HEALTH SCIENCES LIBRARY
LUTHERAN COLLEGE
3024 FAIRFIELD AVE.
FORT WAYNE IN 46807-1697

D1782601

Library of Congress Cataloging-in-Publication Data

Cellular membrane   a key to disease processes / editors, S. Tsuyoshi
   Ohnishi and Tomoko Ohnishi.
      p.   cm. -- (Membrane-linked diseases)
   Includes bibliographical references and index.
   ISBN 0-8493-8091-X
   1. Cell membranes--Abnormalities.   2. Pathology, Cellular.   3. Ion
channels.   I. Ohnishi, S. Tsuyoshi.   II. Ohnishi, Tomoko.
III. Series
   [DNLM: 1. Cell Membrane--pathology.   2. Cell Membrane--physiology.
3. Disease--etiology.   QH 601 C3937]
   RB 152.C45   1992
   616.07--dc20
   DNLM/DLC
   for Library of Congress   92-9530
                                                                                              CIP

This book represents information obtained from authentic and highly regarded sources. Reprinted material is quoted with permission, and sources are indicated. A wide variety of references are listed. Every reasonable effort has been made to give reliable data and information, but the author and the publisher cannot assume responsibility for the validity of all materials or for the consequences of their use.

Neither this book nor any part may be reproduced or transmitted in any form or by any means, electronic or mechanical, including photocopying, microfilming, and recording, or by any information storage and retrieval system, without permission in writing from the publisher.

Direct all inquiries to CRC Press, Inc., 2000 Corporate Blvd., N.W., Boca Raton, Florida, 33431.

© 1993 by CRC Press, Inc.

International Standard Book Number 0-8493-8091-X

Library of Congress Card Number 92-9530
Printed in Mexico   1 2 3 4 5 6 7 8 9 0

Printed on acid-free paper

DEDICATION

To our mentors and teachers who showed us the way and to our families and friends who encouraged us.

CRC Series in Membrane-Linked Diseases

S. Tsuyoshi Ohnishi and Tomoko Ohnishi, Series Editors
Philadelphia Biomedical Research Institute
University of Pennsylvania School of Medicine

Cellular Membrane: A Key to Disease Processes (1992)

Forthcoming Titles
Membrane-Associated Abnormalities in Sickle Cell Disease
Malignant Hyperthermia: A Membrane-Linked Disease
Central Nervous System Trauma
Prostaglandin Derivatives as Membrane Protecting Agents

PREFACE

We, the editors of this series, began our research work in the biological sciences in the late 1950s. It was a time when the significance of membranes in biomedical science was starting to be recognized. However, membranes were considered a "muddy pond" and courage was required to jump into this pond.

Somehow, we jumped in. Together we discovered in 1962 the existence of the muscle protein actin in red cell membranes. Tsuyoshi developed a murexide-dye method to measure kinetics of calcium ion movement across the membrane of the sarcoplasmic reticulum in 1963. He has been studying red blood cells and the sarcoplasmic reticulum ever since. Tomoko prepared intact yeast mitochondria in 1964 and has since dedicated herself to studying oxidative phosphorylation in mitochondrial and bacterial membranes. Her work includes the study of diseases related to the dysfunction of Site-I energy metabolism in mitochondria.

It was only a dream to determine the structure of a membrane 30 years ago. Even SDS-gel electrophoresis was not yet popular. Today, many membrane proteins, receptors, and channels have been isolated and purified. Many of them have been reconstituted in *in vitro* systems to demonstrate their characteristic activities. The primary amino acid sequences were determined and their tertiary structures have been analyzed by X-ray crystallography. Numerous biophysical, biochemical, and genetic technologies have been developed to study membrane phenomena. This development has been so rapid and extensive that study of the structure and function of membranes constitutes a major field in science today. Thus, it is worthwhile at this time to examine and view the future direction of membrane research.

The reason why both of us started studying membranes was simply because they attracted our scientific interest. It is true that "interest" and "curiosity" are always a driving force for scientific discoveries. There are people who believe that the aim of scientists is to find the "truth". However, society demands of medical science that the truth should create "value". Discoveries of truth should result in the development of cures and preventions of disease, and the accumulation of knowledge should help improve the quality of life of mankind. For example, many scientists had observed that a strain of fungi destroyed their bacterial cultures, but A. Fleming pursued this and found that the fungi secrete chemicals which inhibit the formation of the bacterial cell membrane. This led to the discovery of penicillin, which has saved the lives of countless patients.

Thus, a key issue in future membrane research (and in the assurance of its funding) will be to create "value" from the "truth" and to contribute to the elimination of "suffering" and to the enhancement of "wellness" for mankind. Looking back on our own research history, we have gradually incorporated this humanistic mandate. We have recognized that the protection of red cell membranes in sickle cell anemia patients may alleviate pain. We found the cause of a rare but often fatal mishap in operating rooms called "malignant hyperthermia". It is an abnormal genetic response to inhalational anesthetics of calcium release channels in the sarcoplasmic reticulum. We also now know that some myopathies are caused by genetic defects in mitochondrial metabolism.

As described in Chapter 1 of this book, we came to realize that membranes play key roles in many diseases and that they are the target sites for pharmacologic treatment. Thus, we decided to publish a series of books under the title of *Membrane-Linked Diseases* to disseminate this idea.

To accomplish this goal, we set the following aims: (a) to clarify membrane involvement in disease processes, (b) to develop hypotheses on the mechanisms, (c) to design pharmacologic interventions, (d) to encourage readers to build their research on this foundation, and (e) to describe the actual techniques used to help investigators generate new ideas and ease their way.

Volume 1 serves as an overview for the entire series. From Volume 2 on, we will focus on specific subjects and will ask contributors to describe the detailed techniques or "secrets" they use in doing experiments. We believe that techniques always play key roles in research, but that authors do not normally have an opportunity to describe technical details in journal publications. Proposed future titles include: *Membrane-Associated Abnormalities in Sickle Cell Disease; Malignant Hyperthermia: A Membrane-Linked Disease; Central Nervous System Trauma; Prostaglandin Derivatives as Membrane Protecting Agents; Mechanism of Ischemia and Reperfusion Injury; Membranes in Cancer and AIDS;* and *Mitochondria-Linked Diseases.*

It is our sincere hope that this series will help link the most recent discoveries in membrane research to clinical disorders. It is also our hope that this series will stimulate research for the cure and prevention of diseases as well as for improved diagnostic procedures.

Lastly, the editors express their sincere appreciation to all the authors who contributed their excellent articles to meet the ambitious goal of this series.

S. Tsuyoshi Ohnishi
Tomoko Ohnishi

THE EDITORS

S. Tsuyoshi Ohnishi, Ph.D., is Director of the Philadelphia Biomedical Research Institute, King of Prussia, Pennsylvania, and is Visiting Professor of Hahnemann University School of Medicine, Philadelphia, Pennsylvania.

Dr. S. T. Ohnishi graduated from Kyoto University, Kyoto, Japan, in 1954 with a B.S. degree in Physics. He obtained his Master's Degree in Chemistry in 1956 and his Ph.D. from Nagoya University in 1959. He was a Post Doctoral fellow at Nagoya University (Biophysics) from 1960 to 1962, and Visiting Scientist fellow between 1963 and 1965 at the University of Tokyo School of Medicine (Pharmacology). He served as Associate Professor at Waseda University, Tokyo, from 1956 to 1967, Visiting Associate Professor at the Johnson Foundation, University of Pennsylvania, from 1967 to 1968, Assistant Professor at the Medical College of Pennsylvania from 1969 to 1972, Associate Professor at Hahnemann University School of Medicine from 1973 to 1984, and Research Professor at the same university in 1985. He was Director of the Membrane Research Institute from 1985 to 1989, and Director at the Philadelphia Biomedical Research Institute from 1989 to present.

Dr. S. T. Ohnishi was Invited Research Scientist at MIT, Boston, in 1983, Visiting Professor at Hokkaido University, Japan, in 1979, Visiting Professor at the University of Texas, Galveston, in 1981, and Visiting Professor at Mayo Medical School in 1982.

He is a member of the American Society of Biological Chemistry, the Biophysics Society, the Society of Neuroscience, and the American Hematological Society.

Dr. S. T. Ohnishi has made 50 presentations and invited lectures in national and international meetings. He has published 120 papers and is co-author of three books. He has been the recipient of 20 research grants from the National Institutes of Health, National Science Foundation, and other private foundations. His current research interest is to investigate the role of antioxidants in ischemia-reperfusion injury, malaria, cancer, and AIDS.

Tomoko Ohnishi, Ph.D., is Research Professor at the University of Pennsylvania School of Medicine.

Dr. Ohnishi graduated from Kyoto University in 1956 with a B.S. degree in Biochemistry and obtained her Ph.D. in 1962 from Nagoya University. She was a Visiting Scientist fellow between 1961 and 1963 and worked in the Osaka University School of Medicine (Biochemistry). Dr. Ohnishi was Research Associate at the same university between 1963 and 1966. She was Research Associate at Wenner-Gren Institute, University of Stockholm, Sweden, from 1965 to 1966, Visiting Professor at Phillips University, Marburg/Lahn, Germany, in 1966, and Research Associate at Cornell University from

1966 to 1967. Dr. Tomoko Ohnishi was Visiting Assistant Professor at the Johnson Foundation, University of Pennsylvania, from 1967 to 1971, and Assistant Professor at the same Foundation from 1971 to 1977. She served as Research Associate Professor at the University of Pennsylvania Department of Biochemistry and Biophysics from 1977 to 1984, and Research Professor of the same department from 1984 to present.

Dr. Ohnishi received the International Cell Research Award to attend the international conference and workshop in Stockholm in 1965. She was Invited Professor of the Organization of American States Scientific and Technological Program, Buenos Aires, in 1977, Invited Scholar of the Tokyo Metropolitan Institute of Medical Science, Tokyo, in 1982, and Invited Foreign Scholar of CNRS, Paris, from 1983 to 1984.

She is a member of the American Society of Biological Chemistry and the Biophysics Society. She has given 20 invited lectures and 50 presentations in national and international meetings. She has published 161 papers and has been the recipient of many research grants from the National Institutes of Health and the National Science Foundation. Her current research interests relate to the investigation of mitochondrial and bacterial electron transfer and energy coupling, and to the structure-function study of iron-sulphur, quinone, and flavin using EPR spectroscopy.

CONTRIBUTORS

Kiichi Arahata, M.D., Ph.D.
Division of Neuromuscular
  Research
National Institute of Neuroscience
Tokyo, Japan

Takao Asano, M.D., Ph.D.
Department of Neurosurgery
Saitama Medical Center
Saitama, Japan

Lisa Cotterill, M.D., Ph.D.
Section of Surgical Research
Clinical Research Centre
Northwick Park Hospital
Harrow, United Kingdom

Jacqueline Ferrante, Ph.D.
Department of Biochemical
  Pharmacology
State University of New York,
  School of Pharmacy
Buffalo, New York

Tommaso Galeotti, M.D., Ph.D.
Institute of General Pathology
Catholic University
Rome, Italy

Lee A. Goodglick, Ph.D.
Post-doctoral Fellow
Department of Pathology and
  Laboratory Medicine
Brown University
Providence, Rhode Island

Jon Gower, M.D., Ph.D.
Head Office
Medical Research Council
London, United Kingdom

Colin J. Green, M.D., Ph.D.
Section of Surgical Research
Clinical Research Centre
Northwick Park Hospital
Harrow, United Kingdom

Shunichi Kajioka, M.D., Ph.D.
Department of Pharmacology
Faculty of Medicine
Kyushu University
Fukuoka, Japan

**Masahiro Kamouchi, M.D.,
  Ph.D.**
Department of Pharmacology
Faculty of Medicine
Kyushu University
Fukuoka, Japan

Agnes B. Kane, M.D., Ph.D.
Department of Pathology and
  Laboratory Medicine
Brown University
Providence, Rhode Island

Kenji Kitamura, D.D.S., D.D.Sc.
Department of Pharmacology
Faculty of Medicine
Kyushu University
Fukuoka, Japan

Tohru Koide, Ph.D.
Exploratory Research Labs
Chugai Pharmaceutical Co., Ltd.
Chugai, Japan

Motoharu Kondo, M.D., Ph.D.
First Department of Medicine
Kyoto Prefectual University of
  Medicine
Kyoto, Japan

Periannan Kuppusamy, Ph.D.
Department of Medicine,
 Cardiology Division
Johns Hopkins University School of
 Medicine
Baltimore, Maryland

Hiroshi Kuriyama, M.D., Ph.D.
Department of Pharmacology
Faculty of Medicine
Kyushu University
Fukuoka, Japan

Lanfranco Masotti, M.D., Ph.D.
Department of Biochemistry
University of Bologna
Bologna, Italy

Tohru Matsui, M.D., Ph.D.
Department of Neurosurgery
Saitama Medical Center
Saitama, Japan

Masahiko Nakamura, M.D., Ph.D.
Department of Internal Medicine
School of Medicine
Keio University
Tokyo, Japan

Mikio Nakashima, M.D., Ph.D.
Department of Pharmacology
Faculty of Medicine
Kyushu University
Fukuoka, Japan

Masaya Oda, M.D., Ph.D
Department of Internal Medicine
School of Medicine
Keio University
Tokyo, Japan

S. Tsuyoshi Ohnishi, Ph.D.
Philadelphia Biomedical Research
 Institute
King of Prussia, Pennsylvania

Tomoko Ohnishi, Ph.D.
Department of Biochemistry and
 Biophysics
University of Pennsylvania
Philadelphia, Pennsylvania

Yoshio Oosawa, Ph.D.
International Institute for Advanced
 Research
Matsushita Electric Industrial Co.,
 Ltd.
Kyoto, Japan

Ranajit Pal, M.D., Ph.D.
Department of Cell Research
Advanced Bioscience Labs Inc.
Kensington, Maryland

Chester A. Pearlman, M.D.
Department of Psychiatry
Veterans Administration Medical
 Center
Boston, Massachussetts

Marco Ruggiero, M.D., Ph.D.
Department of Molecular Biology
Institute of General Pathology
Firenze, Italy

George G. Somjen, M.D.
Department of Physiology
Duke University Medical Center
Durham, North Carolina

C. A. Stein, M.D., Ph.D.
School of Medicine, Medical
 Oncology Division
Columbia University College of
 Physicians and Surgeons
New York, New York

Hideo Sugita, M.D., Ph.D.
Division of Neuromuscular
  Research
National Institute of Neuroscience
Tokyo, Japan

Yoh Takuwa, M.D., Ph.D.
Department of Clinical Medicine
School of Medicine
University of Tsukuba
Tsukuba, Japan

David J. Triggle, Ph.D.
School of Pharmacy
State University of New York
  School of Pharmacy
Buffalo, New York

Masaharu Tsuchiya, M.D., Ph.D.
Department of Internal Medicine
School of Medicine
Keio University
Tokyo, Japan

Takashi Watanabe, M.D., Ph.D.
Department of Neurosurgery
School of Medicine
Tottori University
Tottori, Japan

Z. Xiong, M.D., Ph.D.
Department of Pharmacology
Faculty of Medicine
Kyushu University
Fukuoka, Japan

Toshikazu Yoshikawa, M.D., Ph.D.
First Department of Internal
  Medicine
School of Medicine
Kyoto Prefectural University
Kyoto, Japan

Jay L. Zweier, M.D.
Department of Medicine,
  Cardiology Division
Johns Hopkins University
Baltimore, Maryland

TABLE OF CONTENTS

PART I: OVERVIEW

Chapter 1
Importance of Biological Membranes in
Disease Processes 3
S. Tsuyoshi Ohnishi

PART II: MEMBRANE TRANSPORT AND MEMBRANE RECEPTORS

Chapter 2
Calcium Ions and Calcium Channel Blockers 23
J. Ferrante and D. J. Triggle

Chapter 3
Features of the ATP-Sensitive Potassium Channel in
Vascular Smooth Muscles 43
**K. Kitamura, S. Kajioka, M. Nakashima,
M. Kamouchi, Z. Xiong, M. Kuriyama, and H. Kuriyama**

Chapter 4
Membrane Ion Channels and Diabetes 79
Yoshio Oosawa

Chapter 5
Microcirculatory Disturbances and Autonomic Nervous
Receptors in Acute Gastric Mucosal Lesion 95
**Masaharu Tsuchiya, Masahiko Nakamura, and
Masaya Oda**

Chapter 6
Membrane Perturbation by Asbestos Fibers and Disease 123
Lee A. Goodglick and Agnes B. Kane

Chapter 7
Membrane Receptors and Signal Transduction in Tumor Cells 141
Marco Ruggiero

Part I
Overview

Chapter 1

IMPORTANCE OF BIOLOGICAL MEMBRANES IN DISEASE PROCESSES

S. Tsuyoshi Ohnishi

TABLE OF CONTENTS

I.	Causes for Illness	4
II.	Evolution of Cell Membrane Structure	4
III.	Features of Cell Membranes	6
IV.	Ion Pumps, Channels, and Membrane Receptors	7
	A. Ion Pumps	7
	B. Calcium Channels	7
	C. Regulation of Pumps and Channels	9
	D. Regulation through Membrane Receptors	10
V.	Pathology of Membranes	11
	A. The Role of Calcium Ions	11
	B. The Role of Oxygen	12
VI.	Pharmacologic Approach to Membrane-Linked Diseases	15
VII.	Diseases Which do not Appear to have Membrane Involvement	16
VIII.	Summary	18
Acknowledgment		18
References		18

ISBN 0-8493-8091-X
© 1993 by CRC Press, Inc.

I. CAUSES FOR ILLNESS

In ancient Chinese medicine, it was believed that there were 404 causes for illness. Precisely, 101 causes for diseases of the "earth" (which means bone and muscle), 101 for "water" (blood), 101 for "fire" (temperature), and 101 for "wind" (respiratory system). Today we describe causative factors almost beyond number. Many different kinds of virus, bacteria, and fungi have been found to cause illness. Some problems are caused by genetic disorders. Some are caused by inappropriate diet. Toxic pollutants or radioactive wastes can cause serious illness or death. Illness can also be caused by psychological factors.

Although there are a wide variety of different causes for illness, there is a common factor: ***All of them damage cells.*** When cells die, tissues and organs die, and finally the whole organism dies. We may consider cell death as a common ultimate factor in diseases. All cells are surrounded by cell membranes, and the cell functions are dependent on highly organized membrane systems (both surface and intracellular membranes). Therefore, the roles of membrane systems are pivotal for any organism.

The disruption of cellular membrane systems may be the cause or result of a disease process. In either case: ***Destruction of membrane systems is synonymous with cell death.***

II. EVOLUTION OF CELL MEMBRANE STRUCTURE

It has been theorized that prior to the existence of life on Earth, the primitive atmosphere was composed of methane, ammonia, carbon dioxide, hydrogen, nitrogen, sulfur dioxide, and water vapor. There was no ozone to shield the surface from strong ultraviolet rays from the sun. Rain, thunderstorms, lightning and volcanic eruptions were common. The Earth's climate was inhospitable. However, these conditions were suitable for synthesizing organic compounds. Scientists have been able to demonstrate that amino acids, nucleotides, and even complicated molecules such as porphyrin can be formed if primitive atmospheric conditions are reproduced in the laboratory and electric spark discharges are applied. Thus, at that time, the ocean was like an organic "soup". In 1929, Oparin developed a theory on the evolution of life in this primordial soup. He proposed that "coacerbates" were first formed. These were small vesicular assemblies surrounded by membranes. He then further suggested that about 4 billion years ago a system with a self-replicating ability emerged.[1,2] Later investigators (in the 1970s) demonstrated that structures similar to cell membranes could be formed from either lipids or amino acids in similar primitive earth conditions.[3-9]

The most important capability of living creatures is their energy metabolism. Without energy, they cannot perform any life activities — growth, movement, reproduction, or maintenance of structure. It is amazing that all living creatures in the world today, be they animals, plants, or even microorganisms, adopt the two identical principles of energy metabolism: (1) Regardless of their size — from 10-μm bacteria to 30-m whales — all living organisms use the same compound, adenosine triphosphate (ATP), as their energy source; and (2) the method of synthesizing ATP is also similar. Almost all life forms oxidatively decompose carbohydrate or fat into CO_2 and H_2 and obtain energy to synthesize ATP (called oxidative phosphorylation), except for certain cases where they synthesize ATP in a non-oxidative way with much lower efficiency (called glycolysis). One of the key mechanisms of oxidative phosphorylation is a carefully controlled step-wise oxidation process involving several cytochromes. All cytochromes have similar protoporphyrin structures. Minor differences in the side-chain structure determine the type of cytochromes. It is fascinating to compare the structure of cytochrome pigments with that of chlorophyls and hemoglobin. They are almost identical except for the metal ions at their reaction center — either iron or magnesium. These two common aspects of energy metabolism strongly suggest that: ***All forms of "life" were created from the primitive environment of the Earth when the situation became appropriate.***

With this knowledge in hand, we could draw a scenario of the evolution of life on Earth as follows: About 3.6 billion years ago, anaerobic organisms with the ability to synthesize ATP through glycolysis appeared. Next, protoporphyrins were synthesized from porphyrin compounds and acquired either iron or copper to become cytochrome pigments. Then, an oxidative phosphorylation mechanism became possible. At that time, creatures used oxygen-containing compounds such as HNO_2, HNO_3, H_2SO_2, H_2SO_3 or fumarate as oxidizing agents, because oxygen gas was not available.

Protoporphyrin also acquired magnesium and produced chlorophyl. Using solar energy, chlorophyl was able to synthesize ATP. Cyanobacteria, the first organism with chlorophyl, is believed to have appeared about 3.5 billion years ago. Then, multicellular organisms such as green algae appeared. The green algae developed the capacity to produce carbohydrates from carbon-dioxide and released oxygen gas. Thus, about 2 billion years ago, oxygen started to accumulate in the sea and in the Earth's atmosphere.

This was a turning point for life, because oxygen changed life on Earth. From oxygen, ozone was formed in the atmosphere, which assembled in the stratosphere to effectively shield the Earth from harmful ultraviolet rays, thus making the planet more habitable for other life forms. With oxygen in the atmosphere, the processs of oxidative phosphorylation (by which bacteria use oxygen as an oxidizing agent) appeared about 1.5 billion years ago. This is called aerobic respiration, and it was much more efficient than any previous method of synthesizing ATP. With the formation of the ozone layer and the appearance of aerobic respiration, the land was able to support its first life forms about 0.4 billion years ago. Primitive forms of green plants and animals appeared on the land, and they incorporated into their cells bacteria which had an ability for oxidative phosphorylation. The relationship between the hosts (plants or animals) and the bacteria is symbionic — the bacteria benefits by being supplied with nutrition, and the host benefits by utilizing the energy produced by the bacteria. These bacteria are believed to be the origin of mitochondria, which are the key energy-transducing microorganisms in higher plants and animals. It is well known that: ***The energy-transducing organnela, chloroplasts, and mitochondria have a highly organized membrane structure.*** By forming a membrane structure, each component of the energy or electron transfer process could occur with high efficiency. The ultra-thin and highly organized structure of the membrane is also advantageous to the reaction. It allows the creation of a high potential gradient and permits energy transfer to take place almost instantly.

The capacity to utilize oxygen thus made the development of higher organisms possible. However, oxygen is like a double-edged sword — it can damage cellular components by oxidation and peroxidation. Life forms had to develop various means to protect against this "oxygen toxicity". Enzymes such as superoxide dismutase, catalase, and glutathione peroxidase as well as many antioxidants, such as glutathione, vitamin C, vitamin E, and carotenes have been developed. In looking back at the history of evolution, we recognize the close interplay between life and its environment: ***Life is created from the environment, but life in turn changes the environment.***

It has been known that genetic abnormality in the structure and function of mitochondria could cause illness, a subject for discussion in Volume 8 of this series.

III. FEATURES OF CELL MEMBRANES

The cells are separated from their environment by a thin cell membrane, which is only about 50 Å thick. There is a potential difference of 60 to 80 mV across the cell membrane. This is equivalent to an electric field of approximately 100,000 V/cm, the level at which dielectric breakdown occurs in most insulators. Thus: ***The dielectric strength of the cell membrane is comparable to that of the best man-made insulators.***

Since there is such a high voltage gradient across the membrane, natural selection evolved the use of a phospholipid bilayer as an insulating material for the membrane. (The phospholipid bilayer is not conductive to electrons, and the point of dielectric breakdown is about 200,000 V/cm.) Since the membrane is embedded in a salt water environment, the nervous system used ions as current carriers instead of electrons. Then, the membrane potential is created by both the selective ion permeability and a difference of ion concentrations between the outside and the inside of the cell.

The cell processes signals through the transient change of the membrane potential, which is known as the "action potential". Animals have developed a nervous system which uses this activity. The ultimate form of the nervous system is the human brain, which processes signals in a more exquisite fashion than even the most sophisticated modern computer.

There are several kinds of ion channels made of protein molecules. They are incorporated into the phospholipid bilayer membrane. When the permeability of the membrane is increased, then ions go through the channels according to the electrochemical potential of each ion. This mechanism functions effectively because of the extremely high potential gradient across cell membranes.

While the cell is alive, the membrane "components" and "wirings" are repaired constantly through biochemical metabolism. Thus, it can be said that: ***The cell membrane is self-maintained.***

Membranes are composed of both lipids and proteins which are mainly made of carbon, hydrogen, nitrogen, and oxygen. Thus, when the cell dies, the membrane is decomposed to H_2O, CO_2, and NH_3. If membranes had been made of silicon or germanium (employing electrons as a current carrier), they would not decompose at death. The surface of the Earth would then have been covered by the remains of membranes of life forms. We are lucky that this was not the case. In short: ***All living systems return to the universe when they die.***

IV. ION PUMPS, CHANNELS, AND MEMBRANE RECEPTORS

A. ION PUMPS

Most cells have an uneven distribution of Na and K ions across the membranes. Perhaps since "life" originated from the sea, the outer environment had a high Na concentration. The primitive form of life separated itself from the outer environment by a thin membrane. It developed a pump system which simultaneously pumps out Na from the inside and pumps K toward the inside. ATP provided energy for this pump. Then it developed Na channels and K channels. These channels are selectively permeable to one species of ion, but not permeable to other ions. The movement of ions through the channels does not require the energy of ATP, for it is driven by the electrochemical potential. The membranes of excitable cells (neurons, axons, synaptosomes, and muscle cells) produce signals by the sequential changes of permeabilities of Na and K channels, which result in a train of "action potentials".

B. CALCIUM CHANNELS

In certain systems, uneven distribution of calcium concentrations could create the membrane potential. In this case, the change of permeability of Ca channels can produce action potentials. Since the sea environment contains a high concentration of calcium ions (on the order of 5 nM), the primitive form of life must have developed a pumping system to pump out calcium from the inside. This eventually decreased the intracellular calcium concentration to the order of 50 nM (a ratio of concentration between outside and inside is 10^5 to 1). Therefore, a change of permeability in the calcium channel could also trigger action potentials.

Since the intracellular calcium concentration becomes very low, the cell acquires the ability to use calcium ions for triggering various physiological functions. When the level of calcium concentration increases above a certain threshold value, muscle contracts, the synapses release neurotransmitters, the cell can start various reactions, and so on. Thus, calcium ions became one of the most important cellular "messengers".

A genetic disease called malignant hyperthermia has been recognized as a potentially lethal mishap in the operating room. A genetically susceptible patient develops high temperature and muscle rigor upon exposure to certain inhalational anesthetics. Death is possible unless proper care is taken immediately. This was found to be caused by an abnormality of the calcium channel in the sarcoplasmic reticulum[10,11] and will be discussed in Volume 3.

The pump which simultaneously moves sodium and potassium ions consists of dimer proteins. Each monomer has a molecular weight of approximately 130,000 Da. The pump to move calcium ions has a similar structure, and its molecular weight is in a similar range. There is a close similarity of the amino acid sequences between these two proteins, suggesting that these pump systems might have developed from the same origin.

The Na, K, and Ca channels are also made of proteins whose molecular weights are approximately 300,000 Da. Again, there is a similarity between these three channels.[12,33]

Besides these voltage-gated channels, there is a potassium efflux channel which is opened by calcium ions at a concentration as low as 1 μM This is called a Ca-activated K channel. There is another membrane-embedded system called the Na-Ca exchanger. It counter-transports Na and Ca ions without requiring the energy of ATP. Figure 1 schematically shows how these pumps and channels are distributed in the phospholipid bilayers of cell membranes.

Which ion pumps (or channels) appeared first, Ca or Na/K? Today, most cells use the difference of concentration in Na^+ and K^+ as a source of the resting potential, and the change of permeability of Na and K channels for signal transduction. However, this does not necessarily mean that the Na/K pump and channels evolved first. I would like to propose that the Ca pump and Ca channel might have evolved first for the following reasons:

1. A number of organic compounds can chelate Ca^{2+}. For example, citric acid, oxalic acid, and other organic acids which have two or more carboxyl groups can bind Ca^{2+} with a high binding constant. On the other hand, it is generally difficult to find naturally occuring systems which can bind Na^+ or K^+ with a high binding constant, or to distinguish Na^+ and K^+. Since a pump or a channel requires a carrier which

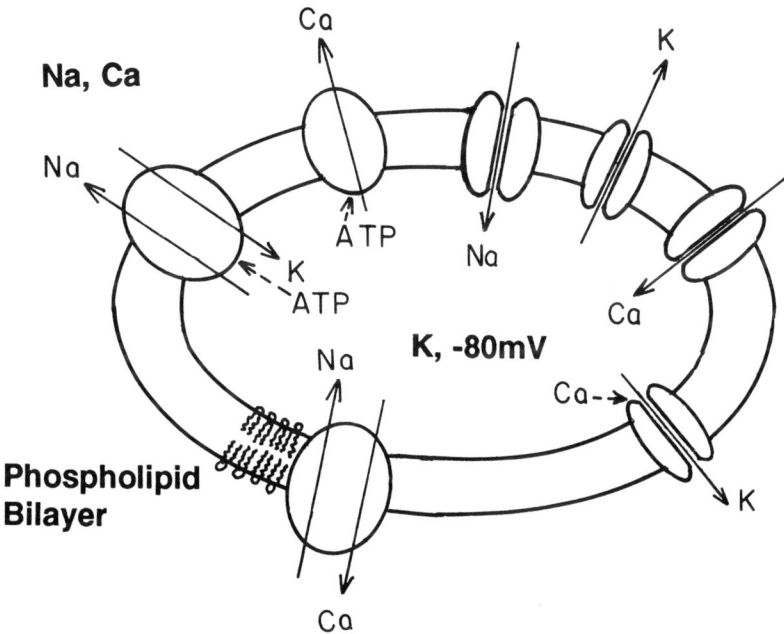

FIGURE 1. Schematic illustration of pumps and channels. From left to right: Na-K-pump, Ca-pump, Na-channel, K-channel, Ca-channel, Ca-activated K-channel, and Na-Ca exchanger.

specifically binds a particular ion, the Ca pump or channel might have appeared first.

2. In the Ca pump, there is a simple stoichiometric relationship between Ca and ATP decomposition. Namely, Ca/ATP = 2. On the other hand, in the Na/K pump, the decomposition of one ATP transports three Na^+ and two K^+. The mechanism of the Ca pump requires one high energy state (E-P), while that for Na/K pump requires two ($E-P_1$ and $E-P_2$). The Na/K pump seems to be more sophisticated than the Ca pump. Thus, the Na/K pump might have evolved from the Ca pump.

3. Many fundamental reactions, such as muscle contraction, neurotransmitter release, and biosynthesis, as well as regulation of many cellular processes, are controlled by Ca^{2+}, though not many reactions are controlled by Na^+ or K^+. Therefore, in the evolution of organisms, the pump and channels which control cellular Ca^{2+} concentration might have appeared first.

C. REGULATION OF PUMPS AND CHANNELS

In the cells of higher organisms, these pumps and channels are regulated in a sophisticated fashion. For instance, the ATP-driven Ca pump is regulated by the phosphorylation of a protein component. The phosphorylation is done

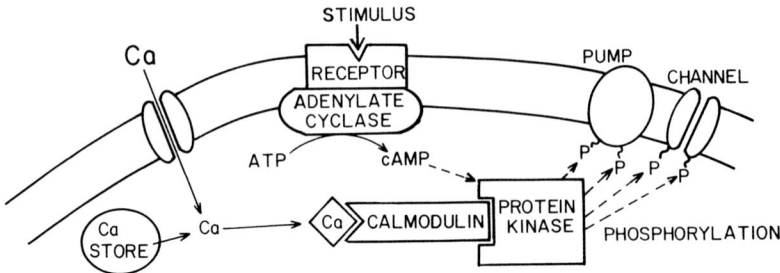

FIGURE 2. Protein phosphorylation by protein kinase is regulated through teamwork of a receptor, cAMP, Ca^{2+}, and calmodulin.

by an enzyme (protein kinase) whose activity is regulated by three factors, i.e., Ca^{2+} ions, a protein called calmodulin (a calcium binding protein with a molecular weight of 20,000) and cyclic AMP (cAMP; another important "messenger"). The intracellular concentration of cAMP is regulated from the external environment through hormones called β-agonists or β-antagonists. When these hormones bind to the receptor (β-adrenergic receptor), the signals either down-regulate or up-regulate the activity of an enzyme (adenylate cyclase) located inside the membrane, which transforms ATP into cAMP (Figure 2).

The permeability of ion channels is also regulated by phosphorylation, which is again dependent on cAMP. It has been suggested that the permeability may also be regulated by a G-protein. When GTP is bound to the G-protein, it is stimulated to regulate the permeability of channels through a cascade of reactions (see next section).

Calmodulin is one of the most ubiquitous intracellular proteins. It can be found in all creatures — bacteria, fungi, plants, and animals. Its binding constant for Ca^{2+} is on the order of 10^6. Thus, it can bind intracellular Ca^{2+} even at a concentration as low as 1 μM. The ubiquitous distribution of calmodulin suggests that Ca^{2+} has been used throughout evolution as a messenger for cellular processes.

D. REGULATION THROUGH MEMBRANE RECEPTORS

As described above, hormone receptors on the membrane play important roles in regulating cell activities. Therefore, if the receptor is defective, it could cause a disease. For example, diabetes and obesity are related to a reduced affinity of the insulin receptor to insulin, and Basedow's disease seems to be related to abnormalities of receptors on the thyroid cell membranes.[13] Some examples of receptor-related diseases are discussed in this volume. In summary, it can be said that: *Life is sustained by membranes, Ca^{2+}, and oxygen.*

V. PATHOLOGY OF MEMBRANES

A. THE ROLE OF CALCIUM IONS

When cells die, the membrane potential disappears. Then, calcium channels open, resulting in the increase of intracellular Ca^{2+} concentration. Since the Ca pump also "dies", the cell cannot pump out Ca^{2+}. Thus, the increase of Ca^{2+} triggers the action of Ca-activated phospholipases and Ca-activated protein-decomposing enzymes. These enzymes attack the cell membrane and cause more Ca^{2+} to enter the cells. This positive feedback reaction quickly destroys the cell membranes and the cell materials.

It is interesting to note that these decomposing enzymes play different roles while the cell is alive. They actually have a maintenance job. For example, it has been suggested that phospholipase A_2 preferentially decomposes peroxidized phospholipids (i.e., phospholipids damaged by oxygen) in order to remove them from the system.[14]

It has been established that these "digestive" enzymes even play a role in the regulation of normal functions. For example, phospholipase A_2 (which is one of major enzymes in the decomposition of cell membranes upon death) is known to produce arachidonic acid (AA) from membrane phospholipids under normal conditions. From AA, regulatory hormones such as prostaglandins, prostacyclins, and thromboxanes are synthesized through an enzyme called "cyclo-oxygenase". Other hormones such as leukotrienes and lipoxins are synthesized from AA through another enzyme called "lipoxygenase". All of them are known to play important roles in regulating a wide variety of cell functions.

Another enzyme called phospholipase C, which decomposes phospholipids in a slightly different manner than phospholipase A_2 does, could also produce AA from membrane phospholipids. Thus, both phospholipases could contribute to the regulation of cell activity by a scheme shown in Figure 3.

Another role for phospholipase C is in the production of diacylglycerol (DG) and inositol-triphosphate (IP_3, a newly discovered "messenger") from the membrane phospholipids. IP_3 stimulates intracellular calcium storage organelles to release Ca^{2+}. Both DG and calcium ions are required to stimulate an enzyme called protein kinase C (PKC). The PKC phosphorylates channel proteins, thereby regulating channel permeability (Figure 4). Therefore, we now know that the same enzymes work both to destroy cell structures upon death and to maintain them when alive. In other words, a lethal effect could become a beneficial effect in a controlled situation. Similarly: *A poison can be used as a medicine.*

In unicellular organisms, the inside of a cell is the world of life, whereas the outside of the cell is the world of death. Thus, membrane enzymes could play roles both in a live state as well as at death. It is difficult to precisely distinguish differences between the live and the dead state. When the system is in good harmony and good coordination, it is alive. But when it is in

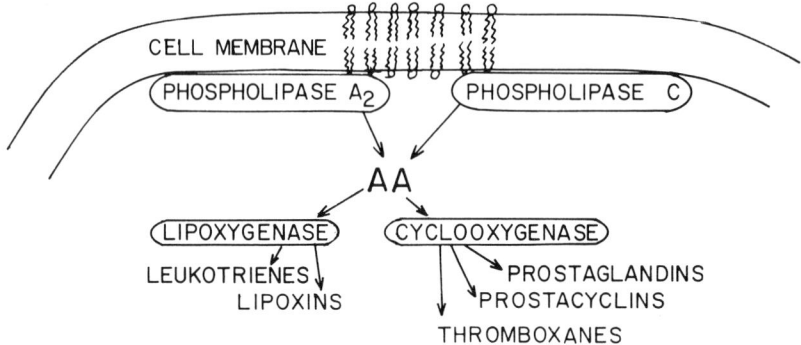

FIGURE 3. Both phospholipase A_2 and phospholipase C decompose the cell membrane to produce arachidonic acid (AA), from which highly active regulators are synthesized.

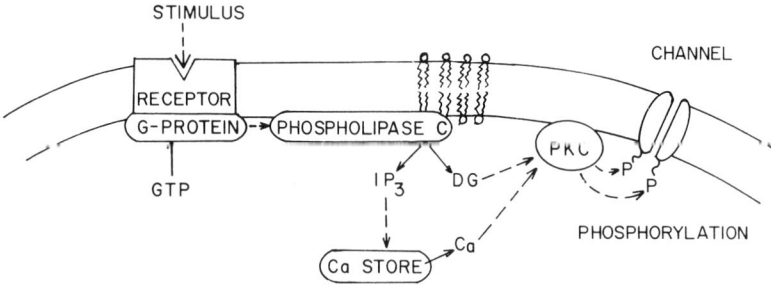

FIGURE 4. A possible pathway of channel regulation. Upon stimulation through a receptor, G-protein binds GTP and activates phospholipase C to produce DG and IP_3 from membrane phospholipids. IP_3 stimulates Ca^{2+} release from the intracellular Ca^{2+} storage organelles. Then, both DG and Ca^{2+} thus released stimulate protein kinase C (PKC), which, in turn, activates channels through protein phosphorylation.

disorder and in chaos, it is dead. In this sense, we can say: *The membrane is the interface between "life" and "death".*

B. THE ROLE OF OXYGEN

Many different pathological disorders can cause hypoxia (lack of oxygen) or ischemia (cessation of blood flow). They reduce the level of ATP and concomitantly increase the intracellular Ca^{2+} concentration. This will abolish the membrane potential. The increase of intracellular calcium concentration activates membrane decomposing enzymes. Weak ischemic damage can be repaired by cells. However, if ischemia continues beyond a certain length of time, then the changes are so great that the membrane sustains irreversible damage and the cell will die.

We all know that oxygen is essential to living cells. However, it has been shown that oxygen may be one of the causes of ischemic cell death. Data

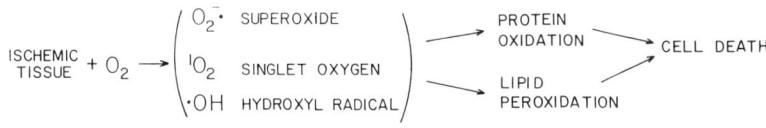

FIGURE 5. Mechanism of membrane damage caused by free radicals produced upon reoxygenation of ischemic tissues.

suggest that lack of oxygen may not be the ultimate cause of cell death. When oxygen is reintroduced to the system after ischemia, it quickly damages the cells. This phenomenon is called "oxygen paradox" or "ischemia-reperfusion injury". It is believed that reoxygenation after deoxygenation produces "oxygen free radicals" such as superoxides, singlet oxygen, and hydroxyl radicals. These radicals have a high reactivity with proteins and lipids, thus they can quickly disrupt membrane functions by forming oxidized proteins and peroxidized lipids. Then, Ca^{2+} enters across the membranes and kills the cells (Figure 5). Two major "killers" in developed countries are stroke and heart attack. Much of the damage resulting from these two pathological states is believed to be from ischemia-reperfusion injury of cell membranes.

Actually, the production of these deleterious oxygen free radicals is not limited to pathological conditions. Even in normal oxygen metabolism of higher plants and animals, a small amount of these radicals is constantly produced by mitochondria. Since creatures developed effective antioxidant systems, these radicals are always scavenged whenever they are produced in the cells. However, in pathological conditions, the production of these radicals exceeds the capacity of natural defense systems and causes serious damage (Figure 6). Thus, oxygen, the most important substance for life, can also kill cells. It can be said that: *A medicine can become a poison.*

As in the case of oxygen, Ca^{2+} could also kill the cell. Figure 7 schematically shows the intracellular Ca^{2+} concentration of a cell. In the resting state, the concentration is less than 0.1 μM. When it reaches the order of 1 μM, it can trigger many cellular reactions, such as muscle contraction, neurotransmitter release, channel regulation and protein synthesis. However, in pathological conditions, such as in ischemia, when Ca^{2+} concentration exceeds 10 μM, the cell is exposed to a danger of self-destroying positive feedback reactions. Thus, we recognize that: *Life is a dynamic equilibrium, and death occurs when equilibrium is lost.*

As to the mechanism of ischemia-reperfusion injury, Masahiro Okuda proposed a slightly different view. His hypothesis can be called an "ATP paradox".[15,16] Cellular ATP is depleted during ischemia, and cellular maintenance is in a state of despair. At this point, if oxygen is reintroduced and the ATP level is suddenly restored, then all cell functions would work at full steam. Okuda hypothesized that the cell cannot tolerate the load and would be damaged.

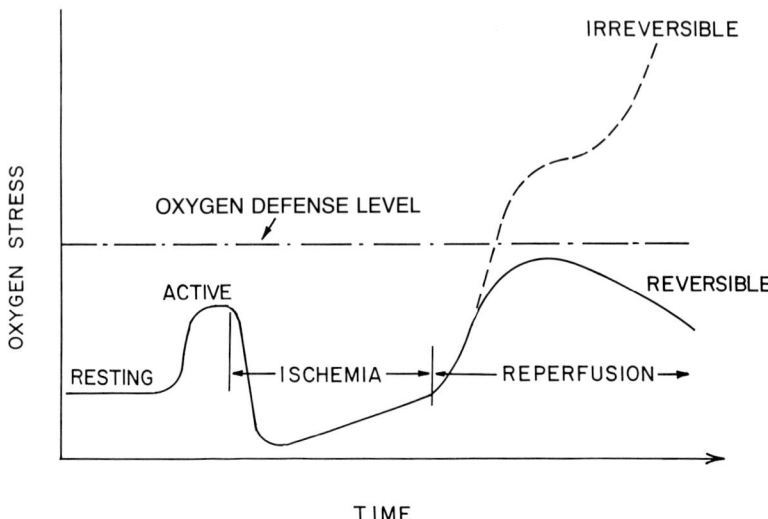

FIGURE 6. Schematic presentation of the level of free radical production (oxygen stress) in a resting state, in an active state, during ischemia, and after reperfusion. Dotted line represents the limit of radical scavenging activity.

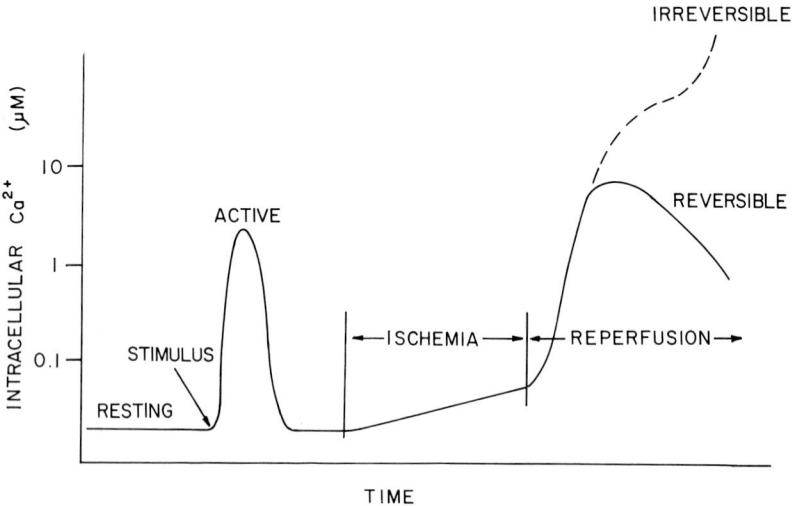

FIGURE 7. Schematic presentation of intracellular Ca^{2+} concentration in a resting state, in an active state (such as muscle contraction, transmitter release, or regulation of other cellular events), during ischemia, and after reperfusion.

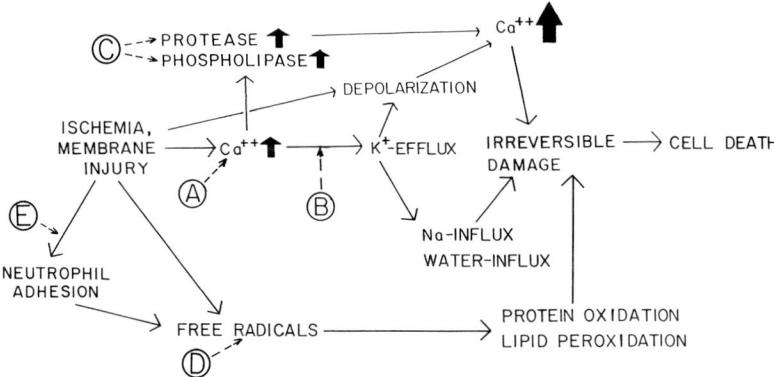

FIGURE 8. Schematic illustration of reactions involved in ischemia-reperfusion injury of cell membranes and its pharmacologic intervention. (A) Ca-blockers, (B) inhibitors for K-efflux, (C) inhibitors for proteases and phospholipases, (D) free radical scavengers, and (E) inhibitors for neutrophil adhesion.

Another topic of interest is the finding that many cells can synthesize a group of proteins called stress proteins in response to attack by oxygen free radicals. The proteins protect these cells and can repair the damage if the insult is not too great. In future volumes of this series, we will discuss the mechanisms of ischemia-reperfusion injury.

VI. PHARMACOLOGIC APPROACH TO MEMBRANE-LINKED DISEASES

Thus, studies of the evolution of cells and membranes suggest that Ca^{2+} and oxygen played key roles in the evolution of life. At the same time, however, organisms are always exposed to the danger of calcium and oxygen toxicity. It is important that the activity of natural defense mechanisms (or repair mechanisms) is always greater than the toxicities of calcium and oxygen. In a sense: *Life is a battle between "evil" (destroying forces) and "good" (repairing forces). When evil wins, diseases occur.*

Thus, interventions against cell death could be provided, based upon (1) protection against calcium activated degenerating reactions and (2) inhibition of oxygen toxicity. Drugs which can inhibit these reactions would have a variety of applications against various diseases. Some of these drugs are discussed in Volumes 1, 4, 5, and 6.

Figure 8 illustrates how these drugs could prevent membranes from suffering ischemia-reperfusion injury. Calcium channel blockers (A) seem to have some effect in blocking calcium entry upon ischemia. However, since these compounds can only block calcium entry through calcium channels, they have a limited efficacy. If the cell membranes are destroyed and calcium ions enter not only through calcium channels but also through ruptures, then calcium channel blockers could not be completely effective. Inhibitors of K-efflux (B) have not been well studied, although some possibilities have been presented.[17,18] Inhibitors of Ca^{2+}-activated phospholipases and proteases (C) could stop the deterioration of membranes. (The efficacy is being studied in animal models.) Antioxidants which could remove oxygen radicals (D) could have a variety of therapeutic effects, because they could work as free radical scavengers. Currently, several antioxidants are proposed to have efficacy against stroke, subarachnoid hemorrhage, heart attack, spinal cord injury, arthritis, diabetes, cancer, and other diseases.

Iron is very important in animals, because both hemoglobin and myoglobin use iron for oxygen binding. When hemorrhage occurs, there is always a chance that free iron can appear. Free iron can cause the production of hydroxyl radicals ($^{\cdot}OH$) according to the Harber-Wise reaction

$$Fe^{2+} + H_2O_2 \rightarrow Fe^{3+} + {}^{\cdot}OH + OH^-$$

The hydroxyl radicals are known to be highly reactive, membrane-toxic free radicals. Therefore, a drug which can remove free iron could also be a candidate as a membrane protecting agent. It has been known that inflammatory response after myocardial ischemia contributes to tissue damage. Entman et al. described how neutrophils play an important role in this reaction. Normally, the role of neutrophils is to kill foreign substances such as bacteria or cancer cells by secreting proteases and producing superoxide radicals (in this immune defense mechanism, the neutrophil uses deleterious superoxide radicals for the benefit of the organisms). However, when tissue and endothelial cells are damaged by ischemia-reperfusion, neutrophils now identify the damaged cells as "foreign substances". Neutrophils are guided to the injury site where they attach to and attack the cells with their "killing weapons". This is an example that: *When things go wrong, a best friend can become a worst enemy.*

Since the adhesion of neutrophils involves membrane receptors, monoclonal antibodies which have been made sensitive to receptor proteins (E) seem to have a specific beneficial effect.[19] Protease inhibitors and free radical scavengers could also inhibit this "self-eating" reaction of neutrophils.

VII. DISEASES WHICH DO NOT APPEAR TO HAVE MEMBRANE INVOLVEMENT

In diseases where membranes may not be directly involved in the pathogenesis, membranes could still be important.

Sickle cell anemia, a genetic disease found mostly in the black population, was found to be caused by genetically abnormal hemoglobin, called sickle hemoglobin.[20] When a patient's blood is exposed to hypoxia, sickle hemoglobin forms bundles of polymer and deforms red blood cells into a rigid "sickle shape" which obstructs blood circulation and triggers their destruction by the spleen. Although abnormal hemoglobin causes the disease, it has been proposed that if we could protect the red cell membrane from subsequent damage, the anemia might be prevented.[21,22] This topic will be discussed in Volume 2.

Cancer is a major "killer" disease in developed countries. It kills almost the same number of people as stroke or heart attack does. It has been known that cancer is caused by uncontrollable cell proliferation, probably caused by an abnormality in protein synthesis. However, the membrane could still play important roles in the development of the disease, metastasis, and in immune responses.

Multidrug resistance is a serious problem in cancer chemotherapy. It has been shown that a glycoprotein with the molecular weight of 170,000 Da (P170) is expressed in the membrane exposed to most anti-cancer agents. Once it is formed, it can pump out not only the exposed agent, but also different agents (even those structurally unrelated) using the energy of ATP.[23,24] Various membrane-active compounds can reverse this resistance, but the toxicity of these compounds has not yet been overcome.[25] The topic of cancer will be discussed in Volume 7.

Malaria is caused by parasites, and in tropical developing countries is a serious health problem. In the 1950s and 1960s, the application of the insecticide DDT and the development of chloroquine seemed to extinguish the disease. However, the mosquitos carrying the parasites became resistant to DDT, and the parasites became resistant to chloroquine. Thus, malaria is back again and over a million die every year. The situation with malaria is worse than it was in the 1950s. The drug resistance of these parasites may be related to that in cancer, because verapamil was found to reverse the drug resistance in both cancer and malaria.[26]

Unfortunately, verapamil is too toxic to be clinically used. Using a mouse malaria model, a safe and effective drug which could reverse the drug resistance is being studied.[27]

Although muscular dystrophy (MD) is an abnormality of muscle fiber, cell membranes of muscle, as well as membrane ATPase activity, are also known to be abnormal. Recent studies have revealed that the membrane of red blood cells is also abnormal in MD patients. This suggests that MD can also be considered a membrane-linked disease.[28]

AIDS is certainly a serious health problem in the world today. Although the disease is caused by an RNA retrovirus (human immunodeficiency virus, HIV), membranes seem to be involved. For example, the attachment of viruses to the receptors of T-cell membranes is the initial reaction in cell infection. At the last stage of cell infection, when matured viruses exit across the cell membrane (budding), a protease activity is involved. A vaccine or agent which could suppress the attachment, or an inhibitor which could inactivate the protease could be a potential therapeutic agent.[29,30]

Psychosomatic diseases also have been seriously affecting a large number of the population. The involvement of receptors in membrane systems has been recognized and studied.[31,32]

VIII. SUMMARY

As our studies unfold the mystery of the cell's functions and its repair mechanisms, the membrane is recognized as an important site in pathogenesis as well as an effective target for therapies. We have a reasonable hope that we can develop potent new tools to help rescue the cell membrane from some lethal threats under pathologic assault. We hope that this new series, "Membrane-Linked Diseases", will shed light on the role of membranes in disease processes and stimulate the development of new therapeutic strategies.

ACKNOWLEDGMENT

The author expresses his appreciation to Professor Tairo Oshima, Tokyo Institute of Technology, for his suggestions.

REFERENCES

1. **Oparin, A. I.**, in *The Origin of Life,* Bernal, J. D., Ed., Weidenfeld and Nicolson, London, 1929.
2. **Oparin, A. I.**, *The Origin of Life,* (translated by Morgulis, S.), Macmillan, New York, 1938.
3. **Felsome, C. F.**, *The Origin of Life,* Freeman and Co., San Francisco.
4. **Kenyon, D. H. and Nissenhaum, A. J.**, *Mol. Evol.,* 7, 245–251, 1976.
5. **Yanagawa, H. and Egami, F.**, Marigranules from glycine and acidic, basic and aromatic amino acids in a modified sea medium, *Proc. Japan Acad.,* Ser. B, 54, 10–14, 1978.
6. **Heinz, B.**, *Naturwissenschaften,* 67, 178–181, 1980.
7. **Fox, S. W.**, *Naturwissenschaften,* 67, 378–383, 1980.
8. **Fox, S. W.**, *Comp. Biochem. Physiol.,* B67, 423–436, 1980.
9. **Fleischaker, G. R.**, Origin of life: an operational definition, in *Origin of Life and Evolution of the Biosphere,* 20, 127–137, 1990.
10. **Ohnishi, S. T., Taylor, S., and Gronert, G. A.**, Calcium-induced calcium-release from sarcoplasmic reticulum of pigs susceptible to malignant hyperthermia: the effect of halothane and dantrolene, *FEBS Lett.,* 161, 103–107, 1983.
11. **Ohnishi, S. T.**, Effects of halothane, caffeine, dantrolene and tetracaine on the calcium permeability of skeletal sarcoplasmic reticulum of melignant hyperthermic pigs, *Biochim. Biophys. Acta,* 897, 261–268, 1987.
12. **Catterall, W. A.**, Structure and function of voltage-sensitive ion channels, *Science,* 242, 50–61, 1988.
13. **Burman, K. D. and Baker, J. R.**, Immune mechanisms in Graves' disease, *Endocrine Rev.,* 6, 183-232, 1985.

14. **van Kuijk, F. J. G., Sevanian, A., Handelman, G. J., and Dratz, E. A.**, A new role for phospholipase A_2: protection of membranes from lipid peroxidation damage, *Trends Biochem. Sci.*, 12, 31–34, 1987.
15. **Okuda, M., Kumar, C., Ikai, I., and Chance, B.**, Effect of respiratory inhibitors on the production of reactive oxygen species in the isolated perfused rat liver, *FASEB J.*, 5, (Abstr.)1278, 1991.
16. **Okuda, M., Lee, H.-C., Kumer, C., and Chance, B.**, Comparison of the effect of a mitochondrial uncoupler, 2,4-dinitrophenol and epinephrine on oxygen radical production in the isolated perfused rat liver, *Acta Physiol. Scand.*, in press.
17. **Tominaga, T., Katagi, K., and Ohnishi, S. T.**, Is Ca^{2+}-activated potassium efflux involved in the formation of ischemic brain edema?, *Brain Res.*, 460, 376–378, 1988.
18. **Ohnishi, S. T., Barr, J. K., and Katagi, C.**, Charybdotoxin improves motor recovery of the rat after spinal cord injury, *Pharmacol. Biochem. Behav.*, 31, 187–191, 1989.
19. **Entman, M. L., et al.**, Inflammation in the course of early myocardial ischemia, *FASEB J.*, 5, 2529–2537, 1991.
20. **Pauling, L., Itano, H. A., Singers, S. J., and Wells, I. C.**, *Science*, 110, 543–545, 1949.
21. **Ohnishi, S. T.**, Inhibition of the *in vitro* formation of irreversibly sickled cells by cepharanthine, *Br. J. Haematol.*, 55, 665–671, 1983.
22. **Ohnishi, S. T., Sadanaga, K. K., Katsuoka, M., and Weidanz, W. P.**, Effects of membrane acting drugs on plasmodium species and sickle cell erythrocytes, *Mol. Cell. Biochem.*, 91, 159–165, 1989.
23. **Gottesman, M. M. and Pastan, I.**, The multidrug-transporter, a double-edged sword, *J. Biol. Chem.*, 263, 12,163–12,166, 1988.
24. **Endicott, J. A. and Ling, V.**, The biochemistry of p-glycoprotein-mediated multidrug resistance, *Annu. Rev. Biochem.*, 58, 137–171, 1989.
25. **Miller, T. P., Grogan, T. M., Dalton, W. S., Sper, C. M., Scheper, R. J., and Salmon, S. E.**, p-Glycoprotein expression in malignant lymphoma and reversal of drug resistance with chemotherapy plus high dose verapamil, *J. Clin. Oncol.*, 9, 17–24, 1991.
26. **Scheibel, L. W., Colombani, P. M., Hess, A. D., Aikawa, M., Atkinson, C. T., and Milhous, W. K.**, Calcium and calmodulin antagonists inhibit human malaria parasites (plasmodium falciparum): implication for drug design, *Proc. Natl. Acad. Sci. U.S.A.*, 84, 7310–7314, 1987.
27. **Chandra, S., Ohnishi, S. T., and Dhawan, B. N.**, Reversal of chloroquine resistance in murine malarial parasites by prostaglandin derivatives, *Am. J. Trop. Med. Hyg.*, to be published.
28. **Arahata, K., and Sugita, H.**, Dystrophin Defect in Duchenne and Becker Muscular Dystrophy, in *Cellular Membrane: A Key to Disease Processes*, Vol. 1, Ohnishi, S. T. and Ohnishi, T., Eds., CRC Press, Boca Raton, FL, 1992, chap. 16.
29. **Stein, C. and Pal, R.**, Anti-HIV compounds with membrane oriented specificity-early results, in *Cellular Membrane: A Key to Disease Processes*, Vol. 1, Ohnishi, S. T. and Ohnishi, T., Eds., CRC Press, Boca Raton, FL, 1992, chap. 14.
30. **Ohnishi, S. T.** Prostaglandin derivatives as chemotherapeutic agents in cancer, AIDS, and malaria, in *Cellular Membrane: A Key to Disease Processes*, Vol. 1, Ohnishi, S. T. and Ohnishi, T., Eds., CRC Press, Boca Raton, FL, 1992, chap. 15.
31. **Plotnikoff, N. S., Murgo, A. J., Faith, R. E., and Wybran, J., Eds.**, *Stress and Immunity*, CRC Press, Boca Raton, FL, 1991.
32. **Briley, M. and File, S. E., Eds.**, *New Concepts in Anxiety*, CRC Press, Boca Raton, FL, 1991.
33. **Heinemann, S. H., Teriau, H., Stuhmer, W., Imoto, K., and Numa, S.**, Calcium channel characteristics conferred on the sodium channel by single mutations, *Nature*, 356, 441-443, 1992.

Part II
*Membrane Transport
and Membrane Receptors*

Chapter 2

CALCIUM IONS AND CALCIUM CHANNEL BLOCKERS

J. Ferrante and D. J. Triggle

TABLE OF CONTENTS

I.	The Regulation of Cellular Calcium	24
II.	Calcium Channels	28
III.	Ca^{2+} Channel Regulation	33
IV.	Summary	37
	Acknowledgments	38
	References	38

ISBN 0-8493-8091-X
© 1993 by CRC Press, Inc.

I. THE REGULATION OF CELLULAR CALCIUM

That Ca^{2+} is critical to cellular excitability was recognized in 1883 by Sidney Ringer, who investigated its role in the maintenance of cardiac contractility.[1] Subsequent investigations have amply demonstrated a fundamental role for Ca^{2+} in stimulus-response coupling processes and in the maintenance of cellular integrity.[2] Although the average human contains approximately 1 kg of Ca^{2+}, it is on the small, approximately 1 to 2%, mobile fraction that the processes controlling cellular integrity, excitability, and responsiveness depend.

This role of Ca^{2+} as a cellular messenger is made possible by a number of factors. The high inwardly directed concentration and electrochemical gradients are appropriate for current carrying function, and a messenger role is made possible by the presence of intracellular Ca^{2+} binding proteins. These proteins, including troponin and calmodulin, serve as Ca^{2+} receptors to translate the information of messenger Ca^{2+}. Additionally, the coordination chemistry of Ca^{2+} makes possible the formation, with polyanionic ligands, of tight complexes with flexible geometry.[3]

For Ca^{2+} to play effectively its multiple roles in structural and excitability control it must be a regulated species. Almost paradoxically, it is the existence of such regulatory processes that separates Ca^{2+} the cellular messenger from Ca^{2+} the cellular toxin. Ca^{2+} out of control is a lethal signal. The regulation of Ca^{2+} is exerted at two levels, the cellular and the organismic. These are not independent processes, but rather are linked intimately both because of relationships between serum and cellular Ca^{2+} and because Ca^{2+} regulating hormones may be involved indirectly and directly in both cellular and organ Ca^{2+} regulation.

The normal resting free intracellular Ca^{2+} concentration of $< 10^{-7} M$ in the face of millimolar concentrations of extracellular Ca^{2+} is made possible by the existence of Ca^{2+} mobilizing and sequestration processes arranged in both series and parallel fashion (Figure 1). These processes permit both the elevation of intracellular Ca^{2+} to the approximately micromolar levels observed during cell excitation and the restoration to the resting levels during the return to resting state. The relative importance of these several processes depends upon the cell type, the stimulus and its time course. Additionally, it is now clear that the distribution of Ca^{2+} during cellular stimulation is both spatially and temporally heterogeneous, perhaps reflecting the necessity to avoid prolonged elevation of intracellular Ca^{2+} during cell signaling processes.[4]

At the organismic level Ca^{2+} absorption, excretion, and mineralization are controlled by a triumvirate of hormones, vitamin D, parathyroid hormone, and calcitonin. These hormones serve to maintain a plasma Ca^{2+} concentration of approximately $2.5 \times 10^{-3} M$ (5.0 meq/l) and are, in turn, regulated by the concentration of Ca^{2+} itself. The reciprocal relationships whereby a de-

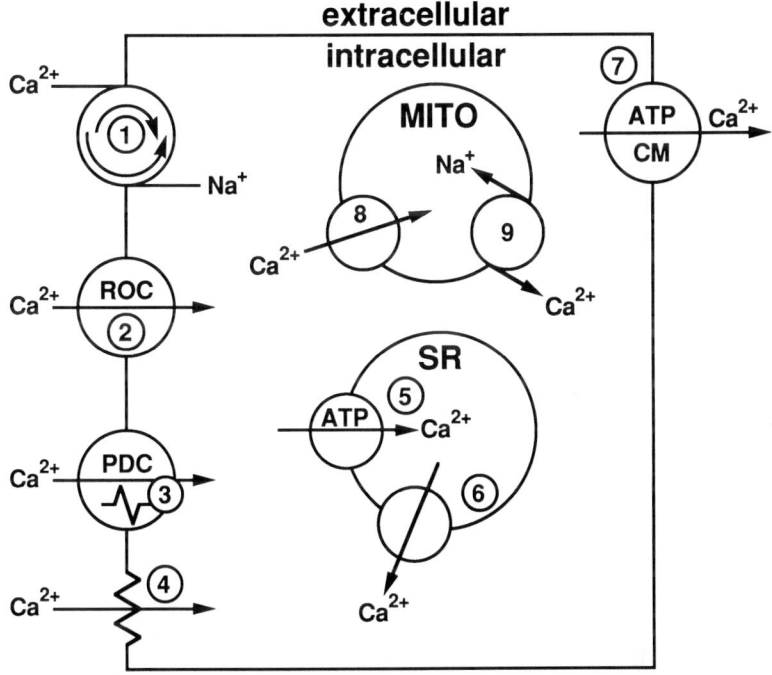

FIGURE 1. Representation of cellular Ca^{2+} regulation with major control loci indicated: (1) $Na^+:Ca^{2+}$ exchange; (2) receptor-operated Ca^{2+} channels; (3) potential-dependent Ca^{2+} channels; (4) "leak" pathway; (5) Ca^{2+} uptake into sarcoplasmic/endoplasmic reticulum; (6) Ca^{2+} release channel; (7) plasmalemmal Ca^{2+}-ATPase; (8) Ca^{2+} uptake into mitochondria; (9) $Na^+:Ca^{2+}$ exchanger of mitochondria.

crease in serum Ca^{2+} is followed by an elevation of PTH with increased Ca^{2+} absorption from the gut, reabsorption from the kidney, and enhanced mobilization from bone are summarized in Table 1. These relationships are likely of importance in at least some categories of essential hypertension. Thus, hypertension, presumed to involve elevated intracellular Ca^{2+} in vascular smooth muscle, is associated in some groups with a dietary Ca^{2+} deficiency, low serum Ca^{2+} levels, and elevated parathyroid hormone levels and may be normalized in both experimental and clinical situations by Ca^{2+} supplementation.[5,6] However, elevated serum Ca^{2+} levels accompanied by vitamin D excess or hyperparathyroidism are also associated with hypertension.[7] The relationship between these changes and the antihypertensive efficacy of Ca^{2+} antagonists is discussed later.

In principle, defects in Ca^{2+} homeostasis can occur at any of the control loci indicated in Figure 1. The ensuing elevation of intracellular Ca^{2+} can be viewed as a critical player in the mediation of the cell and tissue reactivity changes that accompany such disorders as hypertension and bronchial asthma (Figure 2). Thus, a variety of Ca^{2+}-handling changes have been reported to

TABLE 1
Relationship between Serum Calcium and Parathyroid Hormone, Calcitonin, and Calcitriol

	Hormone response		
Plasma signal	Parathyroid hormone	Calcitonin	Calcitriol
Ca^{2+} ↓	↑	↓	↑
Ca^{2+} ↑	↓	↑	
Hormone signal		Plasma Ca^{2+} change	
Parathyroid hormone	↕	↕	
Calcitonin	↕	↕ ↑ ↕	
Calcitriol	↕	↕	

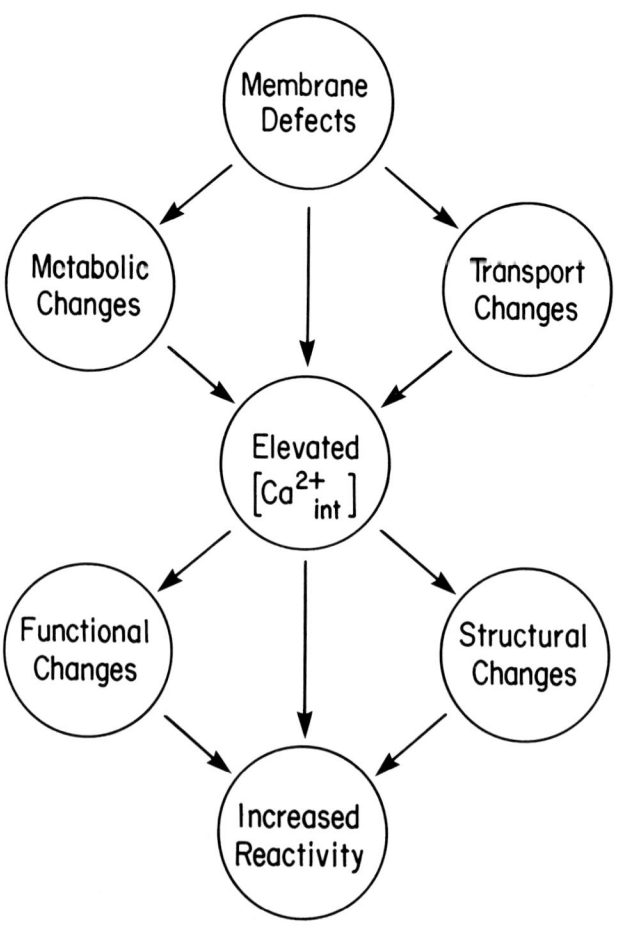

FIGURE 2. Contribution of elevated intracellular Ca^{2+} to generation of hyperreactivity.

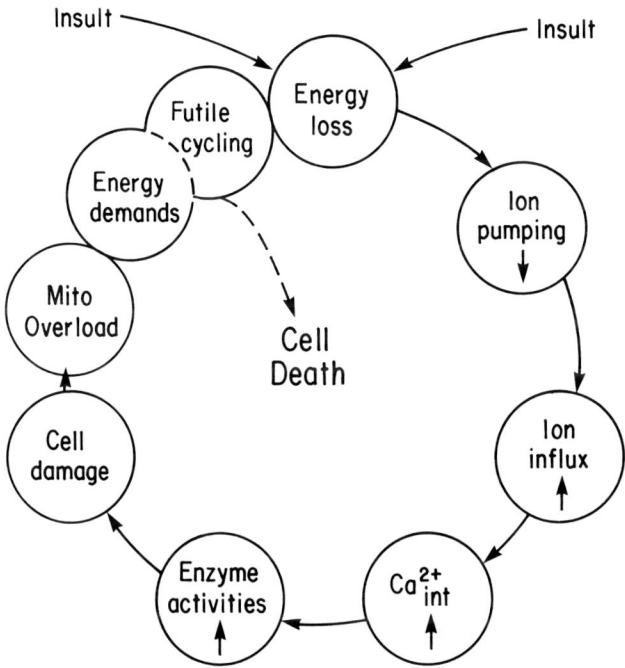

FIGURE 3. A cycle of events in which diverse initial insults cause an initial cell energy loss and Ca^{2+} overload that results in irreversible cell damage and death.

occur in vascular smooth muscle in experimental and clinical situations.[8] Persistent elevation of intracellular Ca^{2+} has long been recognized to be destructive in the "calcium paradox" in which Ca^{2+}-free perfusion or ischemia of the heart produces irreversible cell damage or death.[9,10] The general features of this process, including loss of electrical and mechanical excitability, mitochondrial overload, a loss of high energy phosphates, and a cellular accumulation of Ca^{2+}, are exhibited by many cell types subsequent to diverse noxious physical and chemical stimuli. Given the diversity of cellular Ca^{2+} control processes, it is not necessary to assume that the source of the pathologically elevated intracellular Ca^{2+} is necessarily the same in every cell type. Nonetheless, a general flowsheet of events can be constructed in which a primary event is the loss of cellular ionic homeostasis and energy control (Figure 3). These events cause a rundown of the pumping and exchange capacities of the cell and the resultant mobilization of Ca^{2+} through a variety of pathways and the activation of Ca^{2+}-dependent proteases and phospholipases leading to physical destruction.

In principle, groups of drugs should exist that are selective for each of the several Ca^{2+} mobilization/sequestration processes depicted in Figure 1.[11] In practice this is not so, but one very important group of drugs, the Ca^{2+} channel antagonists, has achieved considerable stature as both therapeutic

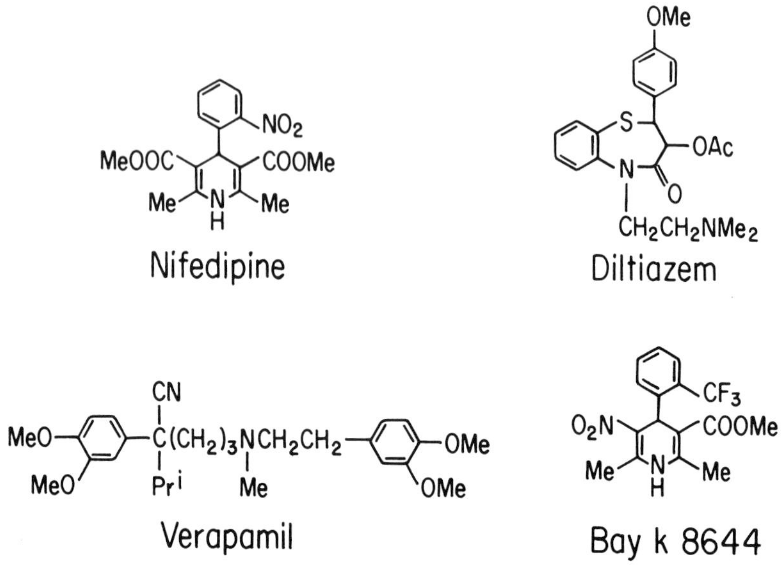

FIGURE 4. Structural formulas of Ca^{2+} channel antagonists (nifedipine, diltiazem, and verapamil) and activators (Bay k 8644).

tools and experimental drugs active at one pathway of Ca^{2+} mobilization. This group, which includes the clinically available verapamil, diltiazem, and nifedipine (Figure 4), is of major importance in cardiovascular medicine with proven effectiveness against angina, hypertension, arrythmias, and some peripheral vascular disorders (Table 2) and with the promise of effectiveness against a variety of other conditions from achalasia to vertigo (Table 3).

II. CALCIUM CHANNELS

Plasmalemmal Ca^{2+} channels represent an important pathway of Ca^{2+} mobilization in response to chemical, physical, and electrical signals. Conventionally, Ca^{2+} channels are considered to be of two major classes — voltage-dependent (VDCCs), activated and inactivated by membrane potential, and receptor-operated channels (ROCs), activated by chemical signals.[12] However, it is clear that VDCCs are modulated by a variety of biochemical inputs and that ROCs are, in turn, modulated by changes in membrane potential.

Receptor-operated channels may be defined by a number of models in which the channel and the pharmacologic receptor are the same protein or oligomeric assembly, where an intermediate messenger (cytosolic or membrane) serves to link receptor and channel, and where the receptor and channel may be linked via G proteins (Figure 5). A particularly interesting example of a receptor-operated Ca^{2+} channel is provided by the N-methyl-D-aspartate

TABLE 2
Therapeutic Uses of Ca^{2+} Channel Antagonists

	Antagonist		
Use	Verapamil* (II)	Nifedipine (II)	Diltiazem (III)
Angina			
Exertional	+++**	+++	+++
Prinzmetal's	+++	+++	+++
Variant	+++	+++	+++
Arrhythmias			
Paroxysmal supraventricular tachyarrhythmias	+++	—	++
Atrial fibrillation and flutter	++	—	++
Hypertension	++	+++	+
Hyperthrophic cardiomyopathy	+	—	—
Raynaud's phenomenon	++	++	++
Cardioplegia	+	+	+
Cerebral vasospasm (post hemmorrhage)	—	+	—

* Classes I, II, and III as defined by the World Health Organization.
** Number of plus signs indicates extent of use: +++, being very common; —, not used.

category of glutamate receptor, activation of which is linked to both memory events and learning behavior and to neurodegeneration initiated by excitotoxins (Figure 6).

Despite increasing interest and definition of ROCs, their potential dependent counterparts remain the better characterized. At least three groups of VDCCs have been recognized, and it is likely that several additional groups and subgroups will be characterized.[12,13] The major channel classes, L, T, and N, are defined by their permeability, electrophysiologic, and pharmacologic characteristics (Table 4). The L channels, of large conductance and long duration, are widely distributed and are of particular importance to the cardiovascular system (Table 2). T channels are activated at negative membrane potentials, rapidly inactivate and are of brief duration. They are likely involved in pacemaking function, but may also contribute to background Ca^{2+} entry. In contrast to the L and T channel classes, N channels appear to be confined to the central and peripheral nervous systems. The pharmacologic characterization of VDCCs is of particular importance. The L channels are represented by the major therapeutically available agents of Figure 4. T channels are less well characterized, but some relatively non-selective agents are available including amiloride, diphenylhydantoin, and the pyrethroid, tetramethrin (Figure 7). N channels are most specifically characterized by ω-toxins, including GVIA and MVIIA (Figure 8) from molluscs of the Conus genus.[14]

TABLE 3
Additional and Potential Uses of Calcium Channel Antagonists

Cardiovascular	Nonvascular smooth muscle
Atherosclerosis	Achalasia
Cardioplegia	Asthma
Cerebral ischemia, focal	Dysmenorrhea
Cerebral ischemia, global	Eclampsia
Congestive heart failure	Esophageal spasm
Hypertrophic cardiomyopathy	Intestinal hypermotility
Migraine	Obstructive lung disorder
Myocardial infarction	Premature labor
Peripheral vascular diseases	Urinary incontinence
Pulmonary hypertension	
Subarachnoid hemorrhage	

Other
Aldosteronism
Cancer chemotherapy
Epilepsy
Glaucoma
Manic syndrome
Motion sickness
Spinal cord injury
Tinnitus
Tourette's disorder
Vertigo

The L class of VDCC is well characterized pharmacologically, and the available drugs serve both as therapeutic agents and as molecular tools with which to characterize and isolate the channel proteins.[11] The chemical and pharmacologic heterogeneity of the Ca^{2+} channel antagonists early suggested that they interacted at discrete sites to modulate channel function. This has been amply confirmed in a variety of tissues through radioligand binding techniques.[15] Discrete binding sites, each exhibiting a specific structure-activity relationship and linked in allosteric fashion one to the other and to the permeation and gating machinery of the channel, exist (Figure 9).[11,15,16] The 1,4-dihydropyridine site is best characterized and, of particular interest, can accommodate potent activator and antagonist species.[11,16] VDCCs can exist in several states or families of states, resting, open, and inactivated (Figure 10). In principle, drugs may interact selectively with, or access preferentially, one or more of these states. Such state-dependent interactions are important to the definition of the selectivity of action of the Ca^{2+} channel antagonists which exhibit higher affinity for the inactivated state which they may access, according to the modulated receptor hypothesis, via hydrophilic or hydro-

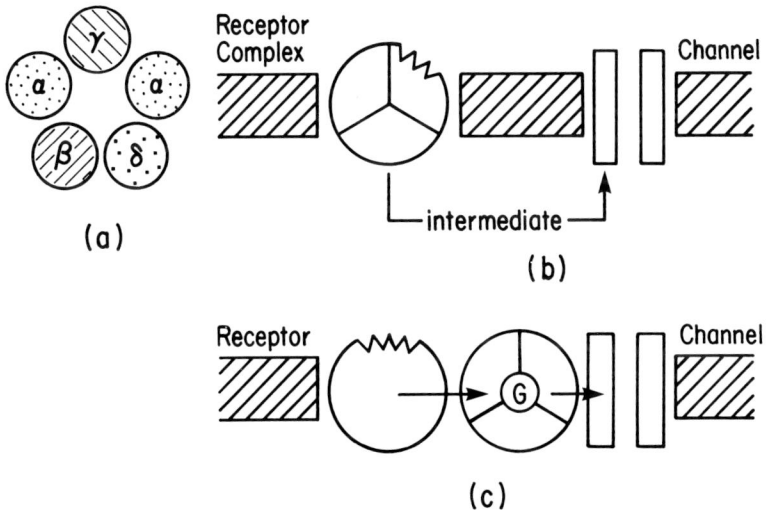

FIGURE 5. Schematic representations of models of receptor-operated ion channels: (a) Arrangement of protein subunits where the receptor and the ion channel functions are carried in the same protein arrangement; (b) arrangement where the receptor and the channel function are discrete entities and communicate via a messenger (cytosolic or membrane); (c) arrangement where the receptor and channel are discrete entities, but communicate directly through a G protein. (From Janis, R. A., Silver, P., and Triggle, D. J., *Adv. Drug Res.*, 16, 309, 1987. With permission.)

phobic pathways (Figure 10). Both electrophysiologic and pharmacologic evidence support the concept that the Ca^{2+} channel antagonists likely interact preferentially with the inactivated channel state favored by membrane depolarization.[17,18] Experimental procedures or conditions that favor this state will increase the apparent affinity of the antagonists. The frequency- and voltage-dependent block exhibited by verapamil and diltiazem and by the 1,4-dihydropyridine antagonists, respectively, underlie, at least in part, their observed pharmacologic and therapeutic selectivity of action.[19]

The existence of both activator and antagonist 1,4-dihydropyridines, and perhaps also of other structures, raises the issue of endogenous ligands, regulatory species for the VDCC which are mimicked in function by these synthetic agents.[20] Such endogenous species remain to be identified unambiguously, although putative factors have been reported,[20-22] but they are theoretically attractive candidates whose aberrant expression may underlie disease states.

The existence of drug binding sites, specific structure-activity relationships, activators, and antagonists and relationships between binding site occupancy and response indicate that VDCCs may be considered as pharmacologic receptors. A further important characteristic of receptors is that they are regulated, homologously and heterologously, under a variety of experi-

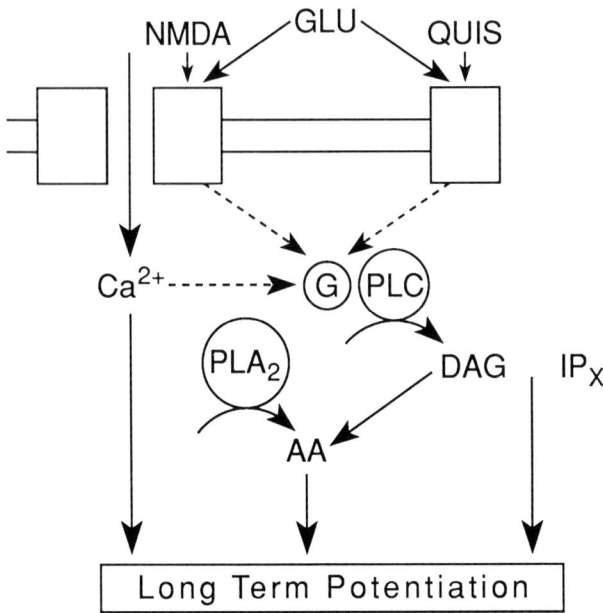

FIGURE 6. Representation of excitatory amino acid receptors for *N*-methyl-*D*-aspartate (NMDA), glutamic acid (GLU), and quisqualate (QUIS) and their linkage to Ca^{2+} movements. The NMDA receptor-channel complex is of particular importance to the movement of Ca^{2+} ions through a receptor-channel complex in the neuron cell membrane. The Ca^{2+} mobilized is involved in a number of processes from long term potentiation to cell death.

TABLE 4
Properties of Plasmalemmal Ca^{2+} Channels[a]

	Channel class		
	L	T	N
Activation range, mV	−10 mV	−70 mV	−30 mV
Inactivation range, mV	−60 to −10 mV	−10 to −60 mV	−120 to −30 mV
Inactivation rate	Very slow	Rapid moderate	
Conductance	25 pS	9 pS	13 pS
Kinetics	Little activation	Brief burst, inactivation	Long burst
Permeation	$Ba^{2+} > Ca^{2+}$	$Ba^{2+} = Ca^{2+}$	$Ba^{2+} > Ca^{2+}$
Cd^{2+} Sensitivity	Sensitive	Insensitive	Sensitive
1,4-DHP sensitivity	Sensitive	Insensitive	Insensitive
ω-Conotoxin sensitivity	Sensitive (neurons) Insensitive (muscle)	Insensitive	Sensitive

[a] Data computed from a variety of sources and are not intended to suggest that these properties are singularly characteristic of each channel class.

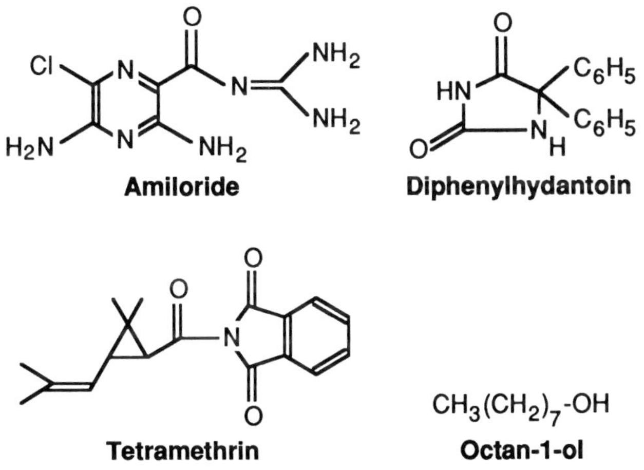

FIGURE 7. Structural formulas of agents active at the T class of Ca^{2+} channels.

①　　　②　　　　③①　　②　　　　③ ㉗
C.K.S.P.G.S.S.C.S.P.T.S.Y.N.C.C.R.S.C.N.P.Y.T.K.R.C.Y

ω - Conotoxin G VI A

①　　　②　　　　③①　　②　　　③㉔
C.K.G.K.A.K.C.S.R.L.M.Y.D.C.C.T.G.S.C.R.S.G.K.C

ω - Conotoxin M VII A

FIGURE 8. Structural formulas of ω-toxins active at the N class of N Ca^{2+} channels.

mental and pathological conditions. The regulation of Ca^{2+} channels is important to the understanding of channel function and development under both physiologic and pathologic conditions.

III. Ca^{2+} CHANNEL REGULATION

The phenomenon of membrane-receptor linked diseases is well documented. Genetic abnormalities, such as familial hypercholesterolemia, autoimmune disorders including myasthenia gravis, drug withdrawal, and dependence phenomena are all attributable to alterations in receptor level or function.[23,24] It is increasingly clear that similar considerations will apply to other membrane components, including Ca^{2+} channels. Considerable evidence now exists to demonstrate that ion channels, including VDCCs, are regulated during disease states and following cell lesions or chronic drug or hormone administration. A selection of such channel alterations is provided in Table 5. A more detailed review is available.[25] Thus far, examples exist

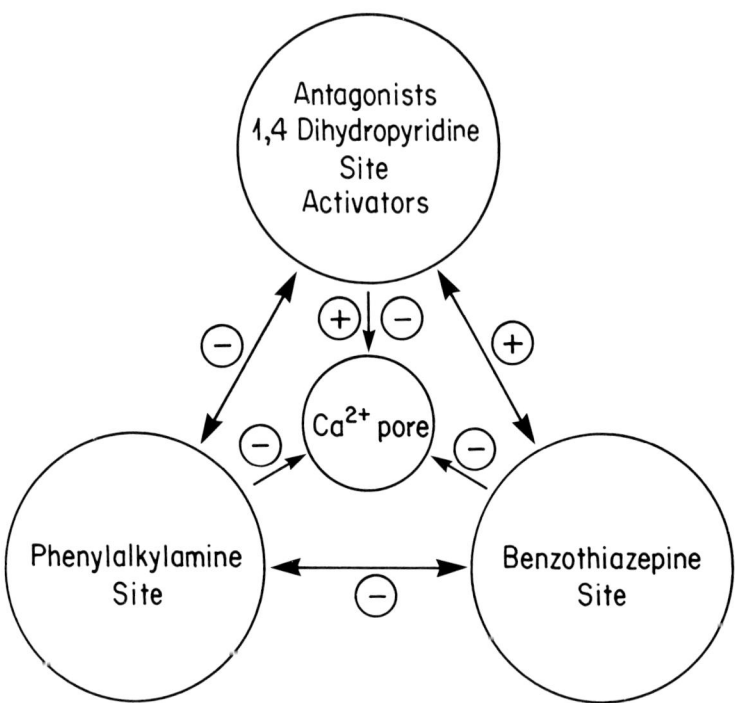

FIGURE 9. Allosteric arrangement of drug-binding sites at the L class of Ca^{2+} channel indicating that the three primary binding sites are linked allosterically one to the other and to the functional machinery of the channel.

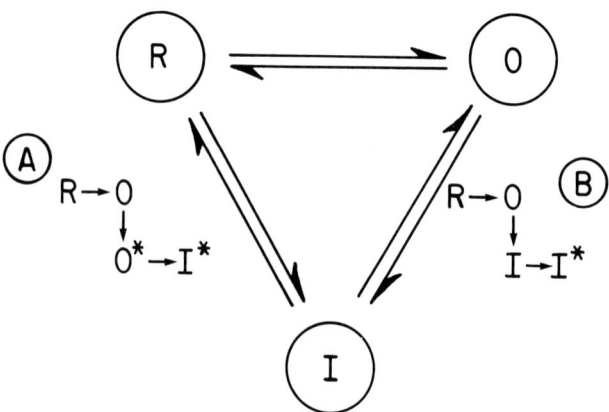

FIGURE 10. The Ca^{2+} channel depicted as cycling through resting, open, and inactivated states. Each of these states may, in principle, offer different affinity and/or access to drugs which thus exhibit state-dependent interactions and apparent affinities which vary according to the distribution of channel states.

TABLE 5
Regulation of Voltage-Dependent Ca^{2+} Channels

Treatment/condition	Species	Tissue	Radioligand effects K_D	B_{max}	Ref.
Homologous regulation					
20 d IV treatment nifedipine	Rat	Nitrendipine Heart, brain	nc* brain nc heart	23% 49%	26
Nifedipine (5 d)	PC 12	PN 200 110	nc	29%	27
Bay K 8644 (5 d)	PC 12	PN 200 110	nc	24%	
Heterologous regulation					
Morphine tolerance	Mice	Nimodipine, brain	nc	60%	28
Phenylephrine (6 d)	Rat	Nitrendipine Heart	nc	32%	29
Hormone Regulation					
Insulin (21 d)	Human	PN 200 110 Muscle	nc	250%	30
Thyroid, hyper hypo	Rat	Nitrendipine Heart	nc nc	42% 26%	31
Estrogen	Rat	Nitrendipine Uterus	nc	96%	32
Neuronal lesions					
6-Hydroxydopamine (IV)	Rat	Nitrendipine Heart	nc	31%	33
6-Hydroxydopamine (IV)	Rat	Nitrendipine Heart	nc	14–23%	34
Reserpine	Rat	Nimodipine Heart	nc	28%	35
Kainic acid (IV)	Rat	Nimodipine Brain	nc	43%	36
Denervation	Rat	Nitrendipine Sk. Mus.	nc	200%	37
Chronic Treatments					
Lead	Rat	Nitrendipine Brain	nc	48%	38
Alcohol (7 d)	Rat	Nimodipine Brain	nc	50%	39
K$^+$ depolarization	PC 12	Nitrendipine	nc	50%	40

TABLE 5 (continued)
Regulation of Voltage-Dependent Ca^{2+} Channels

Treatment/condition	Species	Tissue	Radioligand effects		Ref.
			K_D	B_{max}	
		Disease States			
Hypertension	Rat SHR	Nitrendipine Heart	55%	43%	41
Hypertension	Rat SHR	Nitrendipine Brain	nc	21–40%	42
Ischemia	Guinea pig	Nitrendipine Heart	nc	73%	43
Cardiomyopathy	Human	PN 200 110 Heart	nc	25%	44
Parkinson's disease	Human	Nitrendipine Brain	nc	44-49%	45

* nc = no change.

only for the L type of channel, but it is most unlikely that similar regulation will not be found for other channel classes. It will be of particular interest to determine to what extent channels and their activating receptors are co-regulated and to what extent different coexisting channel classes are co-regulated.

Muscular dysgenesis, a hereditary muscular disorder of mice, is accompanied by a failure of excitation-contraction coupling and slow Ca^{2+} current in skeletal, but not in cardiac, muscle. These functional changes are accompanied by structural changes, including disorganized triads and a loss of 1,4-dihydropyridine binding sites in the transverse tubules.[45,46] This disorder is attributable to alterations in the structural gene coding for the 1,4-dihydropyridine receptor, and microinjection of an expression plasmid into cultured myotubes from these diseased mice restore contractions and 1,4-dihydropyridine binding sites and function.[47]

Alterations in 1,4-dihydropyridine receptors and Ca^{2+} channel function have been implicated in a number of disease states, including experimental and clinical cardiomyopathy. The left ventricle of patients with hypertrophic cardiomyopathy is enlarged and hyperdynamic and such patients respond to Ca^{2+} channel antagonists. The Syrian cardiomyopathic hamster, a model for the human condition, contains elevated numbers of binding sites in heart and other tissues including brain, smooth, and skeletal muscle.[34] The numbers of 1,4-dihydropyridine binding sites are also increased in atrial tissue from human cardiomyopathic individuals and are not associated with changes in β-adrenoceptors and Na^+ channels, consistent with a specificity of association.[44]

Lambert-Eaton myasthenic syndrome (LES) is a neuromuscular disorder characterized by an impairment of evoked acetylcholine release from motor nerve terminals. IgG antibodies from patients with this disorder reduce voltage-dependent Ca^{2+} currents and react with Ca^{2+} channel proteins.[49] Thus, LES appears to be an autoimmune disorder similar to myasthenia gravis, but involving Ca^{2+} channels rather than nicotinic receptors.

A number of observations that 1,4-dihydropyridine binding sites are changed in hypertensive rats in both brain and heart have led to suggestions that VDCCs are altered in this disease. However, the results are not totally consistent and the causality of the events is not established. Thus, increases in the B_{max} and K_D values for [^{3}H]nitrendipine binding occur in cardiac membranes from 24-week SHR, but not from 9-week SHR relative to normotensive controls.[25,41] Both increases and decreases in brain binding sites have been reported according to the hypertensive model.[42,49] However, in venous cells from 1 to 3 day-old rats the proportion of L channel current relative to T channel current was greater in SHR relative to WKY rats.[50]

Ca^{2+} channels are regulated both homologously and heterologously. Chronic *in vivo* administration of verapamil and nifedipine produces down-regulation of brain and heart binding sites,[26] but chronic administration to PC12 cells of nifedipine or S Bay K 8644 resulted in up- and down-regulation, respectively.[27] The differences observed between these *in vivo* and *in vitro* situations may relate both to the different tissues and to the presence of the compensating cardiovascular pathways present in the *in vivo* situation. Chronic morphine or alcohol administration results in the up-regulation of Ca^{2+} channels,[28,39] consistent with observations that withdrawal symptoms from these drugs are reduced by Ca^{2+} channel antagonists.

VDCCs are subject to hormonal regulation. Thus, insulin treatment of cultured human skeletal muscle increases the density of 1,4-dihydropyridine binding sites.[30] Similarly, thyroid treatment causes changes in channel number with increases and decreases of cardiac sites being observed with hypo- and hyperthyroid rats, respectively.[31] Of particular interest, these changes in apparent channel number were accompanied by opposing changes in β-adrenoceptor density. Similarly, estrogen-dominated rat uterus exhibits a significant increase in channel binding sites.[32]

VDCCs are regulated by a variety of neuronal lesions. In rat heart, 6-hydroxydopamine results in an increase in both 1,4-dihydropyridine sites and β-adrenoceptors,[33] and, following intrastriatal injection of kainic acid, a loss of both channel and dopamine receptor-binding sites was observed.[36] A further example of lesion-induced co-regulation of channels and receptors is provided by reserpine, which produces increases in smooth muscle 1,4-dihydropyridine binding sites and α-1-adrenoceptors[51] and a decrease in β-adrenoceptors and an increase in 1,4-dihydropyridine sites in heart.[35]

IV. SUMMARY

VDCCs represent one important pathway for the entry of calcium into the cell. These channels, classifiable in part by the drugs with which they interact, behave similarly to pharmacologic receptors in a number of respects. In particular, they appear to be a regulated species under conditions of chronic occupancy, heterologous receptor influence, and disease states. This is of potential importance to the longterm use of these agents in the treatment of cardiovascular and other disorders for which they are widely prescribed. However, it remains to be established as to what extent these regulatory events are shared by different cell types. Although these agents are in widespread use in cardiovascular medicine, objective reports of withdrawal phenomena are rare.[52-54] More information is needed; however, it is clear that with the recognition that these channels are subject to a diversity of regulatory processes, there will be a continuing impetus to elucidate the fundamental mechanisms involved, to compare channel regulatory processes with those for other receptor types, and to identify further diseases and disorders associated with aberrant channel regulation. Currently, Ca^{2+} channel changes appear to be associated with diverse disorders, including, most recently, malignant hyperthermia.[55] This list will certainly grow. Current attention on Ca^{2+} channel regulation has inevitably focused on the L class of channel, but other channel classes are now available for investigation. Channel regulation and that of associated components, including G proteins,[56] should become a topic of major importance to the general consideration of membrane metabolic diseases in the near future.

ACKNOWLEDGMENTS

Preparation of this work was assisted by grants from the National Institutes of Health (HL 16003 and GM).

REFERENCES

1. **Ringer, S.**, A further contribution regarding the influence of different constituents in the blood on the contraction of the heart, *J. Physiol.*, 4, 29–42, 1984.
2. **Campbell, A. K.**, *Intracellular Calcium: Its Universal Role as Regulator,* John Wiley & Sons, New York, 1983.
3. **Levine, B. A. and Delgarno, D. C.**, The dynamics and function of Ca++ binding proteins, *Biochim. Biophys. Acta,* 726, 187–204, 1983.
4. **Berridge, M. J.**, Cytosolic calcium oscillators, *FASEB J.*, 2, 3074–3082, 1989.
5. **Resnick, L. M.**, Uniformity and diversity of calcium metabolism in hypertension: a conceptual framework, *Am. J. Med.,* 82 (Suppl. 1B), 16–26, 1987.
6. **Young, E. W., Bukoski, R. D., and MacCarron, D. A.**, Calcium metabolism in experimental hypertension, *Proc. Soc. Exp. Biol. Med.,* 187, 123–141, 1988.
7. **Fleckenstein, A., Frey, M., Zorn, J., and Fleckenstein-Grun, G.**, The role of calcium in the pathogenesis of experimental arteriosclerosis, *Trends Pharmacol. Sci.,* 8, 496–501, 1987.
8. **Aoki, K. and Frohlich, E. D.**, (Eds.), *Calcium in Essential Hypertension,* Academic Press, Tokyo, 1988.

9. **Poole-Wilson, P. A., Harding, D. P., Bourdillon, P. D. V., and Jones, M. A.,** Calcium out of control, *J. Mol. Cell. Cardiol.,* 16, 175–187, 1984.
10. **Chapman, R. A. and Tunstall, J.,** The calcium paradox of the heart, *Prog. Biophys. Mol. Biol.,* 50, 67–96, 1987.
11. **Janis, R. A., Silver, P., and Triggle, D. J.,** Drug action and cellular calcium regulation, *Adv. Drug Res.,* 16, 309–591, 1987.
12. **Bean, B. P.,** Classes of calcium channels in vertebrate cells, *Annu. Rev. Physiol.,* 51, 367–384, 1989.
13. **McCleskey, E. W., Fox, A. P., Feldman, D., and Tsien, R. W.,** Different types of calcium channels, *J. Exp. Biol.,* 124, 177–190, 1986.
14. **Gray, W. R., Olivera, B. M., and Cruz, R. J.,** Peptide toxins from venomous Cone snails, *Annu. Rev. Biochem.,* 57, 665–700, 1988.
15. **Triggle, D. J. and Janis, R. A.,** Calcium channel antagonists, new perspectives from the radioligand binding assay, in *Modern Methods in Pharmacology,* Vol. 2, Back, N. and Spector, S., (Eds.), Alan R. Liss, New York, 1984.
16. **Triggle, D. J., Langs, D. A., and Janis, R. A.,** Ca2+ channel ligands: structure-activity relationships of the 1,4-dihydropyridines, *Med. Res. Rev.,* 9, 123–180, 1989.
17. **Sanguinetti, M. C. and Kass, R. S.,** Voltage-dependent block of calcium channel current in the calf cardiac Purkinje fiber by dihydropyridine calcium channel antagonists, *Circ. Res.,* 55, 336–348, 1984.
18. **Bean, B. P.,** Nitrendipine block of cardiac calcium channels: high affinity binding to the inactivated state, *Proc. Natl. Acad. Sci. U.S.A.,* 81, 6388–6392, 1984.
19. **Hondeghem, L. M. and Katzung, B. G.,** Antiarrythmic agents: the modulated receptor mechanism of action of sodium and calcium channel blocking drugs, *Annu. Rev. Pharmacol. Toxicol.,* 24, 387–423, 1984.
20. **Triggle, D. J.,** Endogenous ligands for the Ca^{2+} channel: myths and realities, in *The Calcium Channel: Structure, Function and Implications,* Morad, M. and Nayler, W. G., Eds., Springer-Verlag, Berlin, 1988.
21. **Callewaert, G., Hanbauer, I., and Morad, M.,** Modulation of calcium channels in cardiac and neuronal cells by an endogenous peptide, *Science,* 243, 663–668, 1989.
22. **Janis, R. A., Shrikhande, A. V., Johnson, D. E., McCarthy, R. T., Howard, A. D., Greguski, R., and Scriabine, A.,** Isolation and characterization of a fraction from brain that inhibits 1,4-dihydropyridine binding and L-type calcium channel current. *FEBS Lett.,* 239, 233–236, 1988.
23. **Goldstein, J. L., Brown, M. S., Anderson, R. G. W., Russell, D. W. R., and Schneider, J. W.,** Receptor-mediated endocytosis: concepts emerging from the LDL receptor system, *Annu. Rev. Cell Biol.,* 1, 1–35, 1985.
24. **Hollenberg, M. D.,** Receptor Regulation. I, II, III. *Trends Pharmacol. Sci.,* 6, 242–245, 299–302, 334–337, 1985.
25. **Ferrante, J. and Triggle, D. J.,** Drug- and disease-induced regulation of voltage-dependent calcium channels, submitted.
26. **Gengo, P., Skattebøl, A., Moran, J. F., Gallant, S., Hawthorn, M., and Triggle, D. J.,** Regulation by chronic drug administration of neuronal and cardiac calcium channel, beta-adrenoceptor and muscarinic receptor levels, *Biochem. Pharmacol.,* 37, 627–633, 1988.
27. **Skattebøl, A., Brown, A. M., and Triggle, D. J.,** Homologous regulation of voltage-dependent Ca^{2+} channels by 1,4-dihydropyridines, *Biochem. Biophys. Res. Commun.,* in press.
28. **Ramkumar, V. and El-Fakahany, E. E.,** Increase in (3H)nitrendipine binding sites in the brain in morphine-tolerant mice, *Eur. J. Pharmacol.,* 102, 371–372, 1984.
29. **Gengo, P., Bowling, N., Wyss, V. L., and Hayes, J. S.,** Effects of prolonged phenylephrine infusion on cardiac adrenoceptors and calcium channels, *J. Pharmacol. Exp. Ther.,* 244, 100–105, 1988.

30. **Desnuelle, C., Askanas, V., and Engel, W. K.,** Insulin increases voltage-dependent Ca2+ channels in membranes of aneurally cultured human muscle, *Neurology*, 36 (Suppl. 1), 171–172, 1986.
31. **Hawthorn, M., Gengo, P., Wei, X.-Y., Rutledge, A., Moran, J. F., Gallant, S., and Triggle, J.,** Effect of thyroid status on beta-adrenoceptors and calcium channels in rat cardiac and vascular tissue, *Naunyn-Schmiedebergs Arch. Pharmacol.*, 337, 539–544, 1988.
32. **Batra, S.,** Increase by estrogen of calcium entry and calcium channel density in uterine smooth muscle, *Br. J. Pharmacol.*, 92, 389–392, 1987.
33. **Skattebøl, A. and Triggle, D. J.,** 6-Hydroxydopamine treatment increases beta-adrenoceptors and Ca^{2+} channels in rat heart, *Eur. J. Pharmacol.*, 127, 287–289, 1986.
34. **Wagner, J. A., Reynolds, I. J., Weisman, H. F., Dudeck, P., Weisfeldt, M. L., and Snyder, S. H.,** Calcium antagonist receptors in cardiomyopathic hamster: selective increases in heart, muscle, brain, *Science*, 232, 515–518, 1986.
35. **Ramkumar, V. and El-Fakahany, E. E.,** Selective reduction in the density of (3H)nimodipine binding sites in rat ventricular tissue following reserpine treatment, *Pharmacologist*, 28, 113a, 1986.
36. **Skattebøl, A., Hruska, R. E., Hawthorn, M., and Triggle, D. J.,** Kainic acid lesions decrease striatal dopamine receptors and 1,4-dihydropyridine sites, *Neurosci. Lett.*, 89, 85–89, 1988.
37. **Schmid, A., Kazazoglou, T., Renaud, J., and Lazdunski, M.,** Comparative changes of levels of nitrendipine Ca2+ channels of tetrodotoxin-sensitive Na+ channels and of ouabain-sensitive (Na+K+-ATPase) following denervation of rat and chick skeletal muscle, *FEBS Lett.*, 172, 114–118, 1984.
38. **Rius, R. A., Lucchi, L., Govoni, S., and Trabucchi, M.,** *In vivo* lead exposure alters (^{3}H)nitrendipine binding in rat striatum, *Brain Res.*, 322, 180–183, 1984.
39. **Dolin, S., Little, H., Hudspith, M., Pagonis, C., and Littleton, J.,** Increased dihydropyridine-sensitive calcium channels in rat brain may underlie ethanol physical dependence, *Neuropharmacology*, 26, 275–279, 1987.
40. **DeLorme, E. M., Rabe, C. S., and McGee, R.,** Regulation of the number of functional voltage-sensitive Ca++ channels on PC12 cells by chronic changes in membrane potential, *J. Pharmacol. Exp. Ther.*, 244, 838–843, 1988.
41. **Chatelain, P., Demol, D., and Raba, J.,** Comparison of [^{3}H]nitrendipine binding to heart membranes of normotensive and spontaneously hypertensive rats, *J. Cardiovasc. Pharmacol.*, 6, 220–223, 1984.
42. **Ishii, K., Kano, T., Ando, J., and Yoshida, H.,** Binding of (3H)nitrendipine to heart and cerebral membranes from normotensive and renal, deoxycorticosterone/NaCl and spontaneously hypertensive rats, *Eur. J. Pharmacol.*, 123, 277–278, 1986.
43. **Matucci, R., Bennardini, F., Sciammarella, M. L., Baccare, C., Stendardi, I., Franconi, F., and Giotti, A.,** (3H)Nitrendipine binding in membranes obtained from hypoxic and reoxygenated heart, *Biochem. Pharmacol.*, 36, 1059–1062, 1987.
44. **Wagner, J. A., Sax, F. L., Weisman, H. L., Porterfield, J., McIntosh, C., Weisfeldt, M. L., Snyder, S. H., and Epstein, S. E.,** Calcium-antagonist receptors in the atrial tissue of patients with hypertrophic cardiomyopathy, *N. Engl. J. Med.*, 320, 755–761, 1989.
45. **Nishino, N., Kuno-Noguchi, S. A., Sugiyama, T., and Tanaka, C.,** (3H)Nitrendipine binding sites are decreased in the substantia nigra and striatum of the brain from patients with Parkinson's disease, *Brain Res.*, 377, 186–189, 1986.
46. **Beam, K. G., Knudson, C. M., and Powell, J. A.,** A lethal mutation in mice eliminates the slow calcium current in skeletal muscle cells, *Nature*, 320, 168–170, 1986.
47. **Tanabe, T., Beam, K. G., Powell, J. A., and Numa, S.,** Restoration of excitation-contraction coupling and slow calcium current in dysgenic muscle by dihydropyridine receptor complementary DNA, *Nature*, 336, 134–139, 1988.

48. **Lang, B., Newson-Davies, J., and Wray, D. W.**, The effect of Lambert-Eaton myasthenic syndrome antibody on slow action potentials in mouse cardiac ventricle, *Proc. R. Soc. London, Ser. B,* 235, 103–110, 1988.
49. **Lee, H. R., Watson, M., Yamamura, H. I., and Roeske, W. R.**, Decreased (3H)nitrendipine binding in the brain stem of deoxycorticosterone-NaCl hypertensive rats, *Life Sci.,* 37, 971–977, 1985.
50. **Rusch, N. J. and Hermsmeyer, K.**, Calcium currents are altered in the vascular muscle cell membrane of spontaneously hypertensive rats, *Circ. Res.,* 63, 997–1002, 1988.
51. **Powers, R. E. and Colucci, W. S.**, An increase in putative voltage-dependent calcium channel number following reserpine treatment, *Biochem. Biophys. Res. Commun.,* 132, 844–849, 1985.
52. **Raftery, E. B.**, Cardiovascular drug withdrawal syndromes: a potential problem with calcium antagonists, *Drugs,* 28, 371–374, 1984.
53. **Schroeder, J. S., Walker, S. D., Skallard, M. L., and Hemberger, J. A.**, Absence of rebound from diltiazem therapy in Prinzmetal's variant angina, *J. Am. Coll. Cardiol.,* 6, 174–178, 1985.
54. **Gottlieb, S. O. and Gerstenblith, G.**, Safety of acute calcium antagonist withdrawal studies in patients with unstable angina withdrawn from nifedipine, *Am. J. Cardiol.,* 55, 27E–30E, 1985.
55. **Ervasti, J. M., Claessens, M. T., Mickelson, J. R., and Louis, C. F.**, Altered transverse tubule dihydropyridine receptor binding in malignant hyperthermia, *J. Biol. Chem.,* 264, 2711–2717, 1989.
56. **Rosenthal, W., Hescheler, J., Trautwein, W., and Schultz, G.**, Control of voltage-dependent Ca^{2+} channels by G-protein coupled receptors, *FASEB J.,* 2, 2784–2790, 1988.

Chapter 3

FEATURES OF THE ATP-SENSITIVE POTASSIUM CHANNEL IN VASCULAR SMOOTH MUSCLES

K. Kitamura, S. Kajioka, M. Nakashima, M. Kamouchi, Z. Xiong, and H. Kuriyama

TABLE OF CONTENTS

I.	Introduction	44
II.	Features of Potassium Channels in Vascular Smooth Muscle	46
	A. Macroscopic K Current in Smooth Muscles	47
	B. Unitary K Current in Smooth Muscles Measured Using the Patch-Clamp Procedures	50
	C. ATP-Dependent K Channel in Excitable Cells, Including Vascular Smooth Muscle	52
III.	Effects of Potassium Channel Openers on the ATP-Sensitive Glibenclamide-Sensitive Potassium Channel in Smooth Muscle	54
	A. Changes in Membrane Potential and Ion Fluxes in Smooth Muscle	54
	1. Nicorandil	55
	2. Cromakalim	55
	3. Pinacidil	57
	4. Effects of K-Channel Openers on the SR Membrane	57
	B. Effects of K-Channel Openers on the K Channels in Vascular Smooth Muscle Membranes Measured Using the Voltage- and Patch-Clamp Procedures	58
IV.	Actions of Potassium Channel Openers in Relation to Cardiovascular Diseases	65
V.	Conclusion	66
	Acknowledgment	67
	References	67

ISBN 0-8493-8091-X
© 1993 by CRC Press, Inc.

I. INTRODUCTION

A reduction or cessation of blood flow due to transient spasm or to mechanical narrowing following thickening of the wall of the coronary artery induces ischemia in the heart with pain (angina pectoris), whereas chronic vasoconstriction in resistance vessels induces essential hypertension. The etiology of these states (including the role of heredity) is not yet completely understood. However, pathological changes in calcium (Ca) homeostasis in the cytosol are thought to be closely related to these diseases.

Drugs used for treatment of cardiovascular diseases such as angina pectoris and hypertension include β-adrenoceptor blockers, α-adrenoceptor blockers, nitro-compounds, or angiotensin II converting enzyme inhibitors. Recently, coronary and systemic vascular diseases have been successfully treated with Ca antagonists (Ca channel blockers, Ca entry blockers, or slow channel blockers). These include: (1) the dihydropyridine (DHP) derivatives, nifedipine, nicardipine, nisoldipine, nitrendipine, nimodipine, manidipine, PNP200-110, benidipine, FRC8653, CV-4093, etc.; (2) the phenylalakilamine derivatives (papaverine derivatives), gallopamil (D600), and verapamil; (3) benzothiazepine derivatives, diltiazem, and TA3090, and (4) the piperazine derivative, flunarizine. These Ca antagonists mainly act on the channel which possesses a low threshold, a slow inactivation, and large conductance Ca unitary current (L-subtype), and they reduce the open probability of the channel during depolarization. However, in some vascular tissues, DHP also acts on the channel which possesses a high threshold, a rapid inactivation, and a small conductance Ca unitary current (T-type), though with higher concentrations than that required to inhibit the L-subtype.[1] High concentrations of these agents have been reported to also inhibit the K channel and depolarize the membrane.[2]

Furukawa et al.[3] and Itoh et al.[4] studied the actions of nicorandil (2-nicotineamidoethyl nitrate, SG-75) on the porcine coronary artery and concluded that this drug possesses two actions. It hyperpolarizes the smooth muscle membrane and also prevents the generation of the action potential triggered by Ca influxes. Thus, this agent inhibits activation of the voltage-dependent Ca channel by increasing the threshold at which the action potential is evoked. In addition, this agent has a nitroglycerine-like action and synthesizes cyclic GMP, thus causing vasodilation. In fact, nicorandil has already been introduced as an antianginal agent following *in vivo* and *in vitro* experiments. Uchida et al.[5,6] reported that it has a potency comparable to that of papaverine in the dog and can cause an amelioration of cyclic elevations of the ST segment of the electrocardiogram caused by subtotal occlusion of the left anterior descending coronary artery. In anesthetized open-chest dogs, Taira et al.[7] confirmed the above observations and concluded that nicorandil is a potent antianginal agent without cardiodepressant actions.

In various vascular smooth-muscle tissues, nicorandil has been found to pyherpolarize the membrane in a concentration-dependent manner, though this hyperpolarization occurred only if the concentration of K was 30 mM or less. When the resting membrane was low, as in the mesenteric vein (> -75 mV), the hyperpolarization was less than when it was high, as in the portal vein (resting potential -45 to -50 mV), the hyperpolarization induced by nicorandil depended on the resting membrane potential level. When current-voltage relationships were compared before and after application of nicorandil using the microelectrode method, the lines crossed at about -85 mV, close to the value of the K-equilibrium potential (K potential estimated from the Nernst equation), and nicorandil had reduced the input resistance. This means that nicorandil hyperpolarizes the membrane due to an increase in K permeability.[3,8-16] Nicorandil did not hyperpolarize the membrane in the presence of high K (> 40 mM), yet did relax the tissue in the porcine coronary artery.[4] This agent relaxed vascular tissues with an associated increase in the synthesis of cyclic GMP,[15,17-20] though the relaxing action of nicorandil on agonist-induced contractions was more pronounced than on K-induced contraction.[4]

Ashwood et al.[21] reported that cromakalim (4-(cyclo amido-3,4-dihydropyran) BRL 34915) lowers blood pressure in experimental animals, and this action is thought to result from hyperpolarization of the membrane caused by an increase in its K permeability. These observations were confirmed in measurements of membrane potential, contractions, and ^{86}Rb and ^{42}K effluxes.[14,21-36] Buckingham et al.[37] reported that cromakalim is more potent than nifedipine as a blood-pressure lowering agent and produces less reflex tachycardia. Cromakalim may act as an anti-hypertensive agent through an increase in peripheral blood flow and reduction in after-load via hyperpolarization of the smooth muscle cell membrane.

Pinacidil (N-alkyl-N''-cyano-N'-pyridylguanidine, P1060) has been introduced as an antihypertensive agent.[38,39] Cohen[40] reported that this agent is three times more potent than hydrazine in animal experiments. Following reexamination of this agent after identification of the mechanism of action of nicorandil and cromakalim as K-channel openers, it was realized that this agent also acts as a K-channel opener in causing hyperpolarization of smooth muscle cell membranes.[31,32,34,41]

Minoxidil sulfate (2,4-diamino-6-piperidineylpyridine 3-oxide) also possesses a vasodilating action. Meisheri and Chipkus[42] concluded from indirect evidence that the minoxidil-induced relaxation would not occur in high-K or TEA (tetraethyl-ammonium) containing solution. This agent relaxed the NE-induced contraction except in high-K solution. Following further experiments using ^{42}K efflux, the actions of minoxidil were concluded to be due to its action as a K-channel opener.[43,44] In addition, this agent also induced hypertrichosis in conditions such as alopecia areata.

Diazoxide (7-chloro-3-methyl-2H-1,2,4-benzothiadiazine-1,1-dioxide) causes hyperglycemia due to a reduction in insulin release.[45,46] Recently, it was reported that this agent also has actions as a K-channel opener and vasodilator. However, to induce vasodilation with diazoxide, much higher concentrations were required, and these actions were blocked by glibenclamide.[34,43,47-50] RP 49356 (N-methyl-2-(s-pyridil)-tetrahydrothio-(pyrine-2-carbothio-amide-1-oxide) also has been reported to have a K-channel opening action.[51] In cardiac muscle, this agent acts on the ATP-sensitive K channel from inside the cell membrane rather than from outside.[52] EMD52692 (4-(1,2-dihydro-2-oxo-1-pyridil)-2,2-domethyl-2H-1-benzo-) pyran-6-carbonitrile, also has actions as a K-channel opener in arterial smooth-muscle cells. This agent hyperpolarized the smooth-muscle membrane due to an increase in K conductance, as estimated using the double sucrose-gap method.[53]

Other agents that have been introduced include derivatives of nicorandil (KRN 1391), cromakalim (lemakalim, BRL 38227, BRL 28226, WAY-120, 491, Ro 31-6930, S0121, SR 46142A, CGP42500) and of pinacidil (P1060, LY222675, LY211808, Squibb EP-A-0354553) and also derivatives of minoxidil (Minoxidil sulfate, SKF 11197). Recently, some novel thioformamides (RP 52891, EP-A-0326297, EP-A-0321273, EP-A-0321274) and a dihydropyridine derivative (niguldipine) have been nominated as K-channel openers.[54] However, detailed basic and clinical investigations have not yet been reported.

Nicorandil, cromakalim, pinacidil, and other newly synthesized K-channel openers hyperpolarize the smooth-muscle cell membrane, thus causing relaxation of vascular tissues. Because of the increase in K permeability, these drugs are described as K-channel openers.[55] Nicorandil is now being successfully used in the treatment and prophylaxis of ischemic heart diseases, including angina pectoris, and pinacidil has been applied as an antihypertensive agent. In this chapter, we will briefly introduce the features of K channels, especially the ATP-sensitive K channel, in vascular smooth muscle relevant to the actions of K channel openers, mainly nicorandil, cromakalim, and pinacidil.

II. FEATURES OF POTASSIUM CHANNELS IN VASCULAR SMOOTH MUSCLE

In cardiac muscle, the K channel has been investigated in detail and classified into several subtypes using voltage- and patch-clamp procedures.[56] These are (1) i_{to}, a transient outward current responsible for the early repolarization,[57,58] (2) i_K, the delayed K current important for the formation of a late repolarization,[59-62] (3) i_{Kp}, a channel that carries current during the whole duration of the action potential,[63] and (4) i_{K1}, the channel current responsible for the resting membrane potential and activated by hyperpolarization (an "anomalous rectifying K current".[64] These channel currents are voltage sensitive: either depolarization or hyperpolarization of the membrane activates the channels. In addition, a cytosolic Ca-sensitive K channel, a cytosolic Na-sensitive K channel,[65] and a cytosolic ATP-sensitive K channel (K_{ATP}[66]) have been described in cardiac muscle, as have agonist-sensitive K channels such as an ACh-sensitive channel.[67-71] In visceral smooth muscle, systematic investigations of K channels have not yet been made to the same extent as for cardiac muscle, mainly due to variation in tissue specificity. Therefore, we shall here review the K current in a phenomenological way.

A. MACROSCOPIC K CURRENT IN SMOOTH MUSCLES

The size of the dispersed smooth muscle cells prepared using collagenase differs according to the tissue (5 to 10 μm in width and 120 to 200 μm in length). From individual smooth-muscle cell membranes, more than 10 different types of K currents have been identified using voltage- and patch-clamp procedures.

Using voltage-clamp procedures, the macroscopic current can be recorded on either depolarization or hyperpolarization of the membrane. Using this procedure, the macroscopic current can be classified into at least three types, known as I_t, I_s, and I_{oo}. First, there is a transient outward current (I_t) which occurs just after the generation of a transiently generated inward current (with a holding potential of -60 or -80 mV and a command pulse of -40 mV). The transient inward current was usually a voltage-dependent Ca current which was blocked by application of Ca antagonists or by perfusion of Ca-free solution in the bath, though in some vascular tissues it included a Na current which was blocked by low concentrations of tetrodotoxin.[72a] The transient outward current attenuates with time due to the occurrence of the inactivation process. Subsequently, an outward current (I_s) occurs which shows less attenuation in amplitude even during a long depolarization (slow inactivation). In addition, when the depolarization exceeds -40 mV, oscillatory outward currents (I_{oo}) of irregular amplitude and frequency occur on the I_s.[73-79] Benham et al.[73] called these oscillatory currents spontaneous transient outward currents (STOCs).

The I_t and I_{oo} are closely related to the cytosolic Ca, because the amplitude of the I_t and the frequency and amplitude of the I_{oo} can be modified in an extracellular- and intracellular-Ca dependent manner. By contrast, the I_s is Ca-independent, as shown by using different ionic environments. The I_t has a causal but indirect relation to the influx of Ca induced by activation of voltage-dependent mechanisms (I_{Ca}) since depletion of Ca in the SR following application of ryanodine, caffeine, or the Ca ionophore A23187 diminished the generation of the I_t even following generation of I_{Ca}. Thus, influx of Ca activates the Ca-induced Ca release mechanism (the same Ca channel as the ryanodine- and caffeine-sensitive one) and the released Ca results in an increase in Ca concentration in the cytosol which may activate the Ca-dependent K channel which generates the I_t.[74] This I_t was not sensitive to apamin, a constituent of bee venom (18 amino-acid basic peptide)[80-81] but was inhibited by extracellular application of TEA. This I_t may have a causal relation to the generation of the after-hyperpolarization of the action potential.

The I_{oo} (STOC) is attenuated after depletion of Ca from its store site using A23187, ryanodine, caffeine, or heparin.[75-79] Therefore, release of Ca from the SR may be closely correlated with the I_{oo} and, subsequently, the Ca-activated K channel in the sarcolemma may be activated. Such a postulate is supported by the evidence that the unitary current recorded using the patch-clamp procedure also produced oscillatory enhancement of the channel activities with much the same frequency as the I_{oo} (S. Kajioka, unpublished observations). This oscillation of the outward current may have a causal relation to oscillatory release of Ca from the SR.

Release of Ca from the SR can be measured using three different procedures: (1) Measurement of contraction evoked in skinned muscle tissues, (2) measurement of the Ca transient using aequorin or fura-2, and (3) measurement of the membrane current activated by released Ca. Ca release as estimated from the Ca transient was oscillatory. Let us briefly look at the Ca release from the SR in relation to the generation of the I_{oo}. Using muscle skinned with saponin or α-toxin, accumulated Ca is released by caffeine and inositol, 1,4,5,-trisphosphate (IP_3) as estimated from the amplitude of contraction in skinned muscle tissues. The caffeine-induced contraction could be blocked by procaine[8,82] and the IP_3-induced contraction by heparin.[83] Iino et al.[84] postulated that the sites of IP_3- and ryanodine-induced Ca release may differ, as may the density of their distribution in the SR. The presence of the ryanodine- and IP_3-binding receptors in Ca channels in the SR has been clarified in skeletal muscles,[85,86] and these receptors are distributed differently in different tissues. IP_3 released Ca from the SR in a transient and repetitive manner, causing a rise of cytosolic Ca.[87-90] The sites and mechanisms of the oscillatory release of Ca from the SR are as yet hypothetical. Berridge et al.[87] suggested that since only 30 to 50% of Ca is mobilized by IP_3 in liver cells, there must be different Ca pools which are IP_3-sensitive or -insensitive. The two pools may interact with each other, in both space and time, to generate a complex pattern such as Ca oscillation. Thus, IP_3 would provide a continuous supply of Ca which primes, and ultimately triggers, the IP_3-sensitive pool to release Ca into the cytosol. Once all the stored Ca has been released it is pumped out from the cell and intracellular levels return to near the control level, thus generating a characteristic Ca transient.

In the continued presence of low concentrations of agonists, this transient begins to repeat itself, thus setting up a Ca oscillation (cytosolic oscillator model). On the other hand, Cobbold et al.[88] suggested a receptor control model based on studies of the SR of the rat aorta in primary culture and of the endoplasmic reticulum (ER) in hepatocytes, using the aequorin procedure. According to this model, application of ATP or angiotensin II leads to a transient generation IP_3 which is curtailed at different rates by different rates of inactivation of different receptors or GTP-binding proteins. However,

Wakui et al.[91] opposed the receptor-controlling model on the basis of measurements of the Ca-dependent Cl current activated by IP_3-induced Ca release from the ER. In their experiments, oscillatory Ca release induced by agonists occurred even when concentrations of IP_3 or IP_3S, a non-hydrolyzable IP_3 derivative, were present.

Irvine[90] proposed a different model involving the action of IP_3 together with IP_4 on Ca release from the SR or ER. In this model, IP_3 has two pools, one with the IP_3 receptor and one without. Ca transfer from the IP_3-insensitive pool to the IP_3-sensitive pool would occur via a link triggered by a lowered Ca level in the IP_3-sensitive pool. The function of IP_4 would be to desensitize the negative feedback so that the activation of Ca transfer occurs at high levels of Ca in the IP_3-sensitive pool. Furthermore, if the Ca level in the IP_3-insensitive pool were lower than in the sensitive pool, Ca reuptake rather than release would be predicted.

Ryanodine and heparin do not directly act on the large conductance Ca-dependent K unitary current (Maxi K; Big K) as estimated from the inside-out patch-clamp procedure, but ryanodine modifies the channel current measured using the cell-attached patch.[76] Using the voltage-clamp procedure, continuous application of both agents blocked the generation of I_{oo}. Furthermore, when the membrane was depolarized to -40 mV, the generation of I_{oo} was rare and the frequency of appearance was voltage-dependently enhanced. IP_3 lowered the threshold for the triggering of the oscillatory outward current without depolarization of the membrane (a holding potential of -60 mV). The I_{oo} induced by caffeine was markedly attenuated by pretreatment with IP_3 and vice versa. Furthermore, the heparin-induced inhibition of I_{oo} more strongly affected the agonist-induced I_{oo} than the caffeine-induced one. This suggests that oscillatory release of Ca occurs from at least two pools to produce the I_{oo}. This I_{oo} may result from grouped activations of the large unitary current of the Ca-dependent K channel.

The I_s occurs in the absence of intra- and extracellular Ca, and the amplitude is enhanced in a voltage-dependent manner. Thus, I_s may be a Ca-independent current. Okabe et al.[72b] reported that on prolonged depolarization in a bath containing Mn but absent Ca, the I_s gradually decreases with a long time course, i.e., inactivation occurs with a long time course. Since the reversal potential level remained unchanged, the accumulation of K in the extracellular space may be ignored. This Ca-independent K current was more sensitive to 4-aminopyridine (4-AP) than to TEA, and more potently inhibited at the lower membrane potential than at a higher level, suggesting that 4-AP bound to the channel may be dislodged at the higher level. The action of 4-AP was enhanced at higher pH. Thus neutral 4-AP may act as a channel blocker.

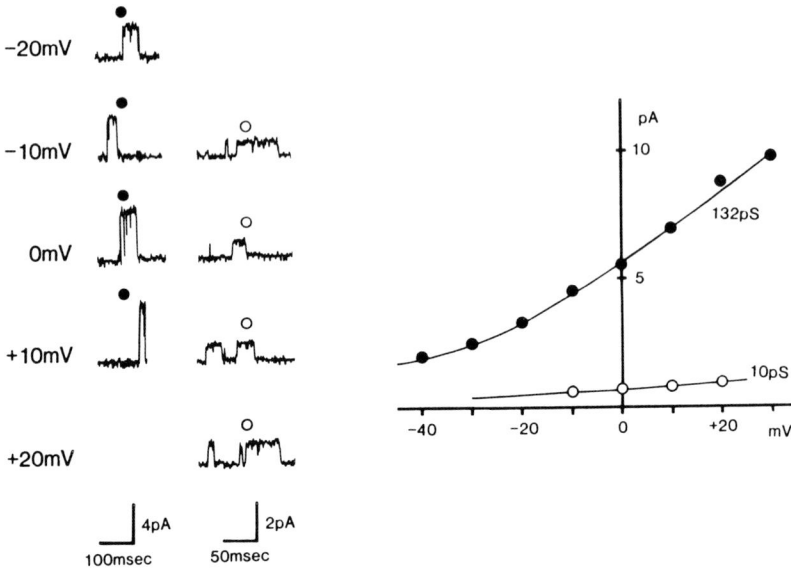

FIGURE 1. Current-voltage relationships of the single K-channel currents recorded in the smooth muscle cells of the rat portal vein. Single-channel currents were recorded at various holding potential levels in outside-out membrane patch configuration. High-K solution with 0.1 mM EGTA filled the pipette and PSS was superfused in the bath. The values of the single channel conductances (132 and 10 pS) were estimated by the slope drawn at holding potentials less negative than −10 mV. Traces of the single-channel currents demonstrated in the left side of the current-voltage relation curves are examples of opening of the 132 (●) and 10 pS (○). K channels recorded at various membrane potentials are indicated. (From Kajioka et al., *J. Pharmacol. Exp. Ther.*, 254, 905, 1990(a). With permission.)

B. UNITARY K CURRENT IN SMOOTH MUSCLES MEASURED USING THE PATCH-CLAMP PROCEDURES

Using the patch-clamp procedures (cell-attached, inside-out, or outside-out), the K unitary current can be classified into several subtypes from the unitary current conductance or from sensitivity to Ca. Inoue et al.[92] found that the Ca-dependent K channel can be subdivided into large and small conductance K channels (200 to 350 pS; big K, BK, or Maxi K, and 30 to 50 pS; small K or SK, measured with high K in the bath and high K in the pipette solution). Subsequently, Inoue et al.[93] found an extracleullar Ca-dependent K channel (middle conductance K channel, 100 to 150 pS) in vascular smooth muscle cells. The extracellular Ca-dependent K channel was less effected by changes in the intracellular Ca buffered with EGTA, as measured using the inside-out patch and cell-attached patch. In addition, the presence of a Ca-independent small conductance K channel has been also reported again in a patch-clamp study.[93] Figure 1 shows the current-voltage relationships of two different Ca-sensitive unitary K currents (SK 10 pS; Maxi

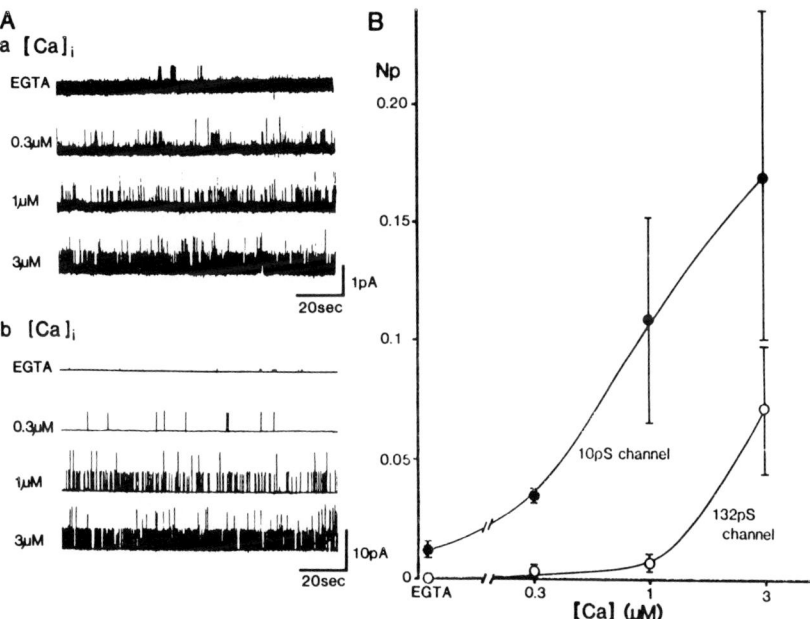

FIGURE 2. Effects of intracellular Ca concentrations on activities of the single-channel currents (rat portal vein). The single-channel currents were recorded with inside-out membrane patch configuration. PSS filled the pipette and high-K solution with various concentrations of free Ca was superfused in the bath. (A) Traces of the 10- (a) and 132-pS (b) K-channels recorded at 0 mV in the absence and presence of various concentrations of free Ca (0.3, 1, and 3 μM). In (a), the 132-pS K channel was blocked by 1 mM TEA. (B) Relationships between the activities of the single K channels and free Ca concentrations of the intracellular side of the membrane (○, the 132-pS K channel; ●, the 10-pS K channel). Np value [(number of channels in the patch membrane) × (mean open probability of individual channel)] was used as a parameter of channel activity. Symbols and bars show the mean values and S.D. (n = 4). (From Kajioka et al., *J. Pharmacol. Exp. Ther.*, 254, 905, 1990(a). With permission.)

K 132 pS) recorded from dispersed smooth muscle cells of the rat portal vein, using the cell-attached patch-clamp procedure, and Figure 2 shows the intracellular Ca sensitivity of these two unitary K currents. The SK was more resistant to reduction in the Ca concentration than the Maxi K.[94]

The Ca-dependent K channel (Maxi K and SK) can also be classified on the basis of factors other than their unitary conductance. For example, the SK can be divided into those that are apamin-sensitive or -insensitive. Apamin (a bee poison) inhibited the SK in cultured skeletal muscles[95] and liver cells.[96] However, apamin had no effect on the intermediate K channel observed in red blood cells[97] and molluscan neurons,[98] or on the Maxi K.[95,99] The apamin-sensitive K channel may have some properties similar to those of the extracellular ATP-sensitive K channel observed in intestinal smooth muscle membranes (the purinergic II-receptor activated K channel[100,101]) This is evidenced by the observation that, in many tissues including intestinal smooth muscle the ATP-induced hyperpolarization was blocked by apamin.[102] The Maxi K, was sensitive to charybdotoxin (a poison from the venom of scorpion *Leiurus quinquestriatus hebraeus*: a highly basic single polypeptide chain of 37 aminoacids and 4353 dalton.[103-107] In addition, the Maxi K was more sensitive to extracellularly applied TEA than was the SK. Thus, 1 mM TEA was enough to inhibit the Maxi K, but a concentration of TEA at least ten times higher was required to inhibit the SK.[94,108,109]

From its voltage-dependency, the K channel can also be classified into a voltage-dependent small conductance Ca-insensitive channel (delayer rectifier K channel, and "A" current generating channel) and voltage-independent K channel (inward rectifying or anomalous rectifying K channel[78]). In some vascular smooth muscle cells, the presence of a funny current (I_f) has been detected (K. Okabe and K. Kitamura, unpublished observations). However, detailed analysis of these currents is still awaited.

Activation of the agonist-receptor induces K channel modification, either activating or inhibiting the receptor-operated K channel. In toad stomach smooth muscle, acetylcholine (ACh) depolarized the membrane and reduced the input resistance. When the effects of ACh were observed using the voltage-clamp method, this agent reduced the outward current and inhibited the open probability of the K channel opening.[110] In intestinal smooth muscle, ACh depolarized the membrane but, in this case, activated mainly Na channels.[93] A different action of ACh was observed when it was applied to cardiac muscle, i.e., this agent accelerated the K channel opening through activations of the α-subunit of GTP-binding proteins.[70,71] Whether these differences in the effects induced by ACh are due to differences in the actions of GTP-binding proteins or in the specificities of the channel correlated with the muscarinic receptor still needs clarification.

Subdivision of the K channel into ATP-sensitive and -insensitive ones is also a means of classification. However, we will review the features of the ATP-sensitive K (K_{ATP}) channel separately, because this channel has a close relation to the actions of K-channel openers and is generally thought to be a target channel both for K-channel openers and for some antidiabetic agents.

C. ATP-DEPENDENT K CHANNEL IN EXCITABLE CELLS, INCLUDING VASCULAR SMOOTH MUSCLE

Since Noma first reported this channel in cardiac muscle membranes in 1983, the regulation of the K_{ATP} channel has been extensively investigated.[111-117] This K channel is normally closed at physiological concentrations of cytosolic ATP but open at lower concentrations. The K_{ATP} channel is slightly inwardly rectifying, and its probability of opening and closing is not greatly dependent on the membrane potential.[118-121] The presence of these channels has been established in many tissues, namely in cardiac muscle,[66,118,120,122] central nerve cells,[123] endocrine glands (pancreatic β-cells,[125-128]) skeletal muscle,[119] and smooth muscle.[30,48,94,109,129]

In cardiac muscle, the biophysical features of the K_{ATP} channel have been investigated in detail.[115] Following the work of Noma,[66] Kakei and Noma,[129] and Kakei et al.[118] reported that this channel was not only sensitive to ATP but also to ADP or GTP. By contrast, in pancreatic β-cells, GTP, GDP, and GTPγS had an accelerating action on the K_{ATP} channel in the presence of Mg.[130] On the action of GTP-binding protein, GTP and GTPγS had an antagonistic action. Thus, in β-cells of the pancreas, activation of the K_{ATP} channel may not require the presence of this binding protein. In cardiac cells, the K_{ATP} channel conductance increased as the 0.24th power of the external K and internal Na decreased the conductance of the single channel current at positive membrane potential levels.[118,121,129] This K_{ATP} channel's kinetic properties depended primarily on the cytosolic ATP concentration.[118] It was sensitive to Mg, and free-ATP more potently inhibited this channel than ATP-Mg.[131,132] Moreover, it was not very sensitive to TEA and 4-AP within the millimolar range.[118]

A new trend of investigation concerning the K_{ATP} channel began with the use of glibenclamide, an antidiabetic sulfonylurea.[133] Since 1985, the K_{ATP} channel has been categorized as a glibenclamide-sensitive channel. It is thought that the hypoglycemic effects of antidiabetic sulfonylureas are directly linked to their ability to block ATP-sensitive K channels in pancreatic β-cells.[134-136] Glibenclamide, however, inhibited this channel in cardiac muscle cells less potently than in pancreatic β-cells.[137,138]

The K_{ATP} channel in the β-cell of pancreatic islets has a unitary current conductance of 20 pS in physiological salt solution and an open probability of 0.1 to 0.2.[135] The distribution of this channel has been calculated to be 10 to 20 channels/mm^2 (a sphere with a diameter of 13 μm[119]), and this value is much higher than that reported for cardiac muscle (0.5/mm[21,22]). With high K in the bath and high K in the pipette, the unitary conductance of the K_{ATP} channel in pancreatic β-cells was estimated to be 50 to 65 pS in the cell-free configuration,[126-128,139] and a similar value was obtained from the inward current using the cell-attached patch-clamp procedure.[125,127,131,140]

Sulfonylureas act principally by decreasing K permeability in the β-cell membrane, thus leading to membrane depolarization, increased Ca influx through activation of the voltage-dependent Ca channel and thus to insulin secretion.[127,141-146] The concentration needed for half-maximum inhibition of ^{86}Rb efflux from β-cells was found to be 0.06 μM for glibenclamide and 40 μM for tolbutamide, another sulfonylurea derivative.[147,148]

In cardiac muscle, glibenclamide acted only on the K_{ATP} channel and not on the instantaneous or delayed rectifying K channels.[147-150] Escande[115] concluded that glibenclamide is about two orders of magnitude more potent in the β-cells of the pancreas than in cardiac muscle cells. Using radiolabeled glibenclamide in microsome of cardiac muscle cell membranes, the potency of radioactive glibenclamide, measured using the displacement test, was much the same as observed using electrophysiological procedures.[147]

Using radiolabeled glibenclamide, the distributions of the K_{ATP} channel has been investigated in the central nervous system. High concentrations of the glibenclamide-binding receptor were found in the substantia nigra, globus, central pallidus, motor cortex, and molecular layer of the cerebral cortex. In the hippocampus, the highest level of binding sites was found in the striatum lucidum of CA_3, where the mossy fibers make synaptic contacts with the proximal part of the apical dendrite of CA_3 neurons. There were high densities of the receptor in CA_3 and intermediate densities in CA_1 and CA_2.[114,151] Bernardi et al.[123] reported that the glibenclamide-sensitive, ATP-sensitive K channel is a protein of 150 kDa, and this value corresponds well with the molecular mass of the receptor of insulinoma cells which is 125 or 140 kDa, estimated by affinity labeling.[152] Glibenclamide altered the response of CA_3 hippocampal neurons to anoxia and this action seems to occur at the presynaptic site on the mossy fiber where the glybenclamide-sensitive K_{ATP} channel is distributed.[124] However, Krnjevic and Loblond[153] had already reported that glibenclamide has no effect on the anoxic response of CA_1 neurons. Ben Ari, therefore, postulated a regional difference in the distribution of the glibenclamide-sensitive receptor. The unitary current conductance of the K_{ATP} channel measured from cultured central neurons of the neonatal rat was reported to be 145 pS.

To judge from investigations in cardiac muscle, pancreatic β-cells, and the central nervous system, the physiological significance of the K_{ATP} channel seems to differ by tissue. In β-cells, hypo- and hyperglycemic conditions regulate K_{ATP} channel activity, opening or closing the channel through changes in the concentration of ATP in the cytosol and thus controlling insulin secretion. However, in cardiac muscle, this channel is activated only in ischemic conditions. ATP is an essential substance for preventing the run-down phenomenon of this channel's activity, but the presence of ATP prevents the opening of the channel. The physiological mechanisms underlying these diverse actions of ATP on the K_{ATP} channel have not yet been identified.

III. EFFECTS OF POTASSIUM CHANNEL OPENERS ON THE ATP-SENSITIVE, GLIBENCLAMIDE-SENSITIVE POTASSIUM CHANNEL IN SMOOTH MUSCLE

A. CHANGES IN MEMBRANE POTENTIAL AND ION FLUXES IN SMOOTH MUSCLE

The membrane potential of vascular smooth muscle cells differs in different regions. For example, the membrane potential in the mesenteric artery is lower than -70 mV, but that in the portal vein is only about -45 mV. Spontaneous membrane activity has been observed only in smooth muscle cells of the portal vein, and whereas many resistance vessels produced action potentials triggered on excitatory junction potentials (e.j.p.s.) evoked by peripheral sympathetic nerve stimulation, some resistance and elastic vessels produced neither action potential nor e.j.p.[154,155]

1. Nicorandil

In the porcine coronary artery, nicorandil hyperpolarized the membrane consistently at K concentrations in the bath between 0.12 and 30 mM. However, a further increase in the concentration of K did not modify the membrane potential. During application of nicorandil, the ionic conductance was increased and current-voltage relation curves observed before and after application of nicorandil crossed at about -85 mV, close to the K equilibrium potential. Such hyperpolarization could be observed in all vascular tissues examined, and the degree of hyperpolarization depended on the resting membrane potential. Thus, smooth muscle cells with a lower (more negative) membrane potential before application of nicorandil were hyperpolarized to a lesser extent than cells with a higher (more positive) membrane potential. These effects of nicorandil were also observed in the mesenteric artery and vein, the hyperpolarization occurring more strongly in the vein than in the artery. In the portal vein, nicorandil blocked spontaneous spike activities as a result of its hyperpolarization of the membrane. The hyperpolarization induced by nicorandil was not inhibited by apamin, and extracellularly applied ATP (via activation of the PII-receptor) had an additive effect on the hyperpolarization. High concentrations of extracellularly applied TEA, 4-AP ($>$ 10 mM), and procaine ($>$ 1 mM) inhibited this nicorandil-induced hyperpolarization. The hyperpolarization induced by nicorandil was weaker in the aorta than in resistance vessels. This agent hyperpolarized intestinal smooth muscle to the same extent as the mesenteric artery but had a smaller effect on tracheal smooth muscle.[3,9-13,15,22,156]

The effects of nicorandil on vascular smooth muscle have been observed using ^{86}Rb efflux procedures, and this agent was found to accelerate ^{86}Rb effluxes only weakly. ^{86}Rb has been commonly and successfully used as a substitute for ^{42}K in studies of the actions of K-channel openers.[14,22,24] However, in rat mesenteric arterioles[157] and in the guinea pig urinary bladder,[158] ^{86}Rb proved to be an unsatisfactory substitute for ^{42}K; in the rat uterus, no detectable increase in either ^{42}K or ^{86}Rb was detected on treatment with K-channel openers.[159]

2. Cromakalim

Cromakalim is a racemate where the substituents in the 3 and 4 positions are in the trans configuration, and the action of cromakalim is stereoselective with the activity residing in the ($-$)-(3S,4R) enantiomer (BRL 38226; ($+$)-S-enantiomer (lemakalim) and BRL 38227; ($-$)-R-enantiomer.[23,37,128,160] Cromakalim hyperpolarized smooth muscle membranes in vascular tissue, in the trachea, and in the urinary bladder, and shifted the membrane potential forward to the K equilibrium potential. This hyperpolarizing effect has been confirmed by ^{86}Rb or ^{42}K efflux measurements in the rat portal vein,[24] guinea pig portal vein,[30,161,162] rat aorta,[22,160] rabbit aorta,[34,35,163] guinea pig trachea,[164,165] guinea pig taenia coli,[141] and guinea pig urinary bladder.[158,166] In general, the potency for activation of the K channel was 30 times or more higher for cromakalim than for nicorandil, though a heterogeneous sensitivity to cromakalim in various vascular tissues has been reported.[167] In the portal vein, the inhibition of spontaneous spike discharges induced by cromakalim was more prolonged than the membrane hyperpolarization, i.e., when the tissue was rinsed with Krebs solution following application of cromakalim, the membrane was rapidly depolarized to the level before application, yet the inhibition of spike generation was still present after complete restoration of the membrane potential.[162] It seems likely that the action of cromakalim is due not only to an increase in K permeability. When the effects of cromakalim or pinacidil were observed on the I_{oo}, the frequency of I_{oo} evoked by a depolarization of -10 mV was reduced under the voltage-clamp condition. Since the depolarization remained the same, the reduction in the frequency of I_{oo} may have been induced by means other than an inhibition of Ca release from the SR (Z. Xiong and K. Kitamura, personal communication).

In the rabbit aorta, norepinephrine (NE) did not depolarize the membrane, though cromakalim hyperpolarized it. NE produced a contraction through the action of synthesized IP_3 on the SR. As a conseqnence, dihydropyridine derivatives had no effect, whereas cromakalim inhibited this contraction.[36,41] On the other hand, NE both depolarized the membrane and produced a contraction in the rat aorta, and this contraction was more potently inhibited by cromakalim than by dihydropyridine.[168] This implies that cromakalim not only activates the K channel but has additional actions that contribute to the production of vasorelaxation. In fact, Okabe et al.[108] have recently reported that cromakalim inhibits the voltage-dependent Ca channel, as observed on application of dihydropyridine derivatives. This agent also partly inhibits IP_3 synthesis (J. Kajikuri and T. Itoh, personal communication). It is also reported that cromakalim inhibited the phasic and tonic component of the NE-induced contraction, and it is postulated that there are two classes of active sites: low (>0.01 μM) and high affinity (0.1 to 0.3 μM). All the actions of cromakalim are abolished in high K.[167]

Cromakalim also acts on other smooth muscle tissues such as those in the trachea, urinary bladder, intestine, and myometrium. In the guinea pig trachea, Allen et al.[164] reported that cromakalim inhibited both the spontaneous tone and the spontaneous slow waves generated in the presence of TEA or procaine. Arch et al.[169] reported that cromakalim is a more potent bronchodilator than nifedipine. In the urinary bladder,[158,166] cromakalim inhibited the spontaneous electrical and mechanical activities. In the guinea pig urinary bladder, cromakalim increased the ^{42}K efflux but not the ^{86}Rb efflux. In addition, in solution containing Rb instead of K, cromakalim produced neither hyperpolarization nor relaxation.[166] In the rat myometrium, Hollingsworth et al.[159,170] reported that this agent acts as a K-channel opener, but the hyperpolarization itself was much smaller than that observed in vascular tissues. Therefore, they concluded that cromakalim may act selectively on the pacemaker cells of the rat myometrium.

In the guinea pig taenia coli, Weir and Weston[14] reported that cromakalim increased the ^{86}Rb efflux and the hyperpolarization of the membrane. However, these actions were not blocked by apamin. In addition, cromakalim increased the synthesis of cyclic AMP in intestinal tissues.[171] Cromakalim acts more potently on the longitudinal muscle layer than on the circular layer, but the actions of cromakalim on intestinal smooth muscles were weaker than on vascular tissues. Much the same conclusion was reached after *in vivo* experiments on the peristalsis reflex.[172]

3. Pinacidil

As briefly described above, the antihypertensive agent pinacidil (a 4-pyridil cyanoguanidine derivative) exerts its vasorelaxing actions by acting as a K-channel opener.[32,34,157,173-175] This agent hyperpolarized vascular smooth muscle and produced vasodilation, mainly through an increase in the open probability of the K_{ATP} channel. However, this action was only one third as powerful as that of cromakalim. Pinacidil has enantiomer, with (−) enantiomer being more potent.[32,176,177] Pinacidil inhibited NE-induced contraction and increased the efflux of ^{86}Rb.[167,178,179]

4. Effects of K-Channel Openers on the SR Membrane

Nicorandil possesses a nitroglycerin-like action and synthesizes cyclic GMP.[13,17,19,30] Cyclic GMP activates the Ca pump mechanism induced by Mg-sensitive ATPase.[180] When nicorandil was applied to skinned muscle tissues prepared with saponin, this agent had no effect on the Ca-tension relationship.[15] Cromakalim had an inhibitory action on the voltage-dependent Ca channel as observed using the voltage-clamp procedure. This action of cromakalim was similar to the action of dihydropyridine derivatives as estimated from changes in the activation and inactivation curves.[108] Nakajima et al.,[181] reported that pinacidil had no effect on the Ca-induced contraction in skinned smooth muscle tissues. However, in the coronary artery, pinacidil enhanced the mechanical response induced by 118 mM K with no change in the cytosolic Ca concentration as estimated using fura-2, whereas it consistently inhibited the contraction and cytosolic Ca concentration activated by NE. The enhancement of the K-induced contraction is unlikely to be due to a reduction in the synthesis of IP_3 (S. Ito and T. Itoh, personal communication), but, on the other hand, the inhibition of the NE-induced contraction does have a close relation with inhibition of IP_3 synthesis (J. Kajikuri and T. Itoh, personal communication). Pinacidil inhibited the hydrolysis of $PI-P_2$ and reduced the amount of synthesized IP_3. This inhibition of IP_3 synthesis was prevented by application of glibenclamide. Therefore, if glibenclamide acts only as an antagonistic agent for K_{ATP} channel opening, hyperpolarization of the membrane induced by pinacidil may inhibit the synthesis of IP_3 in a voltage-dependent manner (T. Itoh and S. Suzuki, personal communication). In fact, cromakalim also inhibits the synthesis of IP_3 (T. Itoh and J. Kajikuri, unpublished observations). This means that cromakalim and pinacidil may also contribute to a reduction in Ca release from the SR and, thus, partly inhibit the mechanical response.

B. EFFECTS OF K-CHANNEL OPENERS ON THE K CHANNELS IN VASCULAR SMOOTH MUSCLE MEMBRANES MEASURED USING THE VOLTAGE- AND PATCH-CLAMP PROCEDURES

Nicorandil, cromakalim, and pinacidil activate the K channel and produce hyperpolarization of vascular smooth muscle cell membranes. To identify the K channel acted on by these K-channel openers, the voltage- and patch-clamp procedures have been applied. Figures 3 to 5 show the effects of cromakalim (Figure 3) and of nicorandil on the membrane currents recorded from the rat portal vein (in the presence of TEA in Figure 4, 4-AP in Figure 5).

Kusano et al.[182] reported that, in the bovine and rabbit aorta, cromakalim opens the Ca-dependent K channel (Maxi K). This result was confirmed by Gelband et al.[183] using the planar membrane which incorporates the K-channel vesicles into an artificial lipid payer and by Trieschmann et al.,[47] using myocytes of human mesenteric artery. Thus, this agent did not modify the unitary channel conductance (about 140 pS), but reduced the closed time of the channel and increased the open probability of the Ca-dependent K channel. Klöckner et al.[184] reported that not only cromakalim, but also diazoxide and (+)-niguldipine, a dihydropyridine derivative, act on the BK. Hermsmeyer[174] reported that, in the presence of 1 mM Ca on the cytosolic side, pinacidil activated the K channel. The above authors concluded that K-channel openers activate the Ca-dependent Maxi K.

On the other hand, Beech and Bolton[185] reported that, in the rabbit portal vein, cromakalim hyperpolarized the membrane by increasing a K current which showed little voltage dependency. This K channel was insensitive to 0.5 mM TEA, which lead Beech and Bolton to posit that this channel may be related to the delayed rectifier K current.

Following the discovery of the K_{ATP} channel and the application of glibenclamide,[147,148] a more detailed analysis on the action of K channel openers was carried out. Standen et al.,[129] using the patch-clamp procedure, reported that the open probability of the unitary BK current recorded from rat and rabbit mesenteric arteries was not modified by high concentrations of cromakalim. However, the Ca-independent large conductance K channel was sensitive both to cromakalim and to pinacidil. This channel was also sensitive to cytosolic ATP (1 mM ATP abolished the channel opening, and cromakalim opened this channel). Moreover, a reduction in the ATP concentration enhanced the open probability of the channel opening, and this action was

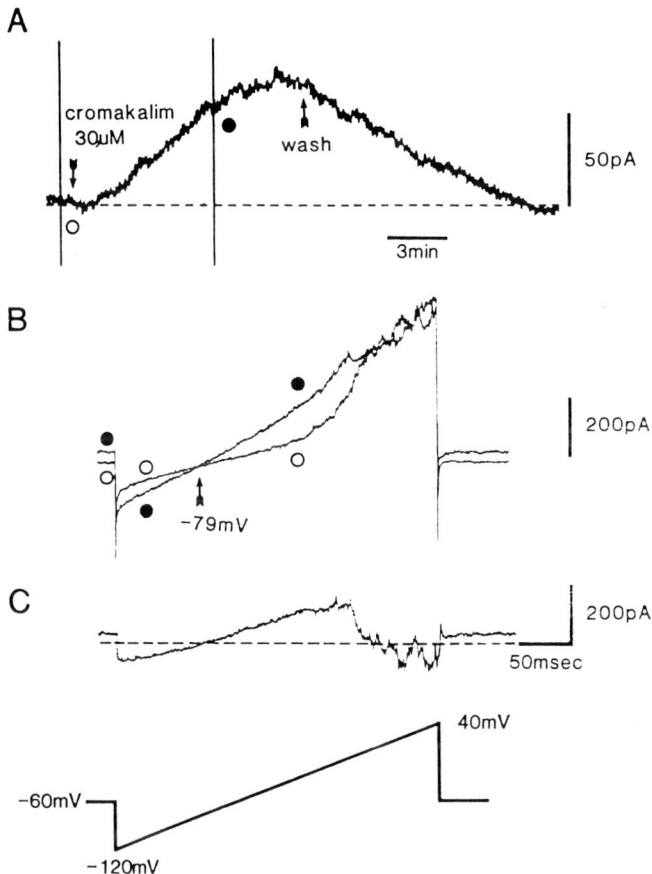

FIGURE 3. Effects of 30 µM cromakalim on the membrane current (A) and current-voltage relationship obtained in the presence or absence of cromakalim (B) in dispersed smooth muscle cells of the rat portal vein. In (A), the membrane potential was kept at 0 mV, and 30 µM cromakalim was superfused (indicated by arrows). The recording pipette contained high K solution (+ 0.3 mM EGTA), with normal PSS in the bath. The dotted line indicates the current level observed in PSS at the holding potential of −60 mV, (holding current). In (B), a ramp voltage pulse (from −120 to +40 mV, 300 ms in duration) was applied at the times marked with vertical lines in (A) (○, control; ●, cromakalim, 30 µM). The arrow in (B) indicates the crossing point of two I-V curves (−79 mV). (C) represents the net membrane current generated by application of 30 µM cromakalim and was obtained by subtraction of the two curves shown in (B). Dotted line indicates the basal current level. (From Okabe et al., *J. Pharmacol. Exp. Ther.*, 250, 832, 1990. With permission.)

prevented by glibenclamide. Furthermore, Standen reported that the hyperpolarization induced by release of endothelium-derived hyperpolarizing factor (EDHF[186,187]) was also blocked by glibenclamide. Thus, the hyperpolarization induced by EDHF is also due to activation of the K_{ATP} channel.

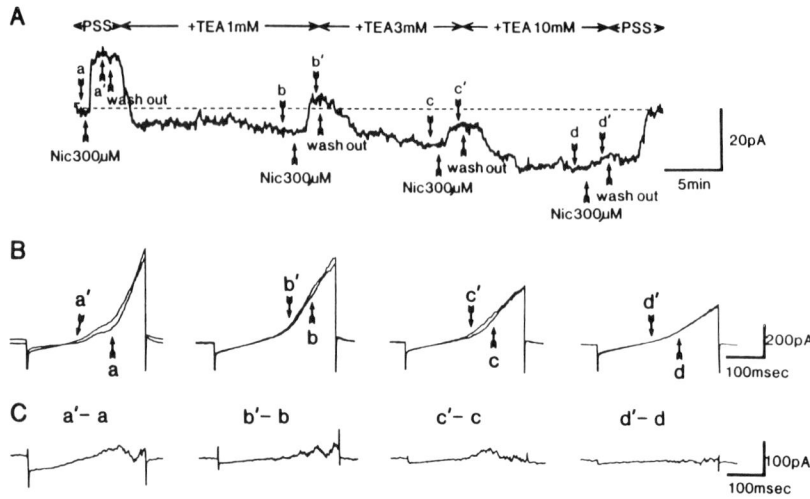

FIGURE 4. Effects of nicorandil (300 μM) on the membrane currents recorded in the absence and presence of various concentrations for TEA (rat portal vein). (A) TEA (1, 3, and 10 mM) were successively superfused in the bath, for period indicated by arrows; ----, represents current level observed before application of nicorandil in the absence of TEA. (B) The current-voltage relationships recorded by application of ramp potential (-120 mV to $+40$ mV) and each letter (a to d, a' to d') attached with current-voltage relation curves corresponds with the same letter indicated in (A). a to d were obtained in the absence of 300 μM nicorandil and a' to d' were in the presence of nicorandil. a and a' were obtained in the absence of TEA. b, b', c, c', d and d' were obtained in the presence of 1, 3, and 10 mM TEA, respectively. (C) Net membrane current induced by nicorandil in the absence and presence of TEA. Net current was obtained by subtracting the membrane current evoked by a ramp potential in the presence of nicorandil (a' to d') by that in the absence of nicorandil (a to d). (From Kajioka et al., *J. Pharmacol. Exp. Ther.*, 254, 905, 1990(a). With permission.)

By contrast, Okabe et al.[108] and Kajioka et al.,[109] using the voltage- and patch-clamp procedures in the rat portal vein, reported that cromakalim and nicorandil act on the same SK channel which is sensitive to cytosolic Ca, ATP, and glibenclamide (unitary conductance of 20 pS with high K in the bath and pipette, but 10 pS with 5.9 mM K in the bath and 140 mM K in the pipette). Figure 6 shows the effects of intracellularly applied ATP on the unitary K current (SK) recorded from dispersed smooth muscle cells of the rat portal vein.

Using the voltage-clamp procedure, cromakalim and nicorandil produced an outward current that was more potently inhibited by 4-AP than by TEA (Figures 4 and 5), and the amplitude of the outward current was markedly attenuated in Ca-free solution containing EGTA and Mn. When a ramp voltage (from -120 mV to $+40$ mV, duration 300 msec) was applied before and in the presence of cromakalim or nicorandil, the current was consistently enlarged in the voltage range -120 to 0 mV. The currents measured before and

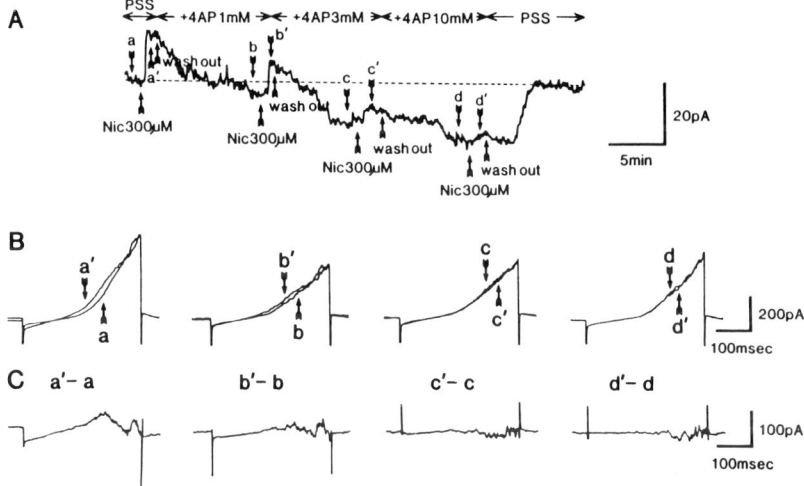

FIGURE 5. Effects of nicorandil (300 μM) on the membrane currents recorded in the absence and presence of various concentrations of 4-AP (rat portal vein). (A) 4-AP (1, 3, and 10 mM) were successively superfused in the bath, for period indicated by arrows. Current level observed before application of nicorandil in the absence of 4-AP. (B) The current-voltage relationships recorded by application of ramp potential (−120 mV to +40 mV) and each letter (a to d, a' to d') attached with current-voltage relation curves corresponds with the same letter indicated in (A). a to d were obtained in the absence of 300 μM nicorandil and a' to d' were in the presence of nicorandil. a and a' were obtained in the absence of 4-AP. b, b', c, c', d, and d' were obtained in the presence of 1, 3, and 10 mM 4-AP, respectively. (C) Net membrane current induced by nicorandil in the absence and presence of 4-AP. Net current was obtained by subtracting the membrane current evoked by a ramp potential in the presence of nicorandil (a' to d') by that in the absence of nicorandil (a to d). From Kajioka et al., *J. Pharmacol. Exp. Ther.*, 254, 905, 1990(a). With permission.)

after application of K-channel openers crossed at −80 mV. This voltage is similar to the K-equilibrium potential. Glibenclamide blocked the generation of the outward current evoked by cromakalim or nicorandil.[94,109] Much the same effects were observed in the case of pinacidil (S. Kajioka, personal communication). Using the cell-attached patch-clamp procedure, when cytosolic Ca was buffered with EGTA to a certain level through the pipette solution, the open probability was increased in a Ca concentration-dependent manner without any change in the unitary current amplitude and conductance. In addition, when the cytosolic ATP was removed this channel activity was enhanced, and at 5 mM ATP the channel opening scarecely occurred (Figure 6). This K_{ATP} channel was not modified by application of charybdotoxin and apamin. Furthermore, the BK was more sensitive to extracellularly applied TEA (blocked by 1 mM) than to 4-AP (blocked by over 10 mM) and the ATP-sensitive Maxi K was more sensitive to 4-AP than to TEA. This means that in the rat portal vein, cromakalim, nicorandil, and pinacidil act on the

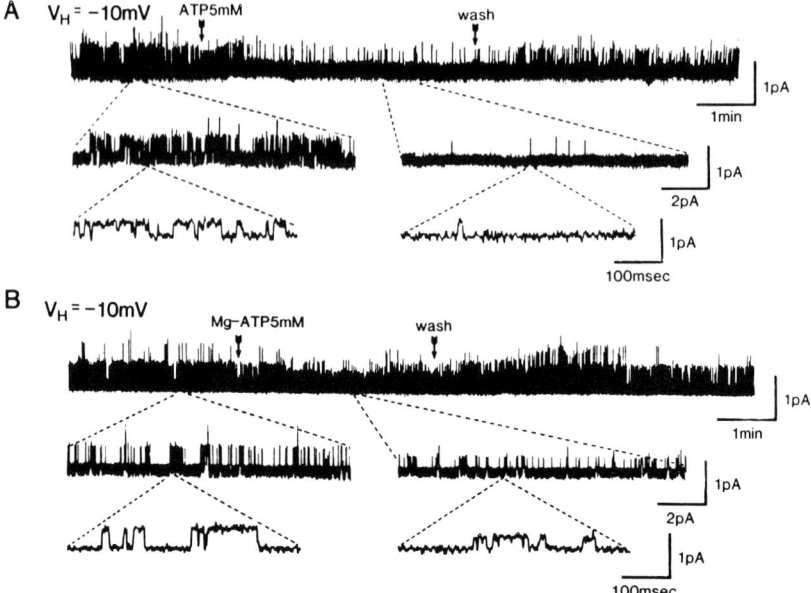

FIGURE 6. Effects of intracellularly applied Na_2ATP (A) and MgATP (B) on the 10-pS K channel (rat portal vein). ATP (as Na_2ATP in A and MgATP in B: 5 mM) was superfused in the bath using the inside-out patch membrane at the holding potential of -10 mV. PSS was filled in the pipette, and the high-K solution with the free Ca concentration of 1 μM (4 mM EGTA) was superfused in the bath. To inhibit the 132-pS K channel, 1 mM TEA was added. The pH of the high-K solution with or without 5 mM ATP was adjusted exactly to 7.25 just before application. Traces were also demonstrated with faster recording speed both in the absence and presence of ATP. (From Kajioka et al., *J. Pharmacol. Exp. Ther.*, 254, 905, 1990(a). With permission.)

same channel. This channel may be the same K_{ATP} channel as described by Beech and Bolton[185] using the same tissue. Figure 7 shows the effects of nicorandil on unitary K currents (Maxi K and SK) recorded from dispersed smooth muscle cells of the rat portal vein, using the cell-attached patch-clamp procedure, and Figure 8 shows the effects of glibenclamide on the accelerated channel opening of unitary K currents (SK) under pretreatment with pinacidil, using the cell-attached patch-clamp procedure.

In cultured cells prepared from the porcine coronary artery, Inoue et al.[186] reported that nicorandil acts on the extracellular Ca- and glibenclamide-sensitive K_{ATP} channel. As described previously, the Ca-sensitive K channel has been classified into cytosolic and extracellular Ca-sensitive K channels.[93] However, the extracellular Ca-sensitive K channel described by Inoue et al.,[93] differed from the K_{ATP} channel in its unitary current conductance and sensitivity. Therefore, the ATP- and glibenclamide-sensitive K channel which is sensitive to K channel openers possesses different natures in different vascular

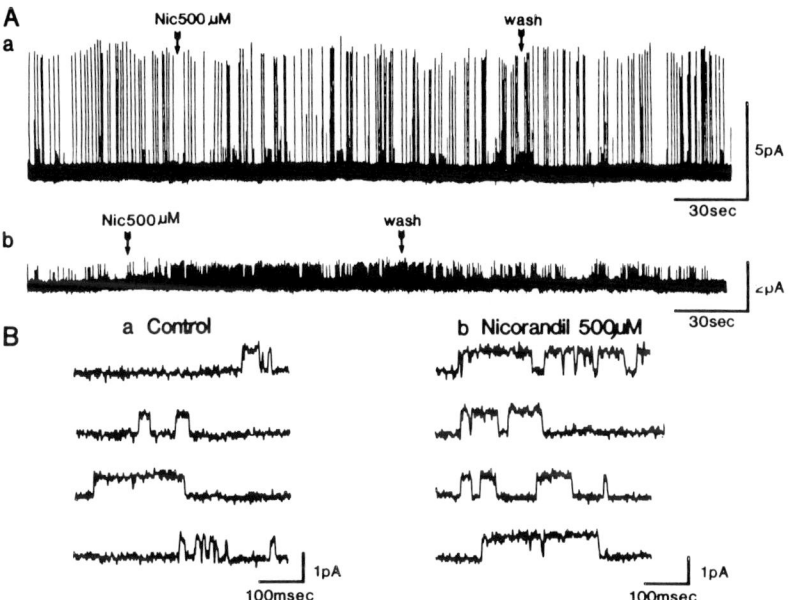

FIGURE 7. Effects of nicorandil (500 μM) on the single channel currents recorded at the membrane potential of +10 mV in the outside-out membrane patch configuration (rat portal vein). (A) Effects of nicorandil (500 μM) on activities of the 132-pS K-channel (a) or 10 pS K channel (b). Nicorandil was applied in the bath (extracellular side of the membrane) during the period indicated by arrows. In (a), opening of both 132-pS and 10-pS K channels was observed. In (b), the 132-pS K channel was blocked by 1 mM TEA. (B) Traces of the 10-pS K-channel current observed in the absence (a) and presence (b) of 500 μM nicorandil. Four typical trace segments (500 ms in duration) were chosen both in the absence (a) and presence (b) of nicorandil. (From Kajioka et al., *J. Pharmacol. Exp. Ther.*, 254, 905, 1990(a). With permission.)

tissues. Thus, the channel current recorded from the guinea pig mesenteric artery following application of K-channel openers is due to activation of the Ca-insensitive, large-conductance K channel,[187] whereas that recorded from bovine and human mesenteric arteries is due to activation of the Ca-sensitive, large conductance K channel.[184] That recorded from rat and guinea-pig portal vein is due to activation of the Ca-sensitive, small conductance K channel[94,108,109] and that recorded from the cultured porcine coronary artery is due to activation of the extracellular Ca-sensitive K channel.[186]

It may be that vascular tissues possess different types of glibenclamide-sensitive K_{ATP} channels. This would parallel the diverse nature of the channels and the effects of K channel openers on cardiac cells and on β-cells of the pancreas. For example, the glibenclamide-sensitive K_{ATP} channel in the β-cell was less effected than that in cardiac cells by K-channel openers, as estimated from ^{86}Rb efflux and only activated after application of 2-ketoiso-capoate or tolbutamide.[34,48,175,188] Diazoxide has a potent inhibitory action on

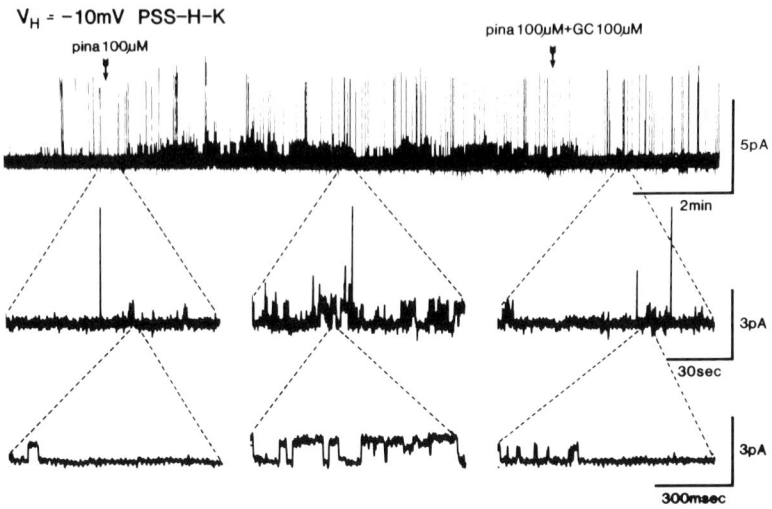

FIGURE 8. Effects of glibenclamide (GC; 100 μM) on unitary K currents under treatment with pinacidil (pina; 100 μM). The K currents were recorded using the cell-attached patch-clamp procedure (dispersed smooth muscle cells of the rat portal vein). PSS was applied in the bath and 140 mM K in the pipette. Note: pinacidil enhanced the channel opening of small unitary current (SK; 10 pS) without any effect on the large unitary current (Maxi K; 132 pS). In the presence of 100 μM pinacidil, glibenclamide blocked the channel opening of the small K unitary current activated by 100 μM pinacidil (Kajioka and Kitamura, unpublished observations).

the glibenclamide-sensitive K_{ATP} channel in the β-cell and increased the release of insulin and produced a lower plasma glucose level, but cromakalim had weaker actions on the release of insulin and hence on the plasma glucose level.[34,48,175] This means that cromakalim acts more potently on the K_{ATP} channel in vascular smooth muscle than in β-cells.

Regional differences in the action of K-channel openers have also been reported using the microelectrode method.[162] Thus, in the mesenteric artery, cromakalim transiently hyperpolarized the membrane, presumably due to rapid desensitization of the channel current, whereas when cromakalim was applied on the mesenteric vein, a long-lasting hyperpolarization, composed of fast and slow components, was recorded. When Mn was added in the solution, the fast component of the hyperpolarization was blocked. These results suggest that the K channel sensitive to K-channel openers in the mesenteric vein was composed of Ca-sensitive and -insensitive K channels.

Minoxidil had no effect on ^{86}Rb-efflux in aortic smooth muscle cells[50] and in this respect resembles cromakalim in the urinary bladder.[166,189] However, the vasorelaxant action of minoxidil was inhibited by glibenclamide.[50] Thus, K-channel openers can be classified into ^{86}Rb-sensitive and -insensitive ones. Thus, different types of ATP-sensitive K channels may be present.

Detailed experiments on the action of diazoxide on the K_{ATP} channel in vascular smooth muscle cells have not been carried out. In the insulin secreting cell line, Kozlowski et al.[190] reported that this agent has a dual action on the K_{ATP} channel, i.e., it causes activation and inactivation of the channel, the former occurring in the presence of Mg-ATP in the cytosol and the latter in the absence of Mg. Presumably, diazoxide may act on the same site as the K channel-opener antagonist, tolubutamide.

IV. ACTIONS OF POTASSIUM CHANNEL OPENERS IN RELATION TO CARDIOVASCULAR DISEASES

Nicorandil is already an established agent for the treatment of angina pectoris[5-7] and is commonly used for this purpose in Japan. Nicorandil has a nitroglycerin-like action and acts as a K-channel opener. In fact, it increased the amount of cyclic GMP,[15,17,18] the synthesis being induced by activation of cytosolic guanylate cyclase. In general, nicorandil acts more potently as a nitrate compound than as a K-channel opener. However, Aizawa et al.[191] clearly demonstrated actions of nicorandil different from those of nitroglycerin on the coronary circulation of anginal patients. These authors reported that nicorandil increases coronary sinus flow and decreases coronary vascular resistance, whereas nitroglycerin exerts the opposite effects. In patients with variant angina, nitroglycerin was a more effective coronary vasorelaxant than nicorandil and, whereas nicorandil relaxed first-order more than second order arteries, nitroglycerin was more effective in second-order vessels. Nicorandil also acts on stable effort angina.[192] These clinical actions of nicorandil may not be completely explained by its actions as a nitrate compound, and an action of nicorandil as a K-channel opener should not be neglected.[193] As a weak K-channel opener, this agent does not induce a reduction in systemic arterial blood pressure.[6]

Pinacidil has been studied extensively as an antihypertensive agent,[35] and only recently identified as a K-channel opener. The hyperpolarizing action of pinacidil is thought to be the main mechanism of its vasodepressant action.[34,35,173] This agent more potently activates the K channel and more strongly hyperpolarizes the membrane than does nicorandil. Side effects of pinacidil are reported to include weight gain, headache, and dyspnea on exertion.[194] This agent inhibited IP_3 synthesis and therefore, its vasodepressing action may not be solely due to K_{ATP} channel activation.

Cromakalim (lemakalim) is not a commercially available medication for clinical use but is undergoing clinical trials as an antihypertensive agent in healthy volunteers and hypertensive patients.[141,193,195-197] This agent also has the properties of a dihydropyridine-type Ca antagonist.[108] In patients, cromakalim has induced headache and edema as side effects caused by peripheral vasodilation. Whether it will be useful for clinical purposes is still unclear.

Diazoxide also possesses properties as an antihypertensive and hyperglycemic agent. In addition, this agent promotes hair growth, as also reported in the case of minoxidil.[167]

V. CONCLUSION

In this chapter, the biophysical features of K channels in vascular smooth muscle have been briefly reviewed. The actions of K-channel openers on the target K_{ATP} channel in vascular smooth muscle tissues were also described in relation to vascular diseases. Of course, this glibenclamide-sensitive K_{ATP} channel is not only distributed on vascular smooth muscle but also on intestine, urinary bladder, myometrium, etc. The actions of glibenclamide, a sulfonylurea, on the K_{ATP} channel were described in terms of its role as an antagonist of K-channel openers. The macroscopic K current in vascular smooth muscle cells was classified phenomenologically into several subtypes (delayed rectifying, inward rectifying (anomalous rectifying), voltage-dependent and -independent, etc. In addition, the channel was also classified according to its sensitivity to cytosolic or extracellular Ca. Furthermore, some agents, such as TEA, aminopyridine, apamin, charybdotoxin, noxiustoxin, dendrotoxin, phallodine, strychnine, etc. act as inhibitors of these different subtypes of the K channel via different mechanisms. The K_{ATP} channels distributed in smooth muscle cell membranes in vascular tissues differ in nature and density, as estimated from channel properties measured using the patch-clamp procedure or the ^{42}K- or ^{86}Rb-efflux measurement. The K channel distributed on smooth muscle membranes possesses features different from those of channels distributed on cardiac and β-cells of the pancreas, as estimated from the actions of ATP, Mg, and the biophysical properties (such as voltage-dependency) of the channel, and also from the actions of K-channel openers and blockers. The common feature is that this channel is a glibenclamide-sensitive K_{ATP} channel in vascular smooth muscle, cardiac muscle, and pancreatic β-cells.

Some K-channel openers and blockers such as diazoxide and glibenclamide would be effective on insulin-secretion-related diseases via the β-cells of the pancreas. In vascular tissues, other K-channel openers such as nicorandil, pinacidil, or chromakalim have been successfully applied or are intended to be applied to cardiovascular diseases. Nicorandil demonstrated weaker potency in activating the K_{ATP} channel than did cromakalim or pinacidil. However, the additional action of nicorandil as a nitro compound would enable it to have a double action in the treatment of angina pectoris, whereas more potent K-channel openers, such as pinacidil and cromakalim, may act as antihypertensive agents for patients if their side effects can be attenuated.

At present, further biophysical analysis of K channels is in progress. It is already clear that glibenclamide-sensitive K_{ATP} channels of different natures are present in different regions of vascular tissue. What is urgently required is clarification as to how many types of K_{ATP} channels are distributed in vascular tissues and how the heterogeneous features of the K_{ATP} channel contribute to the regulation of blood flow and to homeostasis.

ACKNOWLEDGMENT

We are grateful to Dr. R. J. Timms, Birmingham University, for language editing in the pre-production phase. The work was supported by Grants in Aid for Science Research from the Ministry of Education, Science and Culture in Japan.

REFERENCES

1. **Inoue, Y., Xiong, Z., Kitamura, K., and Kuriyama, H.**, Modulation produced by nifedipine of the unitary Ba current of dispersed smooth muscle cells of the rabbit ileum, *Pflügers Arch.*, 414, 534–542, 1989.
2. **Terada, K., Kitamura, K., and Kuriyama, H.**, Different inhibitions of the voltage-dependent K+ current by Ca2+ antagonists in smooth muscle cell membrane of rabbit small intestine, *Pflügers Arch.*, 409, 561–568, 1987.
3. **Furukawa, K., Itoh, T., Kajiwara, M., Kitamura, K., Suzuki, H., Ito, Y., and Kuriyama, H.**, Vasodilating actions of 2-nicotinamidoethyl nitrate on porcine and guinea-pig coronary arteries, *J. Pharmacol. Exp. Ther.*, 218, 248–259, 1981.
4. **Itoh, T., Furukawa, K., Kajiwara, M., Kitamura, K., Suzuki, H., Ito, Y., and Kuriyama, H.**, Effects of 2-nicotinamidoethyl nitrate on smooth muscle cells and on adrenergic transmission in the guinea-pig and porcine mesenteric arteries, *J. Pharmacol. Exp. Ther.*, 218, 260–270, 1981.
5. **Uchida, Y.**, Decreasing potassium conductance — a possible mechanism of phasic coronary vasospasm, *Jpn. Circ. J.*, 49, 128–139, 1985.
6. **Uchida, Y., Yoshimoto, N., and Murao, S.**, Effects of SG-75 (2-nicotinamido ethyl nitrate) on coronary circulation, *Jpn. Heart J.*, 19, 112–124, 1978.
7. **Taira, N., Satoh, K., Yanagisawa, T., Imai, Y., and Hiwatari, M.**, Pharmacological profile of a new coronary vasodilator drug, 2-nicotinamido-ethyl nitrate (SG-75), *Clin. Exp. Pharmacol. Physiol.*, 6, 301–316, 1979.
8. **Itoh, T., Kuriyama, H., and Suzuki, H.**, Excitation-contraction coupling in smooth muscle cells of the guinea-pig mesenteric artery, *J. Physiol.*, 321, 515–535, 1981.
9. **Karashima, T., Itoh, T., and Kuriyama, H.**, Effects of 2-nicotinamidoethyl nitrate on smooth muscle cells of the guinea-pig mesenteric and portal veins, *J. Pharmacol. Exp. Ther.*, 221, 472–480, 1982.
10. **Inoue, T., Ito, Y., and Takeda, T.**, The effects of 2-nicotinamidoethyl nitrate on smooth muscle cells of the dog mesenteric artery and trachea, *Br. J. Pharmacol.*, 80, 459–470, 1983.
11. **Inoue, T., Kammuura, Y., Fujiwara, T., Itoh, T., and Kuriyama, H.**, Effects of 2-nicotineamidoethylnitrate (nicorandil; SG 75) and its derivatives on smooth muscle cells of the canine mesenteric artery, *J. Pharmacol. Exp. Ther.*, 229, 793, 1984.

12. **Kajiwara, M., Droogmans, G., and Casteels, R.,** Effects of 2-nicotinamidoethyl nitrate (Nicorandil) on excitation-contraction coupling in the smooth muscle cells of rabbit ear artery, *J. Pharmacol. Exp. Ther.,* 230, 462–468, 1984.
13. **Allen, S. L., Foster, R. W., Morgan, G. P., and Small, R. C.,** The relaxant action of nicorandil in guinea-pig isolated trachealis, *Br. J. Pharmacol.,* 87, 117–127, 1986.
14. **Weir, S. W. and Weston, A. H.,** Effects of apamin on responses to BRL34915, nicorandil and other relaxants in the guinea-pig taenia caeci, *Br. J. Pharmacol.,* 88, 113–120, 1986.
15. **Sumimoto, K., Domae, M., Yamanaka, K., Nakao, K., Hashimoto, T., Kitamura, K., and Kuriyama, H.,** Actions of nicorandil on vascular smooth muscles, *J. Cardiovasc. Pharmacol.,* 10 (Suppl. 8), 66–75, 1987.
16. **Hamilton, T. C. and Weston, A. H.,** Cromakalim, nicorandil and pinacidil: novel drugs which open potassium channels in smooth muscles, *Gen. Pharmacol.,* 20, 1–9, 1989.
17. **Holzmann, S.,** Cyclic GMP a possible mediator of coronary arterial relaxation by nicorandil (SG-75), *J. Cardiovasc. Pharmacol.,* 5, 364–370, 1983.
18. **Schmidt, K., Reich, R., and Kukovetz, W. R.,** Stimulation of coronary guanylate cyclase by nicorandil (SG-75) as a mechanism of its vasodilating action, *Adv. Cyclic Nucleotide Res.,* 10, 43–53, 1985.
19. **Murad, M., Mittal, C. K., Arnold, W. P., Katsuki, S., and Kimura, H.,** Guanylatecyclase activation by azido, nitrocompounds, nitric oxide, and hydroxyl radical and inhibition by hemoglobin and myoglobin, *Adv. Cyclic Nucleotide Res.,* 9, 131–143, 1978.
20. **Kukovetz, W. R. and Holzmann, S.,** Cyclic GMP in nicorandil-induced vasodilation and tolerance development, *J. Cardiovasc. Pharmacol.,* 10, S25–S30, 1987.
21. **Ashwood, V. A., Cassidy, F., Evans, J. M., Faruk, E. A., and Hamilton, T. C.,** Trans-4-cyclicamido-3,4-dihydro-^{2}H-benzopyran-3-ols as antihypertensive agents, *Proc. VIIIth Int. Med. Chem. Symp. Vol. 1,* (Uppsala), Dahlbom, R. and Nilsson, J. L. G., Eds., Swedish Pharmaceutical Press, Stockholm, 1984, 316.
22. **Weir, S. W. and Weston, A. H.,** The effects of BRL 34915 and nicorandil on electrical and mechanical activity and on ^{86}Rb efflux in rat blood vessels, *Br. J. Pharmacol.,* 88, 121–128, 1986.
23. **Ashwood, V. A., Buckingam, R. E., Cassidy, F., Evans, J. M., Faruk, E. A., Hamilton, T. O., Nash, D. J., Stemp, G., and Willcoks, K.,** Synthesis and antihypertensive activity of 4-(cyclic amido)-2H-1-benzopyrans, *J. Med. Chem.,* 29, 2194–2201, 1986.
24. **Hamilton, T. C., Weir, S. W., and Weston, A. H.,** Comparison of the effects of BRL34915 and verapamil on electrical and mechanical activity in rat portal vein, *Br. J. Pharmacol.,* 88, 103–111, 1986.
25. **Coldwell, M. C. and Howlett, D. R.,** BRL 34915 induced potassium channel activation in rabbit isolated mesenteric artery is not calcium dependent, *Br. J. Pharmacol.,* 88, 443p, 1986.
26. **Coldwell, M. C. and Howlett, D. R.,** Specificity of action of the novel antihypertensive agent, BRL 34915, as a potassium channel activator; comparison with nicorandil, *Biochem. Pharmacol.,* 36, 3663–3669, 1987.
27. **Clapham, J. C. and Wilson, C.,** Anti-spasmogenic and spasmolytic effects of BRL 34915: a comparison with nifedipine and nicorandil, *J. Autonom. Pharmac.,* 7, 1–10, 1987.
28. **Kreye, V. A. W., Gerstheimer, F., and Weston, A. H.,** Effect of BRL 34915 on resting membrane potential and ^{86}Rb efflux in rabbit tonic vascular smooth muscle, *Naunyn-Schmiedebergs Arch. Pharmacol.,* 335, R64, 1987.
29. **Kreye, V. A. W., Gerstheimer, F., and Weston, A. H.,** Effects of the antihypertensive, BRL 34915, on membrane potential and ^{86}Rb efflux in rabbit tonic, vascular smooth muscle, *Pflügers Arch.,* 408, R79, 1987.

30. **Quast, U.**, Effect of the K^+ efflux stimulating vasodilator BRL 34915 on ^{86}Rb efflux and spontaneous activity in guinea pig portal vein, *Br. J. Pharmacol.,* 91, 569–578, 1987.
31. **Southerton, J. S. and Weston, A. H.**, Some effects of Ca^{2+} and K^+-channel-blocking agents on responses to BRL 34915 and pinacidil in the isolated rat portal vein, *J. Physiol.,* 391, 77p, 1987.
32. **Southerton, J. S., Weston, A. H., Bray, K. M., Newgreen, D. T., and Taylor, S. G.**, The potassium channel opening action of pinacidil: studies using biochemical, ion flux and microelectrode techniques, *Naunyn-Schmiedebergs Arch. Pharmacol.,* 338, 310–318, 1988.
33. **Cox, R. H. and Ferrone, R. A.**, Effects of BRL 34915 on ^{86}Rb efflux and active stress in rat vascular smooth muscle, *FASEB J.,* 2, A756, 1988.
34. **Cook, N. S., Quast, U., Hof, R. P., Baumlin, Y., and Pally, C.**, Similarities in the mechanism of action of the two new vasodilator drugs, pinacidil and BRP 34915, *J. Cardiovasc. Pharmacol.,* 11, 90–99, 1988.
35. **Cook, N. S., Quast, U., and Eier, S. W.**, *In vitro* and *in vivo* comparison of two K^+ channel openers: diazoxide and BRL 34915, *Pflügers Arch.,* 411, R46, 1988.
36. **Cook, N. S., Weir, S. W., and Danzeisen, M. C.**, Anti-vasoconstrictor effects of the K+ channel opener cromakalim on the rabbit aorta — comparison with the calcium antagonist isradipine, *Br. J. Pharmacol.,* 95, 741–752, 1988.
37. **Buckingham, R. E., Clapham, J. C., Hamilton, T. C., Longman, S. D., Norton, J., and Poyser, R. H.**, BRL34915, a novel antihypertensive agent: comparison by effects on blood pressure and other haemodynamic parameters with those of nifedipine in animal models, *J. Cardiovasc. Pharmacol.,* 8, 798–804, 1986.
38. **Petersen, H. J., Kaegaard-Nielsen, C., Arrigoni-Martelli, E.**, Synthesis and hypotensive activity of N-alkyl-N''-cyano-N'-pyridylguanidines, *J. Med. Chem.,* 21, 773–781, 1978.
39. **Arrigoni-Martelli, E. and Finucaine, J.**, Pinacidil, in *New Cardiovascular Drugs,* Scriabine, A., Ed., Raven Press, New York, 1985, 133.
40. **Cohen, M. L.**, Pinacidil monohydrate — a novel vasodilator, *Drug Dev. Res.,* 6, 1–10, 1986.
41. **Bray, K. M. and Weston, A. H.**, Effects of K^+ channel openers on the spasmogenic component of caffeine-induced responses in rabbit aorta, *Br. J. Pharmacol.,* 96, 220p, 1989.
42. **Meisheri, K. D. and Cipkus, L. A.**, Minoxidil sulphate acts as a K^+ channel agonist to produce vasodilation. *Fed. Proc.,* 46 (Abstr.), 1383, 1987.
43. **Bray, K. M., Weston, A. H., MacHarg, A. D., Newgreen, D. T., Southerton, J. S., and Duty, S.**, Further studies on the actions of the K-channel-openers cromakalim (BRL34915) and pinacidil in rabbit aorta, *Pflügers Arch.,* 411, R202, 1988.
44. **Meisheri, K. D., Cipkus, L. A., and Taylor, C. J.**, Mechanism of action of minoxidil sulphate-induced vasodilation: a role for increased K^+ permeability, *J. Pharmacol. Exp. Ther.,* 245, 751–760, 1988.
45. **Dollery, C. T., Pentecost, B. L., and Samaan, N. A.**, Drug induced diabetes, *Lancet,* 2, 735–737, 1962.
46. **Talbachnik, I. I. A. and Gulbenkian, A.**, Mechanism of diazoxide hyperglycaemia in animals, *Ann. N.Y. Acad. Sci.,* 150, 204–218, 1968.
47. **Trieschmann, U., Pichlmaier, M., Klöckner, U., and Isenberg, G.**, Vasorelaxation due to K-agonists: single channel recordings from isolated human vascular myocytes, *Pflügers Arch.,* 411, R199, 1988.
48. **Quast, U. and Cook, N. S.**, Moving together: K^+ channels openers and ATP-sensitive K^+ channels, TIPS, 10, 431–435, 1989.

49. **Winquist, R. J., Heaney, L. A., Wallace, A. A., Baskin, E. P., Stein, R. B., Garcia, M. L., and Kaczoroeski, G. J.**, Glyburide blocks the relaxation responses to BRL34915 (Cromakalim), minoxidil sulfate and diazoxide in vascular smooth muscle, *J. Pharmacol. Exp. Ther.*, 248, 149–156, 1989.
50. **Newgreen, D. T., Longmore, J., and Weston, A. H.**, The effect of glibenclamide on the action of cromakalim, diazoxide, and minoxidil on rat aorta, *Br. J. Pharmacol.*, 95, 116p, 1989.
51. **Mondot, S., Mestre, M., Caillard, C. G., and Cavero, I.**, RP49356: a vasorelaxant agent with potassium channel activating properties, *Br. J. Pharmacol.*, Suppl. 95, 8139, 1988.
52. **Thuringer, D. and Escando, D.**, The potassium channel opener RP49356 modifies the ATP-sensitivity of K^+-ATP channels in cardiac myocytes, *Pflügers Arch.*, 414 (Suppl. 1), 175, 1989.
53. **De Peyer, L. E., Lues, I., Gericke, R., and Häusler, G.**, Characterization of a novel $K+$ channel activator, EMD52962, in electrophysiological and pharmacological experiments, *Pflügers Arch.*, 414 (Suppl 1), 191, 1989.
54. **Edward, G. and Weston, A. H.**, Structure-activity relationships of K^+ channel openers, *Trend Pharmacol. Sci.*, 11, 417–422, 1990.
55. **Weston, A. H.**, Some effects of nicorandil on the smooth muscles of the rat and guinea-pig, *J. Cardiovasc. Pharmacol.*, 10 (Suppl. 8), 56–61, 1987.
56. **Carmeliet, E.**, K^+ channels in cardiac cells: mechanisms of activation, inactivation, rectification and K^+_e sensitivity, *Pflügers Arch.*, 414 (Suppl. 1), 88–92, 1989.
57. **Coraboeuf, E. and Carmeliet, E.**, Existence of two transient outward currents in sheep cardiac Purkinje fibres, *Pflügers Arch.*, 392, 352–359, 1982.
58. **Giles, W. R. and Van Ginneken, A. C. G.**, A transient outward current in isolated cells from the crista terminalis of rabbit heart, *J. Physiol.*, 368, 243–264, 1985.
59. **Noble, D. and Tsien, R. W.**, Outward membrane currents activated in the plateau range of potential in cardiac Purkinje fibres, *J. Physiol.*, 200, 205–231, 1969.
60. **Shibasaki, T.**, Conductance and kinetics of delayed rectifier channels in nodal cells of the rabbit heart, *J. Physiol.*, 387, 227–250, 1987.
61. **Tohse, N., Kameyama, M., and Irisawa, H.**, Intracellular $Ca2+$ and protein kinase C modulate $K+$ current in guinea pig heart cells, *Am. J. Physiol.*, 253, H1321–H1324, 1987.
62. **Scamps, F. and Carmeliet, E.**, The effect of external K^+ on the delayed K^+ current in single rabbit Purkinje cells, *Pflügers Arch.*, 414, S169–170, 1989.
63. **Yue, D. T. and Marban, E.**, A novel cardiac potassium channel that is active and conductive at depolarized potentials, *Pflügers Arch.*, 413, 127–133, 1988.
64. **Kurachi, Y.**, Voltage-dependent activation of the inward-rectifier potassium channel in the ventricular membrane of guinea-pig heart, *J. Physiol.*, 366, 365–385, 1985.
65. **Kameyama, M., Kakei, M., Sato, R., Shibasaki, T., Matsuda, H., and Irisawa, H.**, Intracellular Na^+ activates a K^+ channel in mammalian cardiac cells, *Nature*, 309, 354–356, 1984.
66. **Noma, A.**, ATP-regulated K^+ channels in cardiac muscle, *Nature*, 305, 147–148, 1983.
67. **Noma, A. and Trautwein, W.**, Relaxation of the Ach-induced potassium current in the rabbit sinoatrial node cell, *Pflügers Arch.*, 377, 193–200, 1978.
68. **Belardinelli, L. and Isengberg, G.**, Isolated atrial myocytes: adenosine and acetylcholine increase potassium conductance, *Am. J. Physiol.*, 244, 734, 1983.
69. **Kurachi, Y., Nakajima, T., and Sugimoto, T.**, Acetylcholine activation of $K+$ channels in cell-free membrane of atrial cells, *Am. J. Physiol.*, 251, H681–684, 1966.
70. **Yatani, A., Codina, J., Brown, A. M., and Birnbaumer, L.**, Direct activation of mammalian atrial muscarinic potassium channels by GTP regulatory protein G_K, *Science*, 238, 1288–1292, 1987.

71. **Yatani, A., Mattera, R., Codina, J., Graf, R., Okabe, K., Padrell, E., Iyengar, R., Brown, A. M., and Birnbaumer, L.,** The G protein-gated atrial K^+ channel is stimulated by three distinct G_i α-subunits, *Nature,* 336, 680–682, 1988.
72. **Okabe, K., Kitamura, K., and Kuriyama, H.,** The existence of a highly tetrodotoxin sensitive Na channel in freshly dispersed smooth muscle cells of the rabbit main pulmonary artery, *Pflügers Arch.,* 411, 423–428, 1988.
72a. **Okabe, K., Kitamura, K., and Kuriyama, H.,** Features of 4-aminopyridine sensitive outward current observed in single smooth muscle cells from the rabbit pulmonary artery, *Pflügers Arch.,* 410, 69–74, 1987.
73. **Benham, C. D., Bolton, T. B., Lang, R. J., and Takewaki, T.,** The mechanism of action of Ba^{2+} and TEA on single Ca^{2+}-activated K^+ channels in arterial and intestinal smooth muscle cell membrane, *Pflügers Arch.,* 403, 120–127, 1985.
74. **Ohya, Y., Terada, K., Kitamura, K., and Kuriyama, H.,** Membrane currents recorded from a fragment of rabbit intestinal smooth muscle cells, *Am. J. Physiol.,* 251, C335–C346, 1986.
75. **Ohya, Y., Terada, K., Yamaguchi, K., Inoue, R., Okabe, K., Kitamura, K., Hirata, M., and Kuriyama, H.,** Effects of inositol phosphates on the membrane activity of smooth muscle cells of the rabbit portal vein, *Pflügers Arch.,* 412, 382–389, 1988.
76. **Sakai, T., Terada, K., Kitamura, K., and Kuriyama, H.,** Ryanodine inhibits the Ca-dependent K_L current after depletion of Ca stored in smooth muscle cells of the rabbit ileal longitudinal muscle, *Br. J. Pharmacol.,* 95, 1089–1100, 1988.
77. **Bolton, T. B. and Lim, S. P.,** Properties of calcium stores and transient outward currents in single smooth muscle cells of rabbit intestine, *J. Physiol.,* 409, 385–401, 1989.
78. **Hume, J. R. and Leblane, N.,** Macroscopic K+ currents in single smooth muscle cells of the rabbit portal vein, *J. Physiol.,* 413, 49–73, 1989.
79. **Komori, S. and Bolton, T.,** Role of G-proteins in muscarinic receptor inward current and outward currents in rabbit jejunal smooth muscle, *J. Physiol.,* 427, 395–419, 1990.
80. **Vincent, J., Schweitz, H., and Lazdunski, M.,** Structure-function relationships and site of action of apamin, a neurotoxic polypeptide of bee venom with an action on the central nervous system, *Biochemistry,* 14, 2521–2525, 1975.
81. **Herbermann, E.,** Apamin, *Pharmacol. Ther.,* 25, 255–270, 1984.
82. **Itoh, T., Ueno, H., and Kuriyama, H.,** Calcium-induced calcium release mechanism in vascular smooth muscles — assessments based on contractions evoked in intact and saponin-treated skinned muscles, *Experientia,* 41, 989–996, 1985.
83. **Kobayashi, S., Somlyo, A. V., and Somlyo, A. P.,** Heparin inhibits the inositol 1,4,5-trisphosphate-dependent, but not the independent, calcium release induced by guanine nucleotides in vascular smooth muscle, *Biochem. Biophys. Res. Commun.,* 153, 625–631, 1988.
84. **Iino, M., Kobayashi, T., and Endo, M.,** Use of ryanodine for functional removal of the Ca store in smooth muscle cells of the guinea-pig, *Biophys. Biochem. Res. Commun.,* 152, 417–422, 1988.
85. **Takeshima, H., Nishimura, S., Matsumoto, T., Ishida, H., Kangawa, K., Minamino, N., Matsuo, H., Ueda, M., Hanaoka, M., Hirose, T., and Numa, S.,** Primary structure and expression from complementary DNA of skeletal muscle ryanodine receptor, *Nature,* 339, 439–445, 1989.
86. **Furuichi, T., Yoshikawa, S., Miyawaki, A., Wada, K., Maeda, N., and Mikoshiba, K.,** Primary structure and functional expression of the inositol 1,4,5-trisphosphate-binding protein P_{400}, *Nature,* 342, 32–38, 1989.
87. **Berridge, M. J., Cobbold, P. H., and Cuthbertson, K. S. R.,** Spatial and temporal aspects of cell signalling, *Philo. Trans. R. Soc. London Ser.,* B, 320, 324–342, 1988.
88. **Cobbold, P., Daly, M., Dixon, J., and Woods, N.,** Repetitive calcium transient in hormone-stimulated cells, *Biochem. Soc. Transact.,* 17, 69–72, 1989.

89. **Irvine, R. F., Letcher, A. J., Heslop, J. P., and Berridge, M. J.,** The inositol tris/tetrakisphosphate pathway-demonstration of Ins(1,4,5)P$_3$ 3-kinase activity in animal tissues, *Nature,* 320, 631–634, 1986.
90. **Irvine, R. F.,** How do inositol 1,4,5-triphosphate and inositol 1,3,4,5-tetrakisphosphate regulate intracellular Ca^{2+}? *Biochem. Soc. Transact.,* 17, 6–9, 1989.
91. **Wakui, M., Potter, B. V. L., and Petersen, O. H.,** Pulsatile intracellular calcium release does not depend on fluctuations in inositol trisphosphate concentration, *Nature,* 339, 317–320, 1989.
92. **Inoue, R., Kitamura, K., and Kuriyama, H.,** Two Ca-dependent K-channels classified by the application of tetraethylammonium distribute to smooth muscle membranes of the rabbit portal vein, *Pflügers Arch.,* 405, 173–179, 1985.
93. **Inoue, R., Okabe, K., Kitamura, K., and Kuriyama, H.,** A newly-identified Ca^{2+}-dependent K$^+$-channel in the smooth muscle membrane of single cells dispersed from the rabbit portal vein, *Pflügers Arch.,* 406, 138–143, 1986.
94. **Kajioka, S., Oike, M., and Kitamura, K.,** Nicorandil opens a calcium-dependent potassium-channel in smooth muscle cells of the rat portal vein, *J. Pharmacol. Exp. Ther.,* 254, 905–913, 1990.
95. **Blatz, A. L. and Magleby, K. L.,** Single apamin-blocked Ca-activated K$^+$ channels of small conductance in cultured rat skeletal muscle, *Nature,* 323, 718–720, 1986.
96. **Capiod, T. and Ogden, D. C.,** The properties of calcium-activated potassium channels in guinea-pig isolated hepatocytes, *J. Physiol.,* 409, 285–295, 1989.
97. **Burgess, G. M., Claret, M., and Jenkinson, D. H.,** Effects of quinine and apamin on the calcium dependent potassium permeability of mammalian hepatocytes and red cells, *J. Physiol.,* 317, 67–90, 1981.
98. **Hermann, A. and Hartung, K.,** Ca^{2+}-activated K$^+$ conductance in molluscan neurones, *Cell Calcium,* 4, 387–405, 1983.
99. **Romey, G., Hughes, M., Schmid-Antomarchi, H., and Lazdunski, M.,** Apamin; a specific toxin to study a class of Ca^{2+}-dependent K$^+$ channels, *J. Physiol. (Paris),* 79, 259–264, 1984.
100. **Sneddon, P. and Burnstock, G.,** ATP as a co-transmitter in rat tail artery, *Eur. J. Pharmacol.,* 105, 149–152, 1985.
101. **Burnstock, G. and Kennedy, C.,** Purinergic receptors in the cardiovascular system, *Prog. Pharmacol.,* 6(2), 111–132, 1986.
102. **Lazdunski, M., Romey, G., Schmid-Antomarchi, H., Renaud, J. F., Mourre, C., Hugues, M., and Fosset, M.,** The apamin-sensitive Ca2+-dependent K+ channel: molecular properties, differentiation, involvement in muscle diseases, and endogenous ligands in mammalian brain, *Handb. Exp. Pharmacol.,* 83, 135–145, 1988.
103. **Miller, C., Moczydlowski, E., Lattore, R., and Phillips, M.,** Charybtoxin, a protein inhibitor of single Ca^{2+}-activated K$^+$ channels from mammalian skeletal muscle, *Nature,* 313, 316–318, 1985.
104. **Smith, C., Philips, M., and Miller, C.,** Purification of charybdotoxin a specific inhibitor of the high-conductance Ca^{2+}-activated K$^+$ channel, *J. Biol. Chem.,* 261, 14,607–14,613, 1986.
105. **Moczydlowski, E., Lucchesi, K., and Ravindran, A.,** An emerging pharmacology of peptide toxins targeted against potassium channels, *J. Membr. Biol.,* 105, 95–111, 1988.
106. **Gimenez-Gallego, G., Navia, N. A., Reuben, J. P., Katz, G. M., Kaczorowski, G. J., and Garcia, M. L.,** Purification, sequence, and model structure of charybdotoxin, a potent selective inhibitor of calcium-activated potassium channels, *Proc. Natl. Acad. Sci. U.S.A.,* 85, 3329–3333, 1988.
107. **Valdivia, H. H., Smith, J. S., Martin, B. M., Coronado, R., and Possani, L. D.,** Charybdotoxin and noxiustoxin, two homologous peptide inhibitors of the K$^+$ (Ca^{2+}) channel, *FEBS Lett.,* 226, 280–284, 1988.

108. **Okabe, K., Kajioka, S., Nakao, K., Kitamura, K., Kuriyama, H., and Weston, A. H.**, Actions of cromakalim on ionic currents recorded from single smooth muscle cells of the rat portal vein, *J. Pharmacol. Exp. Ther.*, 250, 832–839, 1990.
109. **Kajioka, S., Oike, M., Kitamura, K., and Kuriyama, H.**, Nicorandil opens a Ca-dependent and ATP-sensitive potassium channel in the smooth muscle cells of the rat portal vein, *Jpn. J. Pharmacol.*, 52 (Suppl. 1), 81p., 1990.
110. **Sims, S. M., Singer, J.J., and Walsh, J. V., Jr.**, Cholinergic agonists suppress a potassium current in freshly dissociated smooth muscle cells of the toad, *J. Physiol.*, 367, 503–529, 1985.
110a. **Ashcroft, M. J.**, Adenosine 5'-triphosphate-sensitive potassium channels, *Annu. Rev. Neurosci.*, 11, 97, 1988.
111. **Ashcroft, M. J.**, Potassium channels and modulation of insulin secretion, in *Potassium Channels: Structure, Classification, Function and Therapeutic Potential*, Cook, N. S., Ed., Ellis Horwood, New York, 1990, 300.
112. **Ashcroft, F. M., Kakei, M., Kelly, R. P., and Sutton, R.**, ATP-sensitive K-channels in isolated human pancreatic β-cells, *FEBS Lett.*, 215, 9, 1987.
113. **Ashcroft, F. M., Ashcroft, S. J. H., Berggren, P. O., Betzholz, C., Rorsman, P., Trube, G., and Welsh, M.**, Expression of K channels in xenopus laevis oocytes injected with poly (A^+)mRNA from the insulin-secreting β-cell line, HITT15, *FEBS Lett.*, 239, 185–189, 1988.
114. **DeWeille, J. R., Fosset, M., Mourre, C., Schmid-Antomarchi, H., Bernardi, H., and Lazdunski, M.**, Pharmacology and regulation of ATP-sensitive K^+ channels, *Pflügers Arch.*, 414 (Suppl. 1), 80–87, 1989.
115. **Escande, D.**, The pharmacology of ATP-sensitive K+ channels in the heart, *Pflügers Arch.*, 414 (Suppl. 1), 93–98, 1989.
116. **Petersen, O. H. and Dunne, M. J.**, Regulation of K^+ channels plays a crucial role in the control of insulin secretion, *Pflügers Arch.*, 414 (Suppl. 1), 115–120, 1989.
117. **Rorsman, P. and Trube, G.**, Biophysics and physiology of ATP-regulated K^+ channels (K_{ATP}), in *Potassium Channels, Structure, Classification, Function and Therapeutic Potential*, Cook, N. G., Ed., Ellis Horwood Ltd., New York, 1990, 96–116.
118. **Kakei, M., Noma, A., and Shibasaki, T.**, Properties of adenosine-triphosphate regulated potassium channels in guinea pig ventricular cells, *J. Physiol.*, 363, 441–462, 1985.
119. **Spruce, A. E., Standen, N. B., and Stanfield, P. R.**, Studies of the unitary properties of adenosine-5'-triphosphate-regulated potassium channels of frog skeletal muscle, *J. Physiol.*, 382, 213–236, 1987.
120. **Trube, G. and Hescheler, J.**, Inward-rectifying channels in isolated patches of the heart cell membrane: ATP-dependence and comparison with cell-attached patches, *Pflügers Arch.*, 401, 178–184, 1984.
121. **Zilberter, Y. U., Barnashev, N., Papin, A., Portnov, V., and Khodorov, B.**, Gating kinetics of ATP-sensitive single potassium channels in myocardial cells depends on electromotive force, *Pflügers Arch.*, 411, 584–589, 1988.
122. **Noma, A. and Shibasaki, T.**, Membrane current through adenosine-triphosphate regulated potassium channels in guinea-pig ventricular cells, *J. Physiol.*, 363, 463–480, 1985.
123. **Bernardi, H., Fosset, N., and Lazdunski, M.**, Characterization, purification, and affinity labelling of the brain [^{3}H] glibenclamide-binding protein, a putative neuronal ATP-regulated K^+ channel, *Proc. Natl. Acad. Sci. U.S.A.*, 85, 9816–9820, 1988.
124. **Ben Ari, Y.**, Effect of glibenclamide, a selective blocker of an ATP-K^+ channel, on the anoxic response of hippocampal neurones, *Pflügers Arch.*, 414, 111–114, 1989.
125. **Ashcroft, F. M., Ashcroft, S. J. H., and Harrison, D. E.**, The glucose-sensitive potassium channel in rat pancreatic beta-cells is inhibited by intracellular ATP, *J. Physiol.*, 369, 101p, 1984.
126. **Cook, D. L. and Hales, C. N.**, Intracellular ATP directly blocks K^+ channels in pancreatic β-cells, *Nature*, 311, 271–273, 1984.

127. **Rorsman, P. and Trube, G.**, Glucose dependent K^+ channels in pancreatic β-cells are regulated by intracellular ATP, *Pflügers Arch.*, 405, 305–309, 1985.
128. **Petersen, O. H. and Findlay, I.**, Electrophysiology of the pancreas, *Physiol. Rev.*, 67, 1054–1116, 1987.
129. **Kakei, M. and Noma, A.**, Adenosine 5′-triphosphate-sensitive single potassium channel in the atrioventricular node cell of the rabbit heart, *J. Physiol.*, 352, 265–284, 1984.
130. **Dunne, M. T., Yule, D. I., Gallacher, D. V., and Petersen, O. H.**, Cromakalim (BRL 34915) and diazoxide activate ATP-regulated potassium channels in insulin-secreting cells, *Pflügers Arch.*, 414 (Suppl. 1), 154–155, 1989.
131. **Findlay, I.**, ATP^{4-}- and ATP-Mg inhibit the ATP-sensitive K^+-channel of rat ventricular myocytes, *Pflügers Arch.*, 412, 37–41, 1988.
132. **Ashcroft, F. M. and Kakei, M.**, ATP-sensitive K^+ channels in rat pancreatic β-cells: modulation by ATP and Mg^{2+} ions, *J. Physiol.*, 416, 349–367, 1989.
133. **Loubatieres, A.**, Effects of sulphonylureas on the pancreas, in *The Diabetic Pancreas*, Volk, B. W. and Wellman, K. F., Eds., Bailliere Tindall, London, 1977, 489–515.
134. **Sturgess, N. C., Ashford, M. L. J., Cook, D. L., and Hales, C. N.**, The sulphonylurea receptor may be an ATP-sensitive potassium channel, *Lancet*, 2, 474–475, 1985.
135. **Trube, G., Rorsman, P., and Ohno-Shosaku, T.**, Opposite effects of tolbutamide and diazoxide on the ATP-dependent K^+ channel in mouse pancreatic β-cells, *Pflügers Arch.*, 407, 493–499, 1986.
136. **Nelson, T. Y., Gaines, K. L., Ranjan, A. S., Berg, M., and Boyd, A. E.**, Increased cytosolic calcium: a signal for sulfonylurea-stimulated insulin release from beta cells, *J. Biol. Chem.*, 262, 2608–2612, 1987.
137. **Belles, B., Hescheler, J., and Trube, G.**, Changes of membrane currents in cardiac cells induced by long whole-cell recordings and tolbutamide, *Pflügers Arch.*, 409, 582–588, 1987.
138. **Fosset, M., de Weille, J. R., Green, R. D., Schmid-Antomarchi, H., and Lazdunski, M.**, Antidiabetic sulphonylurea controls action potential properties in heart cells via high affinity receptors that are linked to ATP-dependent K^+ channels, *J. Biol. Chem.*, 263, 7933–7938, 1988.
139. **Misler, D. S., Falke, L. C., Gillis, K., and McDaniel, M. L.**, A metabolite regulated potassium channel in rat pancreatic β cells, *Proc. Natl. Acad. Sci. U.S.A.*, 83, 7119–7123, 1986.
140. **Horie, M., Irisawa, H., and Noma, A.**, Voltage-dependent magnesium block of adenosine-triphosphate-sensitive potassium channel in guinea-pig ventricular cells, *J. Physiol. (London)*, 387, 251–272, 1987.
141. **Ferrer, R., Atwater, I., Omer, E. M., Goncalves, A. A., Croghan, P. C., and Rojas, E.**, Electrophysiological evidence for the inhibition of potassium permeability in pancreatic β-cells by glibenclamide, *Q. J. Exp. Physiol.*, 69, 831–839, 1984.
142. **Gylfe, E., Hellman, B., Schlin, J., and Traljedal, I.-B.**, Interaction of sulphonylurea with the pancreatic β-cell, *Experientia*, 40, 1126–1134, 1984.
143. **Henquin, J. C., Meissner, H. P., and Preissler, M.**, 9-Aminoacridine- and tetraethylammonium-induced reduction of the potassium permeability in pancreatic β-cells, *Biochem. Biophys. Acta*, 587, 579–592, 1979.
144. **Matthews, E. K. and Shotton, P. A.**, The control of ^{86}Rb efflux from rat isolated pancreatic islets by the sulphonylureas tolbutamide and glibenclamide, *Br. J. Pharmacol.*, 82, 689–700, 1984.
145. **Meissner, H. P., Preissler, M., and Henquin, J. C.**, Possible ionic mechanisms of the electrical activity induced by glucose and tolbutamide in pancreatic β-cell, in *Diabetes*, Waldhäsel, W. K., Ed., Excerpta Medica, Amsterdam, 1979, 171–188.
146. **Lambert, D. G., Hughes, K., and Atkins, T. W.**, Insulin release from a cloned hamster β-cell line (HIT-T15): the effects of glucose, amino acid sulfonylurea and colchicine, *Biochem. Biophys. Res. Commun.*, 140, 616–625, 1986.

147. **Schmid-Antomarchi, H., de Weille, J. R., Fosset, M., and Lazdunski, M.**, The receptor for antidiabetic sulfonylureas controls the activity of the ATP-modulated K^+ channel in insulin-secreting cells, *J. Biol. Chem.*, 262, 15,840–15,844, 1987.
148. **Schmid-Antomarchi, H., de Weille, J. R., Fosset, M., and Lazdunski, M.**, The antidiabetic sulfonylurea glibenclamide is a potent blocker of the ATP-modulated K^+ channel in insulin secreting cells, *Biochem. Biophys. Res. Commun.*, 146, 21–25, 1987.
149. **Escande, D., Thuringer, D., Leguern, S., and Cavero, I.**, The potassium channel opener cormakalim (BRL34915) activates ATP-dependent K+ channels in isolated cardiac myocytes, *Biochem. Biophys. Res. Commun.*, 154, 620–625, 1988.
150. **Sanguinetti, M. C., Scott, A. L., Zingaro, G. J., and Siegl, P. K. S.**, BRL 34915 (cromakalim) activates ATP-sensitive K^+ current in cardiac muscle, *Proc. Natl. Acad. Sci. U.S.A.*, 85, 8360–8364, 1988.
151. **Mourre, Ben Ari, Y., Bernardi, H., Fosset, M., and Lazdunski, M.**, Antidiabetic sulfonylurea: localization and binding sites in the brain and effects on the hyperpolarization induced by anoxia in hippocampal slices, *Brain Res.*, 486, 159–164, 1989.
152. **Kramer, W., Oeknomopulos, R., Punter, J., and Summ, H.-D.**, Direct photoaffinity labelling of the putative sulfonylurea receptor in rat β-cell tumor membrane by [^{3}H]glibenclamide, *FEBS Lett.*, 229, 355–359, 1988.
153. **Krnjevic, K. and Leblond, J.**, Are there hippocampal ATP-sensitive K channels that are activated by anoxia?, *Eur. J. Physiol.*, 411, R145, 1988.
154. **Bolton, T. B.**, Mechanisms of action of transmitters and other substances on smooth muscle, *Physiol. Rev.*, 59, 606–718, 1979.
155. **Kuriyama, H., Ito, Y., Suzuki, H., Kitamura, K., and Itoh, T.**, Factors modifying contraction-relaxation cycle in vascular smooth muscles, *Am. J. Physiol.*, 246, H641–H662, 1982.
156. **Sumimnoto et al.**, 1987.
157. **Videbaeck, L. M., Aalkjaer, C., and Mulvany, M. J.**, Pinacidil opens K^+-selective channel causing hyperpolarization and relaxation of noradrenaline contractions in rat mesenteric resistance vessels, *Br. J. Pharmacol.*, 195, 103–108, 1988.
158. **Foster, C. D. and Brading, A. F.**, The effect of potassium channel-antagonists on the BRL34915 activated potassium channel in guinea-pig bladder, *Br. J. Pharmacol.*, 92, 751, 1987.
159. **Hollingsworth, M., Amedee, T., Edwards, D., Mironneau, J., Savineau, J. P., Small, R. O., and Weston, A. H.**, The relaxant action of BRL 34915 in rat uterus, *Br. J. Pharmacol.*, 91, 803–813, 1987.
160. **Hof, R. P., Quast, U., Cook, N. S., and Blarer, S.**, Mechanism of action, systemic and regional haemodynamics of the potassium channel activator BRL34915 and its enantiomers, *Circ. Res.*, 62, 679–686, 1988.
161. **Quast, U. and Baumlin, Y.**, Comparison of the effluxes of $^{42}K^+$ and $^{86}Rb^+$ elicited by cromakalim (BRL34915) in tonic and phasic vascular tissue, *Naunyn-Schmiedebergs Arch. Pharmacol.*, 338, 319–326, 1988.
162. **Nakao, K., Okabe, K., Kitamura, K., Kuriyama, H., and Weston, A. H.**, Characteristics of cromakalim-induced relaxations in smooth muscle cells of guinea-pig mesenteric artery and vein, *Br. J. Pharmacol.*, 95, 795–804, 1988.
163. **Kreye, V. A. W. and Weston, A. H.**, BRL34915-induced stimulation of ^{86}Rb efflux in rabbit aorta and its dependence on calcium, *Proc. Physiol. Soc.*, 347, 36p, 1986.
164. **Allen, S. L., Boyle, J. P., Cortijo, J., Foster, R. W., Morgan, G. P., and Small, R. C.**, Electrical and mechanical effects of BRL 34915 in guinea-pig isolated trachealis, *Br. J. Pharmacol.*, 89, 395–405, 1986.
165. **Morris, J. E. J. and Taylor, S. G.**, Effects of rubidium on relaxant agents in guinea-pig trachea, *Br. J. Pharmacol.*, 96, 232p, 1989.
166. **Foster, C. D., Fujii, K., Kingdom, J., and Brading, A. F.**, The effect of cromakalim on the smooth muscle of the guinea-pig urinary bladder, *Br. J. Pharmacol.*, 97, 281–291, 1989.

167. **Cook, N. L. and Quest, U.**, Potassium channel pharmacology, in *Potassium Channel: Structure, Classification, Function and Therapeutic Potential*, Cook, N. S., Ed., Ellis Horwood, New York, 1980, 181–185.
168. **Buckingham, R. E., Hamilton, T. C., Howlet, D. R., Mootoo, S., and Wilson, C.**, Inhibition of glibenclamide of the vasorelaxant action of cromakalim in the rat skeletal muscle, *Br. J. Pharmacol.*, 97, 57–64, 1989.
169. **Arch, J. R. S., Buckle, D. R., Bumstead, J., Clarke, G. D., Taylor, J. F., and Taylor, S. G.**, Evaluation of the potassium channel activator cromakalim (BRL34915) as a bronchodilator in the guinea-pig: comparison with nifedipine, *Br. J. Pharmacol.*, 95, 763, 1988.
170. **Hollingworth, M., Edwards, D., Rankin, R. J., and Weston, A. H.**, BRL34915 and relaxation of rat uterus-role of K+ channels, *Br. J. Pharmacol.*, 93, 199, 1988.
171. **McHarg, A. D., Southerton, J. S., and Weston, A. H.**, An investigation into the inhibitory action of cromakalim in rabbit isolated mesenteric artery, *Br. J. Pharmacol.*, Suppl. 95, 642p, 1988.
172. **Bechheit, K.-H. and Bertholet, A.**, Inhibition of small intestine motility by cromakalim (BRL34915), *Eur. J. Pharmacol.*, 154, 335–337, 1988.
173. **Bray, K. M., Newgreen, D. T., Small, R. C., Southerton, J. S., Taylor, S. G., Weir, S. W., and Weston, A. H.**, Evidence that the mechanism of the inhibitory action of pinacidil in rat and guinea-pig smooth muscle differs from that of glyceryl trinitrate, *Br. J. Pharmacol.*, 91, 421–429, 1987.
174. **Hermsmeyer, R. K.**, Pinacidil actions on ion channels in vascular smooth muscle, *J. Cardiovasc. Pharmacol.*, 12 (Suppl. 2), 517–522, 1988.
175. **Wilson, C.**, Comparative effects of cromakalim on contractions to noradrenaline and caffeine in rabbit isolated renal artery, *Br. J. Pharmacol.*, 95, 570p, 1988.
176. **Arrigoni-Martelli, E., Nielsen, C. K., Bang Olse, U. B., and Petersen, H. J.**, N''-cyano-N-4-pyridyl-N'-1,2,2-trimethylpropyl-guanidine monohydrate (P1134): a new, potent vasodilator, *Experientia*, 36, 445, 1980.
177. **Cook, N. S., Quast, U., and Manley, P.**, K+ channel opening does not alone explain the vasodilator activity of pinacidil and its enantiomers, *Br. J. Pharmacol.*, 96, 181p, 1989.
178. **Arena, J. P. and Kass, R. S.**, Activation of ATP-sensitive K channels in heart cells by pinacidil: dependence on ATP, *Am. J. Physiol.*, 26, 2092–2096, 1989.
179. **Cavero, I., Mondot, S., Mestre, M., and Escande, D.**, Haemodynamic and pharmacological mechanisms of the hypotensive effects of cromakalim in rats: blocked by glibenclamide, *Br. J. Pharmacol.*, 95, 643p, 1988.
180. **Vrolix, M., Raeymaekers, L., Wuytack, F., Hoffmann, F., and Casteels, R.**, Cyclic GMP-dependent protein kinase stimulates the plasmalemmal Ca^{2+} pump of smooth muscle via phosphorylation of phosphatidylinositol, *Biochem. J.*, 255, 855–863, 1988.
181. **Nakajima, M., Li, Y., Seki, N., and Kuriyama, H.**, Pinacidil indirectly inhibits neuromuscular transmission in the guinea-pig and rabbit mesenteric arteries, *Br. J. Pharmacol.*, 101, 581–586, 1990.
182. **Kusano, K., Barros, F., Katz, G., Garcia, M., Kaczorowski, G., and Reuben, J. P.**, Modulation of K channel activity in aortic smooth muscle by BRL 34915 and a scorpion toxin, *Biophys. J.*, 51, 55a, 1987.
183. **Gelband, C. H., Lodge, N. J., Talvenheime, J. A., and van Breemens, C.**, BRL 34915 increases P_{open} of the large conductance Ca^{2+} activated K^+ channel isolated from rabbit aorta in planar lipid bilayers, *Biophys. J.*, 53, 149a, 1988.
184. **Klöckner, U., Trieschmann, U., and Isenberg, G.**, Pharmacological modulation of calcium and potassium channels in isolated vascular smooth muscle cells, *Drug Res.*, 39, 120–126, 1989.
185. **Beech, D. J. and Bolton, T. B.**, Properties of the cromakalim-induced potassium conductance in smooth muscle cells isolated from the rabbit portal vein, *Br. J. Pharmacol.*, 98, 851–864, 1989.

186. **Inoue, I., Nakaya, Y., Nakaya, S., and Mori, H.**, Extracellular Ca^{2+}-activated K channel in coronary artery smooth muscle cells and its role in vasodilation, *FEBS Lett.*, 255, 281–284, 1989.
187. **Standen, N. B., Quayle, J. M., Davis, N. W., Brayden, J. E., Huang, Y., and Nelson, M. T.**, Hyperpolarizing vasodilators activate ATP-sensitive K^+-channels in arterial smooth muscle, *Science*, 245, 177–180, 1989.
188. **Lebrun, P., Devreuz, V., Hermann, M., and Herchuelz, A.**, Pinacidil inhibits insulin release by increasing K^+ outflow from pancreatic β-cells, *Eur. J. Pharmacol.*, 156, 283–286, 1988.
189. **Foster, K. A.**, Cromakalim activation of potassium channels in guinea-pig trachea: effects of extracellular rubidium, *Br. J. Pharmacol.*, 96, 233p, 1989.
190. **Kozolowski, R. Z., Hales, C. N., and Ashford, M. L. J.**, Dual effects of diazoxide on ATP-K^+ currents recorded from an insulin-secreting cell line, *Br. J. Pharmacol.*, 97, 1039–1050, 1989.
191. **Aizawa, T., Ogasawara, K., and Kato, K.**, Effects of nicorandil on coronary circulation in patients with ischemic heart disease: comparison with nitroglycerin, *J. Cardiovasc. Pharmacol.*, 10 (Suppl. 8), 123–129, 1987.
192. **Kato, K., Asanoi, H., Wakabayashi, C., Hosoda, S., Shiina, A., Hosono, K., Kurita, A., Seki, K., Ishida, K., Kuroiwa, A., and Dukumoto, A.**, Effect of nicorandil on exercise performance in patients with effort angina: a multicenter trial using a treadmill exercise test, *J. Cardiovasc. Pharmacol.*, 10 (Suppl. 8), 98–103, 1987.
193. **Buckingham, R. E.**, *In vivo* studies with drugs which open smooth muscle K^+ channel, in *Potassium Channels: Structure, Classification, Function and Therapeutic Potential*, Cook, N. S., Ed., Ellis Horwood, New York, 1990, 279–299.
194. **Byyny, R. L., Nies, A. S., Lo Verde, M. E., and Mitchell, W. D.**, A double-blind, randomized, controlled trial comparing pinacidil to hyudralazine in essential hypertension, *Clin. Pharmacol. Ther.*, 42, 50–57, 1987.
195. **Dixon, M. S., Thomas, P., Winterton, S. J., and Sheridan, D. J.**, Acute haemodynamic effects of BRL34915, a novel vasodilator, in patients with ischaemic heart disease, *Proc. Br. Pharmacol. Soc. Dublin Meeting,* (Abst.), p.10, 1988.
196. **Vandenburg, M. J., Arr, Woodward, S., Tasker, T., Stewart-Long, P., Fairhurst, G., Stephens, J., and Wood, B.**, Potassium channel activators, hypotensive activity and tolerance: a study with BRL 34915, *Proc. 3rd Euro. Meet. Hypertension,* Milan, June 1987, (Abstr. 586).
197. **Nuyen, P. V., Davis, A., Tasker, T. C. G., and Leenen, F. H. H.**, Effects of BRL34915 on pressor and chronotropic response to i.v. norepinephrine, angiotensin II, and isopreterenol in normal men, *Cardiovasc. Drug Ther.*, 1, 270, 1987.

Chapter 4

MEMBRANE ION CHANNELS AND DIABETES

Yoshio Oosawa

TABLE OF CONTENTS

I.	Introduction	80
II.	Pancreatic Morphology	80
III.	Insulin Secretion	80
IV.	Diabetes	81
V.	Electrophysiology of the Pancreatic β-Cell	82
VI.	Ion Channels in the β-Cell	82
VII.	Sulfonylureas	83
VIII.	Summary	85
IX.	Planar Lipid Bilayer	86
X.	Methods of Bilayer Formation	86
XI.	Ion Channels Incorporated into Planar Lipid Bilayers	87
XII.	Ca^{2+}-Activated K^+ Channel	87
XIII.	ATP-Sensitive K^+ Channel	89
XIV.	Summary	90
Acknowledgments		91
References		91

ISBN 0-8493-8493-X
© 1993 by CRC Press, Inc.

I. INTRODUCTION

Blood glucose homeostasis is controlled by a variety of hormones; of these, only insulin is able to lower the blood glucose concentration and alterations in plasma glucose levels are regulated largely by changes in the rate of insulin secretion. There is substantial evidence to suggest that a derangement of the secretory activity of the β-cell is one of the major factors which produces carbohydrate intolerance and leads to the development of diabetes mellitus. The pancreatic β-cell responds to glucose with a characteristic electrical activity. This electrical activity is produced by ionic fluxes through various ion channels located in the β-cell membrane. The ATP-sensitive K^+ channel is thought to have a key role in the control of β-cell electrical activity. In this review I discuss the role of the β-cell electrical activity and, in particular, that of the ATP-sensitive K^+ channel, in the regulation of insulin secretion from the pancreatic β-cell. Finally, I show that β-cell ion channels can be incorporated into planar lipid bilayers. More general details of pancreatic function and the disorders associated with diabetes mellitus are given in the review by Montague[19] and of electrophysiology of the β-cell in the reviews by Petersen and Findlay;[25] Henquin and Meissner;[10] Ashcroft and Rorsman[4] and of ion channel reconstitution in the review by Miller.[15]

II. PANCREATIC MORPHOLOGY

The pancreas has both exocrine and endocrine functions. The endocrine function is performed by the islets of Langerhans which constitute less than 2% of the wet weight of the adult pancreas. There are several distinct endocrine cell types within the islet, each cell type containing a different hormone stored within secretory granules. The major cell types are the β-cells which contain insulin, α-cells which contain glucagon, δ-cells which contain somatostatin, and the pancreatic polypeptide-containing cells (PP-cells). The relative proportions of the cell types differ in islets located in different parts of the human pancreas. Islets which are located at the head of the pancreas are normally composed of 60 to 70% β-cells, 5 to 10% δ-cells, 20 to 25% α-cells, and 5% PP-cells. However, islets located at the tail of the pancreas are normally composed of 60 to 70% β-cells, 5 to 10% δ-cells, 15 to 20% PP-cells, and only 5% α-cells.

III. INSULIN SECRETION

An increase in the blood glucose concentration above the fasting level of 4 to 5 mM is the major physiological stimulus to insulin release. There is a sigmoidal relationship between the amount of insulin secretion and the extracellular glucose concentration. The most dramatic effects on secretion are observed at glucose concentrations just above the fasting level.

Extracellular calcium is required for the secretion of insulin from the pancreatic β-cell. An increase in the concentration of ionized calcium within the cytoplasmic compartment of the β-cell is thought to trigger the secretory mechanism. Calcium thus acts as an important intracellular signal in the β-cell, linking stimulus recognition events to the secretory process.

The calcium concentration in the cytoplasm of the β-cell is controlled by the relative rates of its influx and efflux across the plasma membrane and by the relative rates of uptake and release from other intracellular compartments, including mitochondria, endoplasmic reticulum, and storage granules. The entry of calcium into the β-cell is thought to involve a number of different routes of which the major path is a specific calcium channel, the voltage-dependent L-type Ca^{2+} channel.

IV. DIABETES

Diabetes mellitus is a disorder of metabolism associated with either a relative or an absolute insulin deficiency. It is one of the major health problems in the developed world. Although the acute and potentially lethal metabolic derangements of diabetes can be controlled with insulin therapy, the long-term complications of the disease, which may involve the cardiovascular, renal, and nervous systems, reduce life expectancy by as much as one third. Clinically, two forms of diabetes mellitus are recognized: Type I or insulin-dependent diabetes mellitus (IDDM) and Type II or non-insulin-dependent diabetes mellitus (NIDDM).

In Type I diabetes the patient has complete lack of functional β-cells in the pancreas. This means that the pancreas cannot produce any insulin, and insulin must be administered to the patient to control glucose homeostasis. The metabolic derangements in Type II diabetes are generally not as severe as those of Type I. This is because the patient has a relatively normal complement of pancreatic β-cells and the fasting plasma insulin levels are normal or may even be elevated. The problem in Type II diabetes is not a lack of biologically effective insulin, but instead appears to result from a reduced secretion or insulin activity. Thus, under some conditions the amount of insulin released is not sufficient to maintain tight control of blood metabolite levels. This is usually because an undefined defect in the β-cell secretory mechanism prevents the β-cell from responding rapidly enough to a secretory stimulus. Blood metabolites can therefore rise to abnormal levels before there is an adequate insulin level in the circulation for nutrient control.

Since there is always a certain amount of insulin present in the circulation there is normally not the same total loss of nutrient homeostasis control like that seen in the Type I diabetic. In the Type II diabetic the major problem in nutrient homeostasis occurs after a meal. Because of the relative lack of biologically effective insulin the blood glucose rises after a meal to a level higher than that encountered in normal individuals and it also takes longer to return to the fasting value. This delay may be so long that the fasting value is not reached before the next meal is consumed and so there is a persistent hyperglycemia.

In the Type II diabetic insulin is not normally required to prevent ketosis, although insulin may be required to prevent hyperglycemia. This therapeutic goal can, however, often be achieved without insulin, by solely strict dietary control or by the use of dietary control plus oral hypoglycemic drugs. The hypoglycemic drugs are mainly sulfonylureas, which currently are employed in the management of Type II diabetes.

V. ELECTROPHYSIOLOGY OF THE PANCREATIC β-CELL

Microelectrode recordings from β-cells within intact islets have shown that glucose produces depolarization and electrical activity in pancreatic β-cells.[10] Similar electrical activity from the single β-cell can be recorded using the perforated patch configuration of the patch clamp technique (Figure 1).[31] The glucose-induced spikes are blocked by $[Ca^{2+}]_o$ removal, treatment with D600 or Mn^{2+}.[6,14,26] These observations indicate that Ca^{2+} is the major ion entering the β-cell during glucose-induced electrical activity.

VI. ION CHANNELS IN THE β-CELL

In the absence of glucose, electrical activity is absent and the β-cell has a stable resting potential of around -70 mV. The patch-clamp technique has allowed identification of the ion channels underlying β-cell electrical activity. The resting potential is mainly determined by the ATP-sensitive K^+ channel. Glucose closes ATP-sensitive K^+ channels and leads to membrane depolarization (Figure 2).[2,18,28] The ability of glucose to close ATP-sensitive K^+ channels is prevented by metabolic inhibitors. This suggests that glucose metabolism generates a second messenger which prevents channel activities. The ATP-sensitive K^+ channel has a unitary conductance of between 50 and 80 pS in symmetrical 140 mM KCl solutions and is highly potassium selective. It is blocked by ATP from the intracellular side of the membrane (Figure 3).[3] Currently, ATP is believed to act as the link between metabolite and membrane events.

Glucose metabolism increases the intracellular ATP and so blocks ATP-sensitive K^+ channels. The membrane K^+-permeability therefore decreases producing membrane depolarization. If this depolarization is sufficient to exceed the threshold for the opening of the voltage-dependent Ca^{2+} channels, electrical activity is initiated.[28] Ca^{2+} enters the cell through Ca^{2+} channels and produces an increase in intracellular Ca^{2+},[1] which causes insulin release. The above model for insulin secretion is schematized Figure 4.

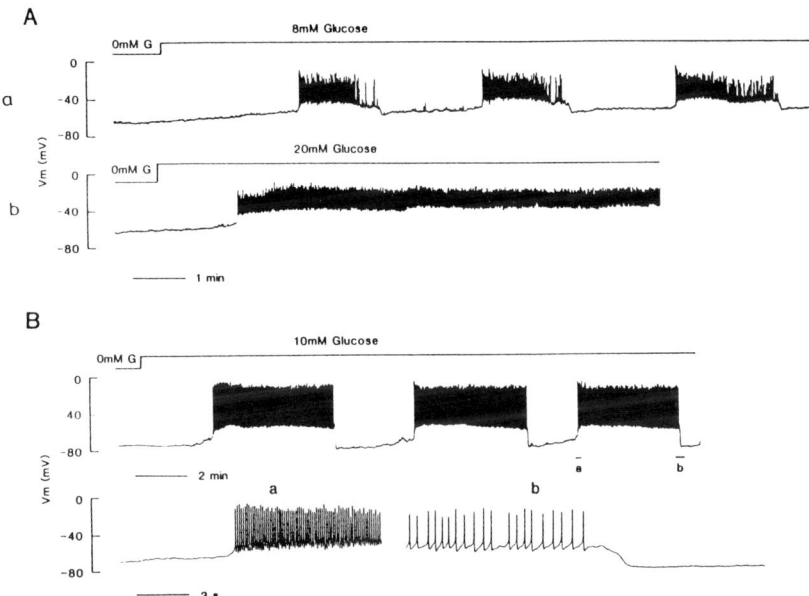

FIGURE 1. Electrical activity elicited by glucose in isolated mouse pancreatic β-cells that were recorded by perforated patch configuration of the patch clamp method.[11] This configuration prevents rapid dialysis of cytosolic constituents from the cell in standard whole-cell configuration. (A) Membrane potential recorded in response to 8 mM (a) and 20 mM glucose (b) from the same β-cell within a cluster of 64 μm diameter (about 5 cells) at 29°C. (B top) Membrane potential recorded in 0 and 10 mM glucose from a single β-cell at 31°C. (B bottom; a, b) Action potentials displayed at a faster time base from the parts of the record marked a and b of the recording above. (From Smith, P. A., Ashcroft, F. M., and Rorsman, P., *FEBS Lett.*, 261, 187–190, 1990a. With permission.)

VII. SULFONYLUREAS

Sulfonylureas are used clinically as hypoglycemic agents in the treatment of Type II diabetes mellitus. Their therapeutic effect is thought to result from their ability to decrease the resting K-permeability of the β-cell and to increase insulin release. The ATP-sensitive K^+ channel is blocked by sulfonylureas, such as tolbutamide and glibenclamide (Figure 5).[3,5,34,35] Sulfonylureas are specific blockers of this channel, as tolbutamide has no effect on any other β-cell channels, such as the voltage-activated K^+ channel,[29] the Ca^{2+} channel,[29] and the Ca^{2+}-activated K^+ channel.[35] Tolbutamide is able to block the ATP-sensitive K^+ channel both from the intra- or extracellular site of the membrane. Further, in cell-attached patches tolbutamide can block the channel even if it is only added to the bath solution and not to the pipette. Based on the fact that tolbutamide is a lipophilic molecule, it has been proposed that the drug may reach its target site by dissolving into the lipid phase of the membrane.[35]

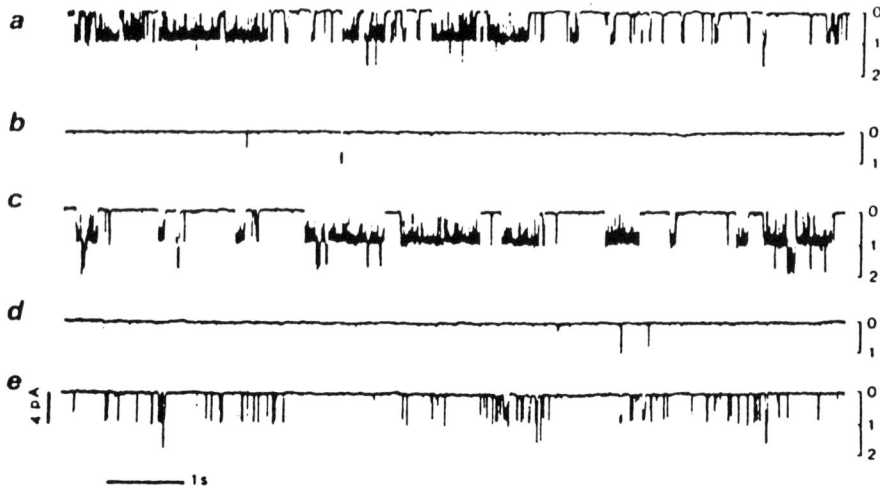

FIGURE 2. Effect of glucose on channel activity: a, in glucose-free solution; b, 7 min after changing to a solution containing 20 mM glucose; c, 12 min after return to glucose-free solution; d, 4 min after perfusion with 20 mM glucose; e, 2 min after perfusion with a solution containing 20 mM glucose plus 20 mM mannoheptulose. Mannoheptulose inhibits glucose metabolism. (From Ashcroft, F. M., Harrison, D. E., and Ashcroft, S. J. H., *Nature*, 312, 446–448, 1984. With permission.)

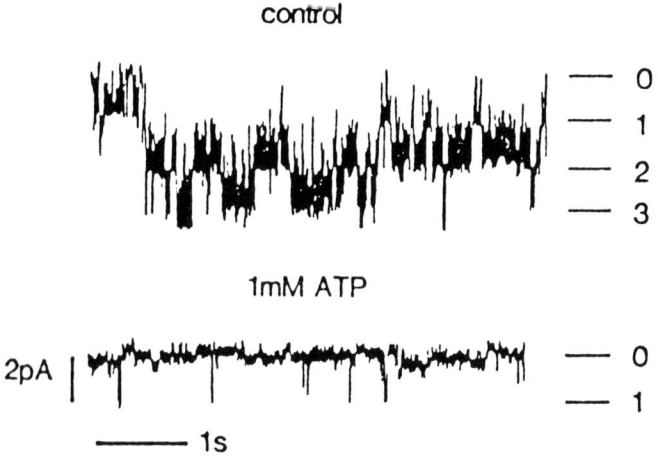

FIGURE 3. ATP inhibits single K$^+$ channel currents. Single channel currents recorded at a membrane potential of −60 mV from an inside-out patch of human β-cell membrane exposed to intracellular solution before (above) and after (below) the addition of 1 mM ATP. The number of channel open is indicated to the right of each trace. (From Ashcroft, F. M., Kakei, M., Kelly, R. P., and Sutton, R., *FEBS Lett.*, 215, 9–12, 1987. With permission.)

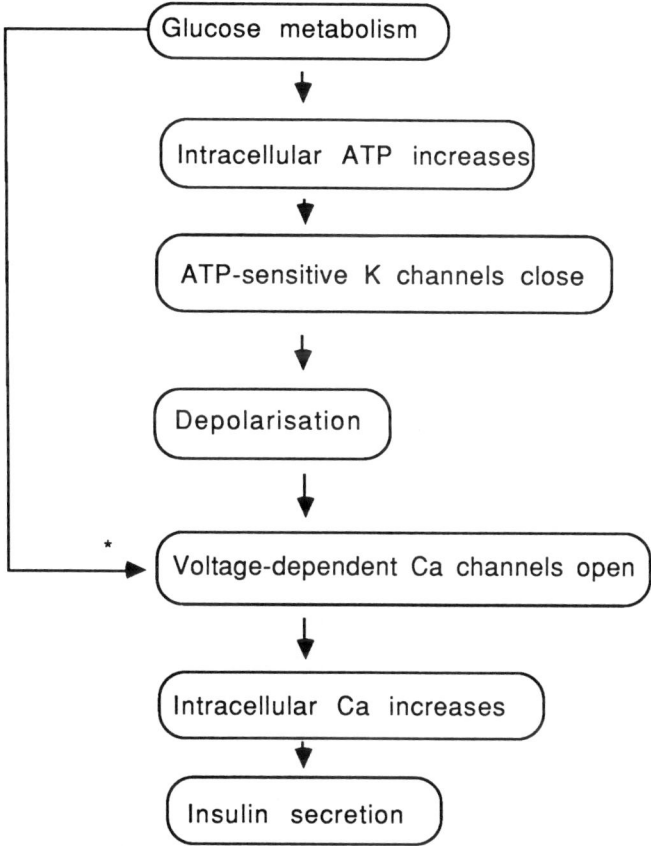

FIGURE 4. The scheme for insulin secretion. *Glucose metabolism modulates Ca^{2+} channel activity.[33]

There is the possibility that Type II diabetes results from a defect in the ATP-sensitive K^+ channel protein or in ATP production. For example, defective glucose uptake or metabolism would result in reduced ATP levels, or alternatively a defect in the channel protein may prevent the channel responding to changes in intracellular ATP. Studies on β-cells isolated from Type II diabetics are required.

VIII. SUMMARY

The β-cell secretes insulin in response to glucose. Associated with this secretion is a characteristic electrical activity which is essential for insulin secretion. This electrical activity is caused by changes in the activities of various ion channels in the β-cell membrane. Among these ion channels, the

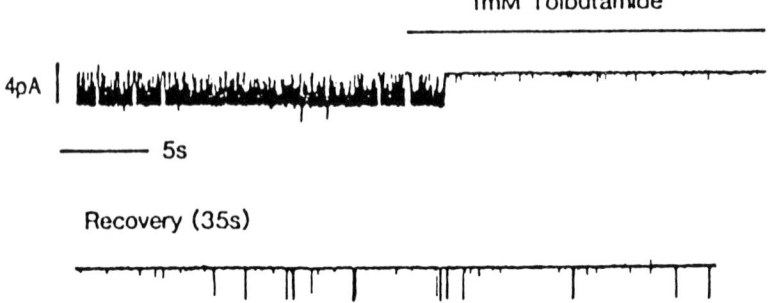

FIGURE 5. Effect of tolbutamide on the ATP-sensitive K^+ channel currents of human β-cell membrane. Single-channel currents recorded at -60 mV from inside-out patches. Tolbutamide was added as indicated by the bar into intracellular solution. The lower trace was obtained 35 s after the upper. (From Ashcroft, F. M., Kakei, M., Kelly, R. P., and Sutton, R., *FEBS Lett.*, 215, 9–12, 1987. With permission.)

ATP-sensitive K^+ channel plays a key role in insulin secretion. The channels are blocked by both physiological stimuli such as glucose and by the hypoglycemic sulfonylurea used to treat Type II diabetes. This initiates electrical activity and insulin secretion. The central role of the ATP-sensitive K^+ channel raises the question of whether defective regulation of this channel may be involved in Type II diabetes.

IX. PLANAR LIPID BILAYER

Besides the patch-clamp technique, ion channel reconstitution is also useful in order to investigate ion channel properties (see Reference 15). In reconstitution studies, an ion channel protein is taken from the native membrane and is transferred to an artificial membrane. It is very useful to investigate ion channel properties under simplified conditions that cannot be obtained in the complicated cellular environment. We tried to incorporate native membrane vesicles from pancreatic β-cell into planar lipid bilayers. A membrane from β-cell line (HIT T15) was used because mass production of membranes from cell lines is easier than collecting many β-cell membranes from animals. This cell line possesses ATP-sensitive K^+ channels, Ca^{2+}-activated K^+ channels, and delayed rectifier K^+ channels.[7,12,21,27]

X. METHODS OF BILAYER FORMATION

Planar lipid bilayers were formed by the painting method.[20,23,24] Figure 6 shows planar lipid bilayer setup. Planar lipid bilayers were formed from a mixture of 50% phosphatidylethanolamine (PE; bovine heart) and 50% phosphatidylserine (PS; bovine brain) dissolved in decane at a concentration of 25 mg/ml. The *cis* chamber (to which the vesicles were added) was voltage

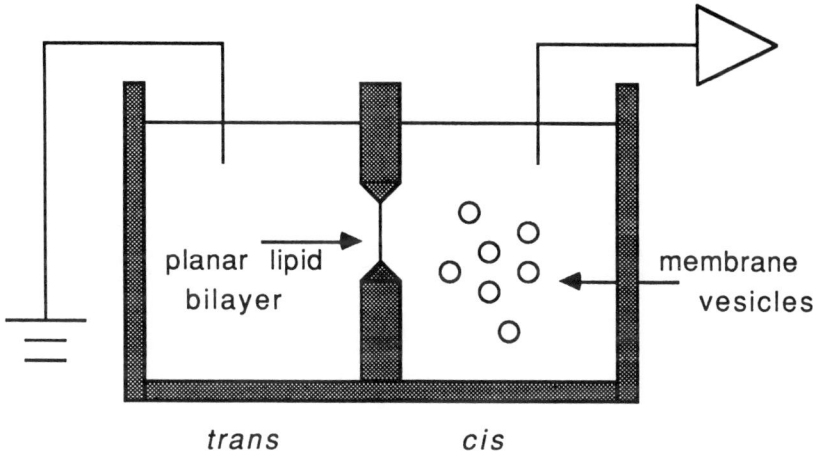

FIGURE 6. Planar lipid bilayer setup.

clamped at various potentials relative to the *trans* chamber. The *trans* chamber was grounded and defined as zero voltage. All potentials refer to the potential of the *cis* chamber with respect to the *trans* chamber (i.e., *cis-trans* voltage). Single channel currents were recorded under voltage clamp using a standard current-to-voltage amplifier and recorded on FM tape or video tape for later analysis. Single channel currents were amplified ($\times 10$ to $\times 50$) and filtered at 200 Hz. They were then digitized at 1 kHz and analyzed using an IBM AT computer and the program PCLAMP. Open probabilities were determined from amplitude histograms constructed from all data points. Free calcium concentrations were adjusted using a Ca-EGTA buffer and the Ca^{2+} concentration calculated using the binding constants of Martell and Smith.[13] Experiments were done at room temperature, 20 to 25°C. Membranes were prepared from HIT T-15 β-cells as described by Gaines et al.[8] with slight modifications.[22] After the addition of 1 mM calcium to the *cis* chamber, the membrane vesicles were then added (final protein concentration 2 to 10 μg/ml).

XI. ION CHANNELS INCORPORATED INTO PLANAR LIPID BILAYERS

We observed two types of K^+ channels in our planar lipid bilayers in KCl solution: a Ca^{2+}-activated K^+ channel and an ATP-sensitive K^+ channel. The latter was observed with a very low frequency.

XII. Ca^{2+}-ACTIVATED K^+ CHANNEL

This channel has large single channel conductance and shows both Ca^{2+}- and voltage-dependency. The effect of membrane potential on single channel currents (A) and channel open probability (B) is shown in Figure 7. Making

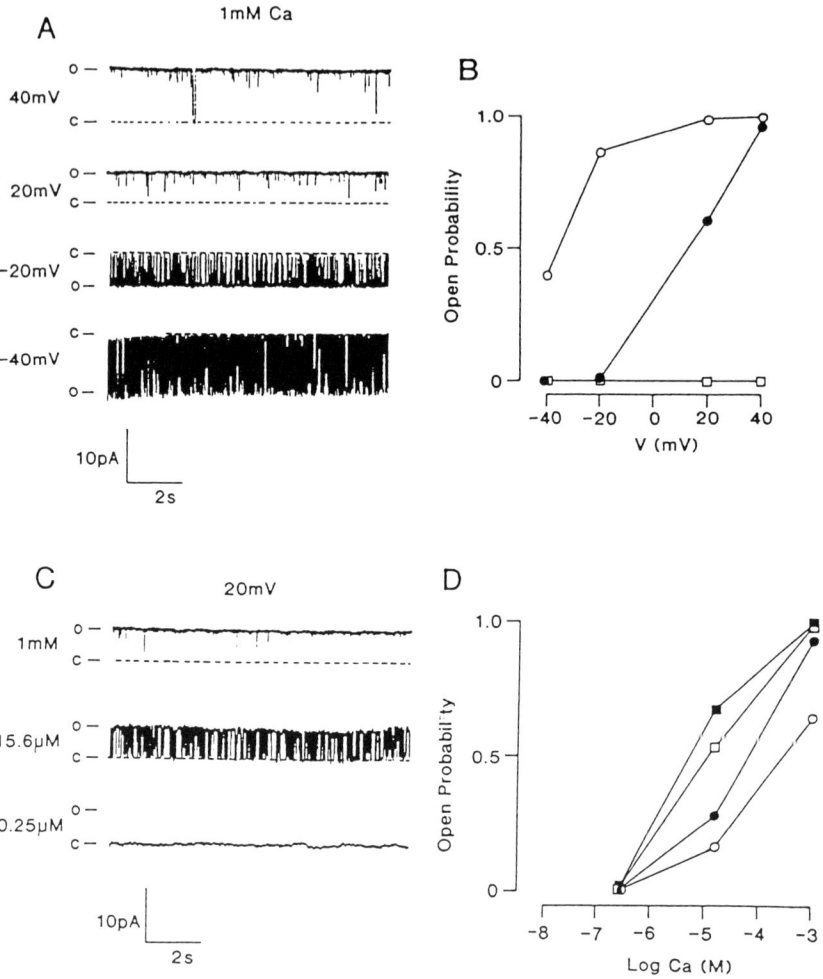

FIGURE 7. Ca^{2+} dependence and voltage dependence of Ca^{2+}-activated K^+ channel. (A) Ca^{2+}-activated K^+ channel currents recorded at different membrane voltages (indicated on the left of each trace). *cis*: 140 mM KCl, 1 mM Ca^{2+}, 10 mM HEPES, pH 7.1; *trans*: 140 mM KCl, 10 mM HEPES, pH 7.1. C and O indicate the closed and open levels, respectively. (B) Relationship between the channel open state probability and membrane potential for the same bilayer at different *cis* calcium concentrations: 1 mM (open circles), 15.6 μM (1 mM $CaCl_2$, 1 mM EGTA) (filled circles), 0.25 μM (1 mM $CaCl_2$, 2 mM EGTA) (open squares). *cis* and *trans*: 140 mM KCl, 10 mM HEPES, pH 7.1. (C) Ca^{2+}-activated K^+ channel currents recorded at different *cis* calcium concentrations at +20 mV. Free calcium concentrations are, from top to bottom, 1 mM, 15.6 μM, 0.25 μM. *cis* and *trans*: 140 mM KCl, 10 mM HEPES, pH 7.1. C and O indicate the closed and open levels, respectively. (D) Relationship between the open state probability and the *cis* side calcium concentrations at four different membrane potentials: +40 mV (filled squares), +20 mV (open squares), −20 mV (filled circles), −40 mV (open circles). (From Ashcroft, F. M. and Rorsman, P., *Prog. Biophys. Molec. Biol.*, 54, 87–143, 1991. With permission.)

the *cis* potential more positive increases the channel open probability: as the *cis* Ca^{2+} concentration is raised, the relationship between open probability and membrane potential shifts to more negative membrane potentials, so that at any given potential channel activity is greater. Channel activity increases as the *cis* Ca^{2+} concentration is raised (Figure 7C). The channel open probability is plotted as a function of *cis* Ca^{2+} concentration in Figure 7D, for four different membrane potentials. At a membrane potential of 0 mV, the open probability is half maximal in ~20 µM Ca^{2+}. Depolarization shifts the relationship between open probability and Ca^{2+} to lower Ca^{2+} concentrations. Conversely, hyperpolarization reduces the Ca^{2+} sensitivity.

The mean single channel current-voltage (I–V) relation measured under symmetrical 140 mM KCl solutions at a *cis* Ca^{2+} concentration of 1 mM is linear, with a slope conductance of 233 pS. The mean single channel I–V relation measured with 250 mM KCl *cis* and 50 mM KCl *trans* shows the reversal potential, -34.5 mV, which was close to the calculated K^+-equilibrium potential (-40 mV) and indicates that the channel strongly selects K^+ over Cl^-. The permeability ratio, P_{Cl}/P_K, was 0.046. Under quasi-physiological ionic conditions (*cis*: 140 mM KCl and *trans*: 135 mM NaCl, 5 mM KCl), the single-channel I–V relation is no longer linear and was better fit by the Goldman-Hodgkin-Katz equation. The best fit was obtained with a K permeability of 3.42×10^{-13} cm^3/s and a P_{Na}/P_K ratio of 0.027. This permeability ratio indicates that the channel is considerably less permeable to Na^+ than K^+. The sensitivity of the channel to *cis* calcium and to *cis* positive potentials suggests that the intracellular end of the protein faces the *cis* solution and the extracellular end faces the *trans* solution.

Charybdotoxin is a potent blocker of high conductance Ca^{2+}-activated K^+ channels.[16,17] Figure 8 shows that 12 nM charybdotoxin greatly reduced the channel open probability when added to the side opposite to that at which Ca^{2+} activated the channel.

Ca^{2+}-activated K^+ channels have already been reported in patch clamp experiments on the insulin-secreting cell lines HIT T15.[7,12] The single channel conductance found in our studies is similar to those reported in previous works. It is now generally believed that the Ca^{2+}-activated K^+ channel contributes to the repolarization of the action potential[30,32] but has little influence on the oscillations in membrane potential (slow waves) upon which they are superimposed.[9]

XIII. ATP-SENSITIVE K^+ CHANNEL

Figure 9 shows single channel currents of an ATP-sensitive K^+ channel before and after *cis* side ATP application in symmetrical 140 mM KCl solutions. The open probability of this channel is not voltage dependent. The mean single channel I–V relation measured with 140 mM KCl on both sides is linear with a slope conductance of 52.6 pS. This channel is blocked by

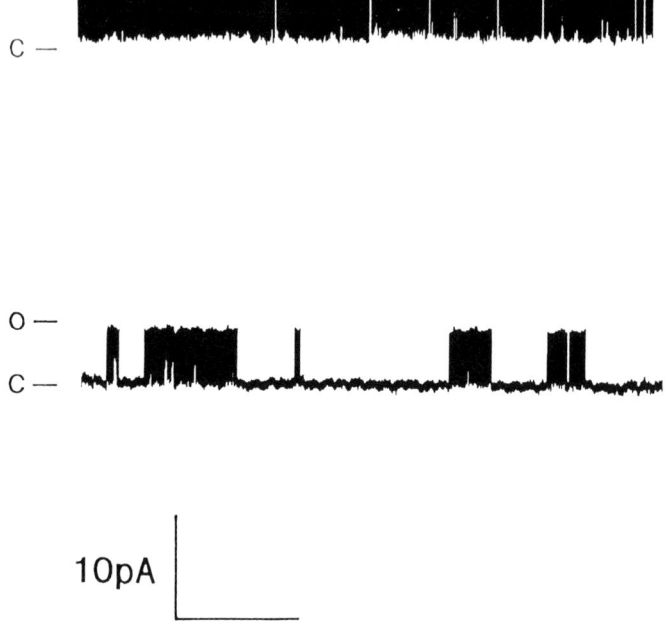

FIGURE 8. Effect of charybdotoxin on the Ca^{2+}-activated K^+ channel. 12 nM charybdotoxin was added to the *trans* side. Before (top trace) and after (bottom trace). Membrane voltage, +20 mV. C and O indicate the closed and open levels, respectively. *cis*: 140 mM KCl, 15.6 μM Ca^{2+}, 10 mM HEPES, pH 7.1; *trans*: 140 mM KCl, 10 mM HEPES, pH 7.1.

0.5 mM ATP completely. The single channel conductance of 52.6 pS found in bilayer study is close to those values found in patch clamp experiments on HIT T15 cells.[7,21]

XIV. SUMMARY

We have shown that both the Ca^{2+}-activated K^+ channels and the ATP-sensitive K^+ channels from HIT T15 cell membranes can be incorporated into planar lipid bilayers. The properties of these channels do not appear to differ from those reported for these channels in native membranes. This means that during the membrane vesicle preparation and the fusion process ion channels of β-cell membranes maintain their properties in the cell.

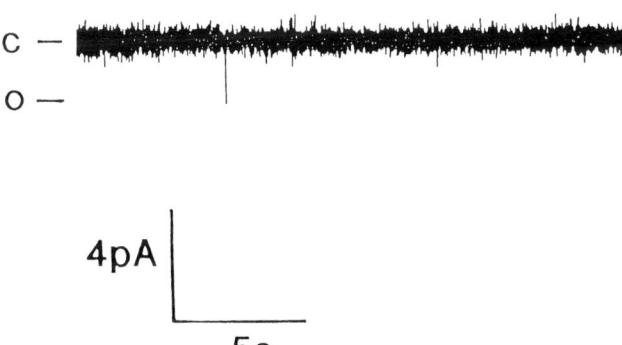

FIGURE 9. Block of the ATP-sensitive K^+ channel by ATP. 0.5 mM ATP added to the *cis* side completely blocked channel activity. Before (top trace) and after (bottom trace). C and O indicate the closed and open levels, respectively. *cis*: 140 mM KCl, 1 mM $CaCl_2$, 10 mM HEPES, pH 7.1; *trans*: 140 mM KCl, 10 mM HEPES, pH 7.1. Membrane voltage, -40 mV.

ACKNOWLEDGMENTS

I thank the Medical Research Council for support. I thank Drs. F. M. Ashcroft (University of Oxford, U. K., at whose laboratory this research was done), S. J. H. Ashcroft, and P. A. Smith for their helpful comments.

REFERENCES

1. **Arkhammar, P., Nilsson, T., Rorsman, P., and Berggren, P.-O.,** Inhibition of ATP-regulated K^+ channels proceeds depolarization-induced increase in cytoplasmic free Ca^{2+} concentration in pancreatic β-cells, *J. Biol. Chem.*, 262, 5448–5454, 1987.
2. **Ashcroft, F. M., Harrison, D. E., and Ashcroft, S. J. H.,** Glucose induces closure of single potassium channels in isolated rat pancreatic β-cells, *Nature,* 312, 446–448, 1984.

3. **Ashcroft, F. M., Kakei, M., Kelly, R. P., and Sutton, R.,** ATP-sensitive K⁺ channels in human isolated pancreatic β-cells, *FEBS Lett.*, 215, 9–12, 1987.
4. **Ashcroft, F. M. and Rorsman, P.,** Electrophysiology of the pancreatic β-cell, *Prog. Biophys. Molec. Biol.*, 54, 87–143, 1991.
5. **Belles, B., Hescheler, J., and Trube, G.,** Changes of membrane currents in cardiac cells induced by long whole-cell recordings and tolbutamide, *Pfluegers Arch.*, 409, 582–588, 1987.
6. **Dean, P. M. and Matthews, E. K.,** Electrical activity in pancreatic islet cells: effect of ions, *J. Physiol. (London)*, 210, 265–275, 1970.
7. **Eddlestone, G. T., Ribalet, B., and Ciani, S.,** Comparative study of K channel behavior in β cell lines with different secretory responses to glucose, *J. Membr. Biol.*, 109, 123–134, 1989.
8. **Gaines, K. L., Hamilton, S., and Boyd, A. E., III,** Characterization of the sulfonylurea receptor on beta cell membranes, *J. Biol. Chem.*, 263, 2589–2592, 1988.
9. **Henquin, J. C.,** Role of voltage- and Ca^{2+}-dependent K⁺ channels in the control of glucose-induced electrical activity in pancreatic β-cells, *Pfluegers Arch.*, 416, 568–572, 1990.
10. **Henquin, J. C. and Meissner, H. P.,** Significance of ionic fluxes and changes in membrane potential for stimulus-secretion coupling in pancreatic β-cells, *Experientia*, 40, 1043–1052, 1984.
11. **Horn, R. and Marty, A.,** Muscarinic activation of ionic currents measured by a new whole-cell recording method, *J. Gen. Physiol.*, 92, 145–159, 1988.
12. **Light, D. B., Van Eenenaam, D. P., Sorenson, R. L., and Levitt, D. G.,** Potassium-selective ion channels in a transformed insulin-secreting cell line, *J. Membr. Biol.*, 95, 63–72, 1987.
13. **Martell, A. E. and Smith, R. M.,** *Critical Stability Constants*, Vols. 1 and 2, Plenum Press, New York, 1974.
14. **Matthews, E. K. and Sakamoto, Y.,** Pancreatic islet cells: electrogenic and electrodiffusional control of membrane potential, *J. Physiol. (London)*, 246, 439–457, 1975.
15. **Miller, C.,** *Ion Channel Reconstitution*, Plenum Press, New York, 1986.
16. **Miller, C.,** Competition for block of a Ca^{2+}-activated K⁺ channel by charybdotoxin and tetraethylammonium, *Neuron*, 1, 1003–1006, 1988.
17. **Miller, C., Moczydlowski, E., Latorre, R., and Phillips, M.,** Charybdotoxin, a protein inhibitor of single Ca^{2+}-activated K⁺ channels from mammalian skeletal muscle, *Nature*, 313, 316–318, 1985.
18. **Misler, S., Falke, L. C., Gillis, K., and McDaniel, M. L.,** A metabolite-regulated potassium channel in rat pancreatic β cells, *Proc. Natl. Acad. Sci. U.S.A.*, 83, 7119–7123, 1986.
19. **Montague, W.,** *Diabetes and the Endocrine Pancreas: A Biochemical Approach*, Croom Helm, Kent, England, 1983.
20. **Mueller, P. and Rudin, D. O.,** Bimolecular lipid membranes: techniques of formation, study of electrical properties, and induction of ionic gating phenomena, in *Laboratory Techniques in Membrane Biophysics*, Passow, H. and Stampfli, R., Eds., Springer-Verlag, Berlin, 1969, 141–156.
21. **Niki, I., Kelly, R. P., Ashcroft, S. J. H., and Ashcroft, F. M.,** ATP-sensitive K-channels in HIT T15 β-cells studied by patch-clamp methods, ⁸⁶Rb efflux and glibenclamide binding, *Pfluegers Arch.*, 415, 47–55, 1989.
22. **Oosawa, Y., Ashcroft, S. J. H., and Smith, P. A.,** Ca^{2+}-activated K⁺ channels from an insulin-secreting β-cell line incorporated into planar lipid bilayers, *J. Physiol. (London)*, 430, 123P, 1990.
23. **Oosawa, Y. and Kasai, M.,** Gibbs-Donnan ratio and channel conductance of *Tetrahymena* cilia in mixed solution of K⁺ and Ca^{2+}, *Biophys. J.*, 54, 407–410, 1988.

24. **Oosawa, Y. and Sokabe, M.**, Cation channels from *Tetrahymena* cilia incorporated into planar lipid bilayers, *Am. J. Physiol.*, 249, C177–C179, 1985.
25. **Petersen, O. H. and Findlay, I.**, Electrophysiology of pancreas, *Physiol. Rev.*, 67, 1054–1116, 1987.
26. **Ribalet, B. and Beigelman, P. M.**, Calcium action potentials and potassium permeability activation in pancreatic β-cells, *Am. J. Physiol.*, 239, C124–C133, 1980.
27. **Ribalet, B., Ciani, S., and Eddlestone, G. T.**, ATP mediates both activation and inhibition of K(ATP) channel activity via cAMP-dependent protein kinase in insulin-secreting cell lines, *J. Gen. Physiol.*, 94, 693–717, 1989.
28. **Rorsman, P. and Trube, G.**, Glucose dependent K^+-channels in pancreatic β-cells are regulated by intracellular ATP, *Pfluegers Arch.*, 405, 305–309, 1985.
29. **Rorsman, P. and Trube, G.**, Calcium and delayed potassium currents in mouse pancreatic β-cells under voltage-clamp conditions, *J. Physiol. (London)*, 374, 531–550, 1986.
30. **Satin, L. S., Hopkins, W. F., Fatherazi, S., and Cook, D. L.**, Expression of a rapid, low-voltage threshold K current in insulin-secreting cells is dependent on intracellular calcium buffering, *J. Membr. Biol.*, 112, 213–222, 1989.
31. **Smith, P. A., Ashcroft, F. M., and Rorsman, P.**, Simultaneous recording of glucose dependent electrical activity and ATP-regulated K^+ currents in isolated mouse pancreatic β-cells, *FEBS Lett.*, 261, 187–190, 1990a.
32. **Smith, P. A., Bokvist, K., Arkhammar, P., Berggren, P.-O., and Rorsman, P.**, Delayed rectifying and calcium-activated K^+ channels and their significance for action potential repolarization in mouse pancreatic β-cells, *J. Gen. Physiol.*, 95, 1041–1059, 1990b.
33. **Smith, P. A., Rorsman, P., and Ashcroft, F. M.**, Modulation of dihydropyridine-sensitive Ca^{2+} channels by glucose metabolism in mouse pancreatic β-cells, *Nature*, 342, 550–553, 1989.
34. **Sturgess, N. C., Ashford, M. L. J., Cook, D. L., and Hales, C. N.**, The sulphonylurea receptor may be an ATP-sensitive potassium channel, *Lancet*, 8453, 474–475, 1985.
35. **Trube, G., Rorsman, P., and Ohno-Shosaku, T.**, Opposite effects of tolbutamide and diazoxide on the ATP-dependent K^+ channel in mouse pancreatic β-cells, *Pfluegers Arch.*, 407, 493–499, 1986.

Chapter 5

MICROCIRCULATORY DISTURBANCES AND AUTONOMIC NERVOUS RECEPTORS IN ACUTE GASTRIC MUCOSAL LESION

Masaharu Tsuchiya, Masahiko Nakamura, and Masaya Oda

TABLE OF CONTENTS

I.	Introduction	96
II.	Distribution of Autonomic Nervous System in the Stomach	96
	A. Cholinergic Nerves	96
	B. Catecholaminergic Nerves	99
	C. Peptidergic Nerves	103
III.	Distribution of Receptors of Vascular Effectors in the Gastric Microcirculatory System	105
	A. Classical Transmitters	105
	1. Muscarinic Acetylcholine Receptors	105
	2. Dopaminergic Receptors	107
	B. Brain-Gut Peptide	107
	C. Endothelin	107
	D. Histamine	108
IV.	Alteration of Autonomic Nervous Activity and Receptor Distribution in Gastric Ulcer Formation	118
V.	Conclusions	118
References		120

ISBN 0-8493-8091-X
© 1993 by CRC Press, Inc.

I. INTRODUCTION

The gastric mucosa is thought to be strongly influenced by physical and psychic stresses, resulting in the formation of the ulcerative lesion, but its pathophysiology remains to be elucidated. Recent histochemical and radioautographic studies have revealed that the gastric mucosa is characterized by the abundant existence of enteric nervous system surrounding the gastric microcirculatory system and gastric glandular cells in comparison with other parts of the gastrointestinal tract. The changes in this nervous activity and the receptor distribution have been found to be closely related to the formation of the gastric mucosal lesion. The purpose of this chapter is to review the current status of our knowledge of the characteristics of the neuroeffector mechanism of the stomach, especially the gastric mucosa.

This chapter is organized as follows. First, the distribution of the autonomic nervous system is described, including the peptidergic and dopaminergic nerves. Next, the distribution of the receptors of neurotransmitters and neuroeffectors is described, mainly obtained from radioautographic studies. Finally, the role of the autonomic nervous system in the formation of the gastric ulcerative lesion is mentioned.

II. DISTRIBUTION OF AUTONOMIC NERVOUS SYSTEM IN THE STOMACH

A. CHOLINERGIC NERVES

The cholinergic nerves in the stomach consist of the intrinsic "enteric" nerves and extrinsic nerve input from the vagal nerve. Vagal input to the enteric nervous system originates in the dorsal motor nucleus of the medulla oblongata.[1] The vagi at the level of the diaphragm contain an average of 56,138 afferent fibers and only 1736 efferent fibers.[2] Most of the cholinergic nerves in the gastric mucosa belong to the enteric nerves. The histochemical localization of the cholinergic nerves was identified by acetylcholinesterase (AChE) activity[3-5] and choline acetyltransferase (ChAc) immunoreactivity.[6] In the submucosal layer of the fundus and antrum, the AChE-positive and ChAc-immunoreactive nerves are recognized near the arterioles (Figure 1). In the mucosal layer, AChE-positive and ChAc-immunoreactive nerves are seen in the lamina propria mucosae between the gastric epithelial cells. By electron microscopic observation, most of the nerve fibers in the mucosal layer contain chiefly small agranular synaptic vesicles, one of the characteristics of cholinergic nerves (Figure 2). These cholinergic nerves were seen near various kinds of fundic and antral glandular cells (Figure 3). In the submucosal and muscular layers, many nerve bundles and ganglion cells positive to AChE histochemistry and ChAc immunoreactivity exist near the smooth muscle cells consisting of the muscularis mucosae and propriae (Figure 4).

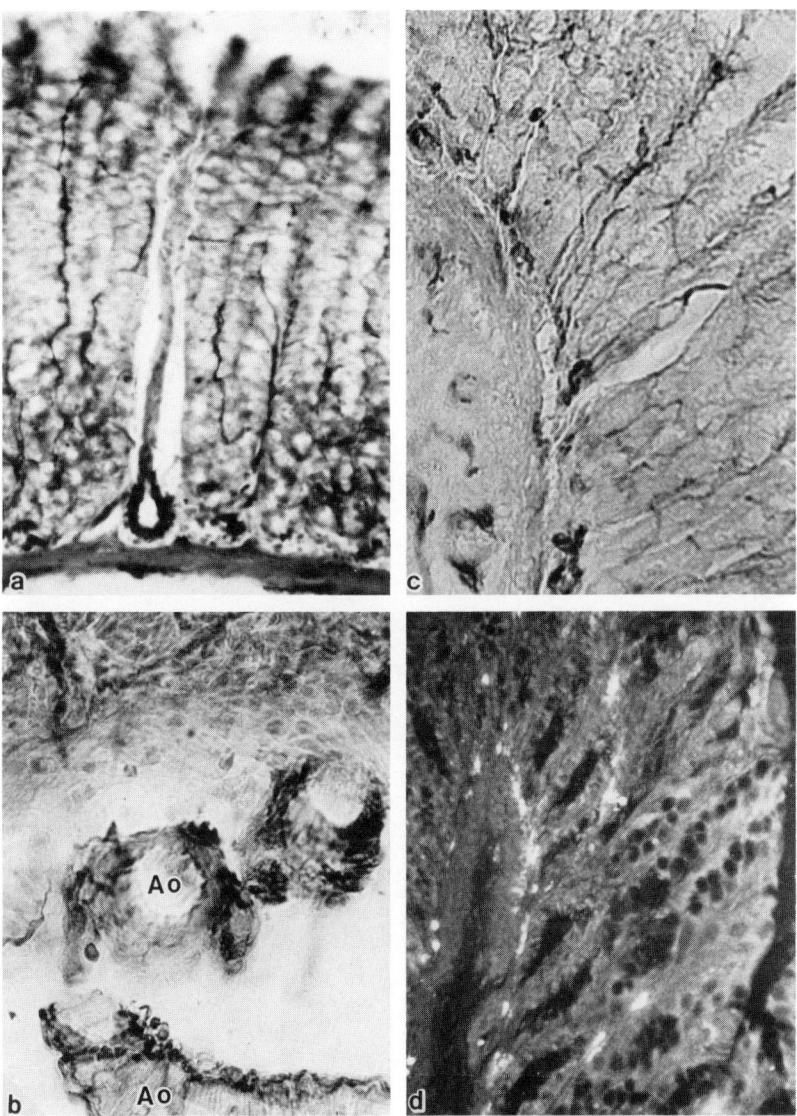

FIGURE 1. AChE histochemistry and ChAc immunohistochemistry of rat stomach. (a) In the fundic mucosa, AChE-positive nerves are seen linearly between the gastric epithelial cells. (Original magnification × 400.) (b) In the fundic submucosal layer, AChE-positive nerves are seen near the arterioles (Ao) and venules. (Original magnification × 2000.) (c) In the antral mucosa, AChE-positive nerves are seen only in the basal half of the mucosal layer. (Original magnification × 400.) (d) In the antral mucosa, ChAc-immunoreactive nerves show almost similar distribution to that of AChE. (Original magnification × 400.)

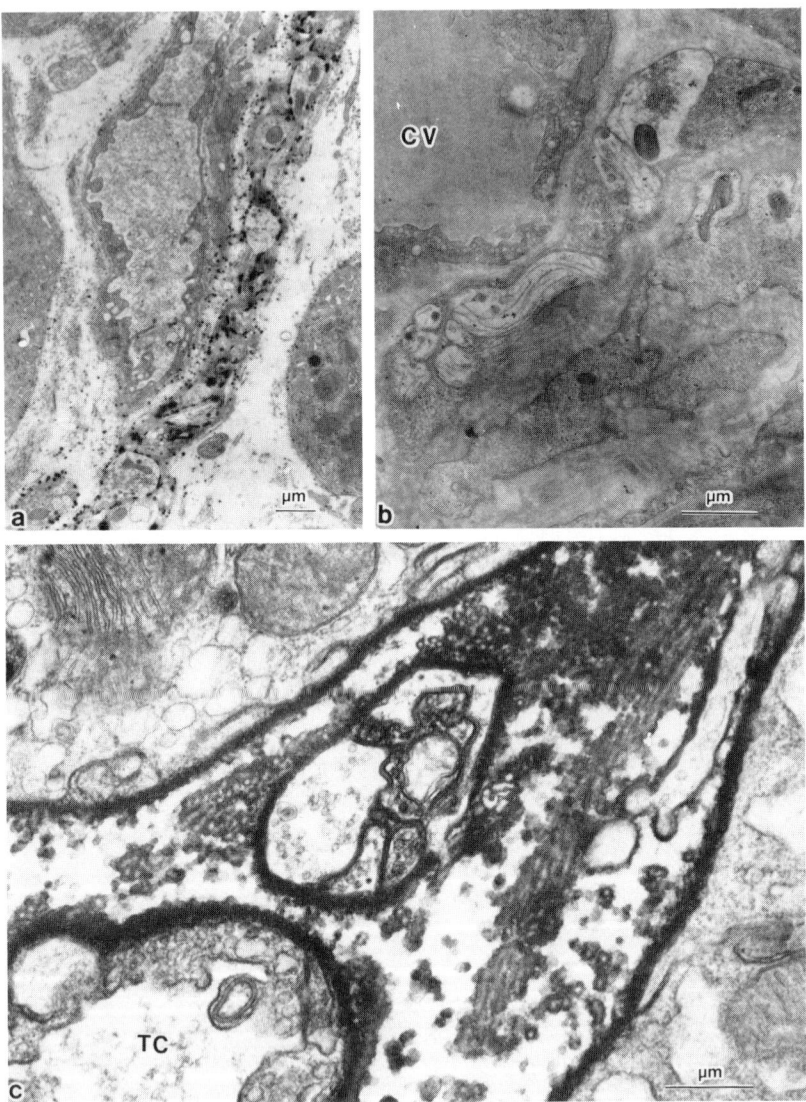

FIGURE 2. Electron micrographs of the unmyelinated nerve fibers of the rat fundic mucosa. (a) By AChE cytochemistry, AChE reaction products are recognized on the axonal membrane of the unmyelinated nerve endings as well as on the basement membrane of the adjacent endothelium and pericyte. (Original magnification × 8000.) (b) Most of the synaptic vesicles of the unmyelinated nerve endings belong to a small, agranular type, corresponding to the cholinergic nerve ending. CV: collecting venule. (Original magnification × 15,000.) (c) By ruthenium red en bloc staining, the unmyelinated nerve endings are clearly seen near the true capillary (TC). (Original magnification × 17,000.)

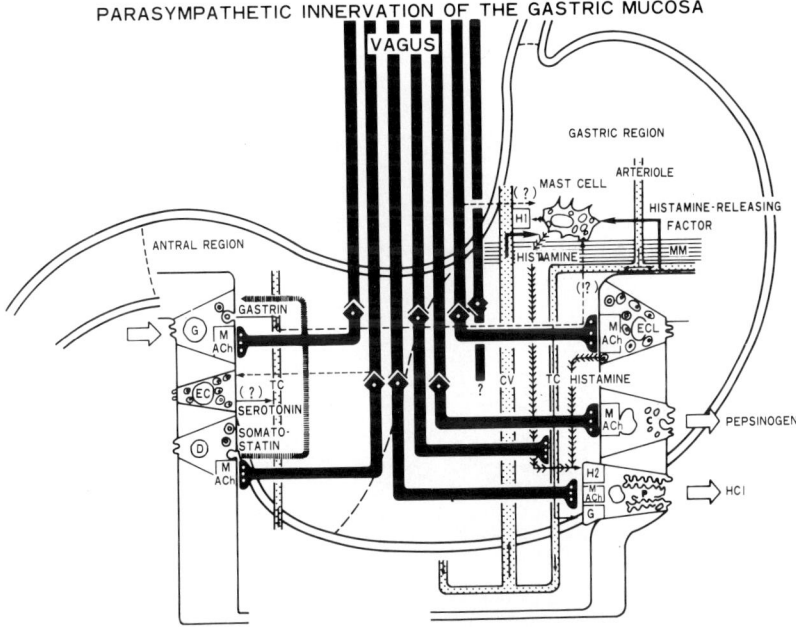

FIGURE 3. Parasympathetic innervation of the gastric mucosa. The cholinergic nerves which form synapses in the myenteric, submucosal, and mucosal plexuses are recognized near the parietal, chief, and ECL (enterochromaffin-like) cells in the fundic mucosa and the mucous, EC (enterochromaffin), and D cells in the antral mucosa as well as on the arterioles and venules.

B. CATECHOLAMINERGIC NERVES

Sympathetic pathways transmit command signals from the central nervous system via spinal nerves terminating in celiac ganglia which function to shut down the gastrointestinal blood flow and motility. From these ganglia, the postganglionic adrenergic fibers reach the stomach as discrete nerves or nerves closely associated with vagal nerves or as nerves accompanying the arterial vessels.

By the formaldehyde-induced fluorescence method, typically Falck-Hillarp's method and its modifications,[7,8] the nerve fibers showing specific noradrenaline fluorescence are located mainly in the perivascular plexuses of the arterioles and slightly in the myenteric and submucosal nerve plexuses (Figures 5 and 6).[9] Electron microscopically, the AChE-negative nerves possessing small electron dense synaptic vesicles clearly demonstrated in the chromaffin or $KMnO_4$-fixed preparations,[10] probably corresponding to the adrenergic fibers, are found mainly along the metarterioles in the basal part of the mucosa as well as along the arterioles in the submucosa, while those in the upper two thirds of the mucosa are almost devoid of adrenergic fibers.

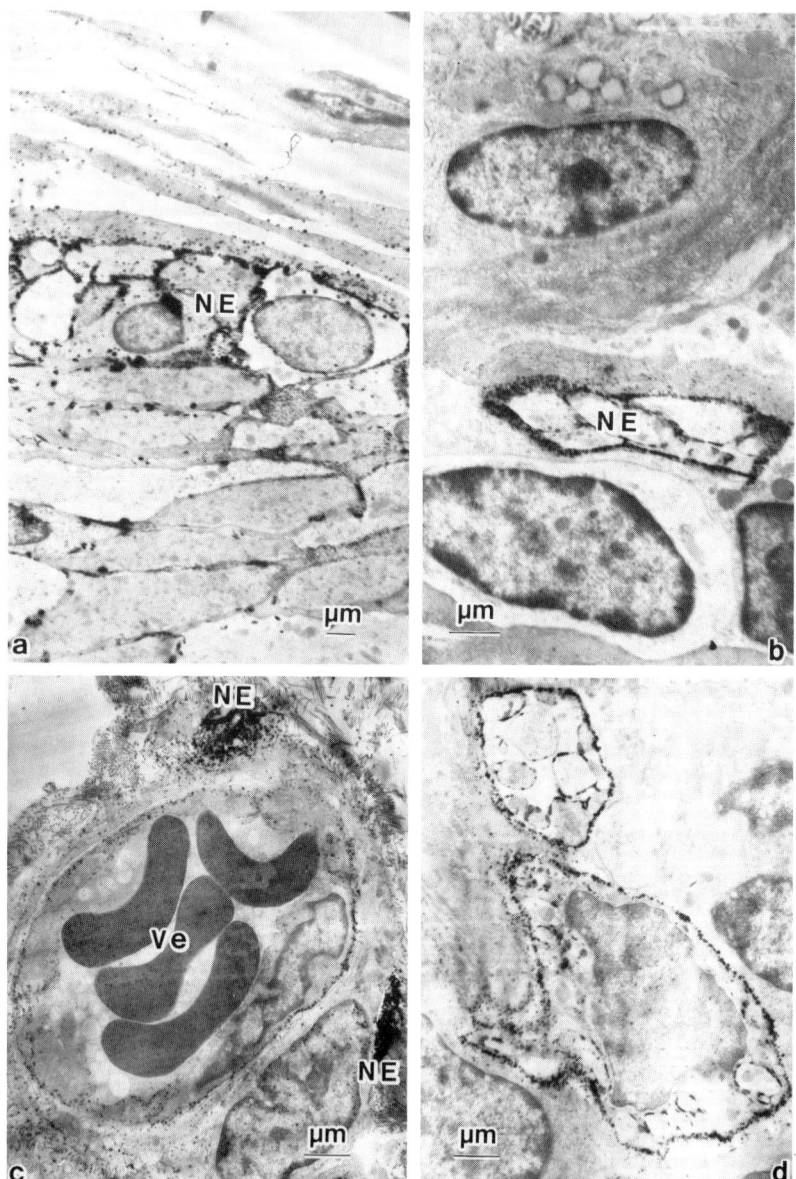

FIGURE 4. AChE cytochemistry of nerve fibers near the muscularis mucosae. (a,b) The AChE-positive nerve fibers are seen near the mucosal side of the muscularis mucosae. The AChE reaction products are seen on the axonal membrane of the unmyelinated nerve endings (NE) and on the basement membrane of the adjacent smooth muscle cells, suggesting the direct cholinergic innervation of the muscularis mucosae. (Original magnification = a: × 4900, b: × 8600.) (c) Some nerve endings are seen near the venules (Ve) passing through the muscularis mucosae. (Original magnification × 8000.) (d) The AChE-positive nerve fibers also exist between the muscle bundles of the muscularis mucosae. (Original magnification × 8000.)

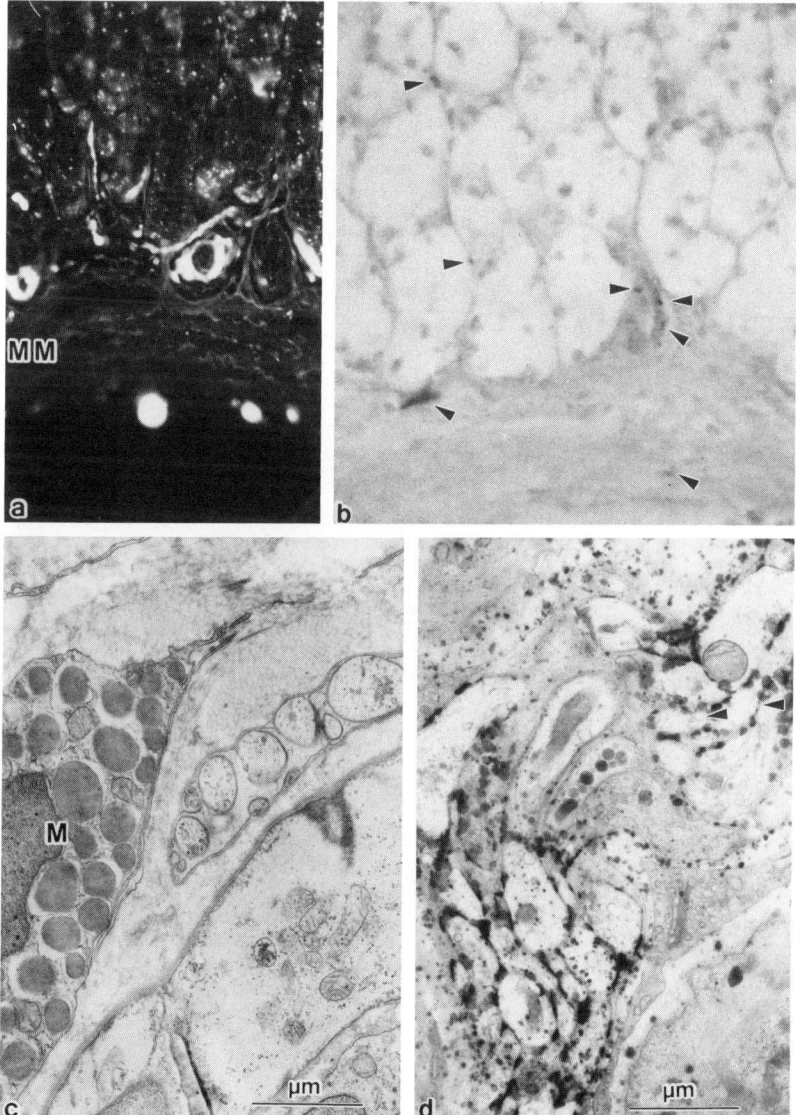

FIGURE 5. Light and electron micrographs showing the catecholaminergic nerves in the basal portion of the gastric mucosa. (a) By Falck-Hillarp's method, adrenergic nerve fibers are recognized surrounding the arterioles and venules and linearly between the lower part of the gastric epithelial cells. MM: muscularis mucosae. (Original magnification × 600.) (b) By tyrosine hydroxylase immunohistochemistry, the catecholaminergic nerves including both adrenergic and dopaminergic nerves are recognized surrounding the arterioles in the basal portion of the gastric mucosa. (Original magnification × 600.) (c) An electron micrograph showing unmyelinated nerve bundles near the arterioles. The nerve fibers exist adjacent to the smooth muscle cells of the arterioles. M: mast cell. (Original magnification × 20,000.) (d) AChE cytochemistry of nerve bundles near the arteriole. Some of the nerve fibers having large electron-dense synaptic vesicles are negative to AChE activity. (Original magnification × 20,000.)

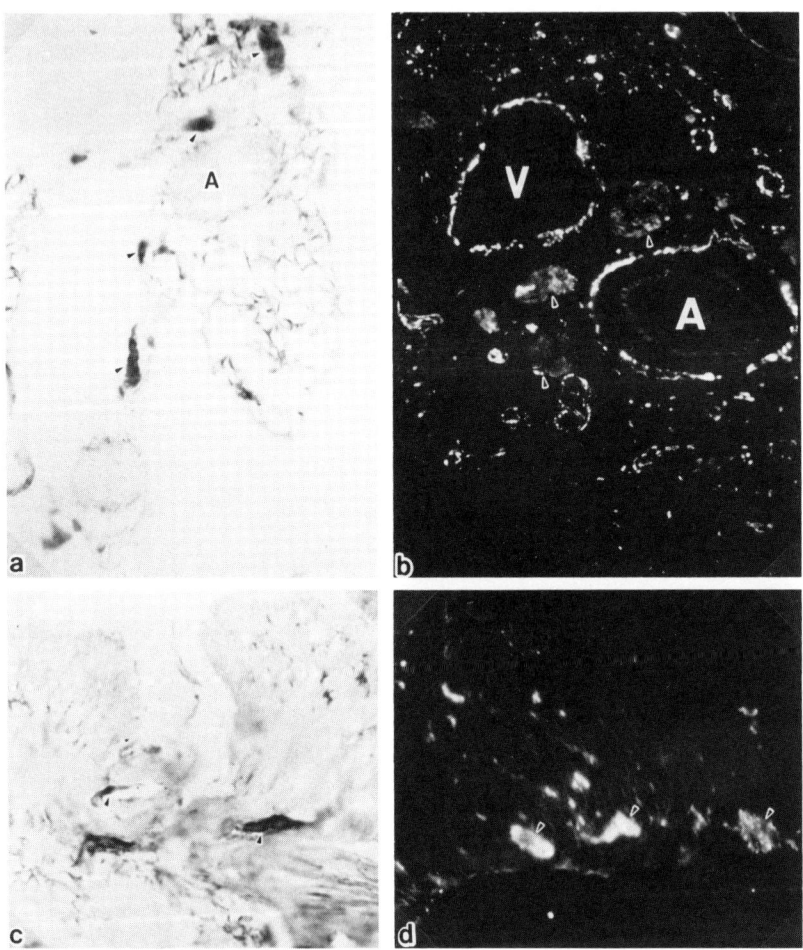

FIGURE 6. AChE histochemistry and noradrenaline fluorescence in the vagal nerve and myenteric plexus. (a) In the rat connective tissues surrounding the vagal nerve, several AChE-positive nerve bundles are recognized near the arteriole (A). (Original magnification × 400.) (b) In the similar area with Part a, noradrenaline fluorescence is seen in the nerve bundles as well as on the perivascular nerve plexus of the artery (A) and vein (V). (Original magnification × 400.) (c) In the proper muscular layer, the AChE-positive nerve fibers are recognized on the myenteric plexus (arrowheads) and nerve fibers in the smooth muscle cells. (Original magnification × 400.) (d) By Falck-Hillarp's method, the monoamine fluorescence is also seen on the myenteric plexus and between the smooth muscle cells. By spectrophotometry, the fluorescence between the smooth muscle cells is found to be mostly derived from dopamine and not noradrenaline. (Original magnification × 400.)

Recently, dopamine, one of the catecholamines, has been shown to be not only the precursor of noradrenaline but the active substance playing a certain role as a neurotransmitter, especially in the kidney.[11] In the stomach, the contraction of the proper smooth muscle cells is also shown to be influenced by dopamine in pharmacological and histochemical studies (Figure 6).[12] The dopaminergic nerves are identified histochemically as dopamine-beta-hydroxylase negative and tyrosine hydroxylase positive nerve fibers. By this criterion, many dopaminergic nerves are found near the arterioles in the muscularis mucosae and propriae in addition to the colocalization with the adrenergic nerves in perivascular plexuses.[13]

C. PEPTIDERGIC NERVES

Various kinds of peptidergic nerves are recognized in the stomach. Vasoactive intestinal polypeptide (VIP), gastrin releasing peptide (GRP), neuropeptide Y (NPY), and calcitonin gene-related peptide (CGRP) are the most abundant peptides in the stomach.

The VIPergic nerves show almost the same distribution as the cholinergic nerves. Electron microscopic studies have shown the coexistence of acetylcholine, VIP, and related peptide-peptide histidine isoleucine (PHI) in one nerve ending.[14] Most of the VIPergic nerves are seen adjacent to the muscularis mucosae and in the proper muscular layer and also in the lamina propria mucosae (Figure 7).[15] The GRP-immunoreactive nerves are also seen abundantly in the whole gastric layers. The GRP-immunoreactive nerves are also found to have the same distribution as the cholinergic nerves.

Recently, CGRP has attracted attention from the viewpoint of the afferent nervous regulation of the gastric circulation, because acute capsaicin treatment was found to prohibit the formation of the gastric mucosal lesion and capsaicin has been found to have a strong effect on the afferent nervous activity mediated by CGRP.[16,17] CGRP-immunoreactive nerves are mainly recognized on the perivascular nerve plexuses of the arterioles and venules as well as in the proper muscular layer.[18]

While most of the brain-gut peptides are thought to dwell in the cholinergic nerves, NPY is thought to be one of the typical brain-gut peptides which coexists with noradrenaline in the adrenergic nerves. By histochemical study, most of the NPY-immunoreactive nerves are seen near the arterioles and venules in the mucosal and submucosal layer, which show the same distribution as the adrenergic nerves.

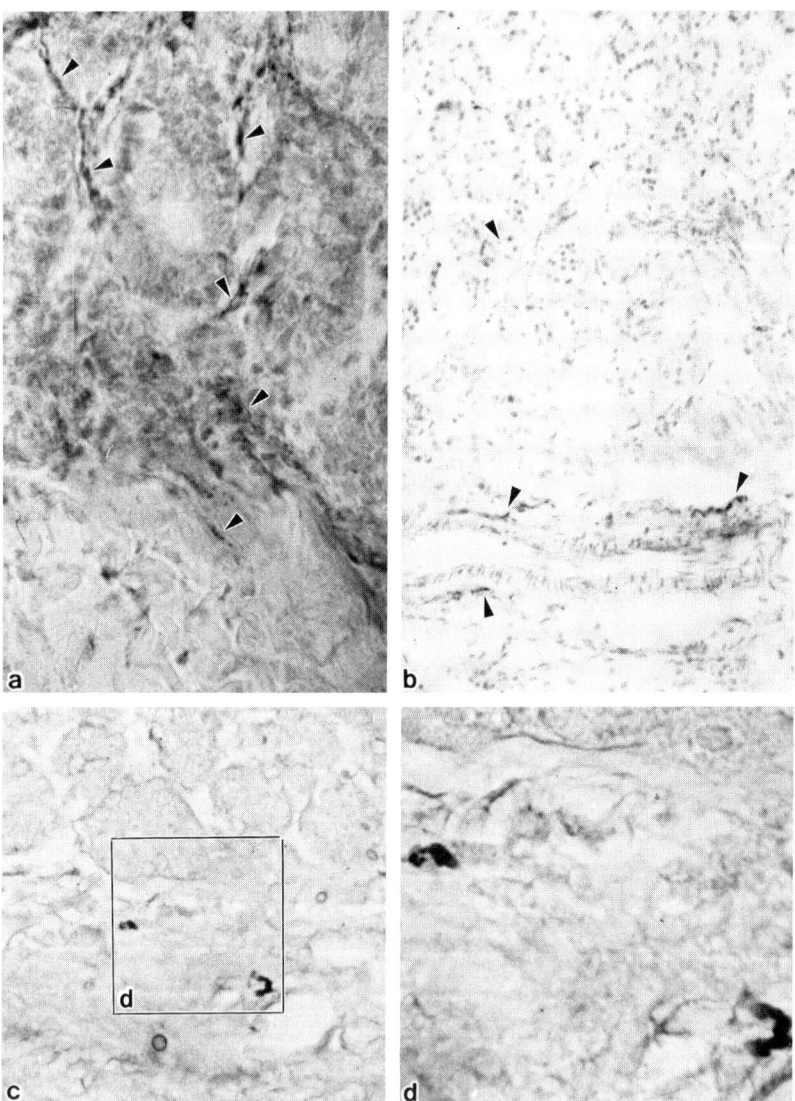

FIGURE 7. Immunohistochemistry of VIP, NPY, and CGRP in the rat fundic mucosa. (a) VIP-immunoreactive nerves are seen linearly between the gastric epithelial cells and adjacent to the smooth muscle cells of the muscularis mucosae. (Original magnification × 800.) (b) CGRP-immunoreactive nerves are mostly corresponding to the perivascular nerve plexuses. (Original magnification × 400.) (c,d) NPY-immunoreactive nerves are recognized in the basal portion of the lamina propria mucosae. (Original magnification = c: × 400, d: × 1200.)

III. DISTRIBUTION OF RECEPTORS OF VASCULAR EFFECTORS IN GASTRIC MICROCIRCULATORY SYSTEM

A. CLASSICAL TRANSMITTERS
1. Muscarinic Acetylcholine Receptors

Radioautographic study using radiolabeled antagonists or agonists is very useful for determining the localization of receptors, since one can obtain a very sensitive measurement of receptors in tissue with a high degree of anatomical discrimination. It is also possible to examine whether the receptor localization obtained reflects a physiological significant locus, by investigating whether the localization can be modified by the application of other agents that are known to affect the interaction between the receptors and their antagonists.

The easiest way to examine the localization of a soluble or diffusible agent is by the *in vitro* application of a radiolabeled ligand to freshly prepared cryostat sections followed by radioautographic examination (*in vitro* method).[19] However, the resolution obtained by this method is poor and it is almost impossible to apply for electron microscopy. Two methods which make it possible to obtain a higher resolution have recently been developed. One of these involves freezing and cryostat sectioning of tissue previously treated with isotope-labeled compounds.[20] The other consists of a freeze drying of isotope-labeled tissue followed by Epon embedding under low pressure.[21] The former method using cryostat sections is very useful at the light microscopic level, but a certain amount of diffusion artifact is inevitable during thawing of the frozen sections. Furthermore, electron microscopic observation is hardly feasible because of the difficulty of making freeze-dried ultrathin sections. Thus, in most of the studies shown in this chapter, the specimens were freeze dried immediately after they were removed to prevent diffusion of the radiolabeled compounds. In addition, the use of Epon embedding in this method permits relatively high resolution to be achieved by light and electron microscopy. The drawback of this method is ice-crystal formation in the tissue, but this problem would be minimized if the freeze-drying process were much quicker and evenly performed at an appropriate low pressure.

By the radioautography of soluble compounds described above, the distribution of the peripheral acetylcholine receptors, i.e., muscarinic acetylcholine receptors (m-AChR), can be visualized using a muscarinic antagonist, ^{3}H-quinuclidinyl benzilate (QNB).[22] The m-AChRs are distributed on the parietal and chief cells in the mucosa and true capillary endothelium as well as on the smooth muscle cells of the arterioles and venules and on the muscular layer. Pharmacological studies have also demonstrated that m-AChRs are localized both on the endothelium and the smooth muscle cells in the microcirculatory system (Figure 8).[23]

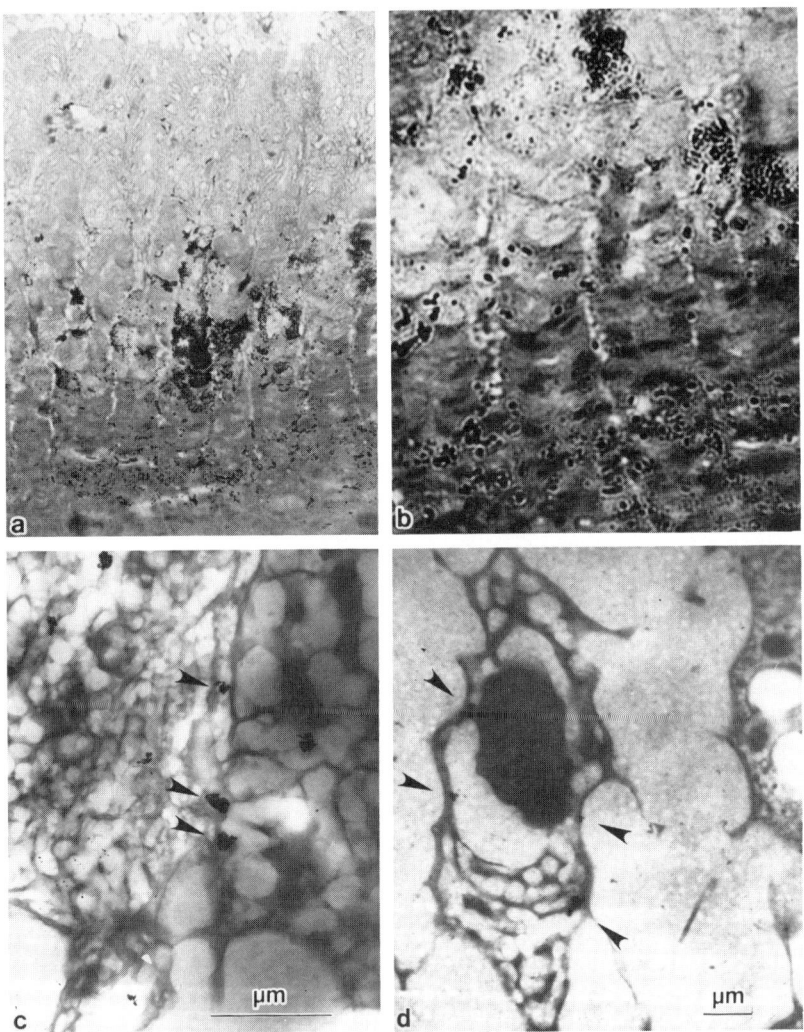

FIGURE 8. Light and electron microscopic radioautographs showing the binding sites of ^{3}H-quinuclidinyl benzilate (QNB) in the rat fundic mucosa (radioautography of soluble compounds). (a,b) By light microscopic radioautographs, the binding sites of QNB are recognized on the epithelial cells, especially on the parietal-cell dominant middle portion of the fundic glands. (Original magnification = a: × 400, b: × 1000.) (c) An electron microscopic radioautograph showing the binding sites of QNB on the parietal cells. Most of the grains are seen near the basolateral membrane of the parietal cells. (Original magnification × 24,000.) (d) An electron microscopic radioautograph showing the binding sites of QNB on the true capillary. Many silver grains, corresponding to the QNB-binding sites, are seen on the endothelial cells. (Original magnification × 10,000.)

Recent pharmacological and molecular biological studies have shown that m-AChR is subclassified into five types.[24,25] Using a muscarinic M_1 antagonist, pirenzepine, the highly specific receptors of pirenzepine, i.e., the M_1 receptors, are seen on the unmyelinated nerve endings and on the endothelium of the true capillary network in the gastric mucosa.[26] With high doses of pirenzepine, both M_1 and classical M_2 or M_3 receptors are visualized on the fundic glandular cells as well as on the unmyelinated nerve endings and on the endothelial cells (Figure 9). Figure 10 schematizes the above-mentioned data concerning the distribution of m-AChR in the stomach.

2. Dopaminergic Receptors

As in other organs, the (α) receptors are localized on the vascular smooth muscle cells (α_1) and perivascular nerve plexus (α_2) (Figure 11). The radioautographic studies revealed that D_1 receptors are located on the arterioles, while D_2 receptors are located on the collecting venules and on the muscularis mucosae. Together with the distribution of dopaminergic fibers in the muscularis mucosae and propriae, the dopaminergic nerves are thought to be very important in the gastric physiology.

B. BRAIN-GUT PEPTIDE

The VIP receptors are mainly seen on the smooth muscle cells of the arterioles and venules as well as on the smooth muscle cells of the muscularis mucosae (Figure 12).[24]

The GRP receptors are also seen on the muscular layer of the stomach.[25,26] GRP was a brain-gut peptide at first thought to be mainly concerned with the secretion of gastrin from the G cell in the antral region of the stomach. However, recent histochemical studies have revealed that the main localization of GRP is the muscular layer.[27]

On the other hand, the NPY-binding sites are seen exclusively on the vascular smooth muscle cells of the arterioles and venules, which coincides with the distribution of NPY-immunoreactive nerves.[28]

CGRP presence and release in the stomach have been recently demonstrated by pharmacological study,[29] but the effector sites of CGRP remain to be elucidated. Our recent histochemical studies on the binding sites of CGRP have revealed the existence of CGRP receptor on the smooth muscle cells of the arteriole and the proper muscular layer (Figure 13).

C. ENDOTHELIN

The endothelium produces vasodilators, collectively called endothelium-derived relaxing factors (EDRFs), as well as vasoconstrictors, one of which has recently been identified and named endothelin. In connection with the gastrointestinal tract, the infusion of endothelin is reported to produce the gastric mucosal injury.[30] The effector sites of endothelin are thought to be

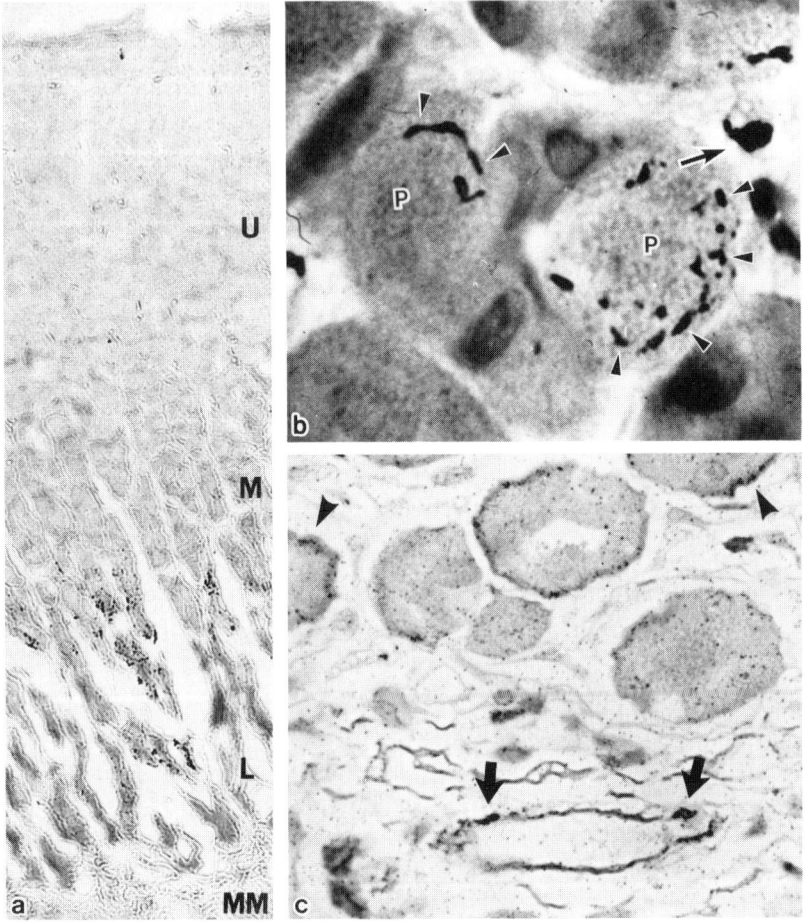

FIGURE 9. Light and electron microscopic radioautographs showing the binding sites of ^{3}H-pirenzepine (PZ) in the rat fundic mucosa (radioautography of soluble compounds). (a) By light microscopic radioautographs, the binding sites of PZ are recognized on the basal one third of the fundic glandular cells. (Original magnification × 300.) (b) In high magnification, the silver grains are seen on the parietal cells (arrowheads) and on the true capillary endothelium. (Original magnification × 3000.) (c) In the basal portion of the gastric mucosa, the silver grains are found on the basal membrane of the chief cells (arrowheads) and on the endothelium of the venules (arrows) and arterioles. (Original magnification × 1500.)

the smooth muscle cells of the arterioles and venules, but recent studies have revealed the interaction of endothelin on the smooth muscle cells of the muscular layers, fibroblasts, and macrophages[31-33] (Figure 14).

D. HISTAMINE

The stomach is one of the richest organs in the storage of histamine, and the most important role of histamine in the gastric mucosa is the regulation

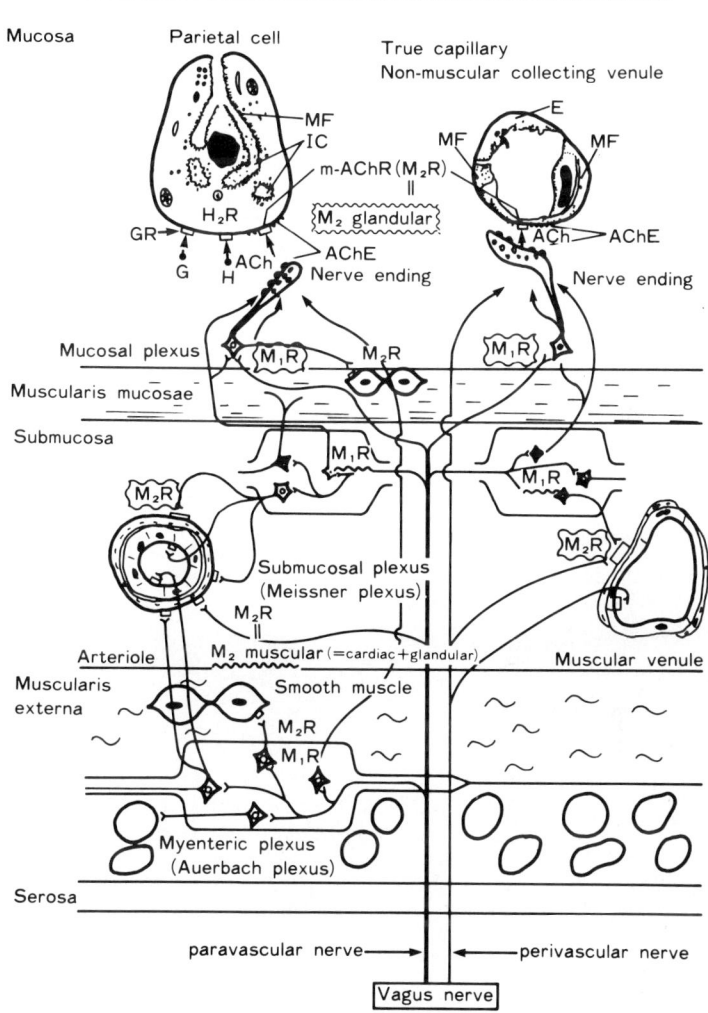

FIGURE 10. Cholinergic innervation of the stomach.

of the acid secretion as shown in Figure 15. The radioautographic studies have shown that in the gastric mucosa histamine also acts as the regulator of the microcirculatory system[34] (Figures 16 and 17). Figure 18 shows the interaction of autonomic nerves and the localization of the corresponding receptors. From the point of microcirculation, using the arteriole as an example, three different regulatory mechanisms exist, i.e., the endothelium, smooth muscle cell, and perivascular nerve plexus (Figure 19).

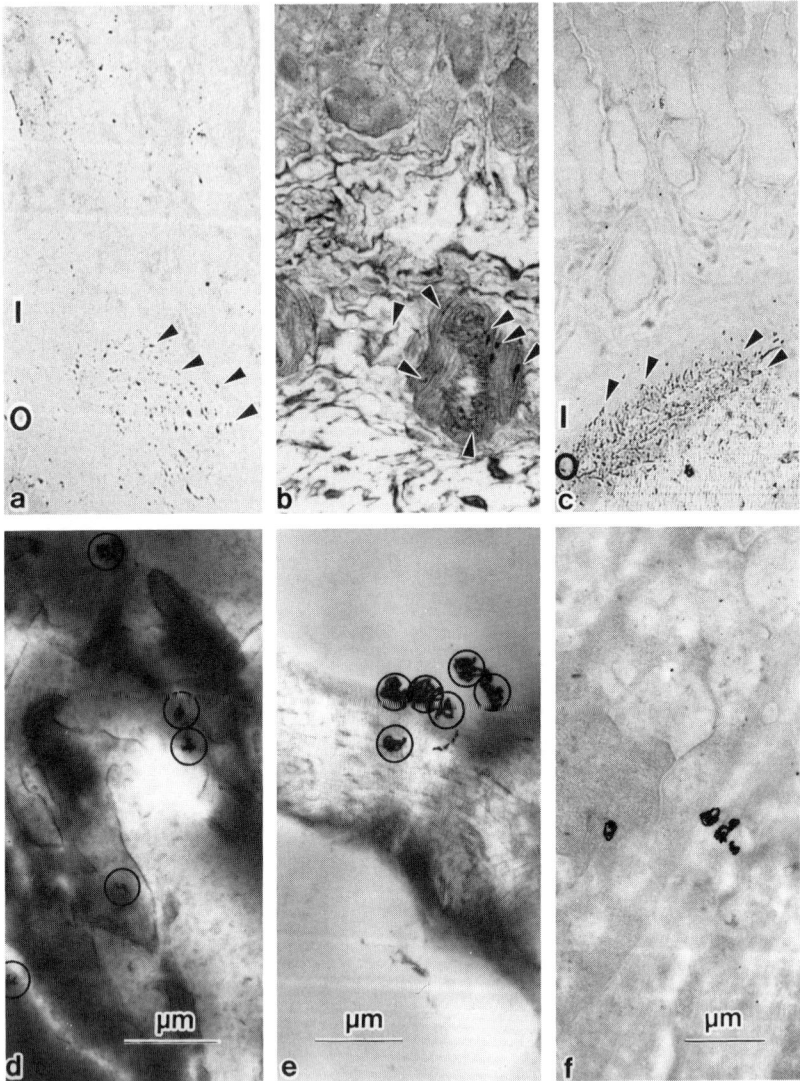

FIGURE 11. Light and electron microscopic radioautographs showing the binding sites of ^{3}H-dopamine, D_1 antagonist, ^{3}H-SCH23390 and D_2 antagonist, ^{3}H-spiperone and ^{3}H-noradrenaline in the rat stomach (radioautography of soluble compounds). (a, d) The binding sites of ^{3}H-dopamine (arrowheads) are found on the outer layer of the muscularis mucosae (O) and on the epithelial cells. The electron microscopic radioautographs show that the silver grains (circles) are scattered on the smooth muscle cells. (Original magnification = a: × 600, d: × 17,000.) (b) The binding sites of SCH23390 (arrowheads) are seen on the vascular smooth muscle cells of the arterioles and the perivascular plexuses. (Original magnification × 600.) (c, e) The binding sites of spiperone (arrowheads) are mainly found on the outer layer of the muscularis mucosae (O). The electron microscopic radioautographs show that the silver grains exist exclusively on the smooth muscle cells. (Original magnification = c: × 460, e: × 15,000.) (f) The binding sites of ^{3}H-noradrenaline (arrowheads) are seen on the unmyelinated nerve endings. (Original magnification × 15,000.)

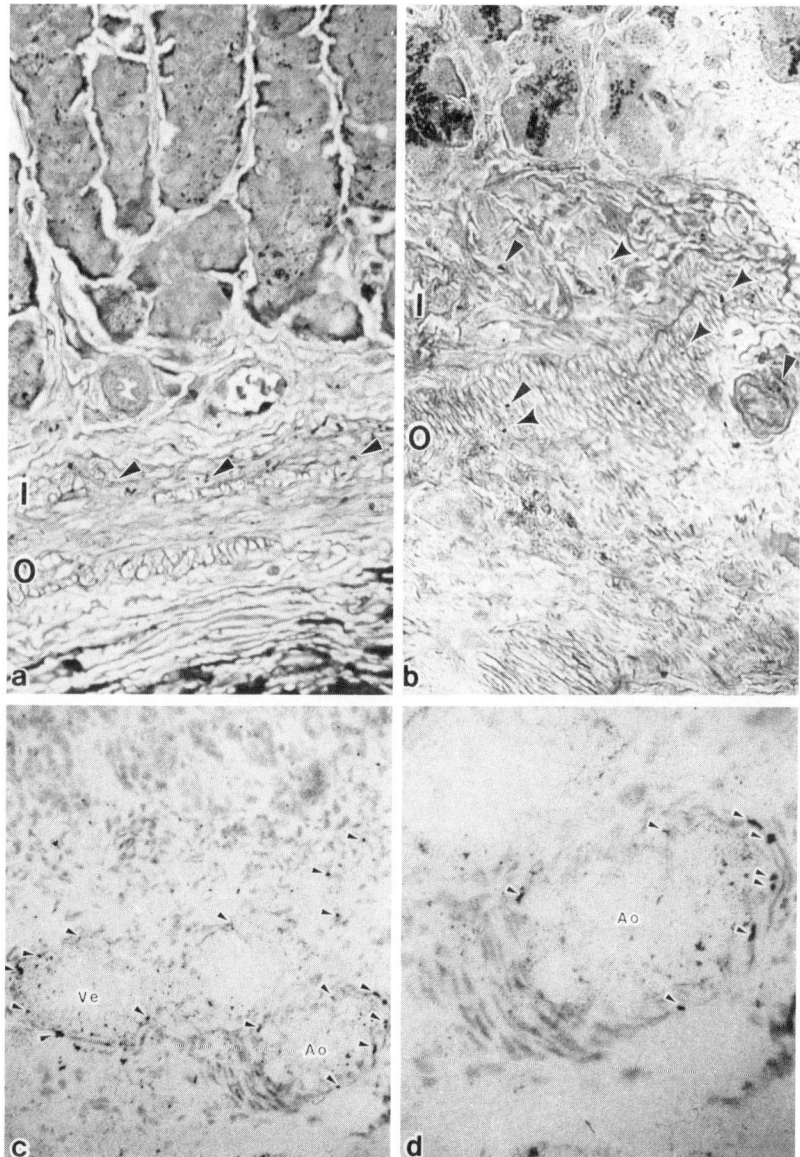

FIGURE 12. Light microscopic radioautographs showing the binding sites of ^{125}I-VIP, ^{125}I-GRP, and ^{3}H-NPY in the rat stomach. (a, b) Radioautography of soluble compounds; (c, d) *in vitro* radioautography. (a) The binding sites of VIP are found on the inner layer of the muscularis mucosae (arrowheads), the endothelia of the arterioles and venules, and the parietal cells. (Original magnification × 600.) (b) The binding sites of GRP in the stomach (arrowheads) are mainly found on the muscularis mucosae as well as on the fundic glandular cells. (Original magnification × 800.) (c, d) The binding sites of NPY are seen on the smooth muscle cells of the arterioles (Ao) and venules (Ve). (Original magnification = c: × 600, d: × 1200.)

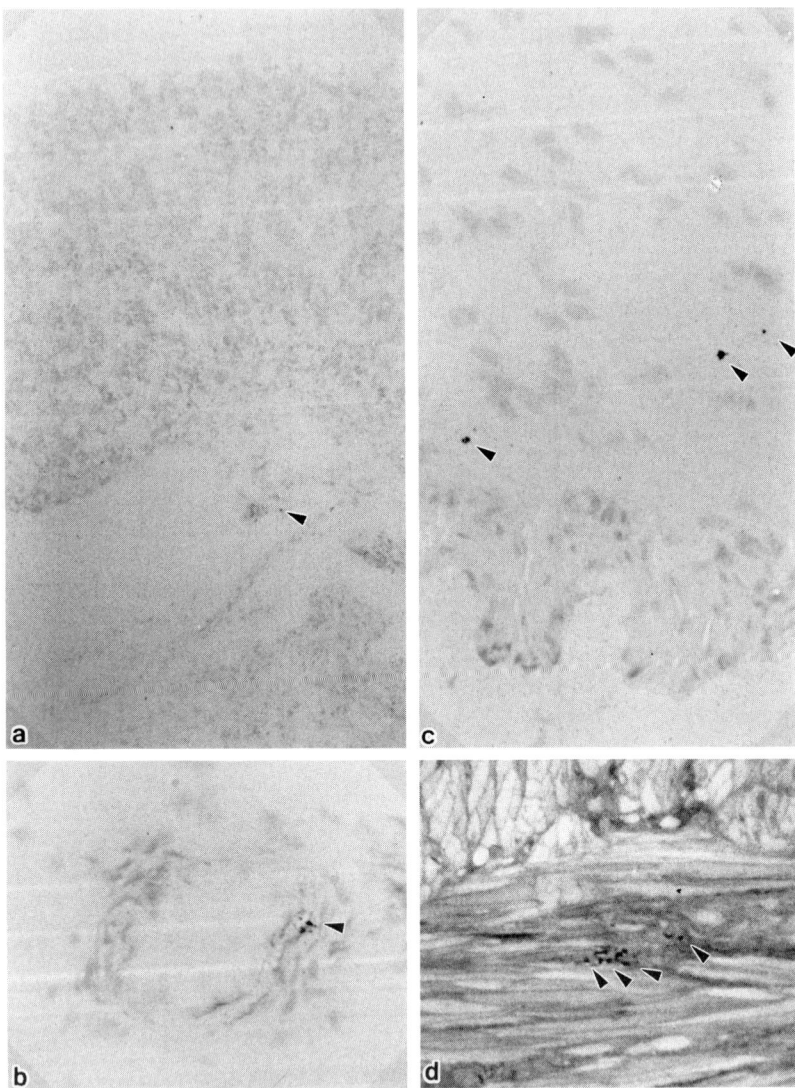

FIGURE 13. Light microscopic radioautographs showing the binding sites of ^{125}I-CGRP in the rat stomach (a-c: *in vitro* radioautography, d: radioautography of soluble compounds). (a, b) The binding sites of CGRP in the mucosal and submucosal layers are mainly seen on the smooth muscle cells of the arteriole. (Original magnification = a: × 400, b: × 1200.) (c, d) The binding sites of CGRP in the muscular layer are seen on the smooth muscle cells (c) and myenteric plexuses (d). (Original magnification = c: × 1200, d: × 1000.)

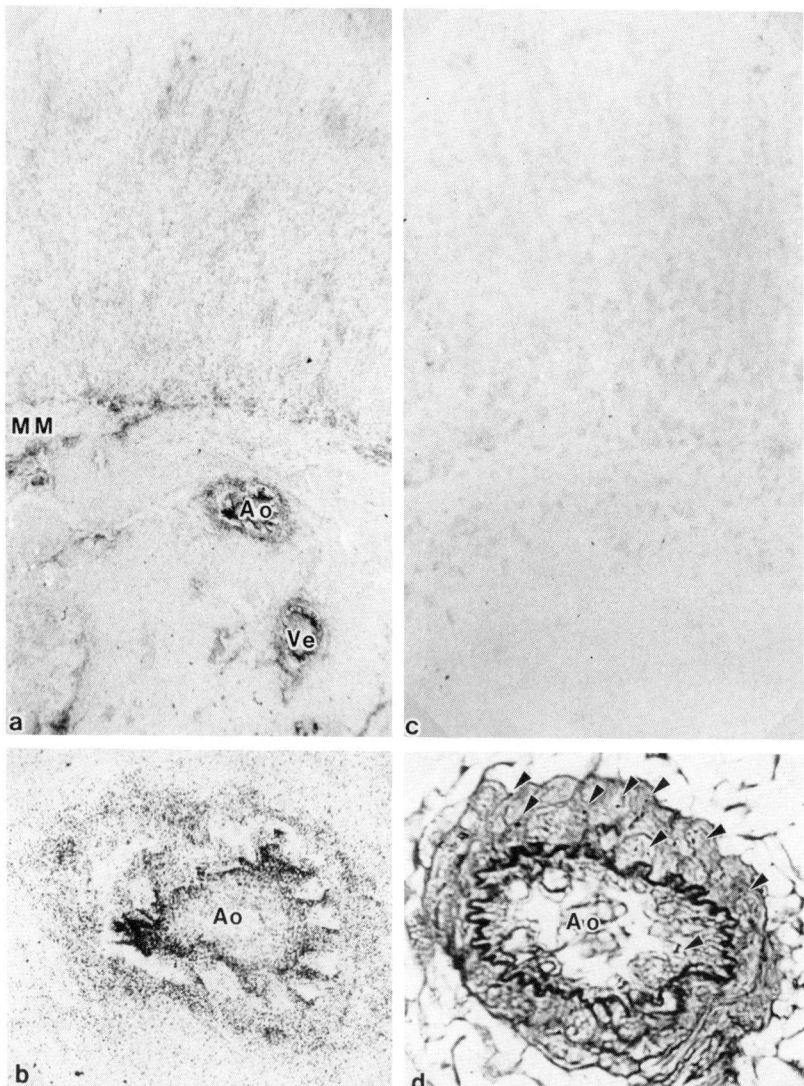

FIGURE 14. Light microscopic radioautographs showing the binding sites of ^{125}I-endothelin-1 in the rat stomach (a-c: *in vitro* radioautography, d: radioautography of soluble compounds). (a, b, d) The bindings of ET-1 on the mucosal and submucosal layer of the rat stomach are recognized on the vascular smooth muscle cells of the arteriole (Ao) and venules (Ve) and slightly on the smooth muscle cells comprising the muscularis mucosae. (Original magnification = a: × 400, b: × 1200, d: × 1200.) (c) By the mixed administration of cold and radiolabeled ligand, the number of the silver grains is significantly decreased. (Original magnification × 400.)

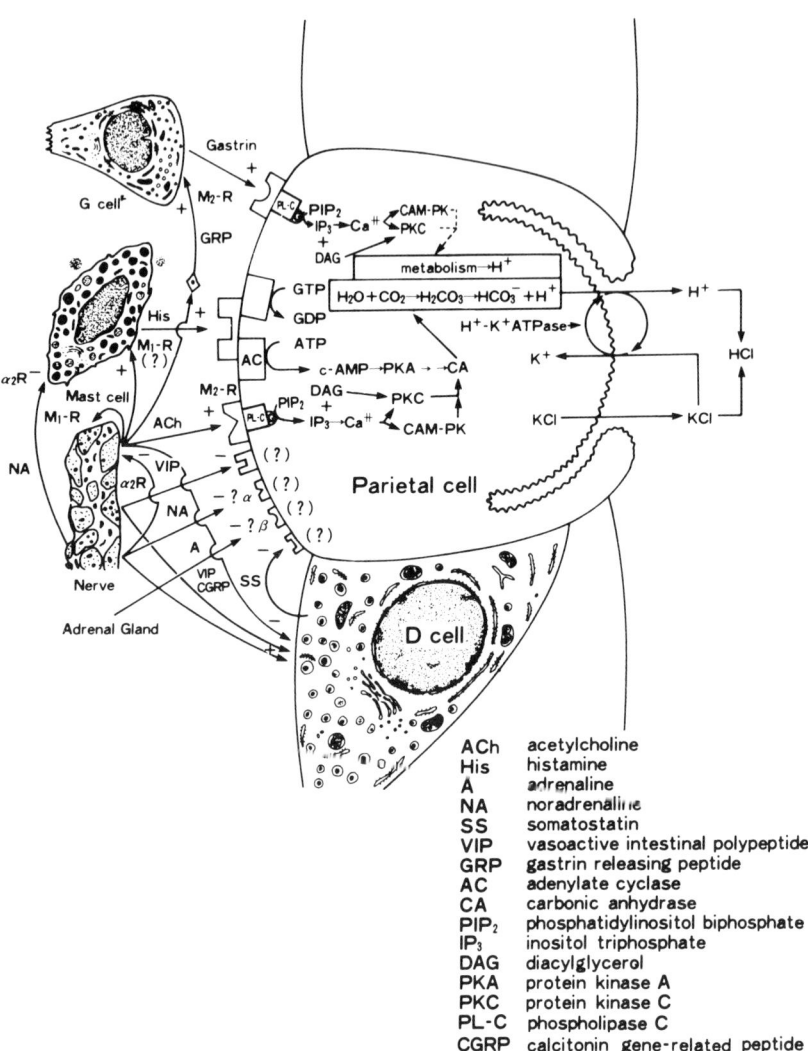

FIGURE 15. Mechanism of acid secretion.

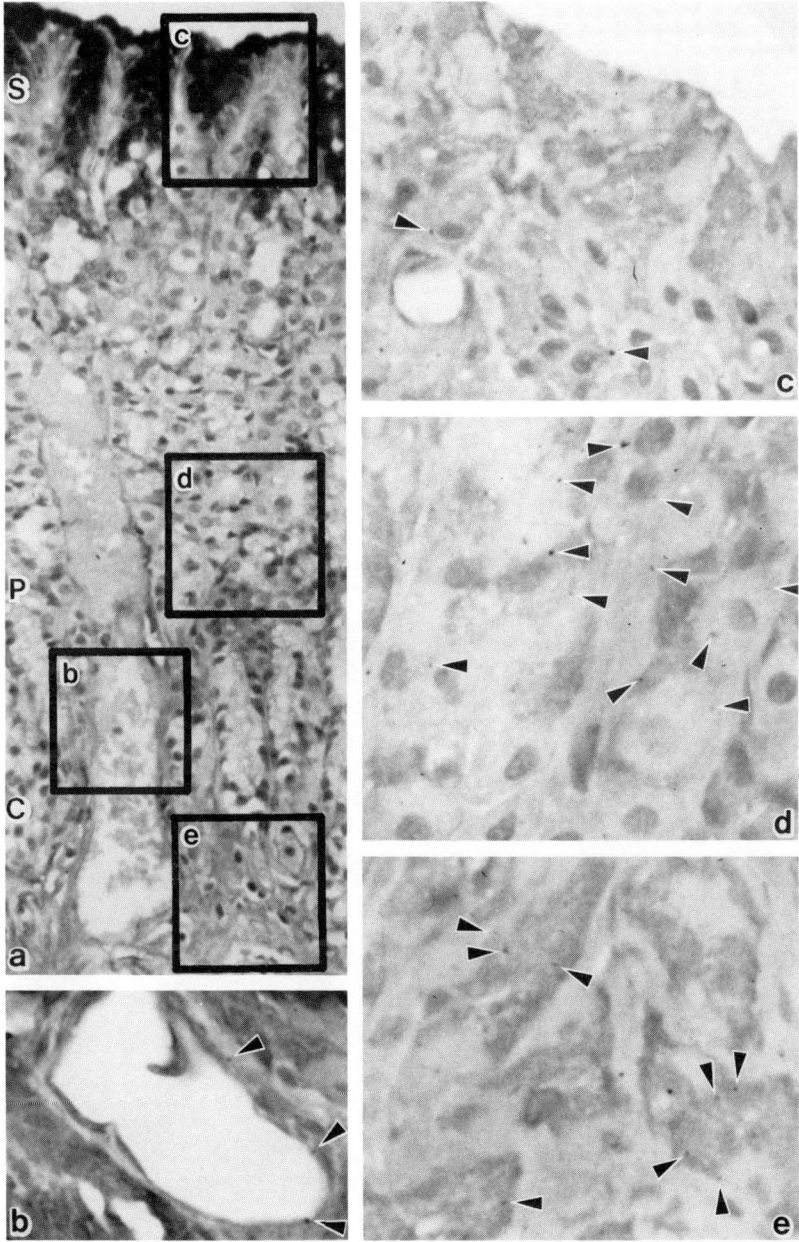

FIGURE 16. Light microscopic radioautographs showing the binding sites of ^{3}H-histamine (radioautography of soluble compounds). The binding sites of histamine are seen on several kinds of epithelial cells and on the endothelial cells of the collecting venules (b). S: surface mucous cell dominant area, P: parietal cell dominant area, C: chief cell dominant area. (Original magnification = a: × 400, b–e: × 1000.)

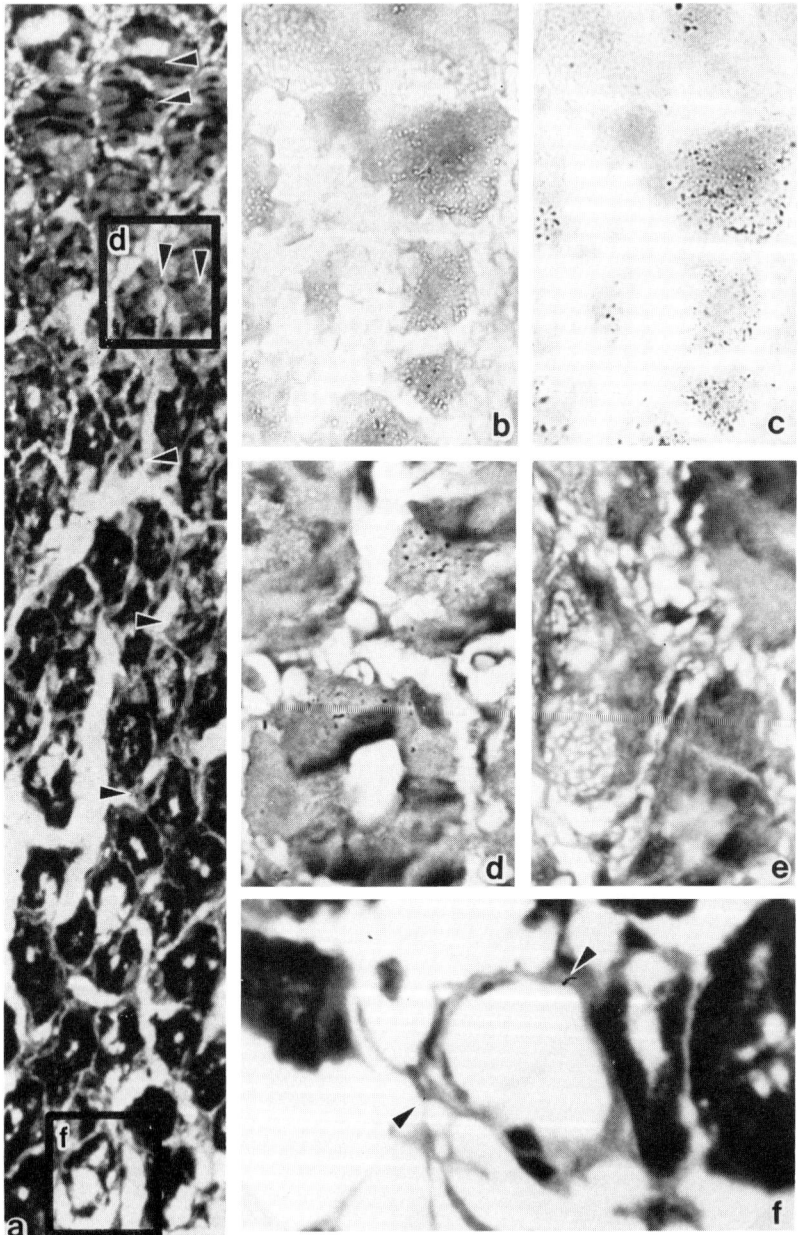

FIGURE 17. Light microscopic radioautographs showing the binding sites of ^{3}H-cimetidine (radioautography of soluble compounds). The binding sites of cimetidine are exclusively seen on the parietal cells and the endothelium of the collecting venules (f). (b, c) No staining; (a, d, e) counterstained with toluidine blue. (Original magnification = a: × 600, b–f: × 1200.)

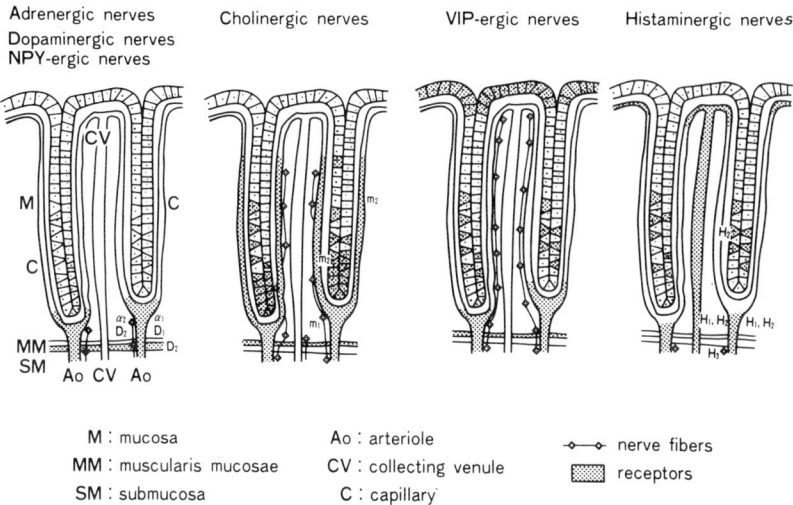

FIGURE 18. Distribution of autonomic nerves and their receptors.

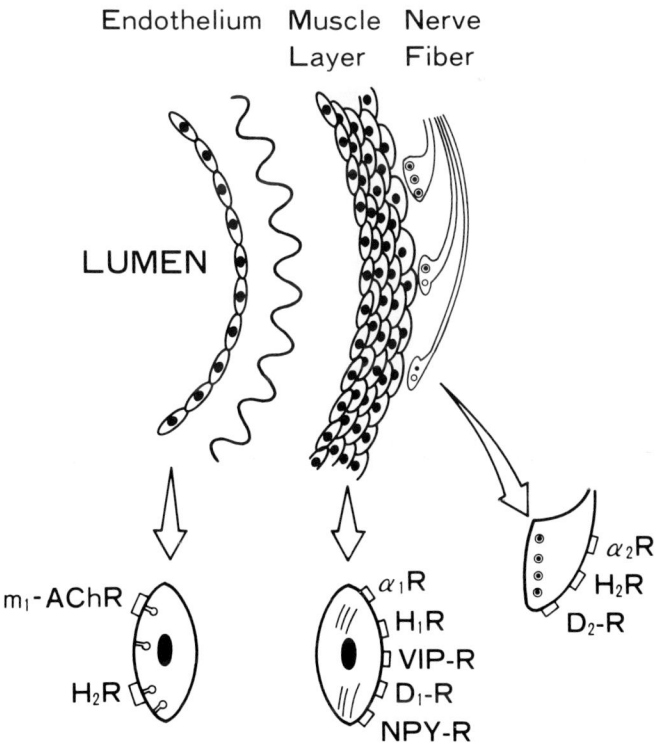

FIGURE 19. Distribution of receptors on arteriole.

IV. ALTERATION OF AUTONOMIC NERVOUS ACTIVITY AND RECEPTOR DISTRIBUTION IN GASTRIC ULCER FORMATION

During the restraint-induced ulcer formation, severe increase of the AChE activity and the decrease of the monoamine fluorescence is recognized, which is followed by the decrease of m-AChR in the gastric mucosa as one of the down regulations of the receptor numbers (Figure 20).

V. CONCLUSIONS

The gastrointestinal tract is shown to be intricately regulated by many kinds of brain-gut peptides (Figure 21). Each neuromodulator and neurotransmitter has a corresponding receptor and their changes are shown to be closely related to the diseased state in experimental animals. The interaction with human disease has not been clarified as yet, so further study is needed to understand the real significance of these mediators in human disease.

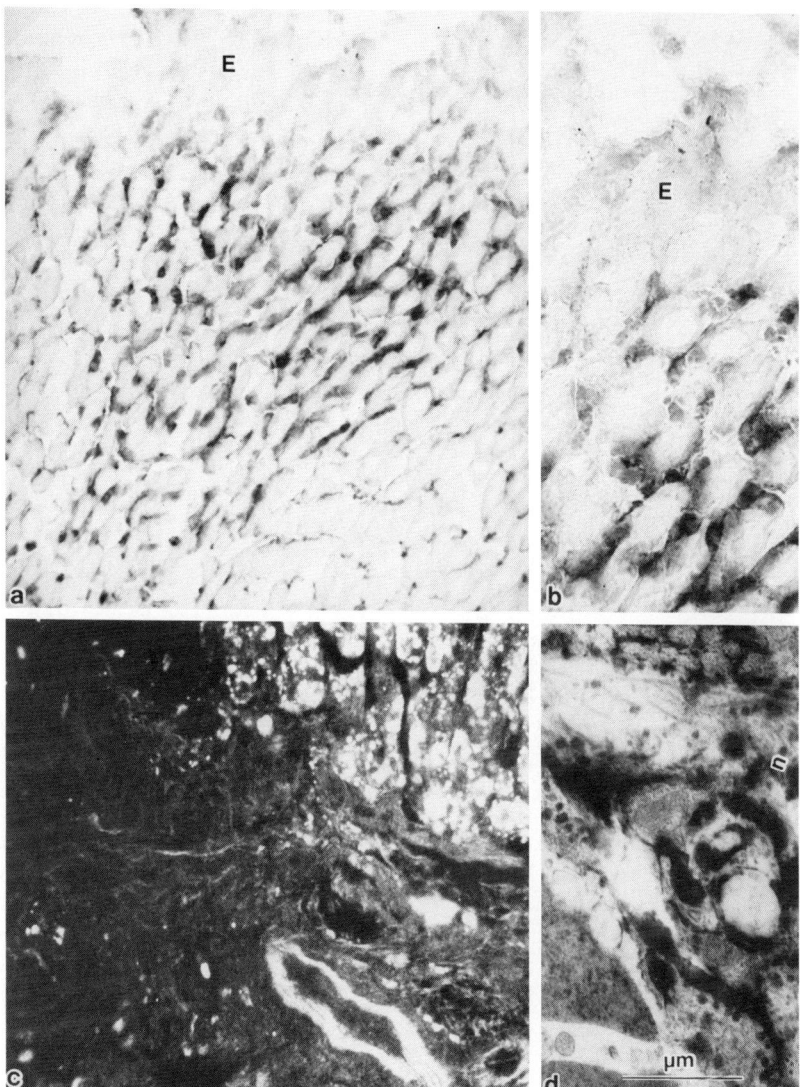

FIGURE 20. Alteration of autonomic nervous activity during experimental ulcer formation. (a, b, d) The AChE histochemical activity is significantly increased just under the erosive lesions (E). By electron microscopic observation, AChE reaction products are accumulated on the axonal membrane and the basement membrane of the adjacent true capillary endothelium. (Original magnification = a: × 600, b: × 1800, d: × 24,000.) (c) By Falck-Hillarp's method, the noradrenaline-specific fluorescence is decreased near the erosive lesion. (Original magnification × 400.)

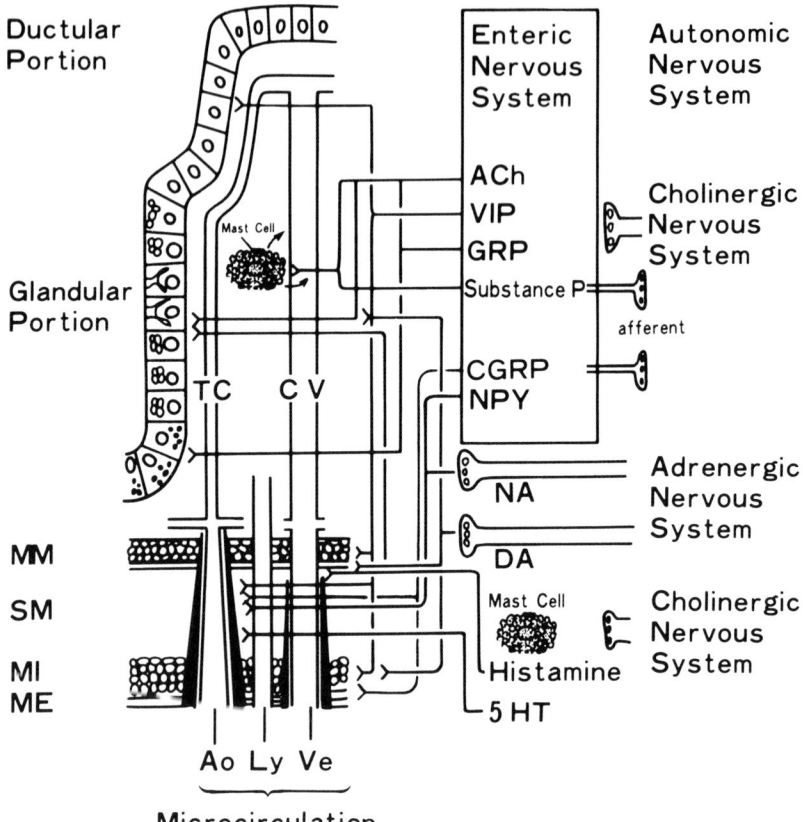

FIGURE 21. Enteric nervous regulation of gastrointestinal tract.

REFERENCES

1. **Satoni, H., Yamamoto, T., Ise, H., and Takatomi, H.,** Origins of the parasympathetic fibers to the cat intestine as demonstrated by the horseradish peroxidase method, *Brain Res.,* 151, 571–578, 1978.
2. **Hoffman, H. H. and Schnitzlein, N. N.,** The number of vagus nerves in man, *Anat. Rec.,* 139, 429–435, 1969.
3. **Bell, C.,** Fine structural localization of acetylcholinesterase at a cholinergic vasodilator nerve-arterial smooth muscle synapse, *Circ. Res.,* 24, 61–70, 1969.
4. **Karnovsky, M. J. and Roots, L.,** A direct coloring thiocholine method for cholinesterases, *J. Histochem. Cytochem.,* 12, 219–221, 1964.

5. **Nakamura, M., Oda, M., Watanabe, N., Tsukada, N., Yonei, Y., and Tsuchiya, M.**, Evidence for direct parasympathetic innervation of parietal cells in the rat glandular stomach — histochemical and electron microscopic cytochemical study, *Okajimas Folia Anat. Jpn.*, 59, 167–180, 1982.
6. **Nakamura, M., Oda, M., Watanabe, N., Tsukada, N., Yonei, Y., Komatsu, H., and Tsuchiya, M.**, Electron microscopic evidence for cholinergic innervation of the gastric mucosal capillaries in rats: a histochemical, electron microscopic and radioautographic study, in *Progress in Microcirculatory Research II*, Courtice, F. C., Garlick, D. G., and Perry, M. A., Eds., Committee in Postgraduate Medical Education, University of NSW, NSW, Australia, 1984, 284–290.
7. **Falck, B. and Hillarp, N.**, Fluorescence of catecholamine and related compounds condensed with formaldehyde, *Histochem. Cytochem.*, 10, 348–354, 1973.
8. **Ajelis, V., Björklund, A., Falck, B., Lindvall, O., Laren, I., and Walles, B.**, Application of aluminium-formaldehyde (ALFA) histofluorescence method for demonstration of peripheral stores of catecholamine and indolamines in freeze-dried paraffine embedded tissue, cryostat sections and whole mounts, *Histochemistry*, 6, 1–15, 1979.
9. **Nakamura, M., Watanabe, N., Tsukada, N., Oda, M., and Tsuchiya, M.**, Demonstration of the adrenergic nerves in the rat gastric mucosa — a histofluorescence and electron microscopic study in comparison with the distribution of the cholinergic nerves, *Okajimas Folia Anat. Jpn.*, 59, 65–86, 1982.
10. **Tranzer, J. P. and Richards, J. G.**, Ultrastructural cytochemistry of biogenic amines in nervous tissue; methodological improvements, *J. Histochem. Cytochem.*, 24, 1178–1193, 1976.
11. **Lockovc, Z. and Relja, W.**, Evidence for a widely distributed dopaminergic system, *Fed. Proc.*, 42, 3000–3004, 1983.
12. **Valenzuela, J. E.**, Dopamine as a possible neurotransmitter in gastric relaxation, *Gastroenterology*, 71, 1019–1022, 1976.
13. **Nishizaki, Y., Oda, M., Nakamura, M., Inoue, J., Morishita, A., Nishida, J., Ishii, K., Yokomori, M., Kaneko, H., Ohya, H., and Tsuchiya, M.**, Significance of 5-hydroxytryptamine in rat gastric mucosa: immunohistochemical and radioautographic observation, in *New Trends in Autonomic Nervous System Research*, Yoshikawa, M., Uono, N., Tanabe, H., and Ishikawa, S., Eds., Elsevier, Amsterdam, 1991, 262–263.
14. **Ekblad, E., Ekelund, M., Graffner, H., Håkanson, R., and Sundler, F.**, Peptide-containting nerve fibers in the stomach wall of rat and mouse, *Gastroenterology*, 84, 241–249, 1985.
15. **Konturek, S. J., Dembinski, A., Thor, P., and Krol, R.**, Comparison of vasoactive intestinal peptide (VIP) and secretin in gastric secretion and mucosal blood flow, *Pfluegers Arch.*, 361, 175–181, 1976.
16. **Holzer, P., Livingston, E. H., Saria, A., and Guth, P. H.**, Sensory neurons mediate protective vasodilatation in rat gastric mucosa, *Am. J. Physiol.*, 260, G363–G370, 1991.
17. **Forster, E. R. and Dockray, G. J.**, The role of calcitonin gene-related peptide in gastric mucosal protection in the rat, *Exp. Physiol.*, 76, 623–626, 1991.
18. **Lee, Y., Shiotani, Y., Hayashi, N., Kamada, T., Hillyard, C. J., Girris, S. I., MacIntyre, I., and Tohyama, M.**, Distribution and origin of calcitonin gene-related peptide in the rat stomach and duodenum: an immunohistochemical analysis, *J. Neural. Trans.*, 68, 1–14, 1987.
19. **Young, W. S., III and Kuhar, M. J.**, A new method for receptor autoradiography: ^{3}H-opioid receptor labeling in mounted sections, *Brain Res.*, 179, 255–270, 1979.
20. **Stumpf, W. E. and Roth, L. J.**, Freeze-drying of small tissue samples and thin frozen sections below $-60°C$. A simple method of cryosorption pumpings, *J. Histochem. Cytochem.*, 15, 243–257, 1967.
21. **Nagata, T., Nawa, T., and Yokota, S.**, A new technique for electron microscopic drymounting radioautography of soluble compounds, *Histochemie*, 18, 241–249, 1969.

22. **Nakamura, M., Oda, M., Watanabe, N., Tsukada, N., Yonei, Y., Komatsu, H., and Tsuchiya, M.,** Demonstration of the localization of muscarinic acetylcholine receptors in the rat gastric mucosa — light and electron microscopic autoradiographic studies using ^{3}H-quinuclidinyl benzilate, *Acta Histochem. Cytochem.,* 17, 297–309, 1984.
23. **Tsukahara, T., Hongo, K., and Kassell, N. F.,** Characterization of muscarinic cholinergic receptors on the endothelium and the smooth muscle of the rabbit thoracic aorta, *J. Cardiovasc. Pharmacol.,* 13, 870–878, 1989.
24. **Goyal, R. K.,** Identification, localization and classification of muscarinic receptor subtypes in the gut, *Life Sci.,* 43, 2209–2220, 1988.
25. **Bonner, T. I., Buckley, N. J., and Young, A. C.,** Identification of a family of muscarinic acetylcholine receptor genes, *Science,* 237, 527–532, 1987.
26. **Nakamura, M., Oda, M., Yonei, Y., Tsukada, N., Komatsu, H., Kaneko, K., and Tsuchiya, M.,** Muscarinic acetylcholine receptors in rat gastric mucosa: a radioautographic study using a potent muscarinic antagonist, ^{3}H-pirenzepine, *Histochemistry,* 83, 479–487, 1985.
27. **Nakamura, M., Oda, M., Kaneko, K., Komatsu, H., Tsukada, N., Honda, K., and Tsuchiya, M.,** Radioautographic studies on the localization of VIP-binding sites in the rat fundic mucosa, *J. Clin. Electron Microsc.,* 19, 484–485, 1986.
28. **Nakamura, M., Oda, M., Kaneko, K., Akaiwa, Y., Tsukada, N., Komatsu, H., and Tsuchiya, M.,** Autoradiographic demonstration of gastrin-releasing peptide-binding sites in the rat gastric mucosa, *Gastroenterology,* 94, 968–976, 1988.
29. **Moran, T. H., Moody, T. W., Hostetler, A. M., Robinson, P. H., Goldrich, M., and McHugh, P. R.,** Distribution of bombesin binding sites in the rat gastrointestinal tract, *Peptides,* 9, 643–649, 1988.
30. **Vigna, S. R., Mantyh, C. R., Giraud, A. S., Soll, A. H., Walsh, J. H., and Mantyh, P. W.,** Localization of specific binding sites of bombesin in the canine gastrointestinal tract, *Gastroenterology,* 93, 1287–1295, 1987.
31. **Nakamura, M., Oda, M., Kaneko, K., Komatsu, H., Honda, K., Azuma, T., Nishizaki, Y., Fujishiro, Y., and Tsuchiya, M.,** Peptidergic and aminergic control of gastric mucosal microcirculation: radioautographic study of receptor localization, in *Microcirculation — An Update,* Vol. 2, Elsevier, Amsterdam, 1987, 229–232.
32. **Inui, T., Chiba, T., Okimura, Y., Morishita, T., Nakamura, A., Yamaguchi, A., Yamatani, T., Kadowaki, S., Chihara, K., and Fujita, T.,** *Life Sci.,* 45, 1199–1206, 1989.
33. **Whittle, B. J. R. and Lopez-Belmonte, J.,** Interactions between the vascular peptide endothelin-1 and sensory neuropeptides in gastric mucosal injury, *Br. J. Pharmacol.,* 102, 950–954, 1991.
34. **Payen, A. N. and Whittle, B. J. R.,** Potent cyclo-oxygenase mediated bronchoconstrictor effects of endothelin in the guinea pig *in vivo, Eur. J. Pharmacol.,* 158, 303–304, 1988.
35. **Takuwa, N., Takuwa, Y., Yanagisawa, M., Yamashita, K., and Masaki, T.,** A novel vasoactive peptide endothelin stimulates mitogenesis through inositol lipid turnover in Swiss 3T3 fibroblasts, 264, 7856–7861, 1989.
36. **Ehrenreich, H., Anderson, R. W., Fox, C. H., Rieckmann, P., Hoffman, G. S., Travis, W. D., Coligan, J. E., Kehrl, J. H., and Fauci, A. S.,** Endothelins, peptides with potent vasoactive properties, are produced by human macrophages, *J. Exp. Med.,* 172, 1741–1748, 1990.
37. **Nakamura, M., Oda, M., Inoue, J., Ito, T., Kurose, I., Fukumura, D., and Tsuchiya, M.,** Binding sites of endothelin-1 and 3 on rat mesentery, stomach and large intestine: radioautographic investigation, in *Microcirculation Annual 1991,* Tsuchiya, M., Asano, M., and Katori, M., Eds., Nihon-Igakukan, Japan, 1991, 71–72.
38. **Nakamura, M., Oda, M., Kaneko, K., Honda, K., Komatsu, H., and Tsuchiya, M.,** Radioautographic characterization of H1 and H2 receptor antagonists binding sites in rat gastric mucosal microcirculatory system, in *Vascular Endothelium in Health and Disease,* Chien, S., Ed., 1988, 151–160.

Chapter 6

MEMBRANE PERTURBATION BY ASBESTOS FIBERS AND DISEASE

Lee A. Goodglick and Agnes B. Kane

TABLE OF CONTENTS

I.	Introduction	124
II.	Entry and Retention of Asbestos Fibers in the Lungs	126
III.	Biology and Function of the Macrophage	127
IV.	Acute Asbestos Cytotoxicity	127
	A. Chemistry of Reactive Oxygen Metabolites	128
	B. Biological Sources of Reactive Oxygen Metabolites	129
	C. Intracellular Defenses Against Reactive Oxygen Metabolites	130
	D. Biological Damage Caused by Reactive Oxygen Metabolites	132
	1. Lipid Peroxidation	132
	2. Protein Oxidation	132
	3. DNA Damage	132
V.	Mechanisms of Asbestos-Induced Diseases	133
	A. Chronic Inflammation and Fibrosis	133
	B. Carcinogenesis	135
	C. Asbestos and Malignant Mesothelioma	136
	D. Asbestos and Lung Cancer	137

Acknowledgment 138

References 138

ISBN 0-8493-8091-X
© 1993 by CRC Press, Inc.

I. INTRODUCTION

The history of asbestos dates back thousands of years to ancient civilizations. The Egyptians, Greeks, and Romans were all intrigued by the properties of asbestos, a mineral fiber that is durable and heat resistant, yet pliable enough to be woven into fabrics. With the beginning of the industrial revolution in the late 19th century, the full potential of asbestos was exploited commercially. By 1973, the consumption of asbestos in the U.S. alone had peaked at approximately 8,000,000 metric tons.[1]

With the increased use of asbestos came a staggering increase in asbestos-related diseases. The first documented case of asbestos exposure associated with fibrous scarring of the lungs (asbestosis) was reported by Dr. H. Montague Murray in 1906 upon performing an autopsy of an asbestos factory worker.[1] In the 1960s, sufficient epidemiologic evidence had been assembled to establish the association between asbestos exposure and the following diseases:[2,3]

1. Scars of the pleural lining (pleural plaques)
2. Scarring or fibrosis of the lungs (asbestosis)
3. Cancer arising within the lungs (bronchogenic carcinoma)
4. Cancer arising from the pleural, peritoneal, or pericardial linings (malignant mesothelioma)

The least serious of these diseases is pleural plaques. These are dense, fibrous scars on the pleural lining surrounding the lung, often located on the diaphragm. This condition is usually diagnosed clinically 10 to 20 years after asbestos exposure when the plaques become calcified and visible on a chest X-ray. Pleural plaques are a marker of prior exposure to asbestos; however, unless they are extensive, lung function is not compromised. A more serious life-threatening complication of asbestos exposure is the development of diffuse interstitial scarring or fibrosis of the lungs (asbestosis) that also develops 10 to 20 years after asbestos exposure. This condition is associated with higher doses or longer exposures to asbestos in the work place. Asbestosis may cause difficulty in breathing and may progress to respiratory failure due to impaired gas exchange.

Inhalation of asbestos has been associated with an increased risk of developing cancer of the lung (bronchogenic carcinoma) and mesothelial linings (malignant mesotheliomas). Cigarette smokers exposed to asbestos risk an 80- to 90-fold increase in bronchogenic carcinomas while the increased risk due to asbestos alone is only fivefold.[4] Malignant mesotheliomas are sarcomas which arise 25 to 50 years following contact with asbestos and develop along the linings of the pleural, pericardial, and peritoneal cavities. Once diagnosed, this tumor usually kills within months. Mesotheliomas rarely occur in the absence of asbestos exposure. A notable exception is the development of

TABLE 1
Classification and Chemical Composition of Asbestos Fibers

Serpentine Asbestos

Chrysotile $\quad$ $Mg_3Si_2O_5(OH)_4$
Fe^{+2}, Fe^{+3}, Mn^{+2}, or Ni^{+2} may substitute for Mg^{+2};
Fe^{+3} or Al^{+3} may replace Si^{+4}

Chrysotile asbestos deposits may be contaminated by amphiboles (tremolite), magnetite, or nemalite which contain iron.

Amphibole Asbestos

Amosite	$(Fe,Mg)_7Si_8O_{22}(OH)_2$
Anthophyllite	Mg_7 or $Fe_7^{+2}Si_8O_{22}(OH)_2$
Actinolite	$Ca_2(Mg,Fe^{+2})_5Si_8O_{22}(OH)_2$
Crocidolite	$Na_2Fe_3^{+2}Fe_2^{+3}Si_8O_{22}(OH)_2$
Tremolite	Ca_2Mg_5 or $Ca_2Fe_5^{+2}Si_8O_{22}(OH)_2$

Modified from Zussman, J., in *Asbestos and Other Fibrous Materials: Mineralogy, Crystal Chemistry, and Health Effects*, Skinner, H. C. W., Ross, M., and Frondel, C., Eds., Oxford University Press, New York, 1988, 45–66.

mesotheliomas in individuals exposed to naturally occurring erionite fibers, a type of zeolite mineral. Unlike bronchogenic carcinomas, there is no increase in malignant mesotheliomas in cigarette smokers. The incidence of these asbestos-related diseases is projected to increase into the 21st century.[3]

The term asbestos refers collectively to a group of naturally occurring hydrated silicate mineral crystals. All types of asbestos have a fibrous morphology with lengths ranging from 0.1 µm to >200 µm and diameters ranging from 0.03 µm to >2.5 µm. By definition, a fiber has a length to diameter ratio (aspect ratio) ≥ 3:1. There are two types of asbestos, serpentines and amphiboles (Table 1). Chrysotile, the only known example of serpentine asbestos, is referred to as a sheet silicate since it is composed of sheets of linked SiO_4 tetrahedrons attached to specific hydrated cations. There are five types of amphiboles: crocidolite, amosite, anthophyllite, tremolite, and actinolite. Amphiboles are composed of double chains of linked SiO_4 tetrahedrons. These chains are stacked one on top of the other, held together electrostatically by the presence of cations, usually (though not exclusively) sodium, iron, magnesium, and calcium. For crocidolite and amosite, for example, iron is the predominant cation. Commercially, the most important types of asbestos are the serpentine, chrysotile, and the amphiboles, crocidolite and amosite.[5] Chrysotile alone accounts for over 90% of all commercially used asbestos. Nevertheless, during World War II, crocidolite was extensively used in the manufacture of gas masks and in shipyards. Approximately 4.5 million people worked in shipyards during World War II.

Asbestos was, and continues to be, an essential resource to industry. Its fibrous morphology, high degree of flexibility, tensile strength, resistance to decomposition by heat and chemicals, and relatively low cost are all unique properties which make asbestos so valuable. Asbestos has been used in thousands of products including heat-protective clothing, electrical and thermal insulation, friction pads for brakes, cement, asphalt, floor and ceiling tiles, paper products, and pipes. Asbestos-related diseases may occur in workers directly or indirectly exposed during mining, milling, manufacture, or application of asbestos fibers. Lung cancer and malignant mesothelioma have also been reported in families of asbestos workers or in people living within 1 or 2 miles of an asbestos factory. No increased incidence of asbestos-related diseases has been documented due to ingestion of low levels of fibers in the water supply. Clusters of malignant mesothelioma have been described in areas of natural deposits of the amphibole, tremolite, in parts of Corsica. With this exception, although asbestos fibers are present in all urban environments at very low levels, there is as yet no reported increased risk of asbestos-related diseases after environmental exposure.[6]

It remains a medical mystery how these mineral fibers can cause this range of malignant and nonmalignant diseases. Several properties of asbestos fibers have been postulated to be relevant to disease:[7]

1. Fibrous geometry
2. Chemical composition
3. Surface reactivity
4. Durability and persistence in the lungs

The entry and deposition of fibers in the lungs, and interaction of fibers with the target cells associated with these asbestos-related diseases, will be described next.

II. ENTRY AND RETENTION OF ASBESTOS FIBERS IN THE LUNGS

Asbestos fibers enter the lungs after inhalation through the nose or mouth. The nose provides the first defense system against inhaled particles. Nasal hairs trap all particles more than 20 μm in diameter and about 50% of particles more than 5 μm in diameter. Fiber length is an important factor determining clearance of fibers trapped at airway bifurcations or deposited in the terminal air spaces. In general, short fibers are cleared more easily, while long fibers (5 to 10 μm) penetrate the alveolar walls into the interstitium and migrate to the pleural and peritoneal linings. The major clearance mechanism is the upward movement of particles and fibers by the mucociliary lining of the tracheobronchial tree to the mouth and nose. Fibers can also be engulfed by phagocytic cells called macrophages, then transported to the mucociliary escalator. Fibers can also be cleared by lymphatics to regional lymph nodes where they are stored in the chest cavity.[8]

If particles escape these routes of clearance, the most important factor that determines persistence of fibers in the lungs is durability. While all types of asbestos fibers are quite resistant to chemicals and heat, the durability of different types of asbestos fibers in the lungs varies depending on their chemical composition. Long, straight amphibole fibers persist unchanged in size, shape, or chemical composition for many years or decades, while curly chrysotile fibers fragment progressively into shorter fibers.[2,3,6]

If foreign particles of any type persist in the lungs, the initial target cell to interact with these particles is the macrophage.[9] The interaction between asbestos fibers and the macrophage is hypothesized to be a critical event that may initiate the development of asbestos-related diseases.[10] First, the normal biology and function of the macrophage will be reviewed. Second, the acute interaction between asbestos fibers and the macrophage will be described, with emphasis on the mechanism of acute cytotoxicity. Finally, the relationship between this acute cytotoxic reaction and the subsequent development of fibrosis and cancer will be discussed.

III. BIOLOGY AND FUNCTION OF THE MACROPHAGE

Macrophages are the first line of defense against any foreign invaders, including microbes, parasites, dust, and mineral fibers. Macrophages also perform routine housekeeping duties including phagocytosis and digestion of aged red blood cells and removal of microbes and dead cells from sites of infection or injury. Macrophages originate from stem cells within the bone marrow and circulate in the blood as monocytes. Monocytes are a subset of white blood cells or leukocytes that clear the blood of foreign invaders and initiate immune reactions. Monocytes leave the blood vessels and enter tissues and organs throughout the body. Here they live for weeks or years and are called resting or resident macrophages. If asbestos fibers persist in the air spaces or become trapped in the walls of the air spaces (interstitium) or migrate to the pleural or peritoneal spaces, the macrophages residing at these sites phagocytize them.[9]

IV. ACUTE ASBESTOS CYTOTOXICITY

The initial interaction between asbestos fibers and macrophages results in injury.[10,11] Until recently, the biochemical mechanism responsible for injury and death of the macrophage target cell was unknown. Macrophages and other cells in culture engulf or phagocytize asbestos fibers as shown by scanning electron microscopy in Figure 1. The process of phagocytosis triggers several metabolic changes in the target cell including ion fluxes, metabolism of arachidonic acid from membrane lipids, and fusion of lysosomes with the phagocytic vacuole which is an invagination of the plasma membrane. For

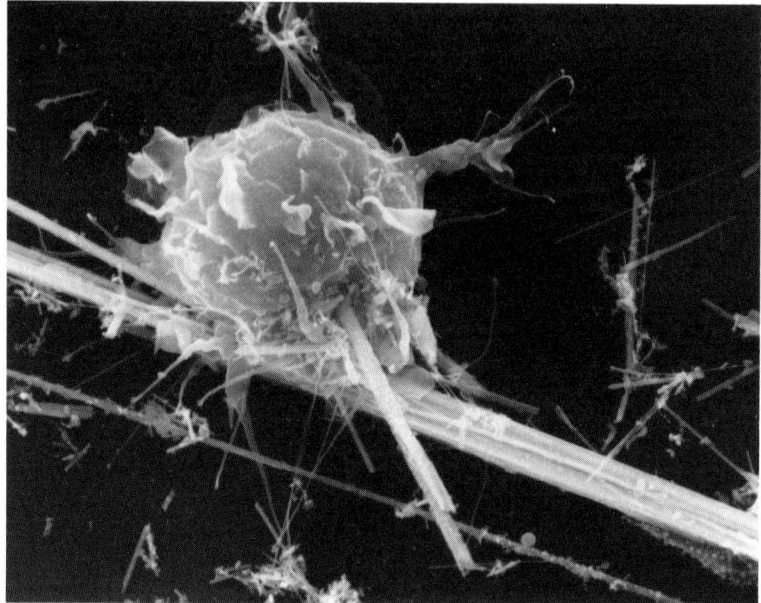

FIGURE 1. Scanning electron micrograph showing a mouse peritoneal macrophage attempting to phagocytize a crocidolite asbestos fiber *in vitro*. Original magnification × 3500.

many years, it was hypothesized that asbestos fibers disrupt lysosomal membranes resulting in intracellular leakage of lysosomal enzymes, autodigestion, and cell death.[10] Experiments conducted by Kane and co-workers suggest that this mechanism is oversimplified. Asbestos fibers also cause peroxidation of membrane lipids[12] and disruption of the mitochondrial membrane potential.[13] It is more likely that damage to multiple cellular membranes is secondary to oxidant-mediated attack rather than the result of direct electrical or mechanical disruption of these membranes by asbestos fibers.

The current hypothesis of acute asbestos cytotoxicity implicates reactive oxygen metabolites as mediators of target cell injury.[14,15] Before presenting the experimental evidence for this hypothesis, it is necessary to review the chemistry and biology of reactive oxygen metabolites.

A. CHEMISTRY OF REACTIVE OXYGEN METABOLITES

Reactive oxygen metabolites are a family of highly unstable compounds arising from the sequential one-electron reduction of molecular oxygen (O_2) as listed in Table 2. Many of these compounds possess an odd number of unpaired electrons and are often called "free radicals". Not all reactive oxygen metabolites, though, are free radicals. Neither hydrogen peroxide (H_2O_2) nor singlet oxygen ($^1O_2$), for example, has extra unpaired electrons.[16]

The chemistry of reactive oxygen metabolites is strongly influenced by

TABLE 2
Reactive Oxygen Metabolites

$^1O_2$
O_2^-
$HO_2^\cdot$
H_2O_2
$HO^\cdot$

their immediate environment, including pH, availability of transition metals, and organic molecules. Transition metals, especially iron or copper, can catalyze redox reactions between reactive oxygen metabolites with the product often being a more potent reactive species. For example, Fe^{+2} or Cu^{+2} can catalyze the formation of the highly unstable hydroxyl radical ($^\cdot OH$) from H_2O_2 and O_2. This reaction is referred to as the metal-catalyzed Haber-Weiss reaction; the second step is called the Fenton reaction:

$$O_2^- + Fe^{+3} \longrightarrow Fe^{+2} + O_2$$
$$H_2O_2 + Fe^{+2} \longrightarrow Fe^{+3} + HO^\cdot + HO^+$$

B. BIOLOGICAL SOURCES OF REACTIVE OXYGEN METABOLITES

Reactive oxygen metabolites are normally produced by all cells and organisms which undergo aerobic metabolism. In eukaryotic cells, the formation of these oxygen metabolites has been detected in the mitochondria, cytosol, endoplasmic reticulum, peroxisomes, nucleus, and plasma membrane. Enzymes which require molecular oxygen as a substrate are called either oxidases or oxygenases. Cytochrome c oxidase, the terminal enzyme complex in the electron transport system of mitochondria, is a good example of an enzyme whose function is to reduce O_2 to H_2O without the oxidation of any other substrate.[16-18]

Macrophages and neutrophils have a unique plasma membrane-bound oxidase. This oxidase is an important mechanism involved in the defense against invading microorganisms. This oxygen-dependent mechanism is referred to as the respiratory burst system and is activated during phagocytosis. Respiratory burst is characterized by an increase in oxygen consumption by the cell, the breakdown of glucose via the hexose monophosphate shunt producing NADPH, and the subsequent utilization of NADPH by a plasma membrane-bound oxidase to reduce molecular oxygen to O_2^-. While O_2^- is the only reactive metabolite directly produced by this oxidase system, subsequent reactions can occur to form other reactive species, primarily H_2O_2, HO_2^- and $^\cdot OH$. Thus, within the phagosome a microorganism can be bombarded by an array of reactive oxygen metabolites. If, however, a foreign substance is too large to be completely engulfed, the macrophage or neutrophil will then spread out along the object and release reactive oxygen metabolites onto its surface.[19] The consequences of triggering the respiratory burst pathway by phagocytosis of inorganic asbestos fibers will be considered after summarizing the normal intracellular defenses against reactive oxygen metabolites.

C. INTRACELLULAR DEFENSES AGAINST REACTIVE OXYGEN METABOLITES

All cells which undergo aerobic metabolism have compounds capable of inactivating reactive oxygen metabolites. The most prominent scavenging enzymes are superoxide dismutase (SOD), catalase, and various peroxidases. Superoxide dismutase and catalase, respectively, catalyze the dismutation of O_2^- and the oxidation of H_2O_2 via the following reactions:

$$2O_2^- + 2H^+ \xrightarrow{SOD} H_2O_2 + O_2$$

$$2H_2O_2 \xrightarrow{Catalase} 2H_2O + O_2$$

Cellular antioxidants include reduced glutathione (GSH), vitamin E, vitamin C, β-carotene, and cysteine. In addition to endogenous cellular defenses, a number of extracellular antioxidants also exist. These include transferrin and albumin (both bind iron), uric acid (radical scavenger), glucose ($^.OH$ scavenger), SOD, and ceruloplasmin (oxidizes iron and detoxifies O_2^-). Ceruloplasmin, which may function in a manner similar to SOD, constitutes the major antioxidant mechanism in blood plasma.[16-19]

A primary purpose of all the scavenging enzymes and antioxidants mentioned above is to protect the cell from the injurious effects of normal levels of reactive oxygen metabolites. As will be discussed, if these cellular defense systems are overwhelmed or depleted, irreversible cell injury may result. The potential role of reactive oxygen metabolites in carcinogenesis will be reviewed subsequently.

In the face of these multiple defense systems against reactive oxygen metabolites, how can these species be responsible for injuring the macrophage phagocytizing asbestos fibers? In the usual reaction involving phagocytosis and killing of invading microorganisms, these reactive oxygen species are consumed by killing the microorganism. Macrophages also engulf inorganic particles and dusts inhaled into the lungs as part of their routine housekeeping functions. In most cases, these dusts are also phagocytized, trigger the respiratory burst activity, and are stored in the phagolysosome without injury to the macrophage or surrounding tissue. What is the unique feature of asbestos fibers responsible for their toxicity to macrophages?

One hypothesis emphasizes the importance of fiber length in cytotoxicity. When macrophages or other cells try to engulf long asbestos fibers as illustrated in Figure 1, it is postulated that "frustrated phagocytosis" causes extracellular leakage of lysosomal enzymes and reactive oxygen species and death of the phagocytic cell.[14] A recent paper that reexamines the role of fiber length in acute asbestos toxicity caused by crocidolite asbestos fibers over a wide range of doses and exposure times argues against this hypothesis.[15]

A second hypothesis also focuses on reactive oxygen metabolites as mediators of acute asbestos cytotoxicity, with emphasis on the importance of chemical composition of fibers, especially iron content. The evidence for this hypothesis is derived from the following experimental results based on several *in vitro* model systems:[14,15]

1. Macrophages exposed to asbestos fibers *in vitro* release O_2^- and H_2O_2.
2. Asbestos toxicity is decreased under hypoxic conditions.
3. Asbestos toxicity is prevented by exogenous superoxide dismutase or catalase.

It is unlikely, however, that the quantities of O_2^- or H_2O_2 generated by the respiratory burst activity of macrophages during phagocytosis of asbestos fibers are directly responsible for cell death. Under similar experimental conditions, macrophages release even more H_2O_2 during phagocytosis of nontoxic mineral particles without any evidence of cell injury. The unique feature of asbestos fibers responsible for their toxicity is postulated to be the availability of Fe^{2+} and Fe^{3+} exposed on the fibers. It is postulated that this exposed iron catalyzes the formation of $^{\cdot}OH$ from O_2^- and H_2O_2 by a modified Haber-Weiss reaction as outlined above or even from dissolved molecular oxygen. The cellular source of O_2^- and H_2O_2 is probably the respiratory burst mechanism of macrophages, although other oxidative reactions involved in arachidonic acid metabolism or purine catabolism may provide these reactive oxygen metabolites. The experimental evidence implicating the hydroxyl radical as the ultimate toxic species responsible for asbestos cytotoxicity is[14,15,20,21]

1. Asbestos fibers coated with the iron chelator deferoxamine show decreased generation of $^{\cdot}OH$ from H_2O_2 in cell-free systems and are less toxic to macrophages, fibroblasts, or a variety of epithelial cells *in vitro*.
2. In the presence of extracellular $FeCl_3$, phagocytosis of inert titanium dioxide particles kills macrophages.
3. In various cell culture models, asbestos toxicity is prevented by the $^{\cdot}OH$ scavenger, dimethylthiourea.

Reactive oxygen species, especially the highly reactive $^{\cdot}OH$, are capable of damaging cells and extracellular matrix. All cellular macromolecules are potentially susceptible to attack; the most vulnerable components are lipids, proteins, and DNA.[16-19]

D. BIOLOGICAL DAMAGE CAUSED BY REACTIVE OXYGEN METABOLITES

1. Lipid Peroxidation

The disruption of cellular membranes due to lipid peroxidation has been shown to have a variety of effects. These include alterations in membrane permeability, fluidity, transport properties, and incorporated proteins. In addition, diffusible breakdown products of lipid peroxidation are also harmful to cellular components. Malonaldehyde, for example, can cross-link proteins. Other lipid oxidation products act as calcium ionophores.

It is unclear whether or not lipid peroxidation is causally related to cell death. Proving causal relationships is indeed difficult, and in most *in vitro* or *in vivo* systems only a temporal correlation exists between lethal cell injury and lipid peroxidation. On the other hand, in model systems of isolated hepatocytes exposed to *t*-butyl hydroperoxide, paraquat, or other lipid peroxidation-inducing compounds, lipid peroxidation could be prevented but cell death still occurred. These results tend to dissociate lipid peroxidation from cytotoxicity in these systems.[22] Similar results were obtained using a model of crocidolite asbestos toxicity to macrophages *in vitro*.[12]

2. Protein Oxidation

Reactive oxygen metabolites can directly damage proteins by disulfide cross-linking, amino acid oxidation, protein strand breaks, and lipid-protein cross-linking. Of these events, amino acid oxidation has been observed most frequently. The amino acids most susceptible to oxidation are those containing sulfur or aromatic rings, methionine, cysteine, phenylalanine, tryptophan, and tyrosine. The oxidation of an amino acid crucial to enzymatic activity or protein structure would lead to disruption of protein function. Furthermore, evidence suggests that damaged proteins are more susceptible to subsequent proteolytic degradation.[16-19]

3. DNA Damage

Reactive oxygen species have been shown to damage cellular DNA and chromosomes. Attack of DNA bases by the hydroxyl radical produces oxidized bases. During the repair process to remove these oxidized bases from DNA, single-strand breaks may be introduced in the DNA backbone. Base substitutions during the repair process may cause mutations in the genetic code. These single-strand breaks, in addition to double-strand breaks, may lead to gross chromosomal changes including sister chromatid exchange, breaks, deletions, and rearrangements. Thus, reactive oxygen species are potentially mutagenic, producing changes in the DNA base sequence, and clastogenic, causing damage to chromosomes.[23]

A classic example of DNA damage induced directly by reactive oxygen species is ionizing radiation. Several oxidizing reagents, including H_2O_2 and benzoyl peroxide, also induce DNA damage. Recently, reactive oxygen metabolites produced by activated phagocytes have been shown to produce DNA strand breaks and chromosomal damage in target cells. Lipid peroxidation breakdown products that are more stable than reactive oxygen metabolites can also induce DNA damage.[16,23]

We have searched for evidence of DNA damage in our model system of acute asbestos toxicity to macrophages *in vitro*. During a 6-h exposure to crocidolite asbestos fibers, there is progressive accumulation of single-strand DNA breaks that occurs in parallel with loss of viability. DNA damage and cell death were ameliorated by adding exogenous catalase to the cultures or by coating fibers with the iron chelator, deferoxamine. Craighead and co-workers have also demonstrated DNA strand breaks in cultures of rat embryo cells after a 12- to 23-h exposure to crocidolite asbestos fibers.[24] It has also been shown in a cell-free system that asbestos fibers in the presence of H_2O_2 produce oxidized bases in DNA.[25] The relationship between DNA damage and cancer induced by exposure to asbestos fibers will be discussed in the next section.

V. MECHANISMS OF ASBESTOS-INDUCED DISEASES

A. CHRONIC INFLAMMATION AND FIBROSIS

Injury to macrophages and neighboring cells at the site of fiber deposition triggers a generalized inflammatory reaction. In the peritoneal cavity, the peak accumulation of monocytes occurs 3 days after a single injection of asbestos fibers.[26] With repeated exposures to asbestos fibers, additional episodes of inflammation ensue. This defense reaction against repeated injury or persistent irritants is called chronic inflammation.[9]

The monocytes recruited to sites of tissue injury and chronic inflammation are called inflammatory or nonspecifically activated macrophages. They are more active metabolically than resident macrophages: they have higher rates of phagocytosis, increased activity of the respiratory burst system, and greater release of arachidonic acid metabolites and lysosomal enzymes. In addition, inflammatory macrophages synthesize and secrete a variety of new enzymes and peptides. Among the peptides synthesized by inflammatory macrophages are growth factors for fibroblasts, endothelial cells, and epithelial cells. These peptide growth factors, listed in Table 3, stimulate the ingrowth of fibroblasts and new blood vessels into the site of tissue damage. This ingrowth is called granulation tissue and characterizes the early phase of wound healing.[27] As an example, clusters of asbestos fibers deposited in the peritoneal lining of a mouse are surrounded by multinucleated macrophages (giant cells), new blood vessels, and fibroblasts as illustrated in Figure 2.

Normally, after a single episode of tissue injury and inflammation, these mediators released from inflammatory macrophages initiate a healing reaction. In epithelial tissues, such as the skin or the lining of the bronchi and alveoli of the lungs, macrophage-derived growth factors may contribute to epithelial

TABLE 3
Peptide Growth Factors Released by Inflammatory Macrophages

Interleukin-1 (IL-1)
Interleukin-6 (IL-6)
Tumor necrosis factor-α (TNF-α)
Colony stimulating factors (CSF)
Transforming growth factor-α (TGF-α)
Transforming growth factor-β (TGF-β)
Insulin-like growth factor-1 (IGF-1)
Platelet-derived growth factor (PDGF)
Basic fibroblast growth factor (bFGF)

Modified from Kelley, J., Am. Rev. Resp. Dis., 141, 765–788, 1990.

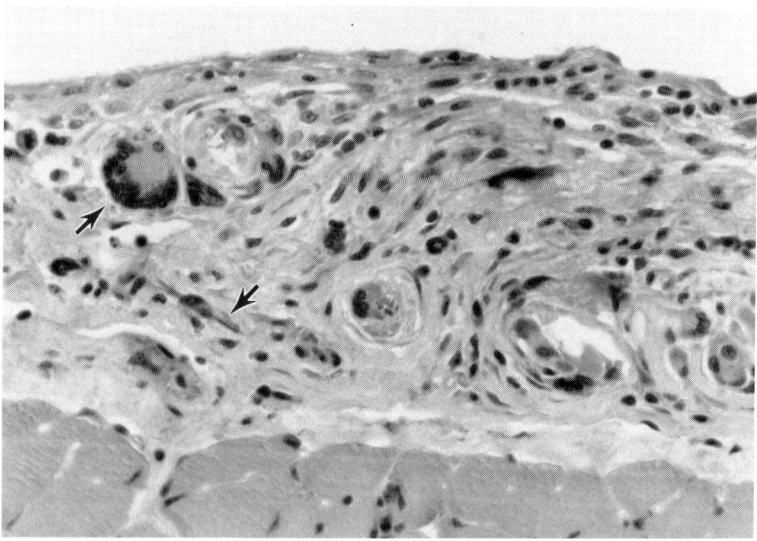

FIGURE 2. Light micrograph showing a histologic section through a cluster of crocidolite asbestos fibers after deposition on the mouse peritoneal lining for 22 weeks. This lesion is covered by a single layer of flat mesothelial cells. Between the surface mesothelial layer and the underlying skeletal muscle, asbestos fibers are seen within macrophages and multinucleated giant cells formed by fusion of macrophages (arrows). The fibers and inflammatory cells are surrounded by fibroblasts and collagen characteristic of a healing wound. Original magnification × 625.

regeneration with restoration of normal tissue architecture. In other organs where the cells are unable to regenerate, such as the heart, or when tissue injury is extensive or repeated, regeneration is limited and healing occurs by extracellular deposition of collagen forming an acellular, fibrous scar. In contrast to self-limiting types of injury, asbestos fibers persist in the lungs and mesothelium for many years and are a source of repeated injury and chronic inflammation. Ultimately, in the lungs this scarring leads to diffuse interstitial fibrosis or asbestosis. In the pleural lining surrounding the lungs, this scarring produces localized pleural plaques or diffuse pleural fibrosis.[2,26]

B. CARCINOGENESIS

The process of carcinogenesis leads to a population of cells with the following properties:

1. Autonomous cell proliferation
2. Altered differentiation
3. Ability to invade surrounding tissue
4. Ability to metastasize or spread to different organs via blood vessels or lymphatic channels

In most cases, the properties of invasive growth and metastases are the reason why cancer is a potentially deadly disease. It is widely accepted that the acquisition of the malignant phenotype is the consequence of a series of cellular changes; that is to say, carcinogenesis is a multistep, dynamic process. In the 1940s, Rous and colleagues identified two early steps involved in chemically induced skin carcinogenesis, "initiation" and "promotion". Subsequently, this two-stage model of carcinogenesis has been applied to other models of liver, mammary gland, brain, kidney, colon, and bladder tumor formation.[28]

Traditionally, initiation is thought to produce an irreversible change in the DNA. This change is often a mutation (i.e., base substitution, frame-shift mutation, deletion, or substitution). Not all initiators are mutagens, however. Initiators may conceivably form DNA adducts, strand breaks, produce transpositions, or act by epigenetic mechanisms. For example, certain carcinogenic agents which do not produce detectable mutations induce aneuploidy. One could imagine mechanisms whereby these agents directly disrupt chromosomes or, alternatively, disturb cellular components, such as microtubules, which are important in chromosomal segregation during mitosis and meiosis.[29] Reactive oxygen metabolites may well function as initiators. In a modified Ames' assay, neutrophils, presumably by releasing reactive oxygen metabolites, are mutagenic. Neutrophils or mononuclear leukocytes coincubated with bacteria, caused a reversion from histidine-requiring to histidine-independent bacteria.[23]

Tumor promoters are agents or stimuli which enhance neoplastic growth following exposure to a low dose of an initiator. Promoters themselves usually lack carcinogenic activity although some have been found to be weak carcinogens. Some initiators, however, are also effective promoters and so are termed complete carcinogens. The concept of tumor promotion was first formulated in the 1930s and 1940s. In classic experiments, carcinomas were induced in the skin of mice or rabbits by the topical application of a subthreshold dose of a carcinogen followed by an application of croton oil. Croton oil comes from the seeds of the *Croton tiglium* plant. The active component of croton oil, phorbol-12-myristate-13-acetate (PMA), was subsequently isolated and has since become one of the most studied promoters both *in vivo* and *in vitro*.[28]

Reactive oxygen metabolites have also been associated with tumor promotion. Peroxides such as H_2O_2, benzoyl peroxide, and lipid peroxides have promoting properties. On the other hand, scavengers of reactive oxygen species and antioxidants often inhibit promotion. SOD can suppress tumor promoter-induced DNA synthesis and mitosis *in vitro,* and CuDIPs, a compound which mimics SOD but which can readily enter cells, inhibits tumor promotion and tumor development in mouse skin.[23,28]

Under what circumstances might reactive oxygen metabolites cause tumor promotion *in vivo*? One important feature of tumor promoters *in vivo* is the elicitation of an inflammatory response. Promoters present at sites of inflammation could further serve to trigger the release of reactive oxygen species from phagocytes. Indeed, most known promoters are inflammatory agents and potent stimulants of the respiratory burst system. Such a situation, it is argued, could function as a promoting environment for initiated cells.[23]

C. ASBESTOS AND MALIGNANT MESOTHELIOMA

A unifying hypothesis that links the acute toxic effects of asbestos fibers mediated by reactive oxygen species with the chronic inflammatory reaction and development of malignant mesothelioma in the pleural and peritoneal linings is diagrammed as follows:[26]

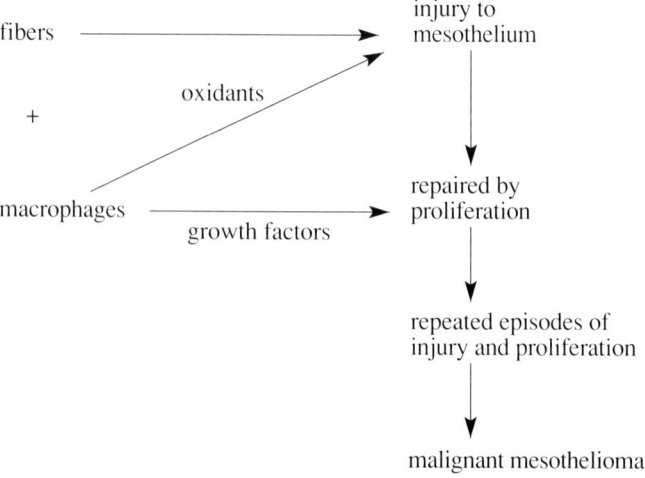

There is some experimental evidence to support this hypothesis based on models of malignant mesotheliomas induced by asbestos fibers in rodents:[13,26,30]

1. When injected directly into the peritoneal cavity, long asbestos fibers are not cleared through lymphatic channels and are trapped at the mesothelial lining.
2. Inflammatory macrophages accumulate around fiber clusters on the mesothelial lining.
3. These inflammatory macrophages release reactive oxygen species at these sites.
4. Adjacent to asbestos fiber clusters, injured macrophages and mesothelial cells are visible 24 h after a single injection of asbestos fibers.
5. This injury is prevented if fibers are coated with the iron chelator, deferoxamine, or by injections of exogenous superoxide dismutase or catalase.
6. After a single injection of asbestos fibers, the mesothelial lining is restored by proliferation of adjacent, uninjured mesothelial cells.
7. Repeated injections of asbestos fibers provoke repeated episodes of injury, inflammation, and mesothelial cell proliferation.
8. Intraperitoneal injection of asbestos fibers induces angiogenesis.

D. ASBESTOS AND LUNG CANCER

The mechanism responsible for the greatly increased incidence of lung cancer, especially in people who smoke cigarettes, is more complex. Cigarette smoke and asbestos fibers interact synergistically in the bronchial epithelium to increase the risk of developing lung cancer.[4] The combination of cigarette smoke and asbestos fibers is postulated to cause:[32,33]

1. Reduced ciliary action with decreased clearance of smoke particles and fibers from the lungs
2. Increased fiber penetration into the bronchial epithelial lining
3. Increased entry of carcinogens into bronchial epithelial cells
4. Increased activation of chemical carcinogens in cigarette smoke

In addition, asbestos fibers have been shown to cause the following changes when added directly to tracheal epithelial cell cultures *in vitro;* many of these changes are characteristic of tumor promoters:[29,31,32]

1. Stimulation of cell proliferation
2. Altered differentiation from epithelial to squamous cells (squamous metaplasia)
3. Increased synthesis of DAG (diacylglycerol)
4. Increased ornithine decarboxylase activity
5. Induction of chromosomal damage by interference with the mitotic spindle

In summary, there are a variety of direct and indirect mechanisms by which asbestos fibers may alter proliferation and differentiation of bronchial epithelial cells that eventually lead to the development of lung cancer. There is a complex interaction between cigarette smoke and asbestos fibers. Thus, asbestos fibers may act as a promoter or co-carcinogen with cigarette smoke to cause lung cancer. In contrast, asbestos fibers appear to act as a complete carcinogen to induce malignant mesotheliomas independently of cigarette smoking. It is postulated that reactive oxygen metabolites released from macrophages at sites of asbestos fiber deposition act during the initiation and the promotion phases in the development of malignant mesotheliomas.

ACKNOWLEDGMENT

The research reported from this laboratory is supported by Grants R01 ES 03721 and R01 ES 03184 from the National Institutes of Health.

REFERENCES

1. **Gross, P. and Braun, D. C.**, Asbestos, in *Toxic and Biomedical Effects of Fibers,* Gross, P. and Braun, D. C., Eds., Noyes Publications, New Jersey, 1984, 9–128.
2. **Churg, A.**, Nonneoplastic diseases caused by asbestos, in *Pathology of Occupational Lung Disease,* Churg, A. and Green, F. H. Y., Eds., Igaku-Shoin, New York, 1988, 213–278.
3. **Churg, A.**, Neoplastic asbestos-induced diseases, in *Pathology of Occupational Lung Disease,* Churg, A. and Green, F. H. Y., Eds., Igaku-Shoin, New York, 1988, 279–326.
4. **Selikoff, I. J., Churg, J., and Hammond, E. C.**, Asbestos exposure and neoplasia, *JAMA,* 188, 22–26, 1964.
5. **Zussman, J.**, The mineralogy of asbestos, in *Asbestos and Other Fibrous Materials: Mineralogy, Crystal Chemistry, and Health Effects,* Skinner, H. C. W., Ross, M., and Frondel, C., Eds., Oxford University Press, New York, 1988, 45–66.
6. **Craighead, J. E. and Mossman, B. T.**, The pathogenesis of asbestos-associated diseases, *N. Engl. J. Med.,* 306, 1445–1446, 1982.
7. **Harington, J. S.**, Fiber carcinogenesis: epidemiologic observations and the Stanton hypothesis, *J. Natl. Canc. Inst.,* 67, 977–989, 1981.
8. **Lippmann, M., Yeates, D. B., and Albert, R. I.**, Deposition, retention, and clearance of inhaled particles, *Br. J. Ind. Med.,* 37, 337–362, 1980.
9. **Sibille, Y. and Reynolds, H. Y.**, Macrophages and polymorphonuclear neutrophils in lung defense and injury, *Am. Rev. Resp. Dis.,* 141, 471–501, 1990.
10. **Harington, J. S., Allison, A. C., and Badami, D. V.**, Mineral fibers: chemical, physicochemical, and biological properties, *Adv. Pharmacol. Chemother.,* 12, 291–402, 1975.
11. **Brody, A. R., Hill, L. H., and Warheit, D. B.**, Induction of early alveolar injury by inhaled asbestos and silica, *Fed. Proc.,* 44, 2596–2601, 1985.
12. **Goodglick, L. A., Pietras, L. A., and Kane, A. B.**, Evaluation of the causal relationship between crocidolite asbestos-induced lipid peroxidation and toxicity to macrophages, *Am. Rev. Resp. Dis.,* 139, 1265–1273, 1989.

13. **Goodglick, L. A. and Kane, A. B.**, Cytotoxicity of long and short crocidolite asbestos fibers in vitro and in vivo, *Cancer Res.,* 50, 5153–5163, 1990.
14. **Mossman, B. T. and Marsh, J. P.**, Evidence supporting a role for active oxygen species in asbestos-induced toxicity and lung disease, *Environ. Health Persp.,* 81, 91–94, 1989.
15. **Goodglick, L. A. and Kane, A. B.**, The role of reactive oxygen metabolites in crocidolite asbestos toxicity to macrophages, *Cancer Res.,* 46, 5558–5566, 1986.
16. **Slater, T. F.**, Free-radical mechanisms in tissue injury, *Biochem. J.,* 22, 1–15, 1984.
17. **Halliwell, B. and Gutteridge, J. M. C.**, Oxygen toxicity, oxygen radicals, transition metals and disease, *Biochem. J.,* 219, 1–14, 1984.
18. **Freeman, B. A. and Crapo, J. D.**, Biology of disease: free radicals and tissue injury, *Lab. Invest.,* 47, 412–426, 1982.
19. **Fantone, J. C. and Ward, P. A.**, Role of oxygen-derived free radicals and metabolites in leukocyte-dependent inflammatory reactions, *Am. J. Pathol.,* 107, 397–418, 1982.
20. **Weitzman, S. A. and Graceffa, P.**, Asbestos catalyzes hydroxyl and superoxide radical release from hydrogen peroxide, *Arch. Biochem. Biophys.,* 228, 373–376, 1984.
21. **Zalma, R., Bonneau, L., Jaurand, M. C., Guignard, J., and Pezerat, H.**, Formation of oxy-radicals by oxygen reduction arising from the surface activity of asbestos, *Can. J. Chem.,* 65, 2338–2341, 1987.
22. **Comporti, M.**, Lipid peroxidation and cellular damage in toxic liver injury, *Lab. Invest.,* 53, 599–623, 1985.
23. **Cerutti, P. A.**, Prooxidant states and tumor promotion, *Science,* 227, 375–381, 1985.
24. **Libbus, B. L., Illenye, S. A., and Craighead, J. E.**, Induction of DNA strand breaks in cultured rat embryo cells by crocidolite asbestos as assessed by nick translation, *Cancer Res.,* 49, 5713–5718, 1989.
25. **Leanderson, P., Soderkvist, P., Tagesson, C., and Axelson, O.**, Formation of 8-hydroxydeoxy-guanosine by asbestos and man made mineral fibres, *Br. J. Ind. Med.,* 45, 309–311, 1988.
26. **Moalli, P. A., Macdonald, J. L., Goodglick, L. A., and Kane, A. B.**, Acute injury and regeneration of the mesothelium in response to asbestos fibers, *Am. J. Pathol.,* 128, 425–445, 1987.
27. **Kelley, J.**, Cytokines of the lung, *Am. Rev. Resp. Dis.,* 141, 765–788, 1990.
28. **Berenblum, I.**, Sequential aspects of chemical carcinogenesis: skin, in *Cancer: A Comprehensive Treatise,* Vol. 1, 2nd ed., Becker, F. F., Ed., Plenum Press, New York, 1982, 451–484.
29. **Hesterberg, T. W. and Barrett, J. C.**, Induction by asbestos fibers of anaphase abnormalities, *Carcinogenesis,* 6, 473–475, 1985.
30. **Branchaud, R. M., Macdonald, J. L., and Kane, A. B.**, Induction of angiogenesis by intraperitoneal injection of asbestos fibers, *FASEB J.,* 3, 1747–1752, 1989.
31. **Mossman, B. T. and Craighead, J. E.**, Mechanisms of asbestos associated bronchogenic carcinoma, in *Asbestos-Related Malignancy,* Antman, K. and Aisner, J., Eds., Grune & Stratton, Orlando, 1987, 137–150.
32. **Hobson, J., Wright, J. L., and Churg, A.**, Active oxygen species mediate asbestos fiber uptake by tracheal epithelial cells, *FASEB J.,* 4, 3135–3139, 1990.

Chapter 7

MEMBRANE RECEPTORS AND SIGNAL TRANSDUCTION IN TUMOR CELLS

Marco Ruggiero

TABLE OF CONTENTS

I.	Introduction	142
II.	The Metabolism of Inositol Lipids	143
III.	Growth Stimuli and Inositol Lipid Metabolism	149
	A. Platelet-Derived Growth Factor (PDGF)	149
	B. Epidermal Growth Factor (EGF)	152
	C. Growth Factors that Stimulate "Alternative" Pathways and Insulin	155
IV.	Oncogenes and Second Messengers	157
V.	Conclusions	165
VI.	Addendum: Recent Developments Concerning Mitogenic Signaling in Normal and Transformed Cells	166
	A. Introduction	166
	B. Cooperation of Inositol Lipid Turnover and Calpain in Mitogenic Signaling	166
Acknowledgments		170
References		170

ISBN 0-8493-8091-X
© 1993 by CRC Press, Inc.

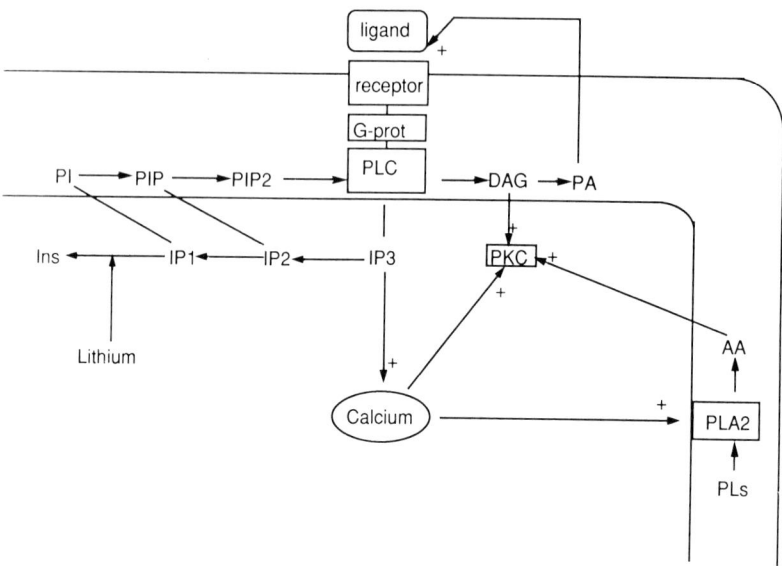

FIGURE 1. Receptor-dependent activation of inositol lipid turnover. Phosphatidylinositol (PI) is phosphorylated to phosphatidylinositol 4-phosphate (PIP), and PIP is phosphorylated to phosphatidylinositol 4,5-bisphosphate (PIP2) by specific kinases. Phospholipase C (PLC) (possibly activated through a GTP-binding protein) hydrolyzes PIP2 to form inositol 1,4,5-trisphosphate (IP3) and 1,2 diacylglycerol (DAG). The hydrolysis of PIP and PI can also give rise to water-soluble inositol phosphates and DAG. IP3 is dephosphorylated to inositol bisphosphates (IP2), inositol monophosphate (IP1), and free inositol (Ins). Lithium inhibits IP1 dephosphorylation and favors the accumulation of IP1. DAG is converted to phosphatidic acid (PA) by DAG kinase. PA can in turn activate the inositol lipid turnover by acting as a growth factor. Protein kinase C (PKC) is activated by DAG and intracellular calcium increase. Calcium mobilization favors the activation of phospholipase A_2 (PLA_2) which liberates arachidonic acid (AA) from membrane phospholipids (PLs). AA can also activate PKC. See the text for details.

a functional relationship between inositol lipid kinases and phosphodiesterases is necessary in order to maintain the pool of PIP2 which is degraded following receptor stimulation or during cell transformation. Quite recently a novel type of PI kinase has been characterized in polyoma middle T-transformed cells.[9] This kinase, which was resolved by immunoprecipitation of middle T/$pp60^{c-src}$, synthesizes phosphatidylinositol 3-monophosphate. This novel inositol phospholipid is a minor component of the PIP in transformed cells; however, the formation of an isomer of PIP during transformation might be correlated to the regulation of mitogenesis.

The crucial event that triggers the cascade of inositol lipid metabolism is the hydrolysis of PIP2 by a specific phosphodiesterase [phospholipase C (PLC), or phosphoinositidase C], to yield 1,2-diacylglycerol (DAG) and inositol 1,4,5-trisphosphate (Ins 1,4,5-P3). Different isozymes of PLC have been characterized in different tissues and cell types.[10] There is general agreement that the enzyme which hydrolyzes PIP2 works in a calcium-independent

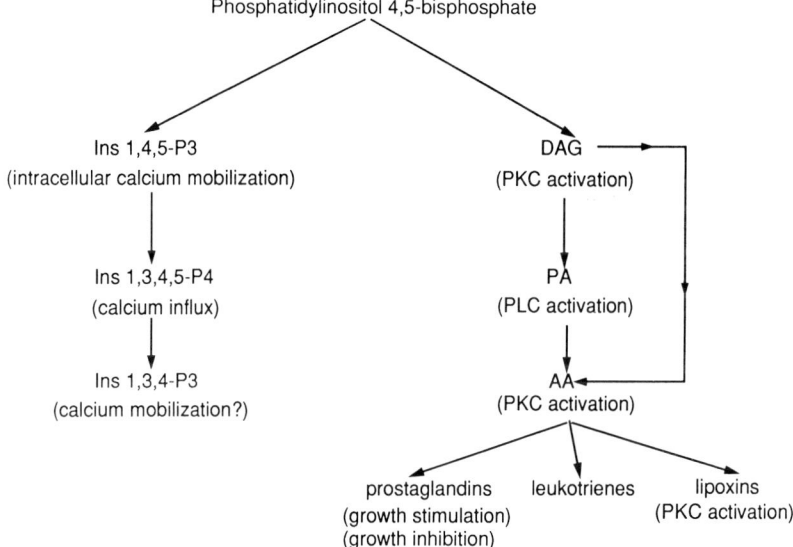

FIGURE 2. Second messengers arising from phosphatidylinositol 4,5-bisphosphate hydrolysis. Inositol 1,4,5-trisphosphate (Ins 1,4,5-P3) and diacylglycerol (DAG) function as intracellular second messengers, mobilizing calcium and activating protein kinase C (PKC), respectively. Furthermore, their metabolism generates several other intra- and extracellular messengers. Abbreviations as in Figure 1. See the text for details.

manner, in that it can be activated at resting calcium concentration. Concerning the mechanism of PLC activation following agonist/receptor interaction, recent evidence points to specific GTP-binding proteins (G-proteins) as couplers between receptors and the enzyme.[11] Since several growth factor receptors possess tyrosine kinase activity that appears to be indispensable for their biological effects,[12] it is tempting to suggest that tyrosine phosphorylation of the enzyme (or of G-proteins or other regulatory factors) might be involved in this activation. Interestingly, functionally distinct G-proteins selectively couple different receptors to inositol lipid turnover in the same cell.[13] Therefore, the nature of the cellular responses to different agonists might be determined by the magnitude of PIPs hydrolysis triggered by G-proteins that differ in their coupling to PLC.

Both products of PIP2 hydrolysis have multiple functions as second messengers (Figure 2): DAG activates protein kinase C (PKC), and Ins 1,4,5-P3 mobilizes calcium from intracellular, non-mitochondrial stores.[2] This represents another peculiarity of the signaling mechanism dependent on inositol lipid turnover: from a single molecule two different second messengers conveying totally distinct information are generated.

Ins 1,4,5-P3 mobilizes intracellular calcium upon binding to its receptor on the endoplasmic reticulum. The phosphate at the 1-position is thought to increase the affinity of the molecule for the receptor, whereas the phosphates at the 4- and 5-positions are necessary for mobilizing calcium.[14] Calcium mobilization is responsible for the activation of other intracellular enzymes such as the calcium-dependent PLC which hydrolyzes PI, phospholipase A_2 (PLA_2), DAG lipase, calcium/calmodulin-dependent kinases, etc. (reviewed in Reference 15). Because of this, PIP and PI are also hydrolyzed in the course of receptor stimulation with the production of DAG, arachidonic acid (AA) and inositol phosphates (InsPs; inositol monophosphate [IP1], and inositol bisphosphate [IP2]). Prolonged formation of DAG from sources other than PIP2 might be important in the mode of action of certain growth factors and/or oncogenes.

Once produced, Ins 1,4,5-P3 is rapidly metabolized along two separate pathways. On one side, it is dephosphorylated by the action of successive inositol phosphatases to IP1 and eventually to free inositol. Interestingly, several of these inositol phosphatases are inhibited by increasing concentration of lithium (for review see Reference 16). Low concentrations (1 mM) primarily inhibit the inositol monophosphatase which dephosphorylates IP1 to free inositol. Higher concentrations (10 mM) inhibit other inositol polyphosphate-specific phosphatases with the resulting accumulation of Ins 1,3,4-P3 and Ins 1,4-P2, but not Ins 1,4,5-P3. It is believed that the dramatic effect of lithium in the treatment of manic depressive patients is due to a disruption of the inositol lipid-related signaling in certain areas of the CNS. Furthermore, lithium stimulates DNA synthesis in different cell types, and this effect too may be ascribed to its action on the catabolism of InsPs. Since lithium favors the accumulation of InsPs following the hydrolysis of PIP2, it is commonly used to detect the stimulation of inositol lipid turnover in a variety of experimental models.

The other pathway involved in the degradation of Ins 1,4,5-P3 consists of phosphorylation in the 3-position to form Ins 1,3,4,5-P4.[17] This inositol tetrakisphosphate is then dephosphorylated to form Ins 1,3,4-P3. These two latter compounds (Ins 1,3,4,5-P4 and Ins 1,3,4-P3) might also be involved in the increase of intracellular calcium, possibly regulating the influx from the outside, or providing sustained elevation of free calcium. Ins 1,4,5-P3 degradation can be influenced by PKC activation, thus providing a negative feedback mechanism that accelerates the inactivation of calcium-mobilizing compounds.[18]

In the course of PIP2 hydrolysis by PLC, cyclic inositol polyphosphates may also be formed.[18] These compounds appear to mobilize calcium, although to a lower extent than Ins 1,4,5-P3. The role of cyclic inositol phosphates remains to be clarified and their involvement in cell proliferation is still an open question.

The other second messenger produced from PIPs hydrolysis is DAG. Unlike InsPs which migrate into the cytoplasm, the apolar DAG molecule remains in the bilayer of the membrane. Its function is to stimulate a calcium/ phospholipid-dependent protein kinase, PKC, and to serve as substrate for the production of other intracellular messengers, phosphatidic acid (PA), AA, and AA metabolites. Furthermore, DAG is a building block in the synthesis of membrane phospholipids, and it can also be formed by pathways other than inositol lipid turnover: DAG can be neosynthesized from the glycolytic pathway,[19] or it can be produced as result of cleavage of other phospholipids by PLC and/or phospholipase D.[20] These alternative routes for DAG formation might be operating in cell transformation, as we shall see in the following sections.

The DAG which is generated from PIP2 hydrolysis contains stearate and arachidonate, and it activates PKC in part by decreasing the apparent K_m for the calcium-mediated activation of the kinase.[21] In this respect, Ins 1,4,5-P3 (which mobilizes calcium) and DAG cooperate in PKC activation, although DAG can activate PKC even at resting calcium concentration.[22]

Without entering into the details of PKC molecular biology, however, it is worth considering that multiple forms of the enzyme exist.[23] These different species of PKC appear to be differently distributed and they may phosphorylate distinct substrates, thus providing a wide range of functional flexibility. It is also emerging that the various isozymes are stimulated differently by messenger molecules formed during receptor activation and inositol lipid turnover. Among activators of PKC other than DAG, AA and lipoxins are worth mentioning, since alterations in the level of these compounds in cell transformation might cause persistent activation of distinct isozymes of PKC. Continuous activation of PKC could then mimic the effect of tumor promoting phorbol esters which induce sustained activation of the enzyme.

Just like Ins 1,4,5-P3, DAG can be metabolized by two separate mechanisms that give rise to additional intra- and extracellular messengers. A specific DAG kinase phosphorylates DAG to form PA which can be utilized for the resynthesis of PI. Interestingly, PA has been shown to act as a growth factor, and it is able to further stimulate inositol lipid turnover.[24,25] Therefore, it may serve as an auto-amplifying messenger that reinforces the signal elicited by growth factors activating inositol lipid metabolism. Recent studies indicate that PA can also be formed by phospholipase D-dependent cleavage of other phospholipids.[20] Furthermore, elevated levels of PA have been observed in cells transformed by some oncogenes[4,5] leading to speculation that PA might participate in the autocrine loop characteristic of uncontrolled cell proliferation. It is likely that in the future PA will be the object of more detailed investigation, since it may be a crucial messenger in the transduction of the mitogenic signal.

Besides being phosphorylated to PA, DAG is the substrate of the calcium-dependent DAG lipase which releases AA. AA is a messenger in itself since it preferentially activates certain PKC isozymes and triggers other processes. In addition, it is rapidly metabolized to form potent biologically active compounds such as prostaglandins, thromboxanes, leukotrienes, and lipoxins. Most of these compounds play a role in the complex network of transmembrane signaling, either amplifying or inhibiting specific responses. To name a few examples, certain prostaglandins (E_2 and F_{2alpha}) are mitogenic;[16] others (prostacyclin, PGI_2, and PGE_1) inhibit receptor-mediated PLC activation;[15] lipoxins can activate specific isozymes of PKC.[23]

The major factor which contributed to the widespread interest in the DAG/PKC pathway is that potent tumor promoters such as phorbol esters, teleocidin, and aplysiatoxin bind and activate PKC.[16] Therefore, activation and/or downregulation of PKC seem to be crucial steps in controlling cell proliferation. Interestingly, in some instances, PKC activation by phorbol esters induces totally opposite effects, determining differentiation instead of growth. To date there is no clear explanation for such an apparent paradox. However, the existence of multiple forms of PKC, the differential phosphorylation of distinct substrates, and/or the differential balance between PKC activation and calcium mobilization in defined cell types might help the understanding of this phenomenon. An analogous situation can be observed in the case of the opposite effects of thrombin on platelets and endothelial cells. In platelets, thrombin induces the release of thromboxane A_2 (TXA_2), an agent that causes platelet aggregation and vasoconstriction. In endothelial cells, however, thrombin promotes the synthesis of PGI_2, a factor which induces vasodilatation and inhibits platelet aggregation. In both cell types thrombin triggers the same signaling mechanism, namely, inositol lipid turnover, calcium mobilization, PKC activation, and release and metabolism of AA. However, since the expression of TXA_2 and PGI_2 synthetases is greatly different in the two cell types, the biological end-result is diametrically opposed.

Given the stimulation of inositol lipid turnover and the formation of a variety of second messengers by mitogenic stimuli and cell transformation, what then are the successive biochemical steps that lead to DNA synthesis? Unfortunately, there are several missing links in the chain of events that culminates in DNA synthesis. However, some of these successive steps in the signaling cascade are currently being elucidated. Elevated cytosolic calcium concentration is typical of cells growing under the influence of growth factors, or transformed by some oncogenes.[11] Increasing calcium concentration is in turn sufficient to induce the transcription of a specific set of early genes (actin, *myc, fos*) that have a role in controlling cell proliferation. Furthermore, calcium is responsible for regulating the phosphorylation of ribosomal protein S6, an integral component of the 40S ribosomal subunit. On the other hand, the DAG/PKC pathway is causally related to key processes such as the activation of the NA^+/H^+ exchanger (with consequent increase of intracellular pH, a prerequisite for DNA synthesis to occur), and the stimulation of topoisomerase II. In addition, several functions of PKC seem to be related to modulating events within the inositol lipid signaling pathway, or connected to other transducing mechanisms (in particular, the cyclic AMP pathway).

3T3 fibroblast cell line with only a few endogenous EGF receptors that shows weak DNA synthesis and no inositol lipid turnover in response to the factor. In EGF receptor-overexpressing NIH/3T3 cells, EGF potently stimulates inositol lipid metabolism and DNA synthesis,[44] to an extent comparable to that observed with PDGF (Figure 4). Thus it appears that the receptors for both factors converge on the same intracellular signaling pathway. Preliminary data seem to indicate that EGF and PDGF stimulate inositol lipid metabolism independently of each other; no "cross talk" between receptors appears to be involved in EGF-induced phosphoinositide turnover (Figure 5). NIH/3T3 fibroblasts are responsive to mitogenic stimuli (PDGF, thrombin, bombesin, etc.) that are known to trigger phosphoinositide turnover; therefore, the transfected EGF receptor might substitute for one of the above-mentioned growth factors in activating the signaling cascade eventually responsible for cell growth. It is tempting to suggest that overexpression of the EGF receptor in normal fibroblasts amplifies EGF-induced inositol lipid turnover up to the point where it can be easily detected by the currently available techniques. This idea is supported by the observation that the ability of EGF to trigger phosphoinositide turnover correlates with its mitogenic effect. Since EGF is a weak mitogen for NIH/3T3 fibroblasts, the EGF-dependent inositol lipid signal might be too low to be detected. Consistently, a low concentration of PDGF which induces only a twofold stimulation of DNA synthesis fails to show appreciable effect on phosphoinositide turnover (Figure 4). Furthermore, a recent report showed that EGF induces similar intracellular calcium redistribution from internal stores in Swiss/3T3 and A431 cells. However, the greatly different number of EGF receptors in the two lines is responsible for a marked difference in the extent of calcium mobilization and, possibly, for the lack of effect on inositol lipid metabolism in Swiss/3T3 fibroblasts.[45]

The EGF receptor has also been overexpressed in the 32D myeloid cell line which is devoid of EGF receptors and totally dependent on interleukin 3 (IL3) for proliferation and survival. Expression of the EGF receptor confers the ability to utilize EGF for transduction of a mitogenic signal.[46] When the transfected cells are propagated in EGF, they exhibit a more mature myeloid phenotype than under conditions of IL3-directed growth. Moreover, exposure to EGF leads to a rapid stimulation of inositol lipid metabolism. These results are of particular interest since IL3 does not stimulate inositol lipid hydrolysis, and so far no other agent able to trigger inositol lipid turnover in 32D cells has been reported. It appears that, in 32D cells, the activated EGF receptor is able to couple to a transducing mechanism that is not routinely used by the cell. This might explain the long adaptation period required by EGF receptor-32D cells to grow in the presence of EGF after switching from IL3 to EGF supplementation.[46]

From these studies it appears that the normal human EGF receptor expressed at high levels is able to couple to inositol lipid metabolism in cell lines that are mitogenically stimulated by EGF. Therefore, the ability to trigger

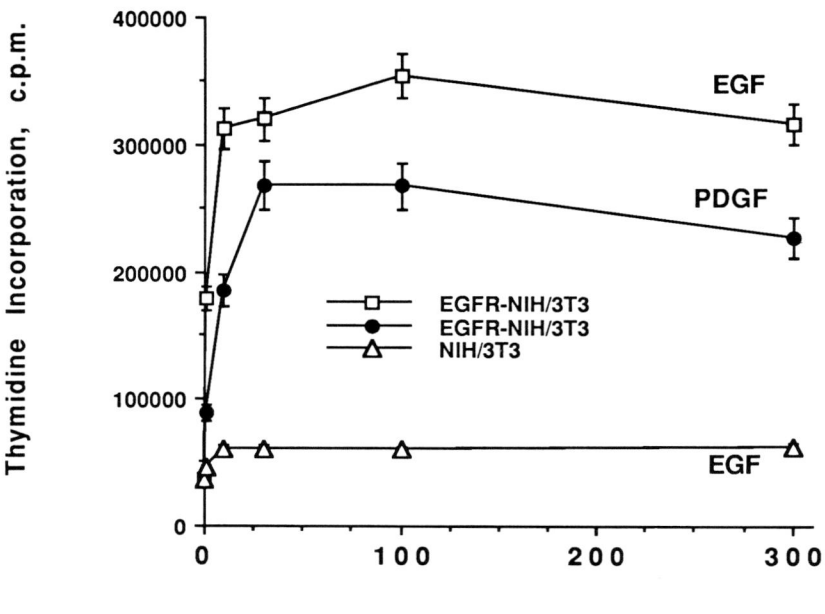
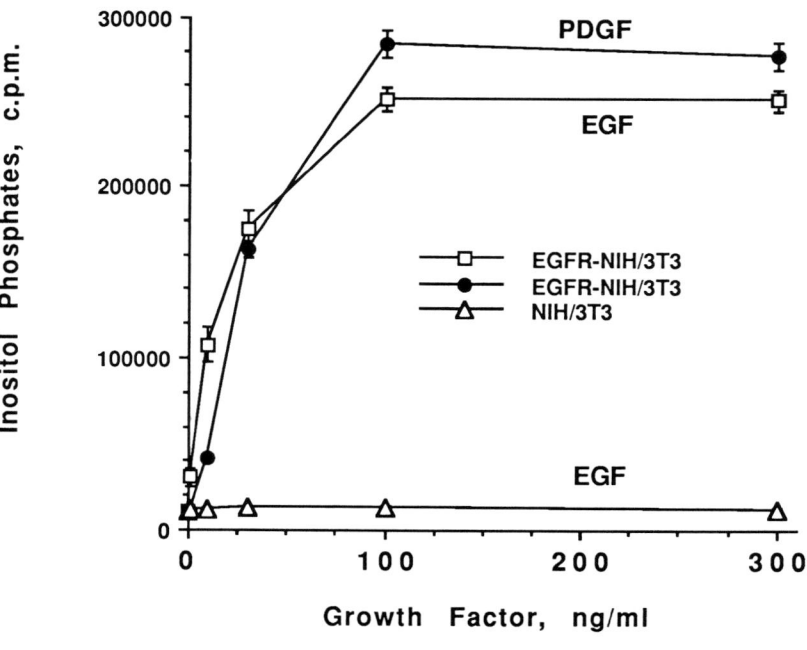

FIGURE 4

PIPs hydrolysis, calcium mobilization, and PKC activation seems to be characteristic of the receptor itself. However, microinjection of antibody to PIP2 does not abolish mitogenesis in response to EGF in wild-type NIH/3T3 fibroblasts, suggesting that inositol lipid metabolism might not be correlated with EGF-induced DNA synthesis in these cells.[30] These results are difficult to interpret, since EGF does not stimulate detectable inositol lipid hydrolysis in wild-type NIH/3T3 cells. It will be interesting to determine whether antibody to PIP2 can block PIPs turnover and DNA synthesis in cells overexpressing the receptor. Thus, although EGF definitely stimulates inositol lipid metabolism concurrently with DNA synthesis in a number of cell lines, there is no clear-cut evidence demonstrating that inositide-derived second messengers actually transduce the EGF mitogenic signal.

C. GROWTH FACTORS THAT STIMULATE "ALTERNATIVE" PATHWAYS AND INSULIN

Several mitogens that induce the hydrolysis of PIP2 have been shown also to cause PC degradation to form DAG and phosphorylcholine.[47] Vasopressin, PDGF, bombesin, and interleukin 1 (IL1) are among the growth factors able to trigger both signaling pathways (reviewed in Reference 20). IL1 represents a distinct example in that it seems to induce different biochemical responses in different cell lines. In T lymphocytes, it stimulates PC hydrolysis without affecting inositol lipid metabolism.[48] However, in these cells, no detectable binding of IL1 can be observed, and the mechanism(s) by which IL1 determines PC hydrolysis has yet to be elucidated. In mouse peritoneal macrophages, on the other hand, IL1 causes the rapid formation of InsPs, indicating that hydrolysis of PIP2 occurs in response to the factor.[49] In human vascular smooth cells, IL1 induces the production of AA metabolites (PGE_1 and PGE_2) that antagonize its growth-promoting effect in the short term, but not during more prolonged exposures.[50] This growth inhibitory effect of prostanoids on growth factors that stimulate inositol lipid hydrolysis has also been reported for the nonapeptide agonist bradykinin.[51] The turnover of PC which generates second messengers such as DAG and AA might then be

FIGURE 4. Dose response of wild-type and EGF receptor overexpressing NIH/3T3 fibroblasts to EGF and PDGF. Upper panel: the growth response was measured by the incorporation of [^{3}H]thymidine into TCA insoluble material, expressed as cpm. Lower panel: cells, prelabeled with [^{3}H]inositol, were stimulated for 30 min at 37°C with the indicated concentration of growth factor in the presence of LiCl. Total inositol phosphates were extracted and separated by anion exchange chromatography. EGFR-NIH/3T3 indicates NIH/3T3 fibroblasts overexpressing the normal human EGF receptor. The relative potencies of the two growth factors appear similar. It is worth noting that for both factors the saturating concentration for inositol phosphate formation was higher than that required for maximum mitogenic stimulation. EGF induced relatively little stimulation of DNA synthesis and no detectable increase of inositol phosphate formation in control NIH/3T3 cells. These results indicate that the ability of EGF and PDGF to stimulate phosphoinositide turnover correlated with their mitogenic effects, thus suggesting a relationship between inositol lipid metabolism and cell growth.

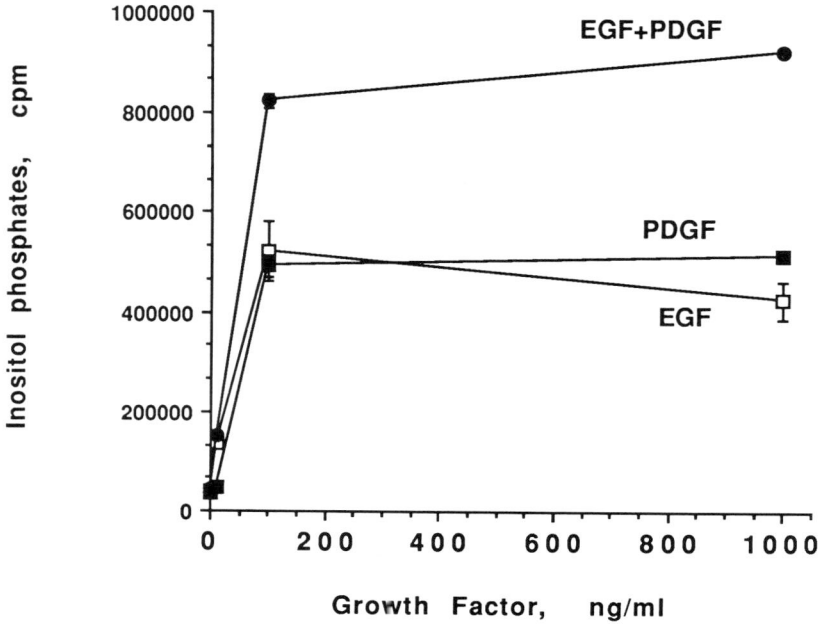

FIGURE 5. Effects of EGF and PDGF on inositol lipid metabolism in EGF receptor overexpressing NIH/3T3 fibroblasts. In this set of experiments, EGF and PDGF were added either separately or together. It is clear that the two growth factors together induced an additive response. This seems to indicate that no "cross talk" between the receptors occurs and that EGF and PDGF stimulate inositol lipid metabolism independently of each other.

involved in the complex balance of growth promoting and inhibiting activities. Since numerous agonists besides growth factors seem to activate PC metabolism, either inducing its degradation or resynthesis,[20] it is possible that remodeling of membrane phospholipids other than PIPs may play a role in signal transduction.

The effect of insulin on signal transduction mechanisms is a complex one, since it is concurrently involved in different biochemical pathways.

One or more G-proteins (tentatively termed G_{ins}) might be modified by insulin receptor-dependent tyrosine phosphorylation, thus activating a PI-glycan-specific PLC. Activation of this enzyme leads to the formation of DAG (whose molecular structure is different than that derived from PIP2 hydrolysis), and water-soluble inositol phosphate-glycan. This in turn would lead to the release of key enzymes (alkaline phosphatase) or compounds (heparan sulfate) that are anchored to the cell membrane through a glycosyl-PI structure.[52] A recent report demonstrates that EGF and insulin-like growth factor-1 also stimulate the hydrolysis of the insulin-sensitive PI-glycan in myocytes.[53] Since all these growth factor receptors exhibit tyrosine kinase activity, it will be interesting to study the relationship between tyrosine phosphorylation and PI-glycan metabolism particularly concerning the factor-specificity of this effect.

It has been proposed that the product of the *ras* oncogene $p21^{ras}$ might be related to the putative G-protein activating the PI-glycan-specific PLC.[52] As we shall see in the following section, *ras*-transformed cells show several biochemical features that are peculiar to insulin action. This suggests that $p21^{ras}$ might mediate some of the insulin effects and/or substitute for them.

In addition to these effects on the PI-glycan pathway, insulin is known to increase DAG content and PKC activity without stimulating PIPs hydrolysis.[54] This result is achieved by biosynthesis of PA and DAG *de novo*, and degradation of PC and phosphatidylethanolamine (PE).[55,56] This pathway for the formation of DAG is of particular interest, since insulin would increase the synthesis of phospholipids (PIPs, PC, PE), by stimulating reactions located between glycerol 3-phosphate and PA in the phospholipid biosynthetic pathway. In this respect, insulin represents a unique example in that it induces phospholipid biosynthesis and degradation at the same time. DAG and PA are crucial components of this transducing mechanism, since they exert second messenger functions and also serve as building blocks for the neosynthesis of phospholipids (including PIPs) that are in turn involved in the generation of intracellular signaling compounds.

IV. ONCOGENES AND SECOND MESSENGERS

Genes capable of inducing neoplastic transformation are termed oncogenes. Most of our knowledge about oncogenes derives from the study of acutely transforming retroviruses. Such viruses, also called transforming retroviruses, possess sequences (viral oncogenes, or v-*onc*) which they capture from the genomes of their host cells. In addition to their importance for understanding oncogenesis, studies on v-*oncs* have provided precious information on normal cellular processes. It has been observed that protooncogene sequences are highly conserved among diverse organisms. These results have led to the suggestion that protooncogenes might perform critical roles in processes such as cell proliferation and differentiation. Strong evidence to support this hypothesis has come from demonstrations that proteins encoded by protooncogenes are components of growth factor-mediated proliferative pathways. In addition to retroviral transduction, other mechanisms are involved in the activation of protooncogenes to transforming oncogenes; they include amplification, mutations, or deletions in the protein coding part of the gene, and deregulation of gene expression. The cellular localization and the properties of several oncogene products suggest that they may cause transformation by substituting for some component of the growth factor-mediated mitogenic pathway (for review on oncogenes see Reference 57).

Activated *ras* oncogenes have been detected in a wide variety of tumor types. The frequency of their detection by DNA transfection techniques ranges from 10 to 20% of human primary malignancies. However, new techniques for detection of point-mutations in eukaryotic cells indicate that *ras* gene activation in tumors occurs even more frequently (see Reference 58 for review on *ras*).

The *ras* genes encode a family of proteins similar in size to one another (M_r 21,000) that appear to be important in the control of cell growth. These p21^{ras} proteins are known to bind GDP, catalyze the hydrolysis of GTP and, in some cases, be autophosphorylated. It has been demonstrated that *ras* proto-oncogenes most commonly acquire malignant properties by single point-mutations affecting either the 12th, 13th, 59th, or 61st codons within the 189 amino acid sequence of the p21 molecule. p21^{ras} proteins share enzymatic properties and distant sequence similarities with G-proteins that might be involved in signal transduction. Since different G-proteins appear to couple receptors for a diverse array of agonsts to distinct signaling pathways (adenylate cyclase, cGMP-specific phosphodiesterase, ion channels, PLC and PLA$_2$), it has been postulated that *ras* genes may play critical roles in pathways central to the regulation of cell growth.

The predicted amino acid sequences of *ras* proteins are highly conserved in species as diverse as humans and yeasts. In yeast, the point-mutated p21^{Ha-ras} protein continuously activates adenylate cyclase, whereas the normal p21^{ras} protein does not show such an effect.[59] However, in mammalian cells (murine fibroblasts), point-mutated p21^{ras} seems to play an opposite role in that it inhibits cAMP formation in response to potent adenylate cyclase agonists such as isoproterenol and PGE$_1$.[60] There is a growing body of evidence indicating that p21^{ras} might be correlated with membrane phospholipid remodeling, ion fluxes, and lipid-derived second messenger formation in mammalian cells (for review see Reference 61).

Several conflicting reports have recently been published concerning the role of phospholipid-derived second messengers in *ras*-induced cell transformation. Essentially, two approaches have been adopted to solve this complex matter: on one side, investigators have studied the effect of different agonists on second messenger formation in *ras*-transformed cells as compared to normal cells. On the other side, the steady-state level of different compounds that may function as second messengers has been measured.

The first evidence concerning *ras* and inositol lipid metabolism was obtained in 1985 by Chiarugi et al. who showed that transformation of 3T3 fibroblasts by a transforming *ras* gene increases the responsiveness of PLC to muscarinic cholinergic stimulation.[60] The following year, the same group provided evidence that this phenomenon occurs in the absence of a change in receptor number and it could be inhibited by the anti-*ras* antibody, Y13-259.[62] Introduction of point-mutated p21^{ras} into normal fibroblasts increases the responsiveness to muscarinic stimulation in terms of inositol lipid metabolism and calcium influx.[62] These results seem to indicate that p21^{ras} may serve in the coupling of membrane receptors to different effector systems. Using a different approach, Wakelam et al.[63] showed that p21$^{N\text{-}ras}$ couples bombesin and PDGF receptors to InsPs production. This observation was extended by Marshall[64] proposing that normal p21$^{Ha\text{-}ras}$ enhances PDGF-induced inositol lipid metabolism, whereas point-mutated p21^{ras} increase PIPs hydrolysis in the absence of agonists. Along the same lines, it has been reported that *ras*-transformed cells show increased responsiveness to bradykinin[65,66] and serotonin.[67] However, in the case of bradykinin, the enhanced stimulation of inositol lipid metabolism is due to an increase of bradykinin receptor number that does not correlate with p21 expression and is not specific for *ras* transformants.[68] The effect on serotonin-induced inositol lipid turnover also seems to be rather nonspecific.[67] To further complicate the matter, other studies showed that *ras*-transformed cells exhibit decreased inositol lipid turnover and phospholipase activity in response to PDGF, serum, and thrombin.[65,67,69]

There are several possible explanations for these apparently conflicting observations. Fundamentally, different experimental conditions (cell culture density, labeling procedures, cell type- or clone-specificity of the observed effects) are most likely responsible for these discrepant observations. This renders experimental data between different laboratories almost impossible to compare.[70] At this point, it can only be proposed that *ras*-induced transformation may lead to subversion of the coupling between different types of receptors and the signal transducing machinery. Agonists such as carbamylcholine, phenylephrine, bradykinin, and vasopressin that induce a short-term inositol lipid signal might have their effects enhanced in *ras* transformants. However, mitogens such as thrombin, bombesin, and PDGF that require longer-lasting accumulation of inositide-derived second messengers might be inhibited (for review see Reference 71). Further studies will reveal whether disruption of receptor-mediated signal transduction is peculiar to *ras*, or is a simple epiphenomenon of neoplastic cell transformation.

Concerning the analysis of the steady-state level of second messengers, there is general agreement that DAG is elevated in *ras*-transformed cells.[72-75] Persistent elevation of DAG would in turn induce partial down-regulation of PKC and/or its substrates, thus desensitizing the DAG/PKC limb of the signaling pathway.[76,77] Discrepancies arise concerning the source of this DAG. According to a number of studies, *ras* transformants show increased basal metabolism of inositol lipids, thus suggesting that DAG is elevated because of persistent enhanced hydrolysis of PIPs.[73,78-82] Other investigators, however, were unable to detect increased inositol lipid turnover in spite of elevated DAG levels.[74-76] Therefore, it has been proposed that DAG in *ras* transformants might be derived either from the hydrolysis of other phospholipids (PC and PE), or from a glycolytic neosynthetic pathway. The constitutive hydrolysis of PC and PE would generate DAG, phosphorylcholine, and phosphorylethanolamine. This conclusion stems from the finding of elevated steady-state levels of phosphorylcholine and phosphorylethanolamine in *ras*-transformed NIH/3T3 fibroblasts prelabeled with [^{3}H]choline or [^{14}C]ethanolamine.[74] However, phosphorylcholine and phosphorylethanolamine are both precursors in the synthesis and products of the hydrolysis of the respective membrane phospholipids. Therefore, the increase in their steady-state level might indicate enhanced *de novo* phospholipid synthesis as well as PLC-mediated hydrolysis. A recent paper of Macara[83] demonstrates that PC hydrolysis is not increased in *ras* transformants, and elevated phosphorylcholine levels are due to enhanced phosphorylcholine neosynthesis.

Concerning the neosynthetic pathway for DAG production, DAG would be neosynthesized from glucose in a manner similar to what happens in response to insulin[55] or glucose.[19] Consistently, *ras*-transformed fibroblasts are not sensitive to insulin in this respect, and insulin is unable to further increase DAG neosynthesis (Figure 6). Indeed some recent studies support the hypothesis of a relationship between insulin and p21^{ras} action. Korn et al.[84] have recently proposed that p21^{ras} might act as mediator of insulin action in *Xenopus* oocytes since anti-p21^{ras} antibodies inhibit oocyte maturation induced by insulin. One or more G-proteins seem to be required for insulin action in vertebrate cells, and it has been suggested that p21^{ras} might be responsible for coupling the insulin receptor to second messenger pathways such as the adenylate cyclase system. Insulin increases the synthesis of phospholipids,[55] and microinjection of p21^{Ha-ras} into *Xenopus* oocytes increases the synthesis of PIP2,[78] just like insulin does in adipocytes. It is worth noting that insulin and p21^{ras} action seem to converge to a common point, namely, *de novo* synthesis of DAG and PA. Consistently, *ras* and phorbol esters activating PKC cooperate in inducing carcinogenesis in fibroblasts and keratinocytes.[85,86] Furthermore, insulin, phorbol esters, DAG, and activated *ras* induce the formation of a key modulator of glycolysis, fructose 2,6-biphosphate through PKC activation (Reference 87 and Bruni et al., unpublished observation). Whatever the case, it is tempting to observe that *ras* transformants show several biochemical features that are peculiar to insulin action. p21^{ras} could represent a multiple coupler of different receptors to specific signaling pathways; consequently, *ras* transforming proteins might selectively interfere with the flow of signaling along diverse biochemical transducing mechanisms. *ras*-induced transformation would then increase PA and DAG neosynthesis with an insulin-like action, and subvert the coupling of neurotransmitter and growth factor receptors to inositol lipid turnover.

However, since almost all tumor cell lines show increased glycolysis, it will be worth investigating whether this insulin-like effect is specific for *ras*, or rather is common to other oncogenes. Indeed, some preliminary experiments seem to suggest that other genes whose products are associated with transducing mechanisms show a similar effect (Figures 7 and 8). Therefore, increased glycolysis might be a common route for DAG production and phospholipid neosynthesis in transformed cells. Alternatively, p21^{ras} might rep-

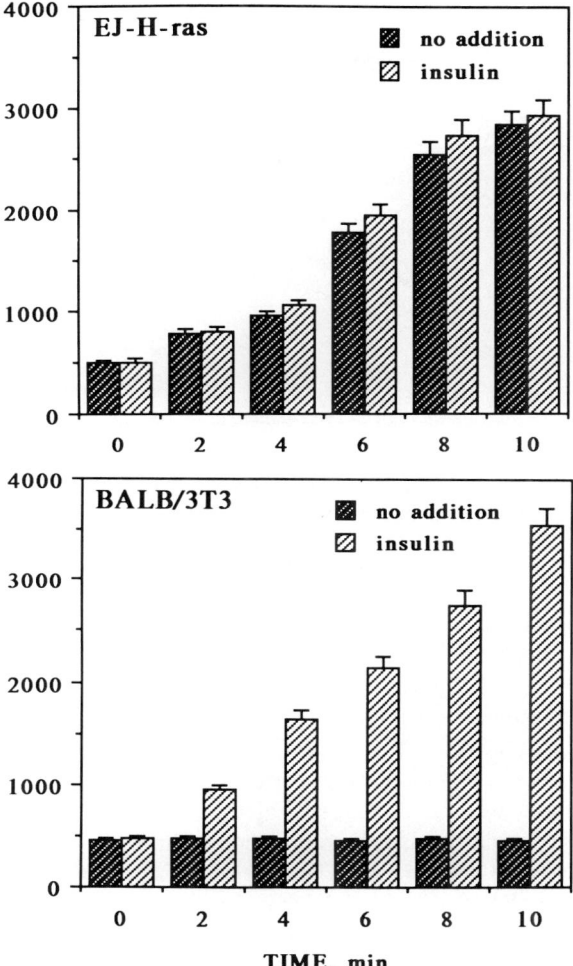

FIGURE 6. Effect of insulin on the acute transfer of ^{14}C from D-[U-^{14}C] glucose into diacylglycerol in normal and *ras*-transformed fibroblasts. Semiconfluent monolayers of normal and EJ-H-*ras*-transformed BALB C/3T3 cells were incubated with 0.5 uCi/ml of [^{14}C]glucose. After addition of the radiolabeled compound, the cells were extracted at different time intervals as indicated on the X axis. Time 0 indicates when the precursor was added. Labeled diacylglycerol (DAG) was rapidly formed from [^{14}C]glucose in *ras*-transformed cells. These results suggest that DAG is synthesized *de novo* from the glycolytic intermediate dihydroxyacetone phosphate and from glycerol 3-phosphate after stepwise acylation to lysophosphatidic acid and phosphatidic acid. Insulin did not modify the incorporation of glucose-derived ^{14}C into DAG in *ras* transformants, whereas it dramatically increased DAG neosynthesis in control fibroblasts. Synthesis of DAG *de novo* might represent the point of convergence of insulin and p21^{ras} action. Since insulin could increase DAG neosynthesis in normal cells but not in *ras* transformants, it is likely that *ras* transformants have maximally activated insulin-like mechanism(s) for DAG formation.

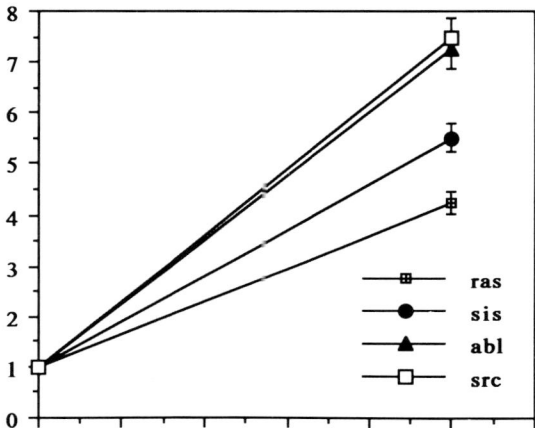

FIGURE 7. Incorporation of glucose-derived ^{14}C into diacylglycerol in normal and transformed fibroblasts. The experimental conditions were as described in the legend to Figure 6. Cells transformed by oncogenes such as *scr*, *abl*, *sis*, and H-*ras* showed a dramatic increase of diacylglycerol (DAG) labeling as compared to normal cells.

resent a necessary *trait-d'union* between the signaling machinery triggered by several oncoproteins (and growth factors) and their respective cellular effectors.

It has been recently demonstrated that G-protein(s) are also involved in the activation of PLA_2.[88] Not surprisingly, p21^{ras} has been proposed as a candidate for PLA_2 activation, mainly because microinjection of p21^{Ha-ras} into fibroblasts stimulates PLA_2.[89] AA (which is released by PLA_2 action on membrane phospholipids) is known to interact with many signaling pathways including guanylate cyclase, adenylate cyclase, PLC, PKC, and calcium mobilization. Therefore, it is possible that some of the alterations of second messengers observed in *ras* transformants might be ascribed to an effect of p21^{ras} on AA release and metabolism.

Although *ras* is the most studied oncogene with respect to signal transduction, other oncogenes have also been investigated. In fact, the products of several other transforming genes seem to be involved in signaling pathways, since they code for a growth factor (*sis*/PDGF), growth factor receptors (*erb*B, *fms*), or membrane-associated protein tyrosine kinases (*src*, *fes*, *abl*). As shown in Table 1, all these oncogenes induce various alterations of inositol lipid metabolism with a resulting derangement of second messenger formation.

Different cell types transformed by the *sis* oncogene (a gene that encodes a PDGF-B homodimer) show elevated steady-state levels of InsPs,[74] DAG,[72] and PI kinase activity.[3] This suggests that *sis*-transformed cells have consti-

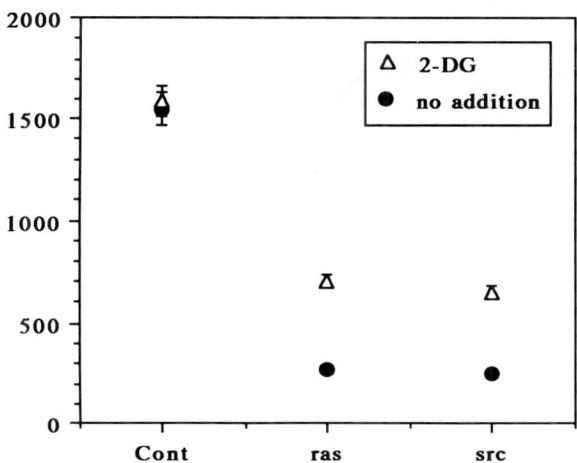

FIGURE 8. Phorbol ester binding to intact control and transformed fibroblasts: effect of 2-deoxy-d-glucose. Semiconfluent monolayers of normal and transformed cells were preincubated with buffer or with the antimetabolite of glucose 2-deoxy-d-glucose (2-DG; 125 mM) for 2 h before addition of phorbol dibutyrate (PDBu). [³H]PDBu (3 nM) was added for 20 min at 37°C and binding was determined. The binding to transformed cells was significantly lower than that to control cells suggesting a decrease of available PKC. Inhibition of glycolysis and DAG neosynthesis by 2-DG could partially restore PDBu binding to transformed cells, whereas it did not affect normal fibroblasts. These results indicate that glucose-derived DAG is responsible for PKC down-regulation in transformed cells.

tutively activated inositol lipid turnover because of persistent stimulation of the PDGF signaling pathway.

The *erb*B oncogene encodes a truncated form of the EGF receptor and it may exhibit a similar signal transduction system leading to cell growth as that resulting from EGF receptor activation. Consistently, chick embryo fibroblasts transformed by *gag*-fused *erb*B gene-carrying virus show enhanced inositol lipid metabolism and subsequent signal transduction. In particular, they show increased activities of PI, PIP, and DAG kinases, higher content of DAG, and elevated PKC activity in the membrane fraction.[4]

The c-*fms* proto-oncogene encodes a product that is related, or possibly identical, to the receptor for the macrophage colony-stimulating factor CSF-1. Like several other growth factor receptors, the product of c-*fms* exhibits tyrosine-specific protein kinase activity, suggesting that the family of viral transforming genes possessing this enzymatic activity (*src, fes, fms*) transforms cells by mimicking the action of cell surface receptors. Cells transformed by the v-*fms* and v-*fes* oncogenes show higher rates of inositol lipid turnover and increased activity of PIPs kinases.[90] Enhanced production of Ins

TABLE 1
The Effects of Oncogenes other than *ras* on Signal Transduction are Reported

Oncogene	Product	Activity	Effect on signal transduction
sis	PDGF-B chain	PDGF agonist	PIPs hydrolysis; PI kinase
erbB	EGF receptor	Tyrosine kinase	PIPs and DAG kinases
fms	CSF-1 receptor	Tyrosine kinase	PIPs hydrolysis; PI kinase
fes	p92^{fes}	Tyrosine kinase	PIPs hydrolysis
src	pp60^{src}	Tyrosine kinase	PIPs hydrolysis; PI kinase
abl	p150^{abl}	Tyrosine kinase	PIPs hydrolysis
dbl	p66	?	Increased bradykinin receptors

Note: "PIPs hydrolysis" indicates phospholipase C-mediated hydrolysis of PIP2, PIP, and/or PI leading to the formation of water-soluble inositol phosphates and diacylglycerol. "PI, PIPs, and DAG kinases" indicate activation of inositol lipid and diacylglycerol kinases with the resulting formation of polyphosphoinositides and phosphatidic acid. See the text for details.

1,4,5-P3 and DAG is probably due to the activation of guanine nucleotide-dependent PIP2-specific-PLC; these findings seem to reinforce the idea that the generation of aberrant hormonally independent signals is associated with cell transformation by genes encoding tyrosine kinases.

Consistent with the previous findings, cells transformed by v-*src* show higher activity of PI kinase[3] and elevated levels of DAG.[76] However, since the comparison of steady-state levels of second messengers in normal and transformed cells has the potential for bias due to different experimental conditions,[70] we used an inducible system to study *src*-related alterations of signal transduction.[91] We examined inositol lipid turnover in a rat fibroblast line infected with a temperature-sensitive mutant of the Rous sarcoma virus. When the infected cells were shifted from the non-permissive to the permissive temperature, rapid and sustained stimulation of PLC was observed along with pp60^{v-src} expression and tyrosine kinase activation. Since pretreatment with Na$^+$ channel inhibitors, or incubation in Na$^+$-free medium prevented temperature shiftdown-induced PLC activation, it is possible that membrane depolarization was responsible for the observed effect. In fact, pp60^{v-src} expression in a very similar system induces membrane depolarization[92] and membrane depolarization itself can stimulate inositol lipid turnover.[93]

Similarly, cells transformed by Abelson murine leukemia virus show constitutively activated PIPs breakdown with increased formation of InsPs.[94]

The human *dbl* oncogene encodes a cytoplasmic protein of about 66 kDa that is equally distributed between cytosol and crude membrane fractions.[95] Its predicted amino acid sequence shows that the *dbl* oncoprotein is highly hydrophilic with no hydrophobic region that would indicate the presence of a potential transmembrane domain or leader sequence.[96] The basis for its partial association with cell membranes remains to be determined and the *dbl* oncoprotein seems to be distinct among known transforming gene products. The normal homolog of the *dbl* oncogene also codes for a novel phosphoprotein of 115 kDa which exhibits some structural similarity with the cytoskeletal protein vimentin.[97] The *dbl* oncogene represents a truncated and rearranged version of its normal homolog and shows significantly higher transforming efficiency as compared to the *dbl* proto-oncogene. The *dbl*-transformed fibroblasts did not show alterations of the basal level of InsPs, PIPs, DAG, or PA. This indicates that the activity of C-type phospholipases, inositol lipid kinases, and DAG kinase was not altered in *dbl*-induced transformation.[68] However, *dbl*-transformed NIH/3T3 cells exhibited increased inositol lipid turnover in response to bradykinin. Further analysis revealed a significantly higher number of bradykinin receptors in *dbl* transfectants as compared to control NIH/3T3. When several clonally derived *dbl* NIH/3T3 transfectants were analyzed, we observed a large variation of their bradykinin receptor number. Cell lines exhibiting increased bradykinin binding, however, failed to show augmented mitogenic response to the peptide agonist. Among other oncogenes, only *ras* showed a similar effect. It can be proposed that increased bradykinin receptor number is a phenomenon common to several cell lines transformed by different oncogenes, and it does not correlate with either enhanced mitogenic responsiveness of transformed cells to the peptide or with the presence of a specific oncogene in the transformant.[68]

V. CONCLUSIONS

The turnover of inositol lipids generates a number of second messengers that are believed to be important in the transduction of mitogenic signals from the membrane to the nucleus. Several growth factors trigger the metabolism of inositol lipids, and there is convincing evidence that this biochemical pathway is one of the major mechanisms by which signals transmitted from cell surface receptors result in DNA synthesis. Because the activation of many growth factor receptors induces tyrosine-specific phosphorylation of certain proteins, it can be proposed that tyrosine phosphorylation and inositol lipid turnover are separate but converging pathways leading to cell proliferation.

Cells transformed by a variety of oncogenes whose products are associated with the plasma membrane show significant alterations of inositol lipid turnover and signal transduction. These findings suggest that a derangement of second messenger production may be associated with the function of certain oncogenes leading to uncontrolled cell growth. In the case of oncogene-associated signal transduction alterations, however, it is not clear whether these changes are specific for some genes and/or whether increased formation of messengers is responsible for cell transformation.

Identification of the biochemical pathways responsible for normal and uncontrolled cell growth will allow the design of specific inhibitors able to block key reactions in the mitogenic signaling machinery. The examples of lithium and aspirin, which disrupt the formation of certain intracellular messengers, are encouraging and have paved the way for the synthesis of specific antagonists/inhibitors of G-proteins, PLC, PKC, PLA_2, etc. It is likely that in the next few years several new drugs will appear, directly affecting the transmembrane signaling pathways involved in cell growth and tumor promotion.

VI. ADDENDUM: RECENT DEVELOPMENTS CONCERNING MITOGENIC SIGNALING IN NORMAL AND TRANSFORMED CELLS

A. INTRODUCTION

This chapter was finished in March 1989. Since then, work in the field of mitogenic signaling and inositol lipid turnover has proceeded at a remarkably high rate: about 200 papers are published each month on inositol lipid metabolism and related subjects. Furthermore, excellent updated reviews appeared in the recent past covering virtually all aspects of transmembrane mitogenic signaling.[99,101,104] Many of the findings reported in this chapter have been confirmed, and new experimental approaches have been developed to gain further insight into the molecular mechanism of mitogenic signal transduction. Fearing the risk of describing results that could be already outdated by the time this chapter will be in press, I decided instead to present a speculative hypothesis about the conversion of short-lasting non-mitogenic signals into long-lasting mitogenic signals that could be involved in the control of transformed cell growth.

B. COOPERATION OF INOSITOL LIPID TURNOVER AND CALPAIN IN MITOGENIC SIGNALING

A major problem in understanding mitogenic signaling concerns the mechanism(s) that convert early, short-lasting signals into late, irreversible responses able to persist even after removal of mitogenic stimuli. The set of intracellular second messengers produced during mitogenic signaling is very similar, if not identical, to that responsible for short-term, reversible responses such as hormonal stimulation and neurotransmission. Since the biochemical events are the same (i.e., protein phosphorylation and inositol lipid turnover) there must be differences in the way the signal is processed within target cells.

In the past few years we studied signal transduction in cells expressing different growth factor receptors or transformed by membrane-associated oncogenes. In particular, we studied mitogenic signaling in null cells into which the normal receptors for EGF, PDGF, and CSF-1 were introduced, and in fibroblasts transformed by *ras, src,* and *dbl* oncogenes.[102,103,105,106,113,122] I came to the conclusion that conversion of short-lived signals into irreversible responses is the crucial event underlying mitogenic signaling both in normal and transformed cells. In this section I would like to focus on the molecular mechanism(s) involved in the generation of irreversible proliferative signals.

Growth factor receptors harbor tyrosine kinase (TK) activity which phosphorylates numerous intracellular substrates including the receptor itself. Growth factor-stimulated tyrosine phosphorylation is the first step in the cascade of events that leads to DNA synthesis; EGF and PDGF receptor mutants in which TK activity was abolished, failed to trigger downstream biochemical passages and DNA synthesis.[108,118] Among the substrates phosphorylated by receptor TKs, three are of particular importance: (1) inositol lipid-specific phospholipase C (PLC); (2) phosphatidylinositol 3-kinase (P13K); and (3) GTPase activating protein (GAP).[98,110,117] While little is yet known about the role of the latter substrate in normal signal transduction, the other two enzymes (PLC and P13K) are key components of the inositol lipid signaling machinery. This complex signaling pathway generates a number of second messengers implicated in the control of cell proliferation.[99,100,111,114,120,126] Studying mitogenic signaling in cells overexpressing PDGF or EGF receptors, we noticed that overall inositol lipid metabolism was continuously stimulated for 6 to 8 h. This time span was critical for growth factor-induced DNA synthesis; removal of growth factors before 6 to 8 h did not allow DNA replication. After 6 to 8 h, however, the mitogens could be removed and growth signaling proceeded resulting in DNA synthesis. Consistently, within 6 to 8 h, growth factor-stimulated inositol lipid turnover ceased as soon as the mitogens were removed either by antibodies or by appropriate washing. These observations indicated that growth factor/receptor complex had to signal for 6 to 8 h in order to trigger irreversible changes able to persist after removal (and/or down-regulation) of the initial stimulus. This concept, however, is difficult to reconcile with the observation that IP3/calcium and DAG/PCK signals are short-lasting in that they are rapidly buffered by efficient intracellular systems. IP3 is both dephosphorylated to IP2 and phosphorylated to inositol (1,3,4,5)tetrakisphosphate (IP4). DAG is deacylated to monoacylglycerol and/or phosphorylated to phosphatidic acid. Calcium is buffered and compartmentalized, and PKC-induced phosphorylation is reversed by phosphatases. Therefore, some additional mechanism must be at work in order to prolong calcium and PKC signals for those critical 6 to 8 h after the initial messengers have gone.

Description of PEST sequences in proteins involved in signal transduction provided a clue in understanding how the mitogenic signal is processed.[128] PEST sequences are regions enriched in proline, glutamic and aspartic acid, serine, and threonine. Since glutamic and aspartic acid are negatively charged, and threonine and serine can be phosphorylated (also by PKC), phosphorylated PEST sequences represent negatively charged areas able to bind and concentrate calcium. Localized concentration of calcium in particular zones of proteins causes the attack of calpain, an intracellular calcium-dependent protease whose calmodulin-like domain binds to phosphorylated PEST sequences. Proteins containing PEST sequences thus become calpain substrates and are rapidly cleaved. It has been noticed that partial hydrolysis of enzymes by calpain leads to deregulation of the enzyme, as if calpain removed inhibitory or regulatory domains.[128] Thus, initial and transient IP3-induced calcium mobilization and PKC-dependent phosphorylation lead to activation of calpain which permanently modifies target proteins. Among these are inositol (1,4,5)trisphosphate 3-kinase (IP3K) and PKC.

IP3K phosphorylates IP3 in the 3-position, thus producing IP4. IP4 cooperates with IP3 in the control of calcium homeostasis by shifting calcium from IP3-insensitive to IP3-sensitive stores.[99] IP4 is formed more slowly than IP3 and persists for much longer; therefore, it is believed to play a role in sustaining calcium mobilization over prolonged periods of time. IP3K is a calcium/calmodulin-dependent enzyme, presumably activated by calcium mobilized through IP3. It can be phosphorylated and modulated by PKC, and, in *src*-transformed cells, might be a target of tyrosine phosphorylation.[107] Like several other calmodulin-dependent enzymes, IP3K contains PEST sequences; in particular, four "strong" and two "weak" PEST regions have been described in the enzyme. Consequently, IP3K is a good substrate for calpain and it can be rapidly cleaved by the protease.[107] In analogy with other calpain enzyme substrates, it is presumable that limited proteolysis gives rise to a deregulated IP3K that would sustain prolonged calcium signaling through formation of IP4. Thus, initial rapid mobilization of calcium by IP3, and phosphorylation by PKC, although transient signals in themselves, would produce permanent modifications in a key enzyme through limited proteolysis; this critical step represents the conversion of a short-term signal into a long-lasting, irreversible response.

A similar mechanism can be postulated for PKC. After it translocates to the membrane, and is activated by DAG, calcium, and phospholipids, PKC is autophosphorylated. Given the presence of PEST sequences in the PKC structure, phosphorylated PKC becomes a substrate for calpain which cleaves the enzyme and removes it from the membrane. This mechanism is also responsible for the phenomenon known as PKC down-regulation which occurs following exposure to tumor promoter phorbol esters.[120] Cleaved PKC translocates back to the cytosol and from there possibly to the nucleus where it can phosphorylate different substrates Furthermore, calpain-cleaved PKC is permanently active and no longer requires cofactors. Different PKC isoenzymes harbor PEST sequences of different strengths,[128] and this would provide them with specific sensitivity to calpain cleavage. This would in turn explain why selective cleavage of one or the other isoenzyme is observed during cell growth and differentiation.[116] As with IP3K, convergence of calcium mobilization, phosphorylation, and calpain activation converts initial short-lived messages into long-term responses.

Thus, mitogenic signaling can be envisaged as a cascade of biochemical events, short-lasting in themselves, which eventually induce permanent modification through limited proteolysis. Calpain would then act as a "converter" of transient signals into long-lived messages. Among the different passages of signaling, calpain-induced proteolysis is the only one truly irreversible. Protein phosphorylation (both tyrosine and non-tyrosine) can be rapidly reversed by the action of intracellular phosphatases. PLC-induced phosphodiesterasic cleavage of phosphoinositides is immediately counteracted by inositol lipid kinases which replenish the pool of inositol lipids. IP3 and DAG are rapidly metabolized, and calcium is buffered and compartmentalized. Only protein hydrolysis by proteases cannot be reversed in the short term, and this renders the "calpain message" a peculiar trait of mitogenic stimulation. Consistent with this role, calpain also cleaves other critical components of the growth signaling pathway such as growth factor receptors, cytoskeletal and structural proteins and, possibly, PLC.[119,128]

If the mitogenic signaling pathway is actually a sequential cascade of definite biochemical steps, three conditions should be respected: (1) inhibition of each step should block the flow of mitogenic signaling and, consequently, DNA synthesis; (2) overexpression, or deregulated function, of each critical passage should be sufficient to induce mitogenesis in the absence of growth factor stimulation; and (3) in normal cells, each passage, taken by itself, should be considered "necessary but not sufficient" for cell growth.

If we take into consideration the enzymes discussed in this section, we see that these three conditions are indeed respected. Inhibition of TK activity blocks growth factor-induced proliferation, whereas overexpressed (or deregulated) TK activity is responsible for cell growth.[112] However, TK activity by itself is insufficient for cell growth if downstream mechanisms are not operating properly.[114,126] Blockade of PLC and inositol lipid turnover inhibits growth factor- and oncoprotein-induced cell growth,[111,114,126] and microinjection of active PLC stimulates proliferation in the absence of other growth stimuli.[127] Inositol lipid turnover alone, however, is definitely not sufficient for cell growth, as demonstrated by the fact that several potent inositol lipid turnover agonists are not mitogens.[100] PKC inhibition prevents proliferation in response to growth factors,[109] whereas PKC overexpression, or point-mutation, leads to abnormal cell growth.[115,121] Stimulation of PKC, however, is not mitogenic per se, as many agents stimulate the kinase without being considered mitogens. Finally, calpain when microinjected leads to mitosis,[124] whereas its inhibition by synthetic compounds or by the natural inhibitor (calpastatin) blocks cell growth.[124,125] Like the above-mentioned enzymes, calpain can be activated in several conditions unrelated to cell proliferation.[119]

From what has been discussed so far it follows that normal mitogenic stimulation only occurs when all the components are sequentially operating, thus transducing the signal from growth factors to the nucleus. Abnormal stimulation of even a single event in this cascade is then sufficient to induce cell growth by triggering downstream effectors. Oncoproteins exploit this mechanism by substituting for growth factors (*sis*), receptors (*erb*B, *fms*), kinases (*src, abl, raf, mos*), and PLC regulating components (*ras*). Detailed knowledge of the mechanism(s) that convert short-lived messages into irreversible responses will allow the design of compounds able to selectively interfere with such critical passages and possibly inhibit transformed cell growth.

ACKNOWLEDGMENTS

The author wishes to thank Dr. S. A. Aaronson for stimulating discussion and continuing support; Janet E. Jones for critically reading this manuscript; the Associazione Italiana per la Ricerca sul Cancro (AIRC), and the Studio Radiologico Ruggiero, Prato, Italy for financial support.

REFERENCES

1. Update: inositol phospholipids and protein kinase C, Update Publishing Company, Research Triangle Park, NC.
2. **Berridge, M. J.**, Inositol trisphosphate and diacylglycerol: two interacting second messengers, *Annu. Rev. Biochem.*, 56, 159–193, 1987.
3. **Kaplan, D. R., Whitman, M., Schaffhausen, B., Pallas, D. C., White, M., Cantley, L., and Roberts, M.**, Common elements in growth factor stimulation and oncogene transformation: 85 kd phosphoprotein and phosphatidylinositol kinase activity, *Cell*, 50, 1021–1029, 1987.
4. **Kato, M., Kawai, S., and Takenawa, T.**, Altered signal transduction in *erb*B-transformed cells, *J. Biol. Chem.*, 262, 5596–5704, 1987.
5. **Huang, M., Chida, K., Kamata, N., Nose, K., Kato, M., Homma, Y., Takenawa, T., and Kuroki, T.**, Enhancement of inositol phospholipid metabolism and activation of protein kinase C in *ras*-transformed rat fibroblasts, *J. Biol. Chem.*, 263, 17975–17980, 1988.
6. **Pike, L. J. and Eakes, A. T.**, Epidermal growth factor stimulates the production of phosphatidylinositol monophosphate and the breakdown of polyphosphoinositides in A431 cells, *J. Biol. Chem.*, 262, 1644–1651, 1987.
7. **Ruggiero, M. and Lapetina, E. G.**, Sustained proteolysis is required for human platelet activation by thrombin, *Thromb. Res.*, 42, 247–255, 1986.
8. **Golden, A. and Brugge, J. S.**, Thrombin treatment induces rapid changes in tyrosine phosphorylation in platelets, *Proc. Nat'. Acad. Sci. U.S.A.*, 86, 901–905, 1989.
9. **Whitman, M., Downes, C. P., Keeler, M., and Cantley, L.**, Type I phosphatidylinositol kinase makes a novel inositol phospholipid, phosphatidylinositol-3-phosphate, *Nature*, 332, 644–646, 1988.
10. **Suh, P. G., Ryu, S. H., Moon, K. H., Suh, H. W., and Rhee, S. G.**, Cloning and sequence of multiple forms of phospholipase C, *Cell*, 54, 161–169, 1988.
11. **Berridge, M. J.**, Inositol lipids and DNA replication, *Phil. Trans. R. Soc. Lond.*, 317, 525–536, 1987.
12. **Moolenaar, W. H., Bierman, A. J., Tilly, B. C., Verlaan, I., Defize, L. H. K., Honegger, A. M., Ullrich, A., and Schlessinger, J.**, A point mutation at the ATP-binding site of the EGF receptor abolishes signal transduction, *EMBO J.*, 7, 707–710, 1988.

13. **Ashkenazi, A., Peralta, E. G., Winslow, J. W., Ramachandran, J., and Capon, D. J.**, Functionally distinct G proteins selectively couple different receptors to PI hydrolysis in the same cell, *Cell,* 56, 487–493, 1988.
14. **Irvine, R. F., Brown, K. D., and Berridge, M. J.**, Specificity of inositol trisphosphate-induced calcium release from permeabilized Swiss-mouse 3T3 cells, *Biochem. J.,* 221, 269–272, 1984.
15. **Lapetina, E. G.**, Inositide-dependent and independent mechanisms in platelet activation, in *Phosphoinositides and Receptor Mechanisms,* Alan R. Liss, New York, 1986, 271–286.
16. **Berridge, M. J.**, Inositol lipids and cell proliferation, *Biochim. Biophys. Acta,* 907, 33–45, 1987.
17. **Irvine, R. F., Letcher, A. J., Heslop, J. P., and Berridge, M. J.**, The inositol tris/tetrakisphosphate pathway-demonstration of Ins(1,4,5)P3-kinase activity in animal tissues, *Nature,* 320, 631–634, 1986.
18. **Majerus, P. W., Connolly, T. M., Deckmyn, H., Ross, T. S., Bross, T. E., Ishii, H., Bansal, V. S., and Wilson, D.**, The metabolism of phosphoinositide-derived messenger molecules, *Science,* 234, 1519–1526, 1986.
19. **Peter-Riesch, B., Fathi, M., Schlegel, W., and Wolheim, C. B.**, Glucose and carbachol generate 1,2-diacylglycerols by different mechanisms in pancreatic islets, *J. Clin. Invest.,* 81, 1154–1161, 1988.
20. **Pelech, S. L. and Vance, D. E.**, Signal transduction via phosphatidylcholine cycles, *Trends Biochem. Sci.,* 1, 28–30, 1988.
21. **Kishimoto, A., Takai, Y., Mori, T., Kikkawa, U., and Nishizuka, Y.**, Activation of calcium and phospholipid-dependent protein kinase by diacylglycerol, its possible relation to phosphatidylinositol turnover, *J. Biol. Chem.,* 255, 2273–2276, 1980.
22. **Nishizuka, Y.**, Studies and perspectives of protein kinase C, *Science,* 233, 305–312, 1986.
23. **Nishizuka, Y.**, The molecular heterogeneity of protein kinase C and its implications for cellular regulation, *Nature,* 334, 661–665, 1988.
24. **Moolenaar, W., Kruijer, W., Tilly, B. C., Verlaan, I., Bierman, A. J., and de Laat, S. W.**, Growth factor-like action of phosphatidic acid, *Nature,* 323, 171–173, 1986.
25. **Murayama, T. and Ui, M.**, Phosphatidic acid may stimulate membrane receptors mediating adenylate cyclase inhibition and phospholipid breakdown in 3T3 fibroblasts, *J. Biol. Chem.,* 262, 5522–5529, 1987.
26. **Ross, R. and Raines, E. W.**, Platelet-derived growth factor; its role in health and disease, *Adv. Exp. Med. Biol.,* 234, 9–21, 1988.
27. **Heldin, C. H., Betsholtz, C., Claesson-Welsh, L., and Westermark, B.**, Subversion of growth regulatory pathways in malignant transformation, *Biochim. Biophys. Acta,* 707, 219–244, 1987.
28. **Matsui, T., Heidaran, M., Miki, T., Popescu, N., La Rochelle, W., Kraus, M., Pierce, J., and Aaronson, S. A.**, Isolation of a novel receptor cDNA establishes the existence of two PDGF receptor genes, *Science,* 243, 800–804, 1989.
29. **Coughlin, S. R., Escobedo, J. A., and Williams, L. T.**, Role of phosphatidylinositol kinase in PDGF receptor signal transduction, *Science,* 243, 1191–1194, 1989.
30. **Matsuoka, K., Fukami, K., Nakanishi, O., Kawai, S., and Takenawa, T.**, Mitogenesis in response to PDGF and bombesin is abolished by microinjection of antibody to PIP2, *Science,* 239, 640–643, 1988.
31. **Fukami, K., Matsuoka, K., Nakanishi, O., Yamakawa, A., Kawai, S., and Takenawa, T.**, Antibody to phosphatidylinositol 4,5-bisphosphate inhibits oncogene-induced mitogenesis, *Proc. Natl. Acad. Sci. U.S.A.,* 85, 9057–9061, 1988.
32. **Westermark, B. and Heldin, C. H.**, Similar action of platelet-derived growth factor and epidermal growth factor in the prereplicative phase of human fibroblasts suggests a common intracellular pathway, *J. Cell. Physiol.,* 124, 43–48, 1985.

33. **Carpenter, G.**, Receptors for epidermal growth factor and other polypeptide mitogens, *Annu. Rev. Biochem.*, 56, 881–913, 1987.
34. **Besterman, J. M., Watson, S. P., and Cuatrecasas, P.**, Lack of association of epidermal growth factor-, insulin-, and serum-induced mitogenesis with stimulation of phosphoinositide degradation in BALB/c 3T3 fibroblasts, *J. Biol. Chem.*, 261, 723–727, 1986.
35. **Ruggiero, M., Pierce, J. H., Fleming, T. P., Matsui, T., Di Fiore, P. P., and Aaronson, S. A.**, Intracellular second messenger formation induced by EGF, IX Int. Washington Spring Symposium (Abstr.), 1989.
36. **Macara, I.**, Activation of $^{45}Ca^{2+}$ influx and $^{22}Na^+/H^+$ exchange by epidermal growth factor and vanadate in A431 cells is independent of phosphatidylinositol turnover and is inhibited by phorbol ester and diacylglycerol, *J. Biol. Chem.*, 261, 9321–9327, 1986.
37. **Hepler, J. R., Nakahata, N., Lovenberg, T. W., DiGuiseppi, J., Herman, B., Earp, H. S., and Harden, T. K.**, Epidermal growth factor stimulates the rapid accumulation of inositol (1,4,5)-trisphosphate and a rise in cytosolic calcium mobilized from intracellular stores in A431 cells, *J. Biol. Chem.*, 262, 2951–2956, 1987.
38. **Takasu, N., Takasu, M., Yamada, T., and Shimizu, T.**, Epidermal growth factor (EGF) produces inositol phosphates and increases cytoplasmic free calcium in cultured porcine thyroid cells, *Biochem. Biophys. Res. Commun.*, 151, 530–534, 1988.
39. **Johnson, R. M. and Garrison, J. C.**, Epidermal growth factor and angiotensin II stimulate formation of inositol 1,4,5- and inositol 1,3,4-trisphosphate in hepatocytes, *J. Biol. Chem.*, 262, 17285–17293, 1987.
40. **Moscat, J., Molloy, C. J., Fleming, T. P., and Aaronson, S. A.**, Epidermal growth factor activates phosphoinositide turnover and protein kinase C in BALB/MK keratinocytes, *Mol. Endocrinol.*, 2, 799–805, 1988.
41. **Olashaw, N. E. and Pledger, W. J.**, Epidermal growth factor stimulates formation of inositol phosphates in BALB/c/3T3 cells pretreated with cholera toxin and isobutylmethylxanthine, *J. Biol. Chem.*, 263, 1111–1114, 1988.
42. **Earp, H. S., Hepler, J. R., Petch, L. A., Miller, A., Berry, A., Harris, J., Raymond, V. W., McCune, B. K., Lee, L. W., Grisham, J. W., and Harden, T. K.**, Epidermal growth factor (EGF) and hormone stimulate phosphoinositide hydrolysis and increase EGF receptor protein synthesis and mRNA levels in rat liver epithelial cells, *J. Biol. Chem.*, 263, 13868–13874, 1988.
43. **Gill, G. N. and Lazar, C. S.**, Increased phosphotyrosine content and inhibition of proliferation in EGF-treated A431 cells, *Nature*, 293, 305–307, 1981.
44. **Pandiella, A., Beguinot, L., Velu, T. J., and Meldolesi, J.**, Transmembrane signalling at epidermal growth factor receptors overexpressed in NIH 3T3 cells, *Biochem. J.*, 254, 223–228, 1988.
45. **Pandiella, A., Malgaroli, A., Meldolesi J., and Vicentini, L. M.**, EGF raises cytosolic Ca^{2+} in A431 and Swiss 3T3 cells by a dual mechanism, *Exp. Cell Res.*, 170, 175–185, 1987.
46. **Pierce, J. H., Ruggiero, M., Fleming, T. P., Di Fiore, P. P., Greenberger, J. S., Varticovski, L., Schlessinger, J., Rovera, G., and Aaronson, S. A.**, Signal transduction through the EGF receptor transfected in IL3-dependent hematopoietic cells, *Science*, 239, 628–631, 1988.
47. **Besterman, J. M., Duronio, V., and Cuatrecasas, P.**, Rapid formation of diacylglycerol from phosphatidylcholine: a pathway for generation of a second messenger, *Proc. Natl. Acad. Sci. U.S.A.*, 83, 6785–6789, 1986.
48. **Rosoff, P. M., Savage, N., and Dinarello, C. A.**, Interleukin-1 stimulates diacylglycerol production in T lymphocytes by a novel mechanism, *Cell*, 54, 73–81, 1988.
49. **Wijelath, E. S., Kardasz, A. M., Drummond, R., and Watson, J.**, Interleukin-one-induced inositol phospholipid breakdown in murine macrophages: possible mechanism of receptor activation, *Biochem. Biophys. Res. Commun.*, 152, 392–397, 1988.

50. **Libby, P., Warner, S. J., and Friedman, G. B.**, Interleukin 1: a mitogen for human vascular smooth muscle cells that induces the release of growth-inhibitory prostanoids, *J. Clin. Invest.*, 81, 487–498, 1988.
51. **Goldstein, R. H. and Wall, M.**, Activation of protein formation and cell division by bradykinin and des-arg-9-bradykinin, *J. Biol. Chem.*, 259, 9263–9268, 1984.
52. **Espinal J.**, Mechanism of insulin action, *Nature*, 328, 574–575, 1987.
53. **Farese, R. V., Nair, G. P., Standaert, M. L., and Cooper, D. R.**, Epidermal growth factor and insulin-like growth factor-1 stimulate the hydrolysis of the insulin-sensitive phosphatidylinositol-glycan in BC3H-1 myocytes, *Biochem. Biophys. Res. Commun.*, 156, 1346–1352, 1988.
54. **Pennington, S. R. and Martin, B. R.**, Insulin-stimulated phosphoinositide metabolism in isolated fat cells, *J. Biol. Chem.*, 260, 11039–11045, 1985.
55. **Farese, R. V., Konda, T. S., Davis, J. S., Standaert, M. L., Pollet, R. J., and Cooper, D. R.**, Insulin rapidly increases diacylglycerol by activating de novo phosphatidic acid synthesis, *Science*, 236, 586–589, 1987.
56. **Farese, R. V., Cooper, D. R., Konda, T. S., Nair, G. P., Standaert, M. L., and Pollet, R. J.**, Insulin provokes coordinate increases in the synthesis of phosphatidylinositol, phosphatidylinositol phosphates and the phosphatidylinositol-glycan in BC3H-1 myocytes, *Biochem. J.*, 256, 185–188, 1988.
57. **Tronick, S. R. and Aaronson, S. A.**, Oncogenes, growth regulation, and cancer, *Adv. Second Messenger Phosphoprotein Res.*, 21, 201–214, 1988.
58. **Barbacid, M.**, ras genes, *Annu. Rev. Biochem.*, 56, 779–827, 1987.
59. **Tamanori, F., Walsh, M., Kataoka, T., and Wigler, M.**, A product of yeast RAS2 gene is a guanine nucleotide binding protein, *Proc. Natl. Acad. Sci. U.S.A.*, 81, 6924–6928, 1984.
60. **Chiarugi, V., Porciatti, F., Pasquali, F., and Bruni, P.**, Transformation of BALB/3T3 cells with EJ/T24/H-*ras* oncogene inhibits adenylate cyclase response to beta-adrenergic agonist while it increases muscarinic receptor dependent hydrolysis of inositol lipids, *Biochem. Biophys. Res. Commun.*, 132, 900–907, 1985.
61. **Chiarugi, V., Ruggiero, M., and Porciatti, F.**, Oncogenes and transmembrane cell signalling, *Cancer Invest.*, 5, 215–229, 1987.
62. **Chiarugi, V. P., Pasquali, F., Vannucchi, S., and Ruggiero, M.**, Point-mutated p21^{ras} couples a muscarinic receptor to calcium channels and polyphosphoinositide hydrolysis, *Biochem. Biophys. Res. Commun.*, 141, 591–599, 1986.
63. **Wakelam, M. J. O., Davies, S. A., Houslay, M. D., McKay, I., Marshall, C. J., and Hall, A.**, Normal p21^{N-ras} couples bombesin and other growth factor receptors to inositol phosphate production, *Nature*, 323, 173–176, 1986.
64. **Marshall, C. J.**, Oncogenes and growth control 1987, *Cell*, 49, 723–725, 1987.
65. **Parries, G., Hoebel, R., and Racker, E.**, Opposing effects of *ras* oncogene on growth factor-stimulated phosphoinositide hydrolysis: desensitization to platelet-derived growth factor and enhanced sensitivity to bradykinin, *Proc. Natl. Acad. Sci. U.S.A.*, 84, 2648–2652, 1987.
66. **Downward, J., DeGunzburg, J., Riehl, R., and Weinberg, R. A.**, p21^{ras}-induced responsiveness of phosphatidylinositol turnover to bradykinin is a receptor number effect, *Proc. Natl. Acad. Sci. U.S.A.*, 85, 5774–5778, 1988.
67. **Seuwen, K., Lagarde, A., and Pouyssegur, J.**, Deregulation of hamster fibroblast proliferation by mutated *ras* oncogene is not mediated by constitutive activation of phosphoinositide-specific phospholipase C, *EMBO J.*, 7, 161–168, 1988.
68. **Ruggiero, M., Srivastava, S. K., Fleming, T. P., Ron, D., and Eva, A.**, NIH/3T3 fibroblasts transformed by the *dbl* oncogene show altered expression of bradykinin receptors, *Oncogene*, in press.
69. **Olinger, P. L. and Gorman, R. R.**, NIH/3T3 cells expressing high levels of the c-*ras* protooncogene display reduced platelet-derived growth factor-stimulated phospholipase activity, *Biochem. Biophys. Res. Commun.*, 150, 937–941, 1988.

70. **Wakelam, M. J. O.**, Inhibition of the amplified bombesin-stimulated inositol phosphate response in N-*ras* transformed cells by high density culturing, *FEBS. Lett.*, 228, 182–186, 1988.
71. **Chiarugi, V. P., Ruggiero, M., and Corradetti, R.**, Oncogenes, protein kinase C, neuronal differentiation and memory, *Neurochem. Int.*, 14, 1–9, 1988.
72. **Preiss, J., Loomis, C. R., Bishop, W. R., Stein, R., Niedel, J. E., and Bell, R. M.**, Quantitative measurement of sn-1,2-diacylglycerols present in platelets, hepatocytes, and *ras*- and *sis*-transformed normal rat kidney cells, *J. Biol. Chem.*, 261, 8597–8600, 1986.
73. **Fleischman, L., Chahwala, S. B., and Cantley, L.**, *Ras*-transformed cells: altered levels of phosphatidylinositol-4,5-bisphosphate and catabolites, *Science*, 231, 407–410, 1986.
74. **Lacal, J. C., Moscat, J., and Aaronson, S. A.**, Novel source of 1,2-diacylglycerol elevated in cells transformed by Ha-*ras* oncogene, *Nature*, 330, 269–272, 1987.
75. **Wolfman, A. and Macara, I. G.**, Elevated levels of diacylglycerol and decreased phorbol ester sensitivity in *ras*-transformed fibroblasts, *Nature*, 325, 359–361, 1987.
76. **Wolfman, A., Wingrove, T. G., Blackshear, P., and Macara, I. G.**, Down-regulation of protein kinase C and of an endogenous 80 kDa substrate in transformed fibroblasts, *J. Biol. Chem.*, 262, 16546–16552, 1987.
77. **Weyman, C. H., Taparowsky, E. J., Wolfson, M., and Ashendel, C. L.**, Partial down-regulation of protein kinase C in C3H 10T 1/2 mouse fibroblasts transfected with the human Ha-*ras* oncogene, *Cancer Res.*, 48, 6535–6541, 1988.
78. **Lacal, J. C., DeLa Pena, P., Moscat, J., Garcia-Barreno, P., Anderson, P. S., and Aaronson, S. A.**, Rapid stimulation of diacylglycerol production in Xenopus oocytes by microinjection of H-*ras* p21, *Science*, 238, 533–536, 1987.
79. **Kamata, T., Sullivan, N. F., and Wooten, M. W.**, Reduced protein kinase C activity in a *ras*-resistant cell line derived from Ki-MSV transformed cells, *Oncogene*, 1, 37–46, 1987.
80. **Hancock, J. F., Marshall, C. J., McKay, I. A., Gardner, S., Houslay, M. D., Hall, A., and Wakelam, M. J. O.**, Mutant but not normal p21^{ras} elevates inositol phospholipid breakdown in two different cell systems, *Oncogene*, 3, 187–193, 1988.
81. **Kamata, T. and Kung, H.**, Effects of *ras*-encoded proteins and platelet-derived growth factor on inositol phospho lipid turnover in NRK cells, *Proc. Natl. Acad. Sci. U.S.A.*, 85, 5799–5803, 1988.
82. **Huang, M., Chida, K., Kamata, N., Nose, K., Kato, M., Homma, Y., Takenawa, T., and Kuroki, T.**, Enhancement of inositol phospholipid metabolism and activation of protein kinase C in *ras*-transformed rat fibroblasts, *J. Biol. Chem.*, 263, 17975–17980, 1988.
83. **Macara, I. G.**, Elevated phosphocholine concentration in *ras*-transformed NIH 3T3 cells arises from increased choline kinase activity, not from phosphatidylcholine breakdown, *Mol. Cell. Biol.*, 9, 325–328, 1989.
84. **Korn, L. J., Siebel, C. W., McCormick, F., and Roth, R. A.**, Ras p21 as a potential mediator of insulin action in Xenopus oocytes, *Science*, 236, 840–843, 1987.
85. **Dotto, G. P., Parada, L. F., and Weinberg, R. A.**, Specific growth response of *ras*-transformed embryo fibroblasts to tumour promoters, *Nature*, 314, 459–462, 1985.
86. **Yuspa, S. H., Kilkenny, A. E., Stanley, J., and Lichti, U.**, Keratinocytes blocked in phorbol ester-responsive early stage of terminal differentiation by sarcoma viruses, *Nature*, 314, 459–462, 1985.
87. **Vasta, V., Bruni, P., and Farnararo, M.**, Mechanism of thrombin-induced rise in platelet fructose 2,6-bisphosphate content. Studies using phorbol myristate acetate, dioctanoylglycerol and ionophore A23187, *Biochem. J.*, 244, 547–551, 1987.
88. **Burgoyne, R. D., Cheek, T. R., and O'Sullivan, A. J.**, Receptor-activation of phospholipase A_2 in cellular signalling, *Trends Biochem. Sci.*, 9, 332–333, 1987.
89. **Bar-sagi, D. and Feramisco, J. R.**, Induction of membrane ruffling and fluid-phase pinocytosis in quiescent fibroblasts by *ras* proteins, *Science*, 233, 1061–1068, 1986.

90. Jakoweski, S., Rettenmier, C. W., Sherr, C. J., and Rock, C. O., A guanine nucleotide-dependent phosphatidylinositol 4,5-diphosphate phospholipase C in cells transformed by the v-*fms* and v-*fes* oncogenes, *J. Biol. Chem.*, 261, 4978–4985, 1986.
91. Chiarugi, V., Porciatti, F., Pasquali, F., Magnelli, L., Giannelli, S., and Ruggiero, M., Polyphosphoinositide metabolism is rapidly stimulated by activation of a temperature-sensitive mutant of Rous sarcoma virus in rat fibroblasts, *Oncogene*, 2, 37–40, 1987.
92. Van der Valk, J., Kerlaan, I., de Laat, S. W., and Moolenaar, W. H., Expression of pp60^{v-src} alters the ionic permeability of the plasma membrane in rat cells, *J. Biol. Chem.*, 262, 2431–2434, 1987.
93. Audigier, S. M., Wang, J. K., and Greengard, P., Membrane depolarization and carbamoylcholine stimulate phosphatidylinositol turnover in intact nerve terminals, *Proc. Natl. Acad. Sci. U.S.A.*, 85, 2859–2863, 1988.
94. Fry, M. J., Gebbhardt, A., Parker, P. J., and Foulkes, J. G., Phosphatidylinositol turnover and transformation of cells by Abelson murine leukaemia virus, *EMBO J.*, 4, 3173–3178, 1985.
95. Srivastava, S. K., Wheelock, R. H., Aaronson, S. A., and Eva, A., Identification of the protein encoded by the human diffuse B-cell lymphoma (*dbl*) oncogene, *Proc. Natl. Acad. Sci. U.S.A.*, 83, 8868–8872, 1986.
96. Eva, A., Vecchio, G., Rao, C. D., Tronick, S. R., and Aaronson, S. A., The predicted DBL oncogene product defines a distinct class of transforming proteins, *Proc. Natl. Acad. Sci. U.S.A.*, 86, 2061–2065, 1988.
97. Ron, D., Tronick, S. R., Aaronson, S. A., and Eva, A., Molecular cloning and characterization of the human *dbl* proto-oncogene: evidence that its overexpression is sufficient to transform NIH/3T3 cells, *EMBO J.*, 8, 2465–2473, 1988.
98. Auger, K. R., Serunian, L. A., Soltoff, S. P., Libby, P., and Cantley, L. C., PDGF-dependent tyrosine phosphorylation stimulates production of novel polyphosphoinositides in intact cells, *Cell*, 57, 167–175, 1989.
99. Berridge, M. J. and Irvine, R. F., Inositol phosphates and cell signalling, *Nature*, 341, 197–205, 1989.
100. Berridge, M. J., Inositol lipids and cell proliferation, *Biochim. Biophys. Acta*, 907, 33–45, 1987.
101. Cantley, L. C., Auger, K. R., Carpenter, C., Duckworth, B., Graziani, A., Kapeller, R., and Soltoff, S., Oncogenes and signal transduction, *Cell*, 64, 281–302, 1991.
102. Chiarugi, V., Bruni, P., Pasquali, F., Magnelli, L., Basi, G., Ruggiero, M., and Farnararo, M., Synthesis of diacylglycerol de novo is responsible for permanent activation and down-regulation of protein kinase C in transformed cells, *Biochem. Biophys. Res. Commun.*, 164, 816–823, 1989b.
103. Chiarugi, V., Porciatti, F., Pasquali, F., Magnelli, L., Giannelli, S., and Ruggiero, M., Polyphosphoinositide metabolism is rapidly stimulated by activation of a temperature-sensitive mutant of Rous sarcoma virus in rat fibroblasts, *Oncogene*, 2, 37–40, 1987.
104. Chiarugi, V., Basi, G., Quattrone, A., Micheletti, R., and Ruggiero, M., The old and the new in transformed cell signalling: glycolysis, diacylglycerol and protein kinase C, *Second Mess. Phosphoproteins*, 13, 69–85, 1990.
105. Chiarugi, V. P., Magnelli, L., Pasquali, F., Basi, G., and Ruggiero, M., Signal transduction in EJ-H-*ras*-transformed cells: de novo synthesis of diacylglycerol and subversion of agonist-stimulated inositol lipid metabolism, *FEBS Lett.*, 252, 129–134, 1989a.
106. Chiarugi, V. P., Pasquali, F., Vannucchi, S., and Ruggiero, M., Point-mutated p21*ras* couples a muscarinic receptor to calcium channels and polyphosphoinositide hydrolysis, *Biochem. Biophys. Res. Commun.*, 141, 591–599, 1986.
107. Choi, K. Y., Kim, H. K., Lee, S. Y., Moon, K. H., Sim, S. S., Kim, J. W., Chung, H. K., and Rhee, S. G., Molecular cloning and expression of a complementary DNA for inositol 1,4,5-trisphosphate 3-kinase, *Science*, 248, 64–66, 1990.
108. Coughlin, S. R., Escobedo, J. A., and Williams, L. T., Role of phosphatidylinositol kinase in PDGF receptor signal transduction, *Science*, 243, 1191–1194, 1989.

109. **Davis, P. D., Hill, C. H., Keech, E., Lawton, G., Nixon, J. S., Sedgwick, A. D., Wadsworth, J., Westmacott, D., and Wilkinson, S. E.**, Potent selective inhibitors of protein kinase C, *FEBS Lett.*, 259, 61–63, 1989.
110. **Downing, J. R., Margolis, B. L., Zilberstein, A., Ashmun, R. A., Ullrich, A., Sherr, C. J., and Schlessinger, J.**, Phospho ipase C-gamma, a substrate for PDGF receptor kinase is not phosphorylated on tyrosine during the mitogenic response to CSF-1, *EMBO J.*, 8, 3345–3350, 1989.
111. **Fukami, K., Matsuoka, K., Nakanishi, O., Yamakawa, A., Kawai, S., and Takenawa, T.**, Antibody to phosphatidylinositol 4,5-bisphosphate inhibits oncogene-induced mitogenesis, *Proc. Natl. Acad. Sci. U.S.A.*, 85, 9057–9061, 1988.
112. **Hunter, T.**, Protein-tyrosine phosphatases: the other side of the coin, *Cell*, 58, 1013–1016, 1989.
113. **Matsui, T., Pierce, J. H., Fleming, T. P., Greenberger, J. S., LaRochelle, W. J., Ruggiero, M., and Aaronson, S. A.**, Independent expression of human alpha or beta platelet-derived growth factor receptor cDNAs in a naive hematopoietic cell leads to functional coupling with mitogenic and chemotactic signaling pathways, *Proc. Natl. Acad. Sci. U.S.A.*, 86, 8314–8318, 1989.
114. **Matsuoka, K., Fukami, K., Nakanishi, O., Kawai, S., and Takenawa, T.**, Mitogenesis in response to PDGF and bombesin is abolished by microinjection of antibody to PIP2, *Science*, 239, 640–643, 1988.
115. **Megidish, T. and Mazurek, N.**, A mutant protein kinase C that can transform fibroblasts, *Nature*, 342, 807–810, 1989.
116. **Melloni, E., Pontremoli, S., Michetti, M., Sacco, O., Cakiroglu, A. G., Jackson, J. F., Rifkind, R. A., and Marks, P. A.**, Protein kinase C activity and hexamethylenebisacetamide-induced erythroleukemia cell differentiation, *Proc. Natl. Acad. Sci. U.S.A.*, 84, 5282–5286, 1987.
117. **Molloy, C. J., Bottaro, D. P., Fleming, T. P., Marshall, M. S., Gibbs, J. B., and Aaronson, S. A.**, PDGF induction of tyrosine phosphorylation of GTPase activating protein, *Nature*, 342, 711–714, 1989.
118. **Moolenaar, W. H., Bierman, A. J., Tilly, B. C., Verlaan, I., Defize, L. H. K., Honegger, A. M., Ullrich, A., and Schlessinger, J.**, A point mutation at the ATP-binding site of the EGF receptor abolishes signal transduction, *EMBO J.*, 7, 707–710, 1988.
119. **Murachi, T.**, Intracellular regulatory system involving calpain and calpastatin, *Biochem. Int.*, 18, 263–294, 1989.
120. **Nishizuka, Y.**, Studies and perspectives of protein kinase C, *Science*, 233, 305–312, 1986.
121. **Persons, D. A., Wilkinson, W. O., Bell, R. M., and Finn, O. Y.**, Altered growth regulation and enhanced tumorigenicity of NIH 3T3 fibroblasts transfected with protein kinase C-1 cDNA, *Cell*, 52, 447–458, 1988.
122. **Pierce, J. H., Ruggiero, M., Fleming, T. P., Di Fiore, P. P., Greenberger, J. S., Varticovski, L., Schlessinger, J., Rovera, G., and Aaronson, S. A.**, Signal transduction through the EGF receptor transfected in IL3-dependent hematopoietic cells, *Science*, 239, 628–631, 1988.
123. **Ruggiero, M., Srivastava, S. K., Fleming, T. P., Ron, D., and Eva, A.**, NIH/3T3 fibroblasts transformed by the *dbl* oncogene show altered expression of bradykinin receptors, *Oncogene*, 4, 767–771, 1989.
124. **Schollmeyer, J. E.**, Calpain II involvement in mitosis, *Science*, 240, 911–913, 1988.
125. **Shoji-Kasai, Y., Senshu, M., Iwashita, S., and Imahori, K.**, Thiol protease-specific inhibitor E-64 arrests human epidermoid carcinoma A431 cells at mitotic metaphase, *Proc. Natl. Acad. Sci. U.S.A.*, 85, 146–150, 1988.
126. **Smith, M. R., Liu, Y., Kim, H., Rhee, S. G., and Kung, H. S.**, Inhibition of serum- and *ras*-stimulated DNA synthesis by antibodies to phospholipase C, *Science*, 247, 1074–1077, 1990.

127. **Smith, M. R., Ryu, S. H., Suh, P. G., Rhee, S. G., and Kung, H. F.**, S-phase induction and transformation of quiescent NIH 3T3 cells by microinjection of phospholipase C, *Proc. Natl. Acad. Sci. U.S.A.*, 86, 3659–3663, 1989.
128. **Wang, K. K. W., Villalobo, A., and Roufogalis, B. D.**, Calmodulin-binding proteins as calpain substrates, *Biochem. J.*, 262, 693–706, 1989.

Part III
*Membrane Injury, Ions,
and Free Radicals*

Chapter 8

EFFECTS OF HYPOXIA ON NEURONAL MEMBRANES

George G. Somjen

TABLE OF CONTENTS

I.	Introductory Remarks and Definitions	182
II.	Some General Consequences of Tissue Hypoxia	183
III.	The Significance of Intracellular Free Calcium Ion Concentration, $[Ca^{2+}]_i$, for Cell Survival	184
IV.	Reversible Arrest of Function in Hypoxic CNS	186
V.	Spreading Depression (SD)-Like Membrane Response of Hypoxic Neurons	188
VI.	Prolonged SD-Like Depolarization Damages Neurons	192
VII.	The Role of Elevated $[Ca^{2+}]_i$ in Neuron Damage	195
VIII.	The Role of NMDA Receptors	196
IX.	Delayed (Post-Ischemic) Neuron Damage is Different from Hypoxic Neuron Damage	197
X.	Summary and Conclusion	198
Acknowledgments		199
References		199

ISBN 0-8493-8091-X
© 1993 by CRC Press, Inc.

I. INTRODUCTORY REMARKS AND DEFINITIONS

Deprived of oxygen or of oxidizable substrate, cells first cease to function and eventually die. It is obvious enough that energy-demanding functions must stop when energy is no longer available, but it is less clear why the cellular machinery cannot be re-started when the supply of energy is resumed. The questions of irreversible cell damage and its prevention lie at the core of the problems of heart attacks and cerebral strokes. The integrity of cell membranes appears to be a key to the survival of cells that are starved of energy.

It is customary to distinguish hypoxia of tissues from ischemia. Hypoxia is defined as the condition in which an organ or an organism with more or less intact circulation lacks oxygen while ischemia means insufficient perfusion with blood. We speak of anoxia when oxygen is totally absent and of hypoxia when the available oxygen is less than what is required for normal function. Ischemia deprives tissues not only of oxygen but also of glucose and other metabolic substrates, as well as of blood-borne regulators such as hormones. Moreover, if the flow of blood is arrested, the internal environment can no longer be controlled and potentially toxic metabolic waste products cannot be removed. Finally, many investigators contend that cells that have survived a period of ischemia may be damaged when the circulation is restored: the so-called reperfusion injury. Still, among the multiplicity of potentially injurious factors operating in ischemia, lack of metabolic energy appears to be the most damaging, while the various other factors are aggravating a bad situation.

All severely hypoxic cells die eventually, but some do so much before others. One might expect that the ability of cells to remain viable (or revivable) in the absence of oxygen is related to their ability to generate high energy phosphates by anaerobic glycolysis. Some tissues in mammals, and the brains of some poikilotherm organisms, indeed survive extended periods of hypoxia in this manner, deriving minimal energy needs from glycolysis.[53,54,86,116,128] In the mammalian brain, however, the capacity for anaerobic glycolysis appears too limited to insure survival by itself.

A much reduced rate of ATP formation may suffice for bare survival, if normal functions that require much energy are temporarily suspended. Hibernation and hypothermia are examples of survival at the expense of reduced function.[130] Experimentally, it has been shown that reducing the workload of perfused kidneys protects the organ against ischemic cell damage.[14] The protection of hypoxic central neurons by lowering the temperature can also be demonstrated experimentally.[4,104]

In some ways energy shortage affects all tissues similarly, but there are also very important differences among cells of various kinds which preclude a uniform solution to the treatment of hypoxia/ischemia of different organs. Rational clinical approaches will require a clear understanding of these differences. This chapter deals mainly with the neuron membranes in the mammalian CNS, but, to provide some background and to place the main topic in context, general reference will be made to other organs as well. More detailed information concerning ischemia of those other organs may be found elsewhere in this book.

II. SOME GENERAL CONSEQUENCES OF TISSUE HYPOXIA

An obvious difference between non-living and living systems is the continuous turnover of the structural constituents of the latter. By continually turning over the stuff of which they are made, live cells avoid the perils of corrosion and metal fatigue that limit the lifetime of inanimate engines. The longevity of living systems is, however, bought at the price of requiring energy not only for doing work, but also for bare survival even when idle. Cell membranes are made of phospholipids and proteins and the biosynthesis of both requires a steady supply of ATP, which serves as a source of both energy and of phosphate groups.[79,111,144] In the absence of oxygen or of oxidizable substrate the supply of ATP is compromised and the molecules that form cell membranes and cytoskeleton can no longer be made and, if their breaking down continues,[27] cells fall apart. The normal half-life of the protein and lipid molecules necessary to maintain the cells' solid structures varies from hours to months.[67,78] Since, however, energy shortage accelerates the breakdown of some of the essential constituents[23] the limit of viability cannot be deduced from measurements made under normal conditions.

The need to replace spent molecules may set the ultimate limit to cell survival, but in several tissues cells become irreversibly injured much before this limit is likely to have been reached. Regrettably, the very organs most needed for the survival of the organism are the most vulnerable. If the delivery of oxygen fails, brain, heart, kidneys, and liver succumb. Neurons of the mammalian central nervous system (CNS) are the most vulnerable of all cells, while glial cells survive much longer. Even among nerve cells there are considerable differences.[55] At normal body temperature some neurons in the brain of gerbils can irreversibly be damaged already after 5 min of ischemia.[28,64,65] By contrast, a small minority of the neurons in the spinal cord of cats appear undamaged after 50 min of experimental ischemia.[102,160] The ventricular muscle of the heart is reportedly revived to full function after 40 to 60 min of total ischemia,[61] and the limit for liver and kidney cells is reportedly similar.[22,80,149] Skeletal muscle can tolerate longer periods, and fibrous connective tissue survives the longest.

Because of their paramount importance, injury of the four vital organs has received much recent attention. Their cells differ greatly in structure and function, and each of these organs is being studied by different research teams. Nevertheless, discussions of hypoxic/ischemic/hypoglycemic cell injury revolved around similar potential causes of cell injury. Energy shortage is believed to trigger a sequence of biophysical and biochemical processes ultimately leading to degeneration, as if the cells of these four organs harbored similar mechanisms for self-destruction.

More than one hypothetical scheme has been proposed to explain the demise of cells during hypoxia/ischemia, and/or during reoxygenation/reperfusion. In a recent review of acute renal failure, Brezis et al.[13] compiled a list of the factors that have been blamed for cell injury, which includes:

1. Energy depletion
2. Cell swelling
3. Intracellular acidosis
4. Mitochondrial dysfunction
5. Phospholipid degradation
6. Elevated intracellular free calcium concentration
7. Depression of membrane-bound Na^+/K^+-ATPase
8. Altered substrate metabolism
9. Oxygen-derived free radicals
10. Lysosomal changes

With minor variations similar lists could have been made for failure of the brain, heart, or liver. Alterations of cell membranes play a central role in several of the proposed mechanisms of cell damage.

III. THE SIGNIFICANCE OF INTRACELLULAR FREE CALCIUM ION CONCENTRATION, $[Ca^{2+}]_i$, FOR CELL SURVIVAL

Numerous authors have suggested that excessive uptake of calcium into cells may cause cell damage.[35,36,62,91,125,132,134,164] Two different versions of the calcium overload hypothesis have been proposed. At first it was suggested that calcium accumulating in mitochondria could hinder oxidative metabolism, thus depriving the cell of its main source of ATP. Taking up Ca^{2+} is a normal function of mitochondria, but an overload could interfere with oxidative metabolism and excessive calcification could cause permanent damage.[164] The role of mitochondrial failure in hypoxic-ischemic cell necrosis has, however, been questioned.[34,35,60,132] The alternative thought is that Ca^{2+} ions stimulate biochemical reactions such as proteolysis, lipolysis, and lipid peroxidation, which, when carried to extreme, could cause the breakdown of cell constituents, especially cell membranes (Figure 1).[22,36,131-134]

It should be clearly kept in mind that the first version, proposing mitochondrial overload, implies an increase of **total** cell calcium content; whereas the second version, postulating accelerated breakdown of macromolecules, implies elevation of **free** cytosolic calcium concentration.

Two sets of key observations support the calcium theory of cell injury.

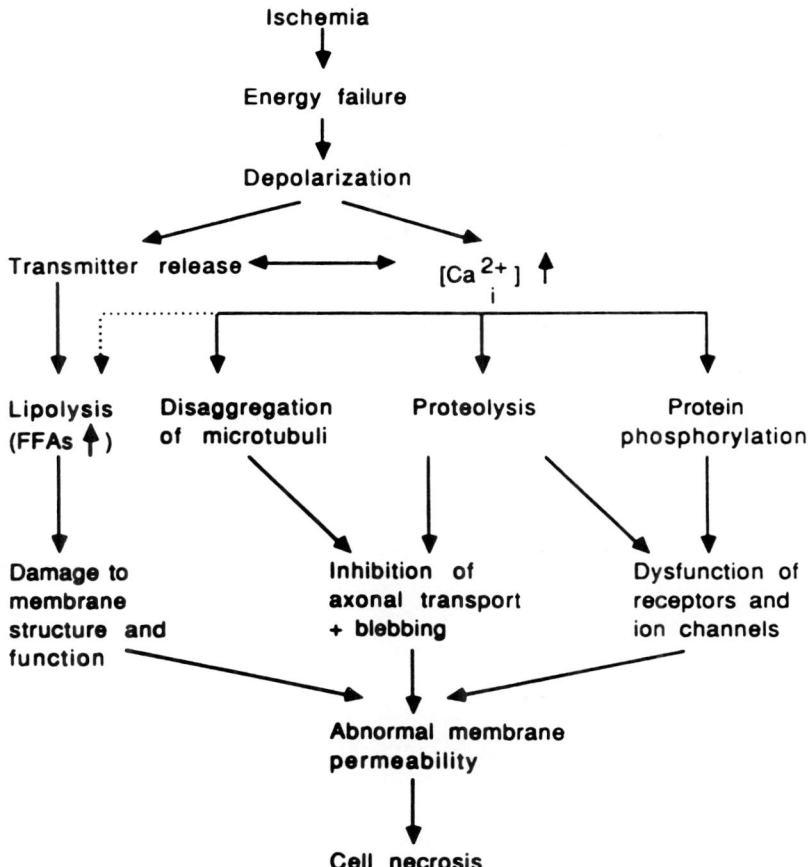

FIGURE 1. B. K. Siesjö's diagram of the pathways of cell destruction caused by metabolic energy shortage. (Siesjö, B. K., *Crit. Care Med.*, 16, 954–963, 1988. With permission.)

The first is that calcium has been shown to accumulate in cells, either during or after hypoxia.[22,35,44,56,57,60,91,135] The early studies relied on estimates of total, not of free, calcium. Only recenty have reliable methods become available to measure free calcium. While in ischemic heart muscle a rise of cytosolic free calcium has indeed been confirmed,[143] in renal and hepatic cells total (bound plus free) calcium increased, but free cytosolic calcium did not change during oxygen deprivation.[80,88,149] The second is that the accumulation of calcium could, of course, be either cause or effect of cell damage. A stronger argument in favor of a causal role is the fact that deleting calcium from the bathing medium can improve the chances for recovery of hypoxic cells from heart, liver, kidney, or brain, while any interference that increases calcium entry into cells worsens the outcome.[7,50,62,115,125,149]

Next we must explore the cause of the rise of $[Ca^{2+}]_i$ during hypoxia. The explanation that comes to mind first is that energy-dependent outward membrane transport of Ca^{2+} must fail during energy shortage.[23,132] In addition, however, it has been proposed that cell swelling causes an inward leaking of the ion.[60,61] Cells indeed swell during hypoxia, for two reasons, the first being that membrane pumps deprived of fuel fail to regulate the internal ion content of the cell which then will gain osmotic equivalents due to Gibbs-Donnan forces. The second reason is that with continuing catabolism abnormal amounts of osmotically active metabolites accumulate in cytoplasm. As long as there is no blood flowing through the tissue, osmotic swelling will be self-limiting as the inward movement of water will raise the extracellular osmotic pressure. Upon reperfusion, however, extracellular water becomes abundant so that cells can suddenly swell excessively. The stretched cell membrane is thought to become incontinent for ions and so large amounts of Ca^{2+} could enter the cytoplasm.[60,61] The concept of a generalized breakdown of the ionic barrier function of cell membranes has, however, been challenged.[29] Neurons differ from inexcitable cells in that calcium can gain access through specialized channels and through unique pathological processes, which are discussed below (see also Figure 7).

IV. REVERSIBLE ARREST OF FUNCTION IN HYPOXIC CNS

If the brain of a human subject is deprived of blood, the subject faints within 7 to 10 s,[118] a condition that is common and, fortunately, completely reversible. The electroencephalogram (EEG) of a fainting subject is dominated by slow waves in the delta frequency range, as described already by the pioneers of EEG.[10,108] Then, in case of complete ischemia, all EEG activity ceases in about 20 s. Spontaneous electrical activity ceases because synaptic transmission is depressed. At the time when the EEG becomes isoelectric, failure of synaptic transmission is not yet complete, for evoked potentials of a lowered amplitude can still be triggered. It is important to realize that the ability to transmit impulses is lost when synaptic potentials become too small to reach threshold, and this occurs much before synaptic potentials themselves become undetectable. Impulse transmission is blocked after 2 to 6 min of severe hypoxia or ischemia in different synapses.[6,15,33,52,82,83] This time, referred to as **arrest latency**[138] or *Lähnungszeit*,[52] may be expected to be longer in anoxia than in ischemia although, to my knowledge, such a difference has never been experimentally demonstrated.

I have recently reviewed the mechanism of the reversible failure of synaptic transmission[138] and will give here only a brief summary.

Many neurons undergo hyperpolarization during the initial phase of hypoxia, and this elevates the threshold of firing.[21,24-26,37,38,40,45,77,94,142] The hyperpolarization is generated by increased permeability of the cell membrane

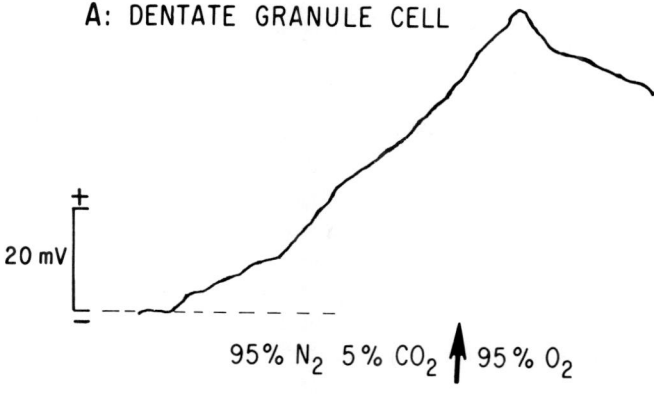

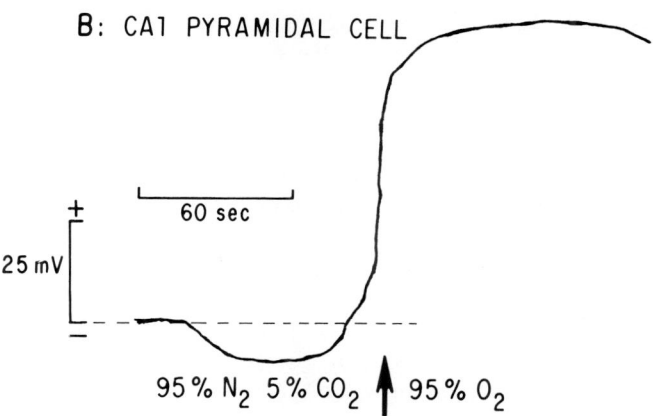

FIGURE 2. Membrane potential changes of neurons in two hippocampal tissue slices during hypoxia. Oxygen was withdrawn a short time before the start of the tracings and readmitted at the times marked by the arrows. Note that the cell in A shows only a gradual depolarization, while the cell in B shows an initial period of hyperpolarization and then SD-like depolarization. (Fast transients have been suppressed to emphasize slow membrane potential change). (From Somjen, G. G., Aitken, P. G., Balestrino, M., and Schiff, S. J., in *Brain Slices: Fundamentals, Applications and Implications,* S. Karger, Basel, 1986, 89–104. With permission.)

to K^+.[44,45,69] Since CO_2 and acidification also cause hyperpolarization,[21,142] it may be expected (but not proven) that it should be more intense in ischemia than in "pure" hypoxia. Hyperpolarization is, however, not observed in all cells during hypoxia (Figure 2),[11,21,37,71,140,142] even though transmission fails at all synapses.

The change that has been seen during hypoxia without exception at all junctions, excitatory and inhibitory, peripheral and central, spinal and cerebral, is a depression of synaptic potentials.[15,31,33,58,70,126,129] The mechanism could be either pre- or postsynaptic, as it could be due to a reduced release of transmitter or to interference with the action of the transmitter on postsynaptic receptors. Available evidence points to a presynaptic site.[20,33,71,113]

Early studies have led to the conclusion that EPSPs in the spinal cord[20,33] as well as EPPs at neuromuscular junctions[58,70] fail because of conduction block in presynaptic terminal arbors. Branch points are sites of low safety factor of propagation, and therefore impulses may be extinguished at such points while they are still normally conducted in the unbranched axon shaft. We have recently found that branch point failure indeed occurs in hypoxic primary afferent fibers.[152] The release of transmitter is, however, likely to be limited for an added reason.[141] In neurons deprived of either oxygen or of glucose inward calcium membrane currents are inhibited.[32,69,152] If calcium cannot enter presynaptic terminals when an impulse arrives there, then the terminal cannot release its transmitter.[165]

V. SPREADING DEPRESSION (SD)-LIKE MEMBRANE RESPONSE OF HYPOXIC NEURONS

Spreading cortical depression was discovered by Leão in the 1940s.[73,74] It consists of a rapid, profound depolarization of all cells, neurons as well as glia, (Figure 3)[18,26] that can be provoked at one point by a variety of strong stimuli, and then can spread slowly to cover the entire cortex of the stimulated hemisphere.[18,90] The onset of SD is often but not always marked by a brief paroxysmal burst of firing, after which all neuronal responses are extinguished. The depolarization can be detected by extracellular electrodes as a large (10 to 40 mV) negative potential shift that is sustained for about 0.5 to 1.5 min, and is sometimes preceded by a small positive wave and is usually followed by another low amplitude but more prolonged positive shift of the extracellular voltage.[18,74,75,90,159] As they depolarize, the cells release large amounts of potassium as well as certain organic compounds, and take up Na^+, Cl^-, Ca^{2+}, and water.[46,68,100,138,155,156,161] The swelling of cells results in a partial obliteration of interstitial spaces.[68,101,158] In the region affected by SD the blood flow increases and oxidative metabolism is stimulated.[46,84,117,159] The interstitial fluid becomes strongly acid, and the acidosis outlasts the negative potential shift.[46,139]

The almost complete loss of membrane potential (Figure 3) and the profound changes in the distribution of ions and other solutes between the in- and outside of cells bespeaks a drastic alteration of membrane properties. The nature of this change is not clear. At first it might seem that the barrier function of cell membranes breaks down completely, if transiently, yet this is not the case. If cell membranes became completely transparent to ions, then the electric resistance (or low-frequency impedance) of the tissue should decrease, but in fact it increases.[76,90,159] The increased electric resistance

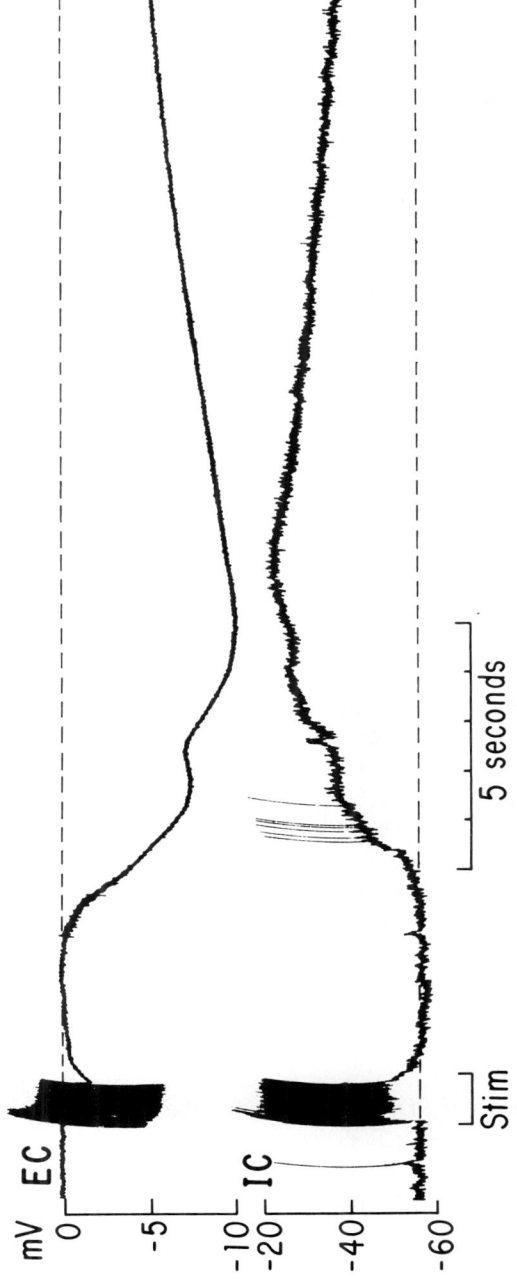

FIGURE 3. Simultaneous extracellular (EC) and intracellular (IC) recording from a hippocampal CA1 pyramidal cell during spreading depression (SD) provoked by an orthodromic stimulus train. Upstrokes of trace are action potentials, reduced in size because of slow frequency response of polygraph pen. Note that the membrane potential change is the algebraic sum of the EC and IC changes. (From Somjen, G. G. and Aitken, P. G., *An. Acad. Bras. Cienc.*, 56, 495–504, 1984. With permission.)

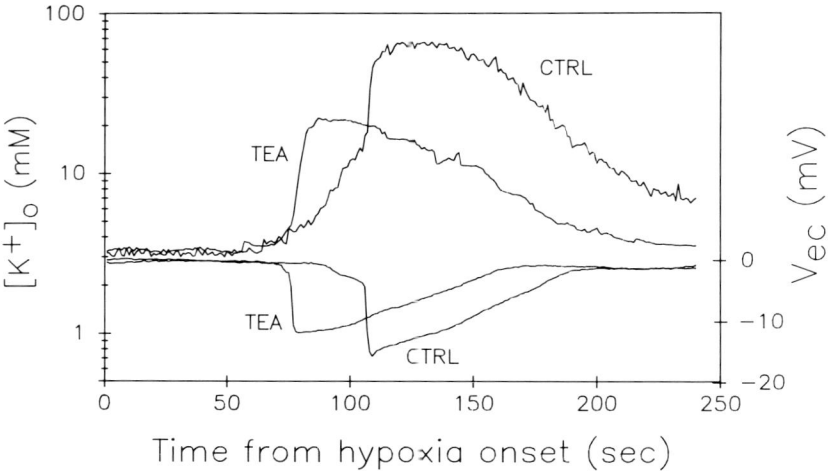

FIGURE 4. Tetra-ethyl-ammonium (TEA) accelerates the onset but reduces the amplitude of hypoxic SD-like depolarization. CTRL: control response; TEA: response of the same slice, in the presence of 10 mM TEA in the bath. Oxygen was returned when SD-like change reached maximum. Upper tracings and left-hand scale show interstitial potassium concentration, lower tracings and right-hand scale show extracellular voltage recorded by the reference barrel of the same electrode. (Unpublished experiment of P. G. Aitken; see also Reference 2.)

indicates restriction of the (extracellular) path for current flow by the swelling of cells whose membrane is still hindering ion currents. Moreover, Phillips and Nicholson[107] have shown that the channels admitting anions during SD, although larger than those acting during synaptic events, have a limited diameter. Moreover, the rise in extracellular potassium, $[K^+]_o$, is reduced in the presence of tetra-ethyl-ammonium (TEA) (Figure 4).[2] This suggests that some of the K^+ exits through normal voltage-sensitive channels[51] for if there had been complete breakdown of membrane barrier function, then TEA would have had no effect. More is not yet known about the membrane change underlying SD.

Among the stimuli that can consistently provoke SD are K^+ ions and glutamate. Both these agents are released from cells during SD. This combination of facts, that the same substances that can initiate SD are also released during SD, suggests a chemically mediated positive feedback mechanism of both the initiation and the spread of SD.[39,153,154] Whether the critical factor is K^+ or glutamate, or a combination of both, or now the one and at other times the other, is not clear. Curtis and Watkins[30] reported that n-methyl-D-asparate (NMDA) is 200 times more potent (in terms of molar concentration) than glutamate in provoking SD, and this has led to the supposition that the NMDA receptor is the specific target for initiating SD. Quisqualate and kainate, specific agonists for the two other known glutamate receptors are, however, also capable of triggering SD in comparable concentrations.[72]

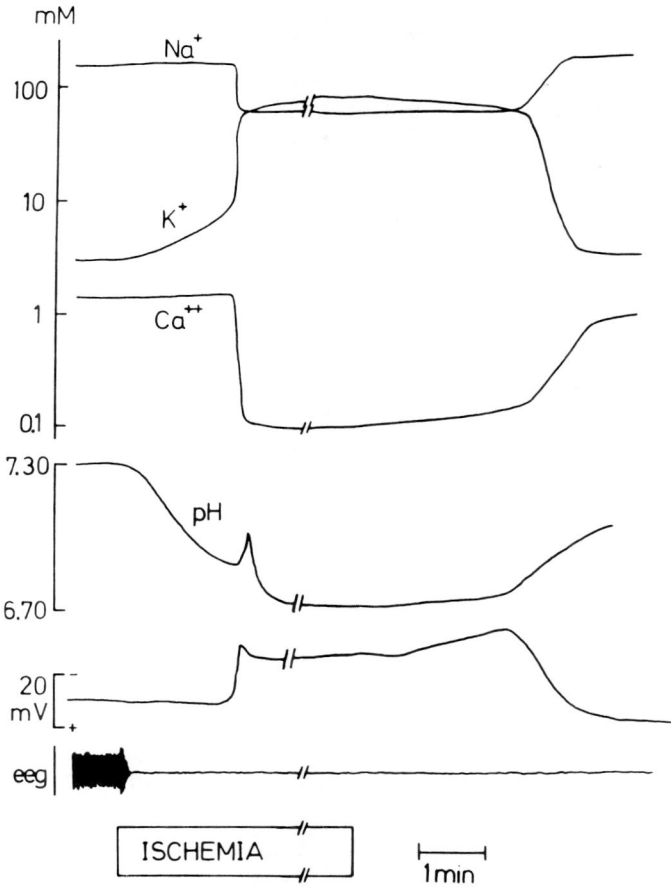

FIGURE 5. Interstitial ion levels and extracellular potential of rat cerebral neocortex during transient ischemia. (From Hansen, A. J. and Lauritzen, M., *An. Acad. Bras. Cienc.*, 56, 457–479, 1984. With permission.)

Leão was also the first to notice that arrest of the blood flow to the cerebral cortex results, in a few minutes, in a shift of potential that is very similar to that seen in SD.[74] Leão as well as Marshall considered the two phenomena, SD and the voltage shift of ischemia, essentially identical.[75,90] In agreement with this supposition, the changes in ion and water distribution are also virtually identical in SD of normoxic CNS and in the SD-like depolarization of hypoxic brain tissue (Figure 5).[44,46,100,101,139] Besides the cerebral neocortex, SD and SD-like hypoxic depolarization can be provoked in other CNS structures, and the likelihood of the one is matched by that of the other.[17] For example, hippocampus is highly susceptible to both SD and hypoxic SD (Figures 2 and 4), whereas the gray matter of the brainstem and the cerebellar cortex are relatively resistant to both.[17] Spinal cord was, until recently, believed to be incapable of generating SD, but it is now clear that, under rather special circumstances, oxygen deprivation can sometimes provoke an SD-like reponse in this structure also.[31]

Although some of the pioneers thought it likely that the mechanisms of the SD of normoxic tissue and the SD-like process during hypoxia are identical,[75,90] others thought of the two as distinct.[16,44,155] It has been pointed out that in "normal" SD the electric responses of neurons are extinguished at the moment when the steep depolarization sets in, whereas in hypoxia or ischemia neuronal responses are suppressed minutes earlier.[17] This, of course, does not negate the identity of the membrane change that brings about the depolarization, it simply stresses the fact that, in hypoxic tissue, a different mechanism interrupts synaptic transmission before the neurons depolarize (see Section IV). It has also been argued that, since in hypoxic tissue the depolarization starts simultaneously in the entire brain, it is not appropriate to call it "spreading" depression. But, as emphasized already by Marshall,[90] this need not be an essential distinction. As an analogy the space-clamped axon comes to mind: the underlying membrane process of the stationary action potential under a space-clamp is identical to that seen in a normally conducting axon.

A more important difference came to light more recently. Several teams found that blockade of NMDA receptors delays the onset of normoxic SD, raises its threshold and slows its spread, but NMDA blocking drugs had no effect on the SD-like depolarization caused by hypoxia or ischemia.[1,47,49,89] From this it seems that normoxic SD is triggered differently from hypoxic SD-like depolarization, but this conclusion does not illuminate the mechanism of the depolarization itself, which still may or may not be the same for both conditions. Because of the persistent doubt concerning the identity of the two mechanisms, many authors use different terms for the two. For example, "anoxic depolarization" (or AD), is frequently found in the literature. But "anoxic depolarization" is too general a term as it could refer to any loss of membrane potential, for example, to van Harreveld's spinal asphyxial potential,[25,155] which is clearly different from SD. For this reason we use "hypoxic SD-like depolarization", which is admittedly cumbersome, but unambiguous.

VI. PROLONGED SD-LIKE DEPOLARIZATION DAMAGES NEURONS

It has been suspected for some time that hypoxic SD-like depolarization is involved in causing neuron damage.[16] We have found that, following transient hypoxia, the chances for a restoration of synaptic function are inversely related to the length of time spent in hypoxic SD-like depolarization.[5-7] If oxygen is temporarily withdrawn from a hippocampal tissue slice preparation, the CA1 region of the slice undergoes SD-like depolarization before the

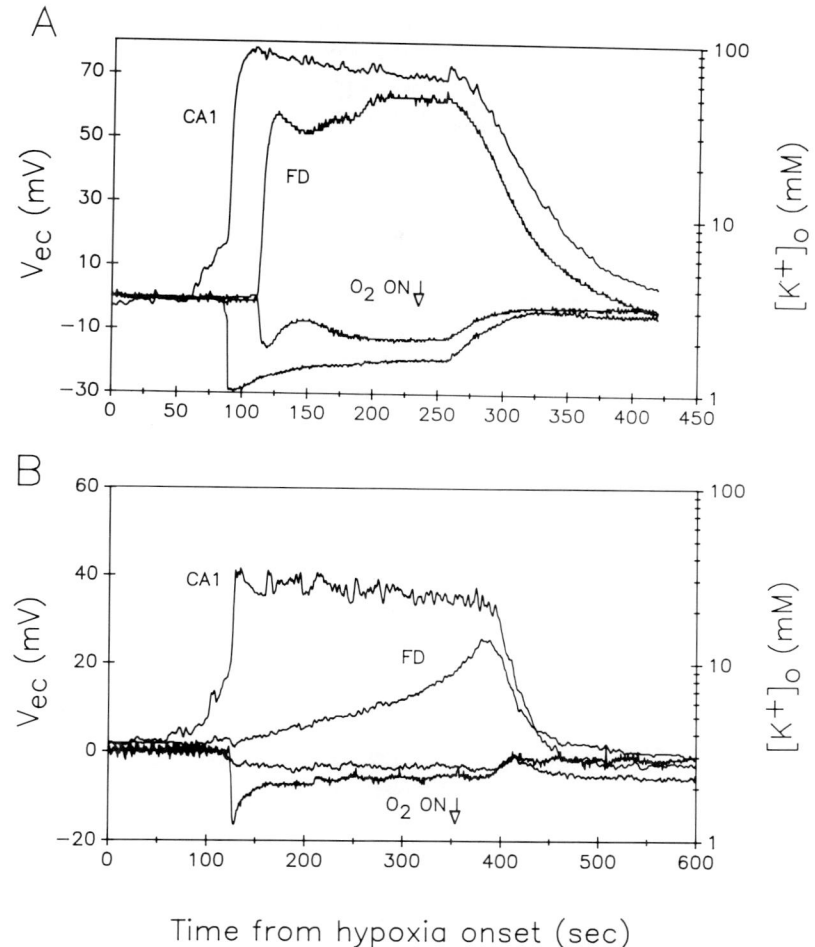

FIGURE 6. Hypoxic changes of CA1 region of hippocampus and fascia dentata (FD) compared. A and B are from two different experiments. Upper tracings and right-hand scales of both graphs show interstitial potassium concentration, lower tracings and left-hand scales show extracellular electric potential. Oxygen withdrawn at time zero, and returned at the time marked by arrows. In A SD-like depolarization started later in FD than in CA1, while in B no SD-like event occurred in FD. (From Balestrino, M., Aitken, P. G., and Somjen, G. G., *Brain Res.*, 497, 102–107, 1989. With permission.)

dentate gyrus (of fascia dentata) of the same slice (Figure 6). Following reoxygenation, synaptic function is more likely to recover in DG than in CA1, and the chances of recovery are inversely correlated with the time spent in the SD-like state in both regions.[6] Drugs such as chlorpromazine[7,8] and gangliosides[5] can protect hippocampal slices against irreversible loss of function to the same degree to which they are able to shorten the time spent in SD-like depolarization.[5,7] By contrast, drugs that block NMDA receptors were not significantly able to delay the onset of hypoxic SD[1] (as also reported by others: References 47, 49, and 89) and, correspondingly, did not offer protection against irreversible loss of function.[1] Rader and Lanthorn[112] tested the effect of depriving hippocampal tissue slices of both oxygen and glucose. Their findings support a relationship between persistent loss of function and prolonged SD-like depolarization but, at variance with others, they found that NMDA antagonists, while not delaying the onset of SD-like depolarization, do prevent the delayed, persistent phase of depolarization and also favor recovery of neuron function.

It has also been shown that brains of juvenile animals are less prone to hypoxic SD than older animals,[42] and it is well known that infantile brains tolerate hypoxia better than older ones.

One observation apparently disagrees with the idea that SD-like depolarization is the main culprit in hypoxic cell damage. Hypoglycemia accelerates, and hyperglycemia retards, the onset of hypoxic SD,[8,43] whereas it has long been known that hyperglycemia increases the risk of brain damage from ischemia.[96,97,131] More recently, however, four different investigating teams have shown that, unlike brain *in situ*, brain slices maintained in a medium containing elevated glucose concentrations tolerate hypoxia better than those in low glucose.[24,103,124,163] The apparently opposite results obtained in experiments on the same tissue *in vitro* and *in situ* is explained by the realization that the deleterious effect of elevated glucose is mediated by increased lactacidosis. Lactic acid content can increase especially if ischemia is incomplete so that glucose continues to trickle into the tissue, and also during reperfusion of the tissue, if ischemia has been transient. The high lactic acid then causes edema and other mischief that damage neurons independently from SD-like depolarization.[96,137] Unlike intact brain, in isolated brain slices edema and blood vessel injury cause no problem, and elevated lactic acid causes no damage.[127]

The idea that SD causes cell damage seems also contradicted by the fact that, in normally oxygenated and perfused cerebral cortex, repeated bouts of SD are well tolerated.[19,98] In these experiments,[19,98] however, the recurrent waves of SD did not last longer than about 1 or 1.5 min. When well-oxygenated hippocampal slices were forced to remain in SD for several minutes by prolonged exposure to very high K^+ concentration, they too suffered lasting loss of function.[50,63]

These observations lead to the conclusions that, while neither normoxic SD nor hypoxic SD-like depolarization cause any lasting damage if terminated within a short enough time, both cause irreversible damage once a critical period of time has been exceeded.

VII. THE ROLE OF ELEVATED $[Ca^{2+}]_i$ IN NEURON DAMAGE

Accepting that prolonged SD and SD-like hypoxic depolarization damage neurons, the next question concerns the mechanism of the damage. We have seen earlier that SD in normoxic tissue greatly stimulates oxidative metabolism.[84,117] Since the supply of energy is denied in hypoxic tissue, the extra demand represented by the depolarized state will accelerate the dwindling of high energy phosphates. Besides, cells swell and the associated stretching of membranes has been suspected of being a lethal factor in and of itself.[120] Finally there is the entrance of large amounts of Ca^{2+} from interstitial fluid into the cytoplasm.

These three factors need not be mutually exclusive, and indeed the one can lead to the other, as we have seen earlier. Convincing experimental evidence points, however, to the rise of $[Ca^{2+}]_i$ as the main damaging event. Deleting Ca^{2+} from the bathing fluid before rendering the tissue hypoxic can protect hippocampal tissue slices against irreversible hypoxic loss of function, while raising external Ca^{2+}, $[Ca^{2+}]_o$, worsens the outcome.[7,62,115] It is important to note that the deficiency of Ca^{2+} does not prevent the SD-like depolarization.[7] If cells do depolarize in hypoxia in the absence of Ca^{2+} as in a normal bath medium, then they are likely to swell also. As Ca^{2+}-deficient cells nonetheless recover function, cell swelling cannot, in itself, cause their demise. There is further evidence implicating $[Ca^{2+}]_i$, in the observation that iontophoretic intracellular injection of Ca^{2+}, but not of Mg^{2+}, irreversibly damages neurons.[5]

A different view has been expressed by Rothman and Olney[120-122] who concluded that the immediate effect of severe hypoxia, what they call "fulminating" neuron damage, is caused by cell swelling and not by elevated $[Ca^{2+}]_i$. To high $[Ca^{2+}]_i$ they attribute a role only in the more slowly evolving delayed neuron damage. It seems likely that there are a number of processes that can injure cells, and the particular one that is set in motion may depend on the experimental conditions. For example, SD-like depolarization requires a certain minimal mass of closely packed cells and therefore it may never occur in tissue culture. For this reason the *coup de grâce* may come from a different event to hypoxic cell cultures and to hypoxic tissue slices. In tissue slices, moreover, the mechanism may be modified by the age of the animal, the temperature, the degree of hypoxia, the composition of the bathing medium, and whether the slice is submerged or exposed at the fluid-gas interface. There is more than one mechanism by which $[Ca^{2+}]_i$ may become elevated (Figure 7), and there are ways other than elevating $[Ca^{2+}]_i$ by which cells can be injured. There may also be differences among clinical syndromes and, clearly, more work will be required to determine which experimental model fits which clinical situation. What our experience does clearly show[5-7] is this: **if** SD-like depolarization occurs in the presence of normal $[Ca^{2+}]_o$, **and** if it is protracted beyond a certain critical time, it **will** injure neurons.

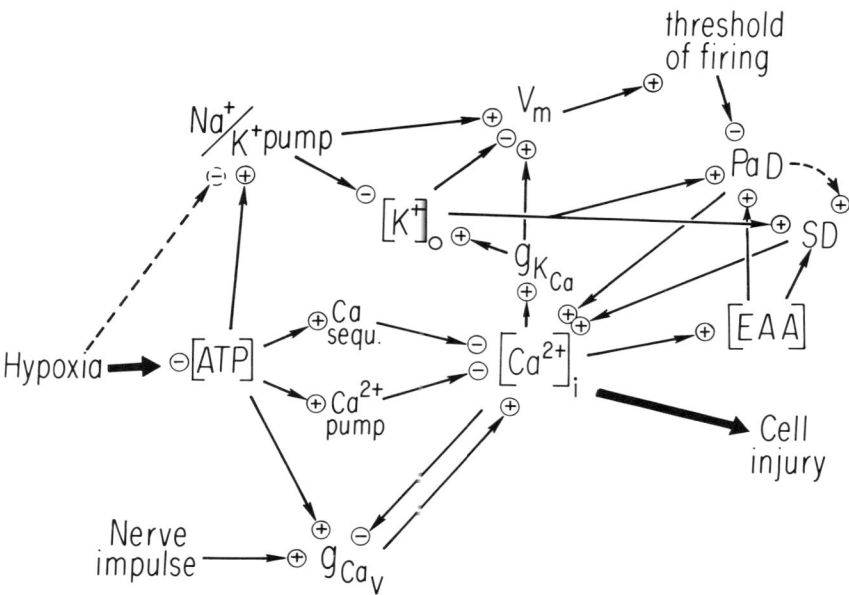

FIGURE 7. Conceptual diagram of the multiple interactive processes relating to intracellular free calcium initiated by hypoxia of neurons. Arrows indicate causal interactions; positive and negative signs indicate enhancement or depression. Hypoxia causes a decrease of available cellular ATP and phosphocreatine ([ATP]). The decline of high energy phosphates results in a depression of active transport of ions (Na^+/K^+ pump: Na K activated membrane ATPase; and Ca^{2+} pump) as well as impaired intracellular calcium sequestration (Ca_{sequ}) and therefore in an increase of both interstitial potassium ($[K^+]_o$) and intracellular calcium ($[Ca^{2+}]_i$). Elevated $[Ca^{2+}]_i$ activates calcium-dependent potassium conductance of neuronal membranes, ($g_{K,Ca}$), causing increase of membrane potential, (V_m), and raising of the threshold of firing. The initial hyperpolarization is then overcome by the accumulating $[K^+]_o$, which now causes depolarization and reduction of threshold. Meanwhile the elevated $[Ca^{2+}]_i$ causes inactivation of voltage-dependent calcium conductance of presynaptic terminals ($g_{Ca,V}$) and hence blockade of synaptic transmission; but later the elevated $[Ca^{2+}]_i$ and the reduced V_m cooperate in causing the abnormal release of excitatory amino acids (EAA) from both synaptic and metabolic transmitter pools. Overflow of EAAs into interstitial fluid, together with the reduced threshold of firing can trigger paroxysmal discharge (PaD) and spreading depression (SD). Both PaD and SD cause further elevation of $[Ca^{2+}]_i$, closing a vicious circle and the eventual activation of the destructive processes depicted in Figure 1. (From Somjen, G. G., in *Mechanisms of Cerebral Hypoxia and Stroke*, Plenum Press, New York, 1988, 447–466. With permission.)

VIII. THE ROLE OF NMDA RECEPTORS

In 1957 Lucas and Newhouse reported that glutamate can be toxic to nerve cells.[85] Over the years this observation has led to the **excitotoxic hypothesis** of neuron damage.[91-93,105,1-9,121,122,157] It is suggested that excessive release of glutamate or related acidic amino acids is responsible for the death of neurons in a number of conditions including status epilepticus, hypoglycemia, ischemia, and hypoxia of the brain. What these conditions have in common is a relative shortage of metabolic energy. In the case of epilepsy this is the result of cells firing "in overdrive" and thus energy demand exceeding metabolic capacity, in spite of vasodilatation.

To explain the toxicity of these compounds, two different ideas are extant. (1) With the opening of ion channels, neurons take up Na^+ and Cl^- and with these ions osmotic water,[120] as has earlier been suggested for cardiac cells.[60,61] (2) Alternatively, it has been suggested that excitotoxic agents induce the uptake of a toxic excess of Ca^{2+}. Of the various known glutamate receptors, it is the NMDA receptor whose action admits Ca^{2+} ions into neurons[110] and, according to some investigators, it is also the NMDA receptor whose excessive activation causes neuron damage.[81,92,136]

At first sight the excitotoxic hypothesis and the idea that prolonged SD causes neuron damage fit well, for glutamate and analog compounds are long known to trigger SD.[72,153] Since SD causes both cell swelling and an increase of $[Ca^{2+}]_i$, it is compatible with both versions of the excitotoxic hypothesis. Some experimental observations do not, however, support that hypoxic SD-like depolarization and hypoxic neuron damage are triggered by the NMDA receptor.

There are three cornerstones to the excitotoxic hypothesis:

1. The demonstrable toxicity of high enough concentrations of excitatory amino acids[85,105,157]
2. The release of glutamate and aspartate into interstitial fluid in ischemic CNS[9,41]
3. Drugs that block the NMDA receptor save neurons subjected to hypoxia or ischemia

It is this last point where we encounter difficulty. For while under certain conditions NMDA blocking drugs did have a protective effect[24,81,112,123,136] under other conditions no protection against hypoxic or ischemic cell death could be seen.[1,12,87,99] Moreover, while these drugs do block or delay SD in normoxic tissue, they have no effect on the SD-like depolarization in hypoxia and ischemia[1,47,49,89] (but see Reference 112). In the next section we will try to resolve these seemingly contradictory findings.

IX. DELAYED (POST-ISCHEMIC) NEURON DAMAGE IS DIFFERENT FROM HYPOXIC NEURON DAMAGE

When blood flow returns to cerebral tissue that has been deprived of blood for a while, the internal environment of the brain can deteriorate so that it becomes hostile to, instead of being supportive of, neurons. This condition can be brought about by malfunction of cerebral blood vessels;[3,95,147,148,162] lactacidosis;[96,97,137] cerebral edema;[137,145] and seizures.[137] Then there is the oft-quoted but controversial formation of free radicals and peroxide when oxygen is readmitted to oxygen-starved cells.[59,66,114,131]

While not denying the threat posed by these various extraneuronal changes, we have recently found that transient ischemia initiates changes in vulnerable neurons that proceed inexorably, independently from environmental factors.[150] This became clear from experiments performed on hippocampal tissue slices prepared from gerbils that previously underwent a brief episode of transient cerebral ischemia. The CA1 region in such tissue slices suffered delayed degeneration even in the well-controlled, supportive environment of a tissue chamber. When post-ischemic tissue slices were placed in a tissue chamber, they first appeared to recover normal function, but within a few hours the CA1 region became excessively excited and thereafter it lost the ability to generate neuronal responses. No SD-like potential shift was seen in these post-ischemic hippocampal slices. When post-ischemic slices were bathed in a medium containing an NMDA antagonist compound, they retained function and did not deteriorate as untreated slices did.[151]

Post-ischemic hyperactivity of neurons has been reported already earlier in experiments on brain *in situ*.[146] Our recent work[150,151] confirms that excitation mediated by NMDA receptors is important in causing delayed post-ischemic neuron damage. Whether or not this form of degeneration is mediated by elevated $[Ca^{2+}]_i$, as is the case with (immediate) hypoxic damage, we cannot yet state.

The seemingly divergent results reported by different teams concerning the protection against ischemic/hypoxic damage by NMDA blocking agents may be reconciled by the just-summarized findings. It seems that NMDA blocking drugs have little or no effect against the damage that neurons suffer **during** hypoxia,[1] but they are very effective in protecting against **delayed** (post-ischemic) neuron degeneration.[151] This indicates that, in cases when NMDA blocking drugs proved protective, the damage occurred in the post-ischemic phase, whereas in cases where the drugs were ineffective, the damage was done already during the hypoxic or ischemic period.

X. SUMMARY AND CONCLUSIONS

When cells of any of the vital organs of mammals are deprived of energy, they first cease to function, and then suffer irreversible damage. The initial, reversible arrest of function is believed to offer some protection against damage by reducing the demand for metabolic energy. Irreversible damage has been attributed to failure of the plasma membrane. Two hypothetical mechanisms of hypoxic/ischemic cell damage are most discussed. One blames stretching of the membrane due to osmotic cell swelling; the other to an inflow of an excess of calcium into cells, leading to a cascade of self-destructive biochemical reactions.

As do other cells, neurons of the central nervous system respond to energy shortage caused by hypoglycemia, hypoxia, or ischemia, by arresting function. The function of CNS neurons is arrested by a failure of synaptic transmission, mediated by two mechanisms: (1) the release of transmitter from presynaptic terminals is curtailed either because impulses fail to propagate past branch points in the terminal arbor of presynapic axons, or because of a blockade of voltage-dependent inward calcium currents; (2) some, but not all, neurons undergo K^+-dependent hyperpolarization, which raises the postsynaptic threshold of firing.

Neurons in the brain are more vulnerable to irreversible hypoxic damage than are other cells of the body, and even among brain cells there is wide variation in tolerance of hypoxia. Selective vulnerability by hypoxic damage appears to be related to susceptibility to a response peculiar to brain cells: the profound depolarization that resembles spreading cortical depression of Leão. Hypoxic SD-like depolarization causes both, cell swelling and elevation of $[Ca^{2+}]_i$. Of these two, the rise of $[Ca^{2+}]_i$ appears to be most injurious. If it persists beyond a critical length of time, neurons are irreversibly injured. Thus, between the early, reversible, and the later, irreversible consequences of hypoxia and ischemia there is a striking contrast. While at first voltage-dependent Ca^{2+}-currents are blocked and synaptic transmitter release is depressed, later large amounts of Ca^{2+} are admitted into cells and excitatory amino acids overflow into interstitial fluid.

Delayed neuron death is distinct from the immediate damage done during hypoxia/ischemia. The delayed damage is caused in part by a deterioration of the internal environment of the brain and in part by intraneuronal processes that proceed independently from environmental changes. SD-like depolarization plays no part in delayed neuron damage. The role of NMDA receptors in immediate hypoxic damage is controversial, but there is good evidence that their activation is important for delayed neuron damage, because drugs that block NMDA receptors have a powerful protective effect.

ACKNOWLEDGMENTS

I thank Drs. P. G. Aitken and L. J. Mandel for critical reading of the draft manuscript and for helpful discussions, and Mrs. M. Andrews for editorial assistance.

REFERENCES

1. **Aitken, P. G., Balestrino, M., and Somjen, G. G.,** NMDA antagonists: lack of protective effect against hypoxic damage in CA1 region of hippocampal slices, *Neurosci. Lett.*, 89, 187–192, 1988.
2. **Aitken, P. G. and Somjen, G. G.,** Tetraethyl ammonium attenuates hypoxia-induced spreading depression in hippocampal slices, *Abstr. Soc. Neurosci.*, 14, 185, 1988.

3. **Ames, A., Wright, R. L., Kowada, M , Thurston, J. M., and Majno, G.**, Cerebral ischemia. II. The no-reflow phenomenon, *Am. J. Pathol.*, 52, 437–453, 1968.
4. **Astrup, J., Rehncrona, S., and Siesjö, B. K.**, The increase in extracellular potassium concentration in the ischemic brain in relation to the pre-ischemic activity and cerebral metabolic rate, *Brain Res.*, 199, 161–174, 1980.
5. **Balestrino, M., Aitken, P. G., Jones, L. S., and Somjen, G. G.**, The role of spreading depression-like hypoxic depolarization in irreversible neuron damage, and its prevention, in *Mechanisms of Cerebral Hypoxia and Stroke*, Somjen, G. G., Ed., Plenum, New York, 1988, 291–301.
6. **Balestrino, M., Aitken, P. G., and Somjen, G. G.**, Spreading depression-like hypoxic depolarization in CA1 and fascia dentata of hippocampal slices: relationship to selective vulnerability, *Brain Res.*, 497, 102–107, 1989.
7. **Balestrino, M. and Somjen, G. G.**, Chlorpromazine protects brain tissue in hypoxia by delaying spreading depression-mediated calcium influx, *Brain Res.*, 385, 219–226, 1986.
8. **Benešová, O., Burešová, O., and Bureš, J.**, Die Wirkung des Chlorpromazins und der Glykämie auf das elektrophysiologisch kontrollierte Überleben der Himrinde bei verschiedenen Körpertemperaturen, *Arch. Exp. Pathol. Pharmakol.*, 231, 550–561, 1957.
9. **Benveniste, H., Drejer, J., Schousboe, A., and Diemer, N. H.**, Elevation of the extracellular concentrations of glutamate and aspartate in rat hippocampus during transient cerebral ischemia monitored by intracerebral microdialysis, *J. Neurochem.*, 43, 1369–1374, 1984.
10. **Berger, H.**, Über das Elektrenkephalogram des Menschen. Neunte Mitteilung, *Arch. Psychiatr. Nervenkr.*, 102, 538–557, 1934.
11. **Bingmann, D., Kolde, G., and Lipinski, H. G.**, Relations between P_{O_2} and neuronal activity in hippocampal slices, *Adv. Exp. Med. Biol.*, 169, 215–255, 1984.
12. **Block, G. A., and Pulsinelli, W. A.**, Excitatory amino acid and purinergic transmitter involvement in ischemia-induced selective neuronal death, in *Mechanisms of Cerebral Hypoxia and Stroke*, Somjen, G. G., Ed., Plenum Press, New York, 1988, 359–365.
13. **Brezis, M., Rosen, S., and Epstein, F. H.**, Acute renal failure, in *The Kidney*, Brenner, B. M. and Rectors, F. C., Eds., W. B. Saunders, Philadelphia, 1986, 735–799.
14. **Brezis, M., Rosen, S., Spokes, K., Silva, P., and Epstein, F. H.**, Transport-dependent anoxic cell injury in the isolated rat kidney, *Am. J. Pathol.*, 116, 327–341, 1984.
15. **Brooks, C. McC. and Eccles, J. C.**, A study of the effects of anaesthesia and asphyxia on the monosynaptic pathway through the spinal cord, *J. Neurophysiol.*, 10, 349–360, 1947.
16. **Bureš, J. and Burešová, O.**, Die anoxische Terminaldepolarisation als Indicator der Vulnerabilität der Grosshirnrinde bei Anoxie und Ischämie, *Pfluegers Arch.*, 264, 325–334, 1957.
17. **Bureš, J. and Burešová, O.**, Susceptibility to spreading depression and anoxia: regional differences and drug control, in *Mechanisms of Cerebral Hypoxia and Stroke*, Somjen, G. G., Ed., Plenum Press, New York, 1988, 253–267.
18. **Bureš, J., Burešová, O., Křivánek, J.**, *The Mechanism and Applications of Leão's Spreading Depression of Electroencephalographic Activity*, Academic Press, New York, 1974.
19. **Burešová, O. and Bureš, J.**, The effect of prolonged cortical spreading depression on learning and memory in rats, *J. Neurobiol.*, 1, 135–146, 1969.
20. **Carregal, E. J. A.**, The site of anoxic block in the spinal monosynaptic pathway, *J. Neurobiol.*, 6, 103–113, 1975.
21. **Caspers, H., Speckmann, E.-J., and Lehmenkühler, A.**, DC potential of the cerebral cortex. Seizure activity and changes in gas pressure, *Rev. Physiol. Biochem. Pharmacol.*, 106, 127, 1987.
22. **Chien, K. R., Abrams, J., Pfau, R. G., and Farber, J. L.**, Prevention by chlorpromazine of ischemic liver cell damage, *Am. J. Pathol.*, 88, 539–558, 1977.

23. **Chien, K. R., Abrams, J., Serroni, A., Martin, J. T., and Farber, J. L.**, Accelerated phospholipid degradation and associated membrane dysfunction in irreversible, ischemic liver cell injury, *J. Biol. Chem.*, 253, 4809–4817, 1978.
24. **Clark, G. D. and Rothman, S. M.**, Blockade of excitatory amino acid receptors protects anoxic hippocampal slices, *Neuroscience*, 21, 665–671, 1987.
25. **Collewijn, H. and van Harreveld, A.**, Intracellular recording from cat spinal motoneurones during acute asphyxia, *J. Physiol. (London)*, 185, 1–4, 1966.
26. **Collewijn, H. and van Harreveld, A.**, Membrane potential of cerebral cortical cells during spreading depression and asphyxia, *Exp. Neurol.*, 15, 425–436, 1966.
27. **Cook, J. S., Will, P. C., Proctor, W. R., and Brake, E. T.**, Turnover of ouabain binding sites and plasma membrane proteins in HeLa cells, in *Biogenesis and Turnover of Membrane Macromolecules*, Cook, J. S., Ed., Raven Press, New York, 1976, 15–36.
28. **Crain, B. J., Westerkam, W. D., Harrison, A. H., and Nadler, J. V.**, Selective neuronal death after transient forebrain ischemia in the mongolian gerbil: a silver impregnation study, *Neuroscience*, 27, 387–402, 1988.
29. **Crake, T. and Poole-Wilson, P. A.**, Evidence that calcium influx on reoxygenation is not due to cell membrane disruption in the isolated rabbit heart, *J. Mol. Cell. Cardiol.*, 18 (Suppl. 4), 31–36, 1986.
30. **Curtis, D. R. and Watkins, J. C.**, Acidic amino acids with strong excitatory actions on mammalian neurons, *J. Physiol. (London)*, 166, 1–14, 1963.
31. **Czéh, G. and Somjen, G. G.**, Hypoxic failure of synaptic transmission in the isolated spinal cord, and the effects of divalent cations, *Brain Res.*, 527, 224–233, 1990.
32. **Duchen, M. R. and Somjen, G. G.**, Effects of cyanide and low glucose on the membrane currents of dissociated mouse primary sensory neurones, *J. Physiol. (London)*, 401, 61P, 1988.
33. **Eccles, R. M., Løyning, Y., and Oshima, T.**, Effects of hypoxia on the monosynaptic reflex pathway in the cat spinal cord, *J. Neurophysiol.*, 29, 315–332, 1966.
34. **Farber, J. L. and El-Mofty, S. K.**, The biochemical pathology of liver cell necrosis, *Am. J. Pathol.*, 81, 237–250, 1975.
35. **Farber, J. L., Chien, K. R., and Mittnacht, S.**, The pathogenesis of irreversible cell injury in ischemia, *Am. J. Pathol.*, 102, 271–281, 1981.
36. **Farber, J. L.**, Membrane injury and calcium homeostasis in the pathogenesis of coagulative necrosis, *Lab. Invest.*, 47, 114–123, 1982.
37. **Fujiwara, N., Higashi, H., Shimoji, K., and Yoshimura, M.**, Effects of hypoxia on rat hippocampal neurones *in vitro*, *J. Physiol. (London)*, 384, 131–151, 1987.
38. **Glötzner, F.**, Intrazelluläre Potentiale, EEG und kortikale Gleichspannung und der Sensorimotorischen Rinde der Katze bei akuter Hypoxie, *Arch. Psychiatr. Nervenkr.*, 210, 274–296, 1967.
39. **Grafstein, B.**, Mechanism of spreading cortical depression, *J. Neurophysiol.*, 19, 154–171, 1956.
40. **Grossman, R. G. and Williams, V. F.**, Electrical activity and ultrastructure of cortical neurons and synapses in ischemia, in *Brain Hypoxia*, Brierly, J. B. and Meldrum, B. S., Eds., Heineman, London, 1971, 61–75.
41. **Hagberg, H., Lehmann, A., Sandberg, M., Nyström, B., Jacobson, I., and Hamberger, A.**, Ischemia-induced shift of inhibitory and excitatory amino acids from intra- to extracellular compartments, *J. Cerebr. Blood Flow Metabol.*, 5, 413–419, 1985.
42. **Hansen, A. J.**, Extracellular potassium concentration of juvenile and adult rat brain cortex during anoxia, *Acta Physiol. Scand.*, 99, 412–420, 1977.
43. **Hansen, A. J.**, The extracellular potassium concentration in brain cortex following ischemia in hypo- and hyperglycemic rats, *Acta Physiol. Scand.*, 102, 324–329, 1978.
44. **Hansen, A. J.**, Effect of anoxia on ion distribution in the brain, *Physiol. Rev.*, 65, 101–148, 1985.

45. **Hansen, A. J., Hounsgaard, J., and Jahnsen, H.,** Anoxia increases potassium conductance in hippocampal nerve cells, *Acta Physiol. Scand.,* 115, 301–310, 1982.
46. **Hansen, A. J. and Lauritzen, M.,** The role of spreading depression in acute brain disorders, *An. Acad. Bras. Cienc.,* 56, 457–479, 1984.
47. **Hansen, A. J., Lauritzen, M., and Wieloch, T.,** NMDA antagonists inhibit spreading depression but not anoxic depolarization, in *Frontiers in Excitatory Amino Acid Research,* Cavalheiro, E. A., Lehman, J., and Turski, L., Eds., Alan R. Liss, New York, 1988, 661–666.
48. **Harris, R. J., Symon, L., Branston, N. M., and Bayhan, M.,** Changes in extracellular calcium activity in cerebral ischemia, *J. Cereb. Blood Flow Metab.,* 1, 203–209, 1981.
49. **Hernández-Cacéres, J., Maciás-Gonzáes, J., Brožek, G., and Bureš, J.,** Systemic ketamine blocks cortical spreading depression but does not delay the onset of terminal anoxic depolarization in rats, *Brain Res.,* 437, 360–364, 1987.
50. **Higashi, H., Sugita, S., Nishi, S., and Shimoji, K.,** The effect of hypoxia on hippocampal neurons and its prevention by Ca^{2+}-antagonists, in *Mechanisms of Cerebral Hypoxia and Stroke,* Somjen, G. G., Ed., Plenum Press, New York, 1988, 205–218.
51. **Hille, B.,** The selective inhibition of delayed potassium current by tetraethylammonium ion, *J. Gen. Physiol.,* 50, 1287–1302, 1967.
52. **Hirsch, H., Euler, K. H., and Schneider, M.,** Über die Erholung und Weiderbelebung des Gehrins nach Ischemie bei Normothermie, *Pfluegers Arch.,* 265, 281–313, 1957.
53. **Hochachka, P. W.,** Inborn resistance to hypoxia and the O_2-dependence of metabolism, in *Mechanisms of Cerebral Hypoxia and Stroke,* Somjen, G. G., Ed., Plenum Press, New York, 1988, 1–8.
54. **Hochachka, P. W. and Somero, G. N,** *Biochemical Adaptation,* Princeton University Press, Princeton, NJ, 1984.
55. **Hossmann, K.-A.,** Post-ischemic resuscitation of the brain: selective vulnerability versus global resistance, *Prog. Brain Res.,* 63 3–27, 1985.
56. **Hossmann, K.-A., Paschen, W., and Csiba, L.,** Relationship between calcium accumulation and recovery of cat brain after prolonged cerebral ischemia, *J. Cereb. Blood Flow Metab.,* 3, 346–353, 1983.
57. **Hossmann, K.-A., Sakaki, S., and Zimmerman, V.,** Cation activities in reversible ischemia of the cat brain, *Stroke,* 8, 77–81, 1977.
58. **Hubbard, J. F. and Løyning, Y.,** The effects of hypoxia on neuromuscular transmission in a mammalian preparation, *J. Physio. (London),* 185, 205–223, 1966.
59. **Imaizumi, S., Tominaga, T., Uenohara, H., Kinouchi, H., Yoshimoto, T., and Suzuki, J.,** Detection of free radicals in cerebral tissue and their relation to cerebral hypoxia/ischemia, in *Mechanisms of Cerebral Hypoxia and Stroke,* Somjen, G. G., Ed., Plenum Press, New York, 1988, 321–335.
60. **Jennings, R. B., Ganote, C. E., and Reimer, K. A.,** Ischemic tissue injury, *Am. J. Pathol.,* 81, 179–194, 1975.
61. **Jennings, R. B. and Reimer, K. A.,** Pathobiology of acute myocardial ischemia, *Hosp. Pract.,* 24, 89–107, 1989.
62. **Kass, I. S. and Lipton, P.,** Calcium and long-term transmission damage following anoxia in dentate gyrus and CA1 regions of the rat hippocampal slice, *J. Physiol. (London),* 378, 313–354, 1986.
63. **Kawasaki, K., Czéh, G., and Somjen, G. G.,** Prolonged exposure to high potassium concentration results in irreversible loss of synaptic transmission in hippocampal tissue slices, *Brain Res.,* 457, 322–329, 1988.
64. **Kirino, T.,** Delayed neuronal death in the gerbil hippocampus, *Brain Res.,* 239, 57–69, 1982.
65. **Kirino, T. and Sano, K.,** Selective vulnerability in the gerbil hippocampus following transient ischemia, *Acta Neuropathol. (Berlin),* 62, 201–208, 1984.
66. **Kogure, K., Arai, H., Abe, K., and Nakano, M.,** Free radical damage of the brain following ischemia, *Prog. Brain Res.* 63, 237–259, 1985.

67. **Kotyk, A., Janacek, K., and Koryta, J.,** *Biophysical Chemistry of Membrane Function,* Wiley, Chichester, 1988.
68. **Kow, L.-M. and van Harreveld, A.,** Iron and water movements in isolated chicken retinas during spreading depression, *Neurobiology,* 2, 61–69, 1972.
69. **Kmjević, K. and Leblond, J.,** Changes in membrane currents of hippocampal neurons evoked by brief anoxia, *J. Neurophysiol.,* 62, 15–30, 1989.
70. **Kmjević, K. and Miledi, R.,** Presynaptic failure of neuromuscular propagation in rats, *J. Physiol. (London),* 149, 1–22, 1959.
71. **Langmoen, I. A. and Berg-Johnsen, J.,** Intracellular recording from neurones in rat cerebral cortex during hypoxia, *Acta Neurochirurg.,* Suppl. 43, 168–171, 1988.
72. **Lauritzen, M., Rice, M. E., Okada, Y., and Nicholson, C.,** Quisqualate, kainate and NMDA can initiate spreading depression in the turtle cerebellum, *Brain Res.,* 475, 317–327, 1988.
73. **Leão, A. A. P.,** Spreading depression of activity in the cerebral cortex, *J. Neurophysiol.,* 7, 359–390, 1944.
74. **Leão, A. A. P.,** Further observations on the spreading depression of activity in the cerebral cortex, *J. Neurophysiol.,* 10, 409–414, 1947.
75. **Leão, A. A. P.,** The slow voltage variation of cortical spreading depression of activity, *Electroencephalogr. Clin. Neurophysiol.,* 3, 315–321, 1951.
76. **Leão, A. A. P. and Martins-Ferreira, H.,** Alteração da impedancia electrica no decurso de depressão alastrante da atividade do córtex cerebral, *An. Acad. Brasil. Cienc.,* 25, 259–266, 1953.
77. **Leblond, J., and Krnjević, K.,** Hypoxic changes in hippocampal neurons, *J. Neurophysiol.,* 62, 1–14, 1989.
78. **Lee, T. C., Stephens, N., Moehl, A., and Snyder, F.,** Turnover of rat liver plasma membrane phospholipids, *Biochem. Biophys. Acta,* 291, 86–92, 1973.
79. **Lehninger, A. L.,** *Principles of Biochemistry,* Worth, New York, 1982, 583–614, 615–644, 888–890.
80. **Lemasters, J. J., DiGuiseppi, J., Nieminen, A.-L., and Herman, B.,** Blebbing, free Ca^{2+} and mitochondrial membrane potential preceding cell death in hepatocytes, *Nature,* 325, 78–81, 1987.
81. **Lipton, P., Raley, K., and Lobner, D.,** Long-term inhibition of synaptic transmission and macromolecular synthesis following anoxia in the rat hippocampal slice: interaction between Ca^{2+} and NMDA receptors, in *Mechanisms of Cerebral Hypoxia and Stroke,* Somjen, G. G., Ed., Plenum Press, New York, 1988, 229–249.
82. **Lipton, P. and Whittingham, T. S.,** The effect of hypoxia on evoked potentials in the in vitro hippocampus, *J. Physiol. (London),* 287, 427–438, 1979.
83. **Lipton, P. and Whittingham, T. S.,** Reduced ATP concentration as a basis for synaptic transmission failure during hypoxia in the in vitro guinea pig hippocampus, *J. Physiol. (London),* 325, 51–65, 1982.
84. **Lothman, E., LaManna, J., Cordingley, G., Rosenthal, M., and Somjen, G.,** Responses of electrical potential, potassium levels and oxidative metabolic activity of the cerebral neocortex of cats, *Brain Res.,* 88, 15–36, 1975.
85. **Lucas, D. R. and Newhouse, J. P.,** The toxic effect of sodium-1-glutamate on the inner layers of the retina, *A.M.A. Arch. Ophthalmol.,* 58, 193–201, 1957.
86. **Lutz, P. L., Rosenthal, M., and Sick, T. J.,** Living without oxygen: turtle brain as a model of anaerobic metabolism, *Mol. Physiol.,* 8, 411–425, 1985.
87. **Magnuson, K., Gustafsson, I., Westerberg, E., and Wieloch, T.,** Neurotransmitter modulation of neuronal damage following cerebral ischemia: effects of protein ubiquination, in *Mechanisms of Cerebral Hypoxia and Stroke,* Somjen, G. G., Ed., Plenum Press, New York, 1988, 309–319.

88. **Mandel, L. J., Takano, T., Soltoff, S. P., Jacobs, W. R., LeFurgey, A., and Ingram, P.**, Multiple roles of calcium in anoxic-induced injury in renal proximal tubules, in *Cell Calcium and the Control of Membrane Transport*, Mandel, L. J. and Eaton, D. C., Eds., Rockefeller University Press, New York, 1987, 277–293.
89. **Marranes, R., De Prins, E., Willems, R., and Wauquier, A.**, NMDA antagonists inhibit cortical spreading depression, but accelerate the onset of neuronal depolarization induced by asphyxia, in *Mechanisms of Cerebral Hypoxia and Stroke*, Somjen, G. G., Ed., Plenum Press, New York, 1988, 303–304.
90. **Marshall, W. H.**, Spreading cortical depression of Leão, *Physiol. Rev.*, 39, 239–279, 1959.
91. **Meldrum, B. S.**, Metabolic effects of prolonged epileptic seizures and the causation of epileptic brain damage, in *Metabolic Disorders of the Nervous System*, Rose, F. C., Ed., Pitman, London, 1981, 175–187.
92. **Meldrum, B. S.**, Possible therapeutic applications of antagonists of excitatory amino acid transmitters, *Clin. Sci.*, 68, 113–122, 1985.
93. **Meldrum, B. S., Evans, M., and Swan, J.**, Excitatory amino acid transmission and protection against brain damage, in *Mechanisms of Cerebral Hypoxia and Stroke*, Somjen, G. G., Ed., Plenum Press, New York, 1988, 349–358.
94. **Misgeld, U. and Frotscher, M.**, Dependency of the viability of neurons in hippocampal slices on oxygen supply, *Brain Res. Bul.*, 8, 95–100, 1982.
95. **Moskalenko, Y. E., Weinstein, G. B., Parfenov, V. E., Bodó, M., and Gaidar, B. V.**, Cerebral blood flow and its responsiveness to CO_2 after traumatic and ischemic brain injuries, in *Mechanisms of Cerebral Hypoxia and Stroke*, Somjen, G. G., Ed., Plenum Press, New York, 1988, 135–136.
96. **Myers, R. E.**, A unitary theory of causation of anoxic and hypoxic brain pathology, *Adv. Neurol.*, 26, 195–213, 1979.
97. **Myers, R. E.**, High lactic acid, not reduced ATP: cause of brain injury from oxygen deprivation, *Exc. Med. Int. Congr. Ser.* 532, 231–236, 1981.
98. **Nedergaard, M. and Hansen, A. J.**, Spreading depression is not associated with neuronal injury in the normal brain, *Brain Res.*, 449, 395–398, 1988.
99. **Nelgård, B., Gustafson, I., Hansen, A., Lauritzen, M., and Wieloch, T.**, MK-801, a noncompetitive NMDA receptor antagonist, does not protect against neuronal damage in the brain following cerebral ischemia *Abstr. Soc. Neurosci.*, 15, 43, 1989.
100. **Nicholson, C. and Kraig, R. P.**, The behavior of extracellular ions during spreading depression, in *The Application of Ion-Selective Electrodes*, Zeuthen, T., Ed., Elsevier, Amsterdam, 1981, 217–238.
101. **Nicholson, C., Phillips, J. M., Tobias, C., and Kraig, R. P.**, Extracellular potassium, calcium and volume profiles during spreading depression, in *Ion-Selective Microelectrodes and the Use in Excitable Tissues*, Vyklický, L., Hník, P., and Syková, E., Eds., Plenum Press, New York, 1981, 211–223.
102. **Niechaj, A. and van Harreveld, A.**, The nature of postasphyxial rigidity examined by intracellular recording from motoneurons, *Exp. Neurol.*, 18, 68–78, 1967.
103. **Okada, Y.**, Reversibility of neuronal function of hippocampal slice during deprivation of oxygen and/or glucose, in *Mechanisms of Cerebral Hypoxia and Stroke*, Somjen, G. G., Ed., Plenum Press, New York, 1988, 191–203.
104. **Okada, Y., Tanimoto, M., and Yoneda, K.**, The protective effect of hypothermia on reversibility in the neuronal function of the hippocampal slice during long lasting anoxia, *Neurosci. Lett.*, 84, 277–282, 1988.
105. **Olney, J. W.**, Brain lesions, obesity and other disturbances in mice treated with monosodium glutamate, *Science*, 164, 719–721, 1969.
106. **Peters, T.**, Calcium in physiological and pathological cell function, *Eur. Neurol.*, 25 (Suppl. 1), 27–44, 1986.

107. **Phillips, J. M. and Nicholson, C.**, Anion permeability in spreading depression investigated with ion selective microelectrodes, *Brain Res.*, 173, 567–571, 1979.
108. **Prawdicz-Neminsky, W. W.**, Zur Kenntnis der elektrischen und der Innervationsvorgänge in den funktionellen Elementen und Geweben des tierischen Organismus. Elektrocerebrogram der Säugetiere, *Pfluegers Arch.*, 209, 362–382, 1925.
109. **Pulsinelli, W. A.**, Selective neuronal vulnerability: morphological and molecular characteristics, *Prog. Brain Res.*, 63, 29–38, 1985.
110. **Pumain, R., Kurcewicz, I., and Louvel, J.**, Ionic changes induced by excitatory amino acids in the rat cerebral cortex, *Can. J. Physiol. Pharmacol.*, 65, 1067–1077, 1987.
111. **Rabinowitz, M.**, Control of metabolism and synthesis of macromolecules in normal and ischemic heart, *J. Mol. Cell. Cardiol.*, 2, 277–292, 1971.
112. **Rader, R. K. and Lanthorn, T. H.**, Experimental ischemia induces a persistent depolarization blocked by decreased calcium and NMDA antagonists, *Neurosci. Lett.*, 99, 125–130, 1989.
113. **Rader, R. K., Lanthorn, T. H., and Lipton, P.**, Effects of hypoxia on responses to acidic amino acids in the *in vitro* hippocampal slice, *Soc. Neurosci. Abstr.*, 13, 1495, 1987.
114. **Rehncrona, S., Westerberg, E., Åkeson, B., and Siesjö, B. K.**, Brain cortical fatty acids and phospholipids during and following complete and severe incomplete ischemia, *J. Neurochem.*, 38, 84–93, 1982.
115. **Roberts, E. L. and Sick, T. J.**, Calcium-sensitive recovery of extracellular potassium and synaptic transmission in rat hippocampal slices exposed to brief anoxia, *Brain Res.*, 456, 113–119, 1988.
116. **Rosenthal, M., Feng, Z.-C., and Sick, T. J.**, Brain vulnerability and survival during anoxia: protective strategies of hypoxia-resistant vertebrates, in *Mechanisms of Cerebral Hypoxia and Stroke*, Somjen, G. G., Ed., Plenum Press, New York, 1988, 9–21.
117. **Rosenthal, M. and Somjen, G. G.**, Spreading depression, sustained potential shifts and metabolic activity of cerebral cortex of cats, *J. Neurophysiol.*, 36, 739–749, 1973.
118. **Rossen, R., Kabat, H., and Anderson, J. P.**, Acute arrest of cerebral circulation in man, *Arch. Neurol. Psychiatr.*, 50, 510–528, 1943.
119. **Rothman, S.**, Synaptic release of excitatory amino acid neurotransmitter mediates anoxic cell death, *J. Neurosci.*, 4, 1884–1891, 1984.
120. **Rothman, S. M.**, The neurotoxicity of excitatory amino acids is produced by passive chloride influx, *J. Neurosci.*, 5, 1483–1489, 1985.
121. **Rothman, S. M. and Olney, J. W.**, Glutamate and the pathophysiology of hypoxicischemic brain damage, *Ann. Neurol.*, 19, 105–111, 1986.
122. **Rothman, S. M. and Olney, J. W.**, Excitotoxicity and the NMDA receptor, *Trends Neurosci.*, 10, 299–302, 1987.
123. **Rothman, S. M., Thurston, J. H., Hauhart, R. E., Clark, G. D., and Solomon, J. S.**, Ketamine protects hippocampal neurons from anoxia in vitro, *Neuroscience*, 21, 673–678, 1987.
124. **Roufa, D. K., Lanthorn, T. H., Rader, R. K., Rapp, S. R., and Contreras, P. C.**, Protection of hippocampal neurons from "ischemic" insult *in vitro* by acidic amino acid antagonists, in *Mechanisms of Cerebral Hypoxia and Stroke*, Somjen, G. G., Ed., Plenum Press, New York, 1988, 367–376.
125. **Schanne, F. A. X., Kane, A. B., Young, E. E., and Farber, J. L.**, Calcium dependence of toxic cell death: a final common pathway, *Science*, 206, 700–702, 1979.
126. **Schiff, S. J. and Somjen, G. G.**, Hyperexcitability following moderate hypoxia in hippocampal tissue slices, *Brain Res.*, 337, 337–340, 1985.
127. **Schurr, A., Dong, W.-Q., Reid, K. H., West, C. A., and Rigor, B. M.**, Lack of adverse effect of lactic acid on hypoxic neuronal tissue in vitro, *Neuroscience*, 22 (Suppl. S744), 1987.

128. **Sick, T. J., Rosenthal, M., LaManna, J., and Lutz, P. L.,** Brain potassium ion homeostasis, anoxia and metabolic inhibition in turtles and rats, *Am. J. Physiol.*, 243, R281–R288, 1982.
129. **Sick, T. J., Solow, E. L., and Roberts, E. L.,** Extracellular potassium activity and electrophysiology in the hippocampal slice: paradoxical recovery of synaptic transmission during anoxia, *Brain Res.*, 418, 227–234, 1987.
130. **Siebke, H., Breivik, H., Rød, T., and Lind, B.,** Survival after 40 minutes submersion without cerebral sequelae, *Lancet*, 1, 1275–1277, 1975.
131. **Siesjö, B. K.,** Cell damage in the brain: a speculative synthesis, *J. Cereb. Blood Flow Metab.*, 1, 155–185, 1981.
132. **Siesjö, B. K.,** Calcium and ischemic brain damage, *Eur. Neurol.*, 25 (Suppl. 1), 45–56, 1986.
133. **Siesjö, B. K.,** Mechanisms of ischemic brain damage, *Crit. Care Med.*, 16, 954–963, 1988.
134. **Siesjö, B. K. and Wieloch, T.,** Molecular mechanisms of ischemic brain damage: Ca^{2+}-related events, in *Cerebrovascular Diseases*, Plum, F. and Pulsinelli, W., Eds., Raven Press, New York, 1985, 187–197.
135. **Simon, R. P., Griffiths, T., Evans, M. C., Swan, J. H., and Meldrum, B. S.,** Calcium overload in the selectively vulnerable neurons of the hippocampus during and after ischemia: an EM study in the rat, *J. Cereb. Blood Flow Metab.*, 4, 350–361, 1984.
136. **Simon, R. P., Swan, J. H., Griffiths, T., and Meldrum, B. S.,** Blockade of n-methyl-d-aspartate receptors may protect against ischemic damage in the brain, *Science*, 226, 850–852, 1984.
137. **Smith, M.-L. and Siesjö, B. K.,** Acidosis-related brain damage: immediate and delayed events, in *Mechanisms of Cerebral Hypoxia and Stroke*, Somjen, G. G., Ed., Plenum Press, New York, 1988, 57–71.
137b. **Somjen, G. G.,** Basic mechanisms in cerebral hypoxia and stroke: background, review and conclusions, in *Mechanisms of Cerebral Hypoxia and Stroke*, Somjen, G. G., Ed., Plenum Press, New York, 1988, 447–466.
138. **Somjen, G. G.,** Mechanism of the reversible arrest of function during transient cerebral hypoxia and ischemia, in *Cerebral Ischaemia and Resuscitation*, Schurr, A., Ed., CRC Press, Boca Raton, FL, 1990, 301–317.
139. **Somjen, G. G. and Aitken, P. G.,** The ionic and metabolic responses associated with neuronal depression of Leão's type in cerebral cortex and hippocampal formation, *An. Acad. Bras. Cienc.*, 56, 495–504, 1984.
140. **Somjen, G. G., Aitken, P. G., Balestrino, M., and Schiff, S. J.,** Uses and abuses of in vitro systems in the study of the pathophysiology of the central nervous system, in *Brain Slices: Fundamentals, Applications and Implications*, Schurr, A., Teyler, T. J., and Tseng, M. T., Eds., S. Karger, Basel, 1986, 89–104.
141. **Somjen, G. G., Schiff, S. J., Aitken, P. G., and Balestrino, M.,** Forms of suppression of neuronal function: Leão's depression, hypoxia and hyperthermia, in *Inactivation of Hypersensitive Neurons*, Chalazonitis, N. and Gola, M., Eds., Alan R. Liss, New York, 1987, 137–145.
142. **Speckmann, E.-J., Caspers, H., and Sokolov, W.,** Aktivitätsänderungen spinaler Neurone während und nach einer Asphyxie, *Pfluegers Arch.*, 319, 122–138, 1970.
143. **Steenbergen, C., Murphy, E., Levy, L., and London, R. E.,** Elevation of cytosolic free calcium concentration early in myocardial ischemia in perfused rat heart, *Circ. Res.*, 60, 700–707, 1987.
144. **Stryer, L.,** *Biochemistry*, 3rd ed., W. H. Freeman, New York, 1988, chaps. 23 and 30.
145. **Suzuki, R., Yamaguchi, T., Kirino, T., Orzi, F., and Klatzo, I.,** The effects of 5-minute ischemia in mongolian gerbils. I. Blood-brain barrier, cerebral blood flow and local cerebral glucose utilization changes, *Acta Neuropathol. (Berlin)*, 60, 207–216, 1983.

146. **Suzuki, R., Yamaguchi, T., Li, C.-L., and Klatzo, L.,** The effect of 5-minute ischemia in mongolian gerbils. II. Changes of spontaneous neuronal activity in cerebral cortex and Ca1 sector of hippocampus, *Acta Neuropathol. (Berlin),* 60, 217–222, 1983.
147. **Symon, L.,** The concept of intracerebral "steal", in *International Anaesthesiology Clinics, Cerebral Circulation,* McDowall, G., Ed., Little, Brown, Boston, 1969, 597–615.
148. **Symon, L.,** Physiological aspects of brain ischemia in the experimental primate and man, in *Mechanisms of Cerebral Hypoxia and Stroke,* Somjen, G. G., Ed., Plenum Press, New York, 1988, 91–107.
149. **Takano, T., Soltoff, S. P., Murdaugh, S., and Mandel, L. J.,** Intracellular respiratory dysfunction and cell injury in short-term anoxia of rabbit renal proximal tubules, *J. Clin. Invest.,* 76, 2377–2384, 1985.
150. **Urban, L., Neill, K. H., Crain, B. J., Nadler, J. V., and Somjen, G. G.,** Postischemic synaptic physiology in area CA1 of the gerbil hippocampus studied *in vitro, J. Neurosci.,* 9, 3966–3975, 1989.
151. **Urban, L., Neill, K. H., Crain, B. J., Nadler, J. V., and Somjen, G. G.,** Effects of NMDA antagonist on postischemic physiology of CA1 hippocampal pyramidal cells, *Abstr. Soc. Neurosci.,* 15, 357, 1989.
152. **Urbán, L. and Somjen, G. G.,** Reversible effects of hypoxia on neurons in mouse dorsal root ganglia in vitro, *Brain Res.,* 520, 36–42, 1990.
153. **van Harreveld, A.,** Compounds in brain extracts causing spreading depression of cerebral cortical activity and contraction of crustacean muscle, *J. Neurochem.,* 3, 300–315, 1959.
154. **van Harreveld, A.,** Two mechanisms for spreading depression in the chicken retina, *J. Neurobiol.,* 9, 419–431, 1978.
155. **van Harreveld, A.,** *Brain Tissue Electrolytes,* Butterworths, Washington, D.C., 1966.
156. **van Harreveld, A. and Fifkova, E.,** Glutamate release from retina during spreading depression, *J. Neurobiol.,* 2, 13–29, 1970.
157. **van Harreveld, A. and Fifkova, E.,** Light- and electron microscopic changes in central nervous tissue after electrophoretic injection of glutamate, *Exp. Mol. Pathol.,* 15, 61–81, 1971.
158. **van Harreveld, A. and Ochs, S.,** Cerebral impedance changes after circulatory arrest, *Am. J. Physiol.,* 187, 180–192, 1956.
159. **van Harreveld, A. and Ochs, S.,** Electrical and vascular concomitants of spreading depression, *Am. J. Physiol.,* 189, 159–166, 1957.
160. **van Harreveld, A. and Schadé, J. P.,** Nerve cell destruction by the asphyxiation of the spinal cord, *J. Neuropathol. Exp. Neurol.,* 21, 410–423, 1962.
161. **Vyskočil, F., Kříž, N., and Bureš, J.,** Potassium selective electrodes used for measuring the extracellular brain potassium during spreading depression and anoxic depolarization in rats, *Brain Res.,* 39, 255–259, 1972.
162. **Welsh, F. A.,** Role of vascular factors in regional ischemic injury, *Prog. Brain Res.,* 63, 19–27, 1985.
163. **West, C. A., Schurr, A., Reid, K. H., and Shields, C. B.,** Protection against hypoxia by high glucose: a study using the in vitro hippocampal slice, *Abstr. Soc. Neurosci.,* 12, 1526, 1986.
164. **Wrogemann, K. and Penna, S. D. J.,** Mitochondrial calcium overload: a general mechanism for cell necrosis in muscle diseases, *Lancet,* 1, 672–673, 1976.
165. **Young, J. N. and Somjen, G. G.,** Suppression of presynaptic calcium currents by hypoxia in hippocampal tissue slices, *Brain Res.,* 573, 70–76, 1992.

Chapter 9

THE PATHOGENETIC MECHANISM UNDERLYING CEREBRAL VASOSPASM

Takao Asano, Tohru Matsui, Takashi Watanabe, Tohru Koide, and Yoh Takuwa

TABLE OF CONTENTS

I. Introduction 210

II. Overview of the Past Research on the Pathogenetic Mechanism of Vasospasm 212

III. Lipid Peroxidation and Cerebral Vasospasm 215

IV. Interaction between Lipid Hydroperoxides and Lipoxygenase Pathway 221

V. Protein Kinase C-Mediated Contraction of Smooth Muscles: Deus ex Machina? 228

VI. Phorbol Ester-Induced Contraction of the Canine Basilar Artery 231

VII. Relevance of the PKC-System to Late Spasm 233

VIII. Conclusion 237

References 238

I. INTRODUCTION

Subarachnoid hemorrhage (SAH) due to rupture of an intracranial aneurysm usually leads to the occurrence of two types of ischemic brain damage. The first, which occurs in the acute stage of SAH, is due to the extravasation of arterial blood within the cranial cavity, causing a local destruction of the brain parenchyma through formation of hematomas or a diffuse brain ischemia through an enormously elevated intracranial pressure. The second type is more insidious in onset, usually starting several days after SAH, and its presenting symptoms are either or both of the focal and generalized neurological deteriorations. This secondary deterioration in the neurological status has been designated as 'delayed ischemic neurological deficits' (DINDs) in discrimination from the acute, direct brain damage.[23] While the direct and immediate brain damage ensuing SAH defies all the therapeutic possibilities, the secondary neurological deterioration has been a matter of great concern, particularly to neurosurgeons, because it occurs in a considerable percentage of patients (about 35% of total morbidity and mortality),[33] posing difficult problems as to the timing of and the selection of patients for surgical interventions. It has been established that the occurrence of DIND is causally related to the angiographic narrowing of major intracranial arteries,[23,24,55,56,78] the underlying pathomechanism of which has been one of the major subjects of research in neurosurgery.

The angiographic narrowing of major cerebral arteries following SAH is called "cerebral vasospasm" or "cerebral arterial spasm" because it persists as long as a month once it gets started. As shown in Figure 1, the narrowing may be segmental in an artery, or may involve many arteries causing diffuse narrowing of all the arteries visualized in the angiogram. Vasospasm preferentially occurs in the artery harboring the ruptured aneurysm, but not rarely in remote arteries. Since the advent of CT scan, it has clearly been demonstrated that cerebral vasospasm occurs where there is a thick subarachnoid clot, and vice versa (Figure 1).[24,56] Thus, the subarachnoid clot has been regarded as the primary cause of vasospasm. The other eminent clinical feature of vasospasm is its protracted time-course, with a delayed onset. This peculiar time-course of vasospasm in human SAH had been obscured by the fact that in the animal model of SAH, vasospasm runs a biphasic course, namely, the acute and late spasm.[14] Whether or not vasospasm in humans has an acute phase compatible to that of animal models was once a subject for debate, but the issue has been settled since series of angiographic studies have shown that vasospasm is not seen during the first 3 days in the vast majority of patients, and that the peak incidence is on day 6 to 7.[55,78]

Also of importance are the pathological changes of the arterial wall accompanying vasospasm. The earliest changes consisted of swelling and thickening of the intimal layer with adhesion of platelets and leukocytes (1 to 9 days after SAH), which then proceeded to the stage of subintimal cellular proliferation (4 to 28 days). The most severe changes, being observed later

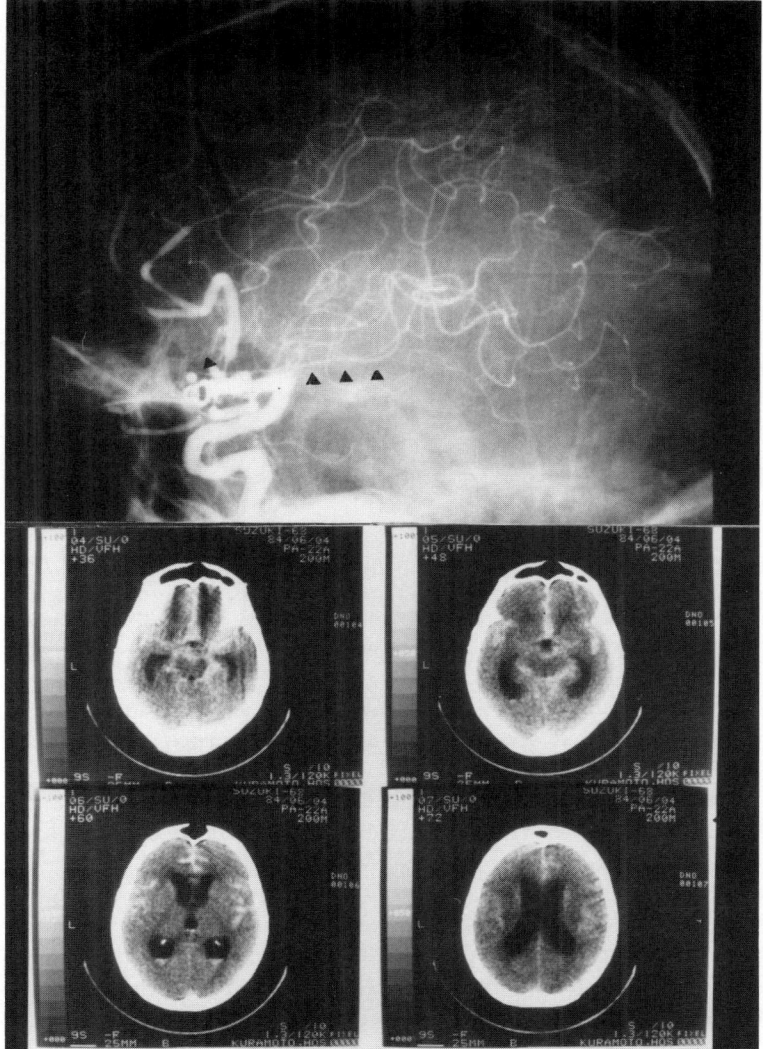

FIGURE 1. Above: A postoperative angiogram showing severe, diffuse vasospasm. ▲, Segmental narrowing of the anterior cerebral artery, ▲▲▲, diffuse narrowing of the middle cerebral artery. Below: The preoperative CT scan of the above patient. Note the diffuse, thick clot involving all the subarachnoid cisterns.

on (16 to 40 days) consisted of prominent subintimal cellular proliferation, rupture and splitting of the internal elastic membrane, and intramural hemorrhage with fibrosis and necrosis (myonecrosis) of the medial layer of the vessel (Figure 2).[66,67] Thus, the topographical association with the subarachnoid clot, the delayed onset, the protracted time-course, and the organic

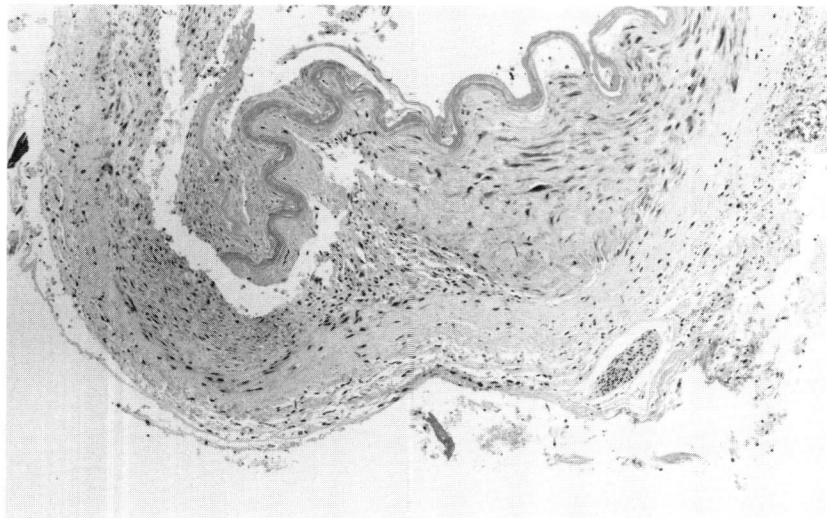

FIGURE 2. The organic changes in the arterial wall associated with severe vasospasm (the internal carotid artery, day 8). Note accumulation of cells in the fold of the endothelium, the cellular proliferation in the medial layer, the cell infiltration in the medial and the adventitial layers, and the edema in the subintimal and the medial layers.

changes in the arterial wall are the eminent clinical features of cerebral vasospasm, which should serve as guidelines for the research on the underlying pathomechanism.

II. OVERVIEW OF THE PAST RESEARCH ON THE PATHOGENETIC MECHANISM OF VASOSPASM

The research on the pathogenetic mechanism of cerebral vasospasm started from the viewpoint that it represents an active contraction of arterial smooth muscles, as implicated by the term 'spasm'. This presumption still seems to be true since we neurosurgeons often observe that the artery showing a significant narrowing in the preoperative angiogram is thin and whitish in color, presenting the appearance of nothing but a constricted artery (Plate 1)*. Taking the occurrence of organic changes into consideration, the train of events as follows has been presumed as a basic pathomechanism leading to the occurrence of vasospasm: SAH → liberation of vasocontractile substances into the CSF → prolonged contraction of arterial smooth muscle → secondary organic changes in the arterial wall presumably due to nutritional derangements.[16,30]

---

* Plate 1 follows page 304.

TABLE 1
List of Representative Putative Spasmogens Suggested

Proteins

Angiotensin II
Bradykinin
Fibrin degradation products
Oxyhemoglobin and its degradation products
RBC ghosts
Thrombin
Vasopressin
Unidentified polypeptides

Amines

Acetylcholine
Dopamine
Epinephrine
Norepinphrine
Histamine
Serotonin
Kynurenine
Tryptamine

Eicosanoids

Arachidonic acid
Prostaglandins E_2, F_2
Thromboxane A_2
Hydroperoxides such as 15- and 12-HPETEs
Leukotrienes C_4 and D_4

Inorganic Substances

Calcium
Potassium
Magnesium deficiency
Hydrogen peroxide

Based on the above thesis, numerous studies have been conducted in search of the vasocontractile substance(s), i.e., "spasmogens". Table 1 shows only a part of spasmogens hitherto suggested.[79] Nevertheless, there are at least several conditions which a putative spasmogen must fulfill before it is deemed the cause of vasospasm: (1) its concentration in the CSF or in the cerebral artery must be high enough to elicit smooth muscle contraction; (2) it must explain the delayed onset and the protracted time-course of vasospasm; (3) an antagonist to it, or an inhibitor of its synthesis must prevent or ameliorate the occurrence of vasospasm; (4) in animal models, the cisternal infusion of

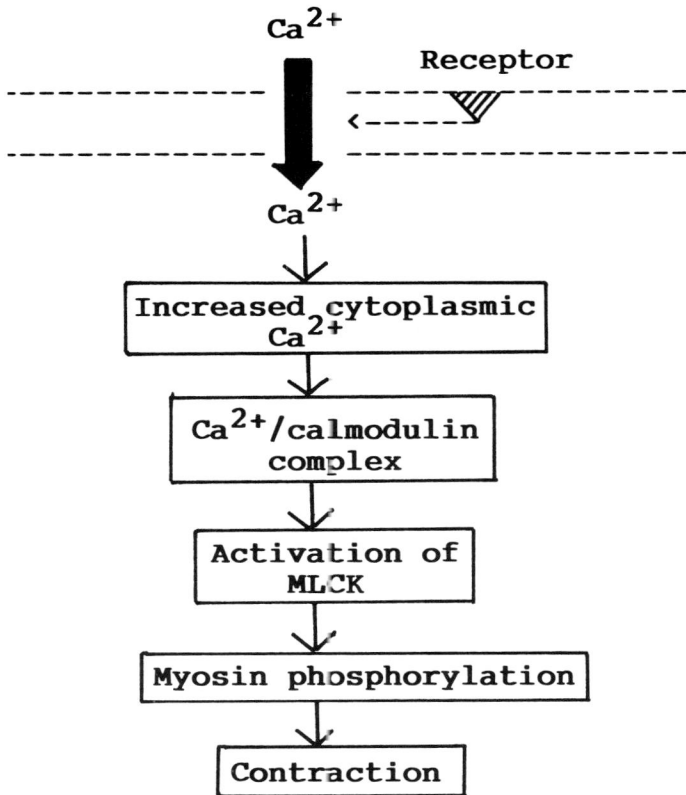

FIGURE 3. The Ca^{2+}/CaM/MLCK system leading to myosin phosphorylation and tension development.

the substance must induce a prolonged contraction of cerebral arteries, etc. Unfortunately, none of the proposed spasmogens has ever completely satisfied the above criteria. Moreover, attempts to prevent vasospasm by administration of vasodilators invariably failed, regardless of the type of the agent.[79] This failure culminated in the trials using calcium antagonists.

It has been believed that an agonist induces smooth muscle contraction by causing an intracellular influx of external calcium (Ca^{2+}) via the Ca^{2+} channel.[27] The subsequent formation of Ca^{2+}/calmodulin (CaM) complex activates myosin light chain kinase (MLCK), leading to phosphorylation of myosin light chain (MLC) and force generation owing to actin-myosin interaction (Figure 3).[32] Assuming Ca^{2+} influx is the final common path of smooth muscle contraction, the inhibition of the Ca^{2+} channel must effectively prevent or reverse vasospasm, no matter what the spasmogen may be.[3] This line of thinking led many investigators to evaluate the therapeutic or preventive effects of various Ca^{2+} antagonists on vasospasm. However, conducted clinical and experimental studies on the whole indicate that these agents do not directly dilate the already spastic arteries, although they may bring some improvements in the neurological outcome.[3,8,20,40,42,49,51,73] Since the spasmogen has never been identified, and since the spastic artery turned out to be unresponsive to Ca^{2+} antagonists, the presumption that vasospasm represents an agonist-induced contraction of the arterial smooth muscle has become difficult to maintain.

A possible alternative approach to pathogenesis of vasospasm has been to ascribe the arterial narrowing not to smooth muscle contraction but to the organic changes in its wall such as the intimal thickening or the decreased distensibility.[13,16,66,67] Although the organic changes in the arterial wall may in some way contribute to the occurrence of vasospasm, it does not necessarily follow that they represent the primary cause of vasospasm. In the first place, the intimal thickening due to subintimal fibrosis sufficient to cause narrowing of the arterial lumen is rather late in its occurrence, lagging far behind the period of the maximal incidence of angiographical spasm. Hence it may better be regarded not as the cause but as the result of vasospasm. Second, the decreased distensibility of the wall of the spastic artery, a well documented fact, has been conjectured to be the cause of the luminal narrowing as well as the decreased blood flow by itself.[13] As to the cause of this decreased distensibility, however, an antecedent active contraction of the arterial smooth muscle was considered to be responsible.[13] Therefore, the discussion obviously becomes circular as to the cause of the "antecedent" vasospasm.

Thus, the current state-of-the-art may be summarized as follows: the concept that vasospasm is due to smooth muscle contraction has not been substantiated in past research, but it is still congruous to the experience of neurosurgeons. The inflammatory changes, or other organic changes within the arterial wall, may play some role in the occurrence of vasospasm, but there has been no convincing evidence showing that these pathological changes per se are the primary cause of vasospasm. Clearly, research on the pathogenetic mechanism of vasospasm has migrated into a 'cul de sac', from where a breakthrough should be sought.

III. LIPID PEROXIDATION AND CEREBRAL VASOSPASM

Whereas the vasocontractile substances have been the main target of research on the pathogenesis of vasospasm, we started our study from an entirely different angle. The idea was formed through a simple clinical observation at surgery (before the advent of CT scan) that the cerebral artery in spasm is always surrounded by a subarachnoid clot undergoing hemolysis. Presuming that the processes involved in hemolysis may participate in the occurrence of vasospasm, we surveyed relevant literature, noticing the fol-

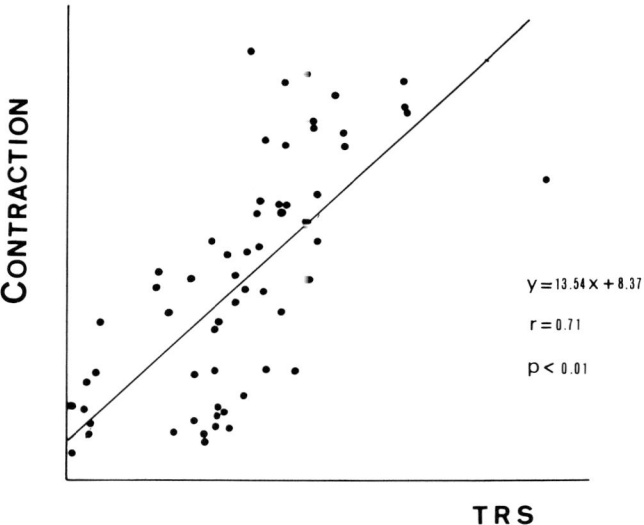

FIGURE 4. The parallel developments of he thiobarbituric acid-positive substances (TRS) and the vasocontractile capacity in the incuba ed RBCs.

lowing issues: (1) among various components of the whole blood, the hemolyzed red blood cells caused the most prominent inflammatory reactions (aseptic meningitis) when intracisternally injected in dogs;[31] (2) lipid peroxidation induced by superoxide anion (O_2^-), liberated from oxyhemoglobin (oxyHb) in its conversion to methemoglobin, participates in the biochemical processes of hemolysis;[34,46,81] (3) the degradation products of oxyHb have potent catalytic actions to lipid peroxidation;[11,72,80] (4) while all the structures facing the subarachnoid space are exposed to degradation products of oxyHb, deposition of iron or degradation products of hemoglobin is recognized within the cerebral artery.[16,41] With the above knowledge, it was rather easy to speculate that the degradation products of oxyHb induce lipid peroxiation, which exerts some toxic effects on the cerebral artery, leading to the occurrence of vasospasm.[5]

Thus, we started out to search for evidence showing the occurrence of lipid peroxidation following SAH, and its relation to vasospasm. The obtained results are briefly described below, together with some pertinent data reported by other authors.

Incubation of RBCs — Centrifuged, canine RBCs mixed with the same amount of Krebs buffer were incubated in an aseptic condition at 37°C and the changes in its content of thiobarbituric acid (TBA)-reactive substance (TRS) and the vasocontractile capacity *in vitro* were examined. The TRS remarkably increased 2 to 3 days after the start of incubation, with a parallel increase in its vasocontractile capacity (Figure 4).[5]

The vasocontractile action of a lipid hydroperoxide, 15-HPETE — A hydroperoxide of arachidonic acid, 15-hydroperoxyeicosatetraenoic acid (15-HPETE) was prepared using soybean lipoxygenase. *In vitro,* the agent showed a potent vasocontractile capacity comparable to that of $PGF_{2\alpha}$ (Figure 5), which was inhibited by pretreatment with free radical scavengers such as vitamin E, hydroquinone, and AVS.[38]

Intracisternal injection of 15-HPETE in dogs — Intracisternal injection of 15-HPETE (2 mg) induced a prolonged angiographical constriction of the canine basilar artery, which was composed of early and late phases.[61] Ultrastructural changes in the arterial wall such as degenerative changes in the endothelial and medial smooth muscle cells were associated with the prolonged arterial constriction.

The PGI_2 synthetic capacity of the canine basilar artery exposed to SAH — The change in the PGI_2 synthetic capacity of the basilar artery following SAH was studied using the canine SAH model since PGI_2 synthase has been known to be inhibited by lipid peroxides.[82] As shown in Figure 6, a significant diminution in the PGI_2 synthetic capacity of the basilar artery was revealed following SAH.[62]

The TRS in the CSF of SAH patients — A time-coursed analysis with the CSF of SAH patients disclosed a prominent increase in the CSF content of the TRS during the first 10 days after SAH. A significant difference was revealed between the patients who developed vasospasm and those who did not (Figure 7).[5,60] It is noteworthy that the duration of the increase in TRS was long enough to overlap the beginning of vasospasm.

The free radical scavengers in the CSF — Since lipid peroxidation is strictly regarded by the activity of free radical scavengers, the vitamin E content and the activity of glutathione peroxidase were studied using the CSF of SAH patients. Both showed a transient increase immediately after SAH and then gradually decreased (Figure 8 A, B).[60,77] The superoxide dismutase (SOD) and catalase activities in the CSF of SAH patients are shown in Figure 8 C, D.[57] The gradual diminution in the activities of these free radical scavengers seems to be compatible with the reciprocal occurrence of lipid peroxidation. The cause of a reverse increase in the catalase activity is unclear, but it may well represent a biological reaction to an enhanced generation of lipid hydroperoxides. It may thus be inferred that the delayed onset of vasospasm is related to the gradual diminution in the activity of the free radical scavenging system in the CSF.

The therapeutic effects of free radical scavengers — Using various animal SAH models, the effects of free radical scavengers such as AVS (nicaraven),[6,76] glutathione,[76] AA861,[83] and U74006F[74] on cerebral vasospasm have been studied. Significant inhibition of spasm development has been reported with each of the above drugs.

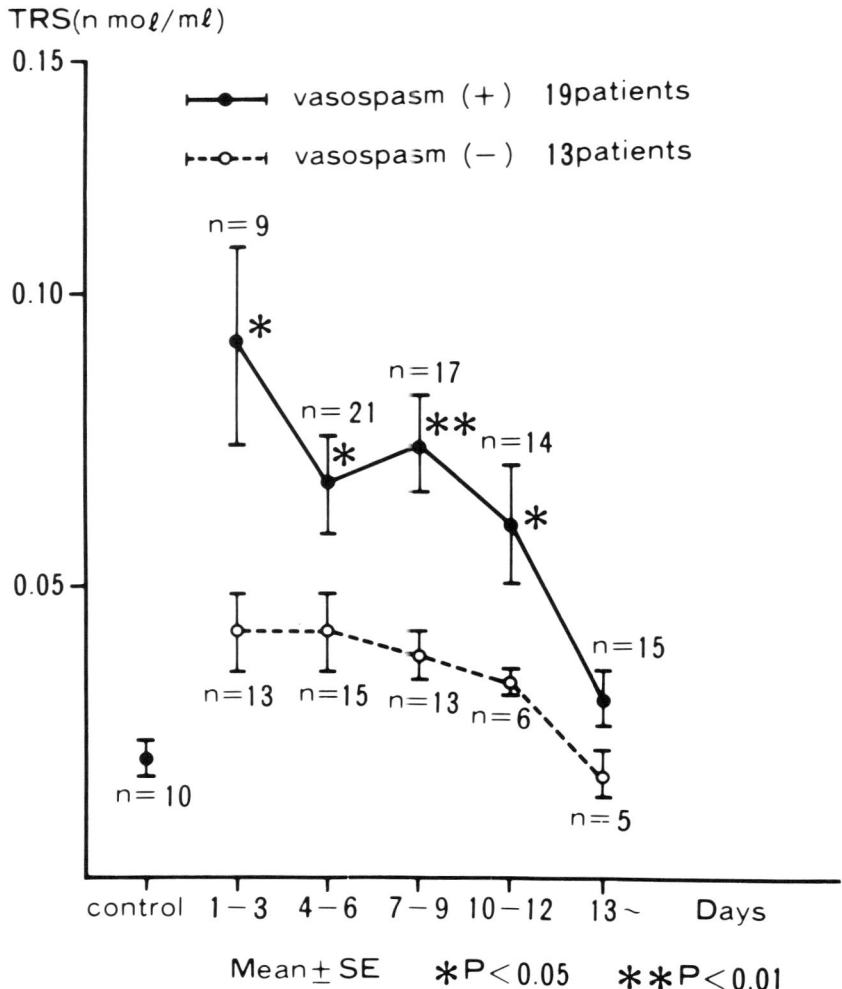

FIGURE 7. The CSF content of TBA-reactive substances (TRS) in SAH patients. The normal value is shown at the left, below.

in significant amounts only in the clot (Table 2).[76] However, any other HETEs were not detected either in the CSF or the basilar artery.

The lipoxygenase activity in the canine basilar artery — On incubation in the presence of arachidonic acid and a calcium ionophore A23187, the canine basilar artery exposed to SAH exhibited 5-lipoxygenase activity, producing 5-HETE and leukotrienes (LTs) B_4 and C_4 (Figure 9).[64] This enhancement in the 5-lipoxygenase activity in the basilar artery following SAH may be ascribed to the action of lipid hydroperoxides, since 15-HPETE was shown to activate lipoxygenases of the canine basilar artery (Figure 10).[38,64]

Intracisternal injection of 12-HPETE in dogs — As the preceding study showed that 12-HETE was the predominant product in the subarachnoid clot, we examined whether or not its precursor, 12-HPETE would cause vasospasm (paper in preparation). Single intracisternal injection of 12-HPETE (0.5 mg), newly prepared from porcine leukocytes, induced a prolonged constriction of the canine basilar artery (Figure 11A), which was similar to the effect of 15-HPETE. Of particular interest is the fact that 12-HPETE was not detected in the CSF even 5 min after injection, whereas its stable metabolite, 12-HETE gradually disappeared from the CSF (Figure 11B). It is surprising that single injection of 12- or 15-HPETE caused a long-lasting constriction of the cerebral artery, in spite of the fact that these agents were very rapidly cleared from the CSF.

IV. INTERACTION BETWEEN LIPID HYDROPEROXIDES AND THE LIPOXYGENASE PATHWAY

Results obtained as described above provide ample evidence to suspect that lipid peroxidation is involved in the pathogenesis of cerebral vasospasm. Although this thesis seems to have been fortified by the recent reports showing the therapeutic effects of various free radical scavengers on vasospasm[74,83] and the increase in the TRS content within the canine basilar artery exposed to SAH,[58] further substantiation, particularly in terms of the radical species initiating free radical reactions, is obviously required.

Results of our recent studies concerning lipoxygenase products raise a possibiilty that there is an interaction between lipid peroxidation which is due to nonenzymatic free radical reactions and the lipoxygenase pathway which is a part of the system of enzymatic peroxidation of free arachidonic acid, i.e., the arachidonate cascade.[4,7] In this regard, it has been shown that trace amounts of oxygen free radicals and/or lipid hydroperoxides are needed to start the key enzymes of the arachidonate cascade, i.e., cycloxygenase and lipoxygenases.[39] Thus the activities of those enzymes are under the control of the ambient level of free radicals.[39] Our own study also showed that 5-lipoxygenase of the canine basilar artery is markedly activated by 15-HPETE. In regard to SAH, therefore, it may be conjectured that lipid peroxidation triggered by degradation products of hemoglobin leads to a rise in the level of ambient hydroperoxides, which then stimulates the arachidonate cascade within the adjacent tissues (Figure 12).[4] The products of thus stimulated arachidonate cascade may differ depending on the kind of the tissue, because each tissue has its own preferential pathway. For example, the preponderant lipoxygenase of the canine RBCs is known to be 12-lipoxygenase,[37] which fits our data. The main product of the cerebral artery would be LTs, since it was shown to possess only 5-lipoxygenase activity. The prominent increases in the CSF levels of prostaglandins, as well as LTs following SAH,[9,75] are

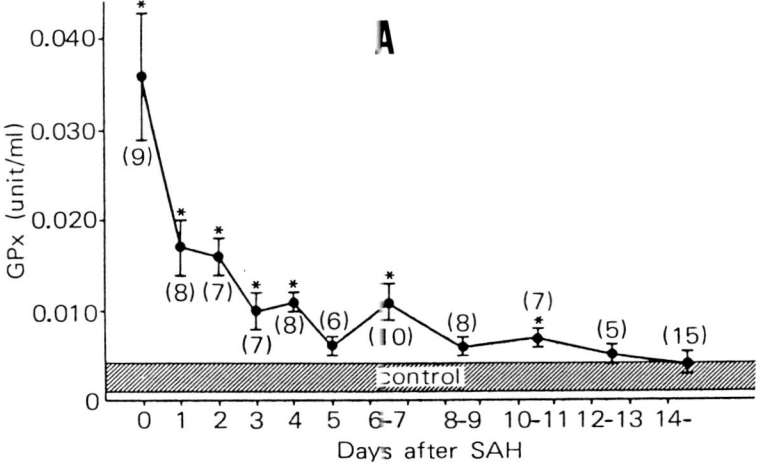
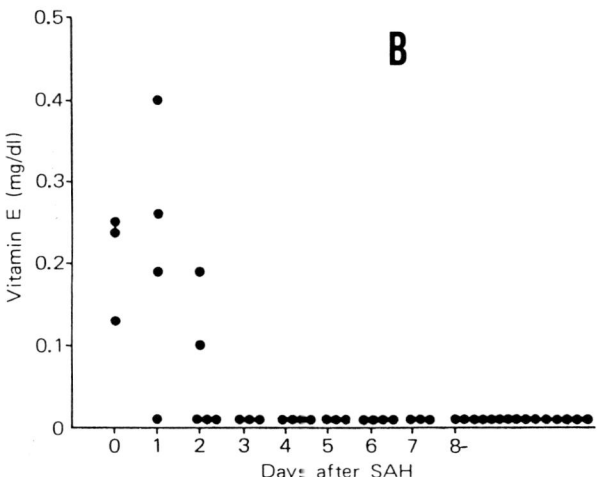

FIGURE 8. The alteration of the CSF free radical scavenging system following SAH. (A) The activity of glutathione peroxidase (GPx);[77] (B) vitamin E content;[77] (C) the activity of superoxide dismutase (SOD);[57] (D) the activity of catalase.[57] Figures 9C and D with permission of the author. Note the rapid decrease in the GPx activity and the vitamin E content (A, B). Whereas the SOD activity was significantly lower in patients with vasospasm than in those without, the activity of catalase revealed a reverse relationship (D).

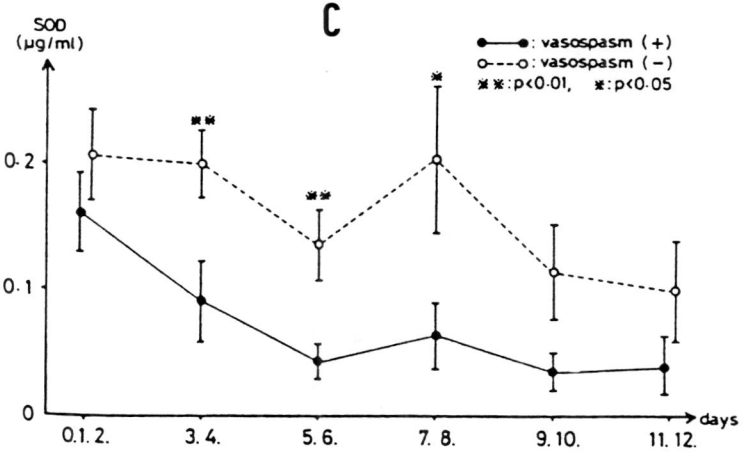
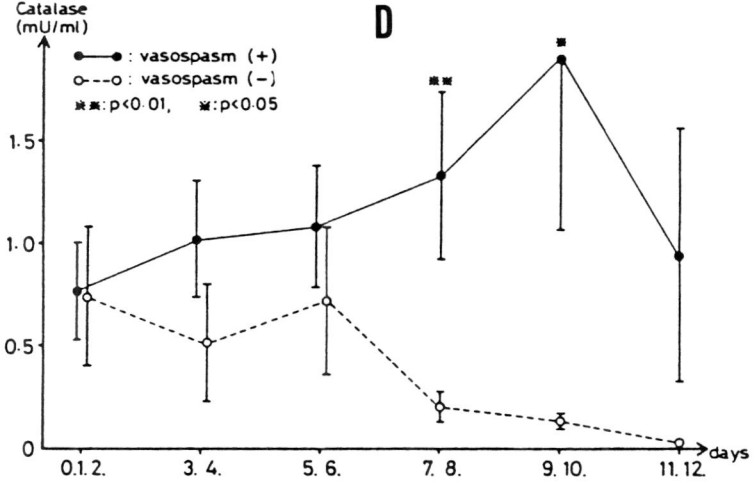

FIGURE 8 (continued)

thus likely to be owing to a generalized enhancement of the preferential pathway of the arachidonate cascade in every tissue in contact with the subarachnoid clot. In support of this view, we have shown that administration of free radical scavengers (glutathione or AVS) not only ameliorated vasospasm but also reduced the activity of 5-lipoxygenase in the basilar artery in the canine two-hemorrhage model.[76]

TABLE 2
The HETE Content in Each of the CSF, Subarachnoid Clot, and the Basilar Artery

	Control	Day 1	Day 2	Day 3	Day 5	Day 8	Recovery (%)
CSF		ND (n = 5)	1.88 ± 2.00 (n = 5)	1.29 ± 0.60 (n = 5)	176 ± 2.43 (n = 5)	ND (n = 5)	98.4 ± 8.2 (n = 10)
Subarachnoid clot (12-HETE, nmol/g tissue)			ND (n = 5)	ND (n = 5)	ND (n = 5)	1.89 ± 2.16 (n = 6)	89.6 ± 13.3 (n = 21)
Basilar artery	ND (n = 5)					ND (n = 6)	97.2 ± 10.7 (n = 26)

Note: ND = not detected.

Assay of Arachidonate 5-Lipoxygenase in Basilar Artery

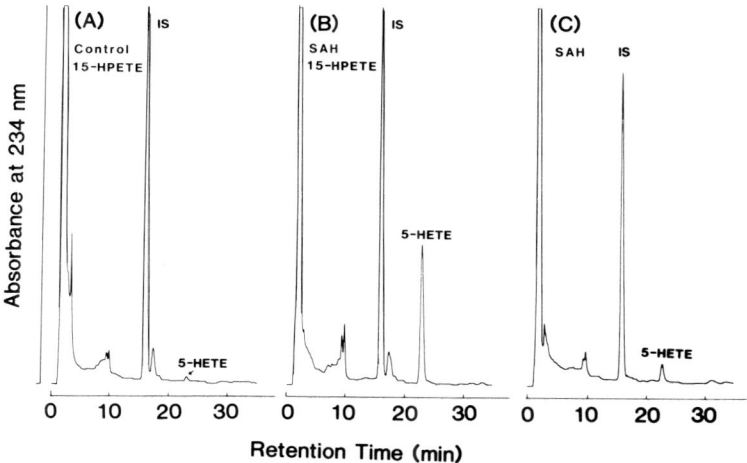

LT Formation in Basilar Artery after SAH

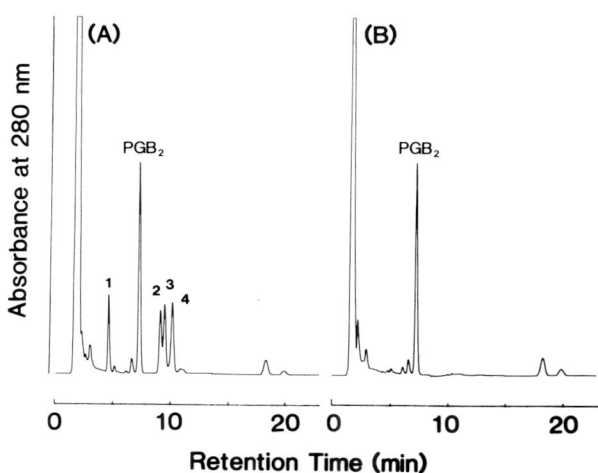

FIGURE 9. The 5-lipoxygenase activity of the canine basilar artery following experimental SAH. Above: the normal basilar artery produced a minute amount of 5-HETE on incubation with arachidonic acid and A23187 (A), which became greater following SAH (C), and further enhanced by additon of 15-HPETE (B). IS: internal standard. Below: Leukotriene (LT) formation in the normal basilar artery (B) and in the artery exposed to SAH (A). 1, LTC_4; 2, 3, and 4, 6-*trans*-LTB_4, 12-epi-6-*trans*-LTB_4, and LTB_4, respectively.

Activation of 5-Lipoxygenase by Hydroperoxy Acids

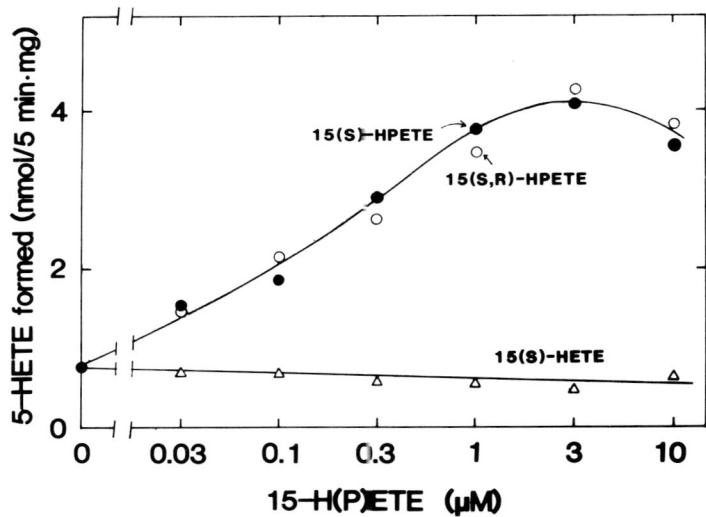

FIGURE 10. Activation of 5-lipoxygenase of the canine basilar artery by 15-HPETE and 15-HETE. Note the similar activation by either of 15(S)- and 15(S,R)-HPETEs. 15(S)-HETE showed no action.

Inferred from the above view is that each product of the arachidonate cascade, the generation of which is enhanced, may play some role in the occurrence of vasospasm, since eicosanoids are known to have potent biological actions, particularly in terms of smooth muscle contraction and inflammatory reactions.[28,59,82] While this surmise remains to be verified, it seems to be an interesting avenue to explore, because organic changes, presumably of an inflammatory nature, are invariably seen in the wall of the spastic artery. The apparent associations between the known action of each lipoxygenase product and each feature of the organic changes in vasospasm are depicted in Figure 13.

In the face of the rather rapid disappearance from the CSF of 12-HPETE and its metabolite 12-HETE, the effect of 12-HPETE causing a prolonged basilar artery contraction is bewildering. Whereas 12-HPETE has a direct vasocontractile action like its isomer, 15-HPETE, 12-HETE has none (unpublished data). However, since the involvement of the direct vasocontractile effect of 12-HPETE is inconceivable because of its rapid disappearance from the CSF, other effects of the agent must be sought to explain the occurrence of the prolonged vasoconstriction. In this respect, it seems possible that 12-HPETE is metabolized within the cerebral artery to some more stable metabolite(s) that has a potent vasocontractile action. Considering the time course

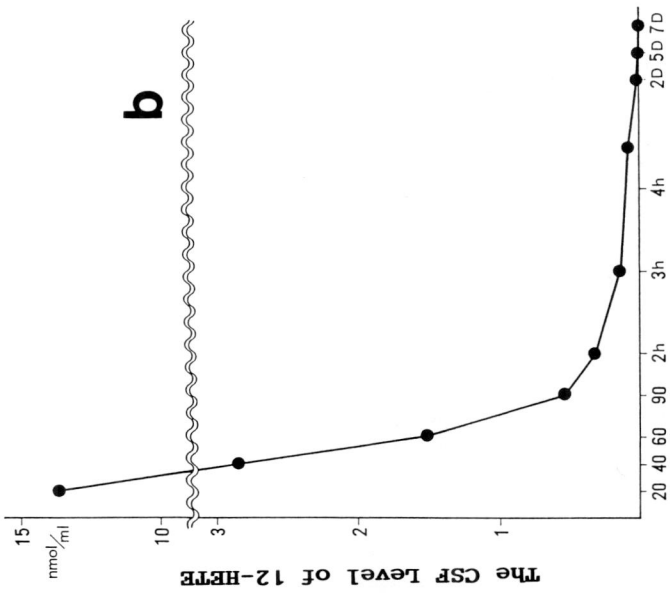
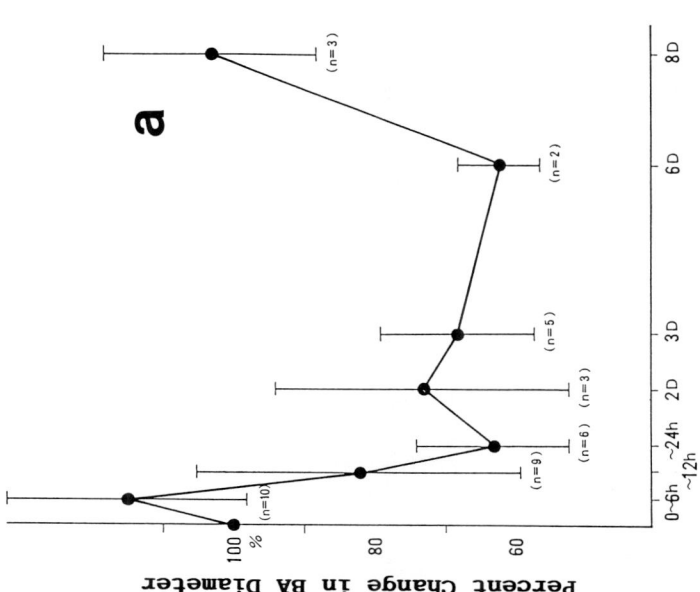

FIGURE 11. (a) The change in the angiographical diameter of the basilar artery following the intracisternal injection of 12-HPETE. Note the transient dilation followed by prolonged contraction. (b) The time-course of the CSF content of 12-HETE following the intracisternal injection of 12-HPETE.

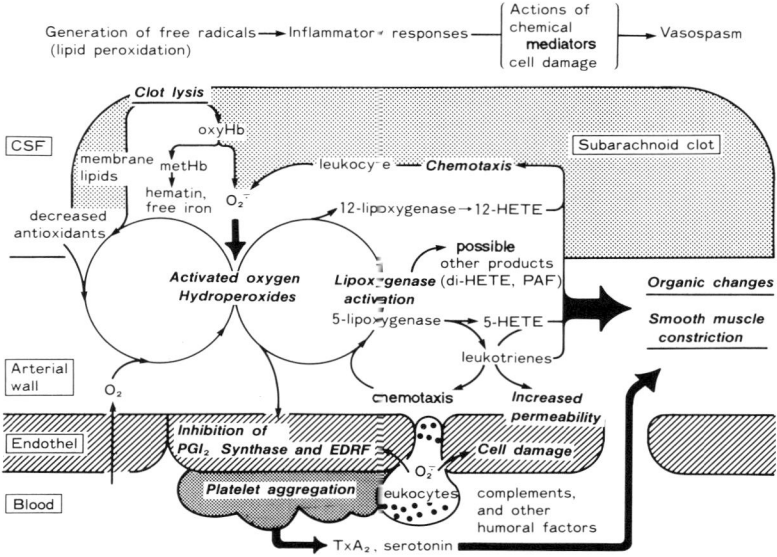

FIGURE 12. The possible interactions between lipid peroxidation and the arachidonate cascade occurring in tissues and cells, being in contact with the subarachnoid clot.

of the arterial constriction, however, this seems rather unlikely. An alternative possibility would be that 12-HPETE and/or 12-HETE exert an irreversible toxic effect on the arterial smooth muscle, leading to a prolonged constriction. To explore this possibility, we surveyed the literature in the hope of finding a clue to elucidate the seemingly distant relationship between 12-HPETE and prolonged arterial constriction.

V. PROTEIN KINASE C-MEDIATED CONTRACTION OF SMOOTH MUSCLES: DEUS EX MACHINA?

That uncertainties remain as to the mechanism operating in the sustained phase of agonist-induced smooth muscle contraction was disclosed by a survey of relevant literature. The contraction of smooth muscle in response to various agonists is usually composed of a fast, "phasic", and a slower, sustained "tonic" component. Inasmuch as the thesis that the influx of external Ca^{2+} heralds the train of events leading to phosphorylation of MLC and force generation (Figure 13) has been widely accepted, recent studies disclosed discordant findings, particularly concerning the mechanisms operating in the sustained contraction. Namely, it has been shown that both the agonist-induced increases in the intracellular concentration of Ca^{2+} and in the amount of phosphorylated MLC are transient, whereas the force is maintained during the sustained contraction.[19,32,70] After the initial force generation, therefore,

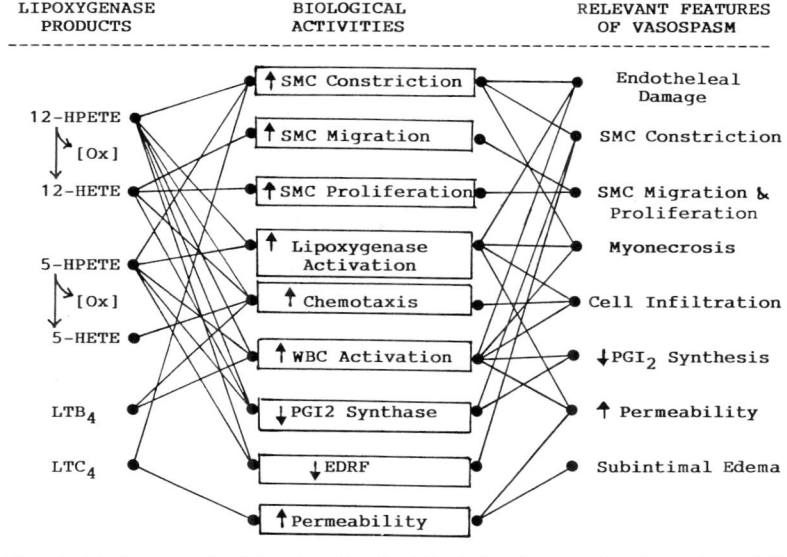

FIGURE 13. Possible relationships between each lipoxygenase product, its biological activities, and the feature of organic changes associated with vasospasm.

the Ca^{2+}/CaM/MLCK system is considered to cease its function, and it must be replaced by some other mechanism whereby the force is maintained in the succeeding tonic phase. As such a mechanism, Aksoy et al. conjectured that MLC phosphorylation leads to formation of latch bridges, which are involved in the maintenance of tension with reduced cross-bridge cycling rates.[1] While the formation of latch bridges has not been substantiated as yet, Rasmussen et al. put forth a novel hypothesis that the maintenance of force during the sustained phase of contraction mainly depends on the phosphorylation of cytosolic contractile proteins, owing to the action of protein kinase C.[54,69,70,71]

Briefly, while agonists activate phospholipase C (PLC) leading to stimulation of the phosphatidylinositol (PI) cycle, subsequently released inositol-1,4,5-triphosphate (IP3) in turn stimulates the release of Ca^{2+} from the intracellular store site, rendering the Ca^{2+}/CaM-dependent protein kinases such as MLCK fully active.[12] On the other hand, the simultaneously released 1,2-diacylglycerol (DAG) stabilizes the membrane-bound form of protein kinase C (PKC) to the cell membrane, thereby markedly enhancing its affinity to Ca^{2+}. Hence, the enzyme becomes fully activated even without a net increase in the cytosolic Ca^{2+} concentration.[36,47,48] Although the intracellular release of IP3 and DAG formation are transient, the effect of PKC stimulation is rather long-lasting because the resulting phosphoproteins owing to PKC stimulation are relatively resistant to the action of phosphatases.[36] Thus, these

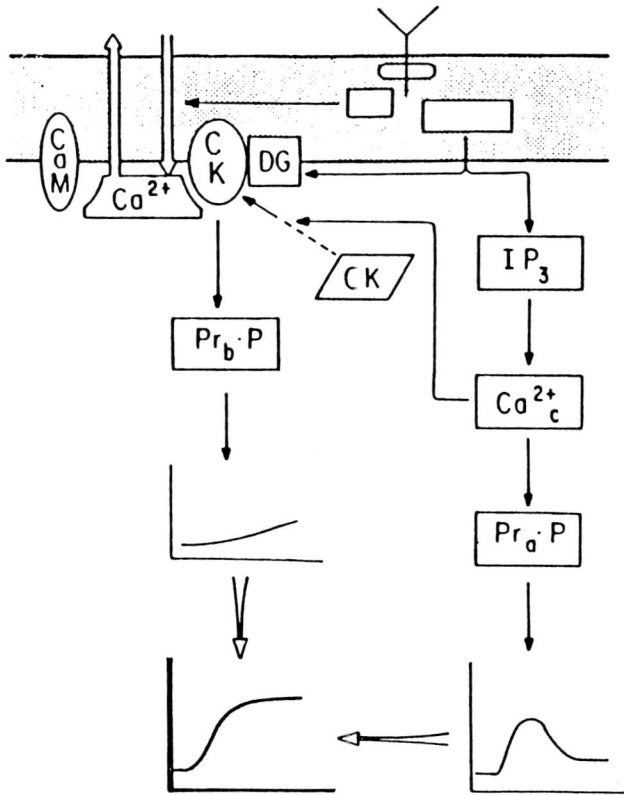

FIGURE 14. Two pathways of smooth muscle contraction. $Pr_a \cdot P$: phosphorylation of proteins (myosin by the Ca^{2+}/CaM/MLCK system leading to a phasic contraction. $Pr_b \cdot P$: phosphorylation of contractile proteins other than myosin by the protein kinase C (CK) system leading to a slowly developing, sustained contraction. The agonist-induced contraction is considered as a sum of the above two responses.

authors conjectured that there are two pathways by which agonists activate contraction: a Ca^{2+}/CaM pathway which initiates the response, and a PKC pathway which, along with the Ca^{2+}/CaM pathway, sustains contraction (Figure 14).

This hypothesis, taken in conjunction with the fact that smooth muscle is rich in C-kinase,[45] triggered a surge of interest in the role of PKC in eliciting contraction of arterial smooth muscles, and it has so far been supported by findings that phorbol esters induce sustained contraction in various types of smooth muscles, which is partially dependent on external Ca^{2+} [10,15,17,18,22,25,26,35,43,44,50,53,65,68,71] and accompanied by phosphorylation of proteins other than MLC.[15,65,71]

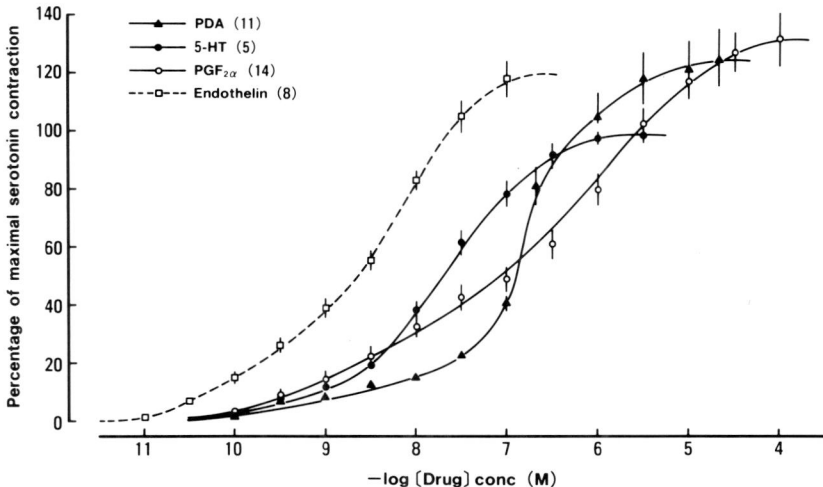

FIGURE 15. The contractile response of the canine basilar artery to each of phorbol-1,2-acetate (PDA), serotonin (5-HT), $PGF_{2\alpha}$, and endothelin.

More than being attractive, the above thesis seems especially pertinent to the pathogenesis of vasospasm as it primarily concerns the mechanism of sustained smooth muscle contraction. The partial dependence of the PKC-system on external Ca^{2+} is a likely explanation as to why vasospasm is unresponsive to calcium antagonists. It may be further speculated that activation of PKC may be the mechanism whereby 12- and 15-HPETE induce prolonged vasoconstriction, since various lipids including lipoxygenase products are known to directly activate PKC.[21] Of particular interest in this respect is the report showing that 12- and 15-HETE inhibit DAG kinase, leading to an increased intracellular level of DAG, an intrinsic activator of PKC.[63]

Thus, we undertook studies to see whether or not the PKC system as shown above exists in the canine cerebral artery, playing a significant role in the occurrence of vasospasm following experimental SAH.[85]

VI. PHORBOL ESTER-INDUCED CONTRACTION OF THE CANINE BASILAR ARTERY

In the *in vitro* study according to the method of Allen,[2] a phorbol ester, phorbol-1,2-diacetate (PDA) was shown to induce a slowly developing contraction of the beagle basilar artery segment. The dose-response curve to PDA in comparison to those of other agonists is shown in Figure 15.[85] In the Ca^{2+}-free medium, there was a significant (the ED_{50} increasing by one order) rightward shift of the dose response curve to PDA, while the maximal tension developed was unchanged. Thus, the PDA-induced contraction in the canine basilar artery was shown to be only partially dependent on external Ca^{2+}.[85]

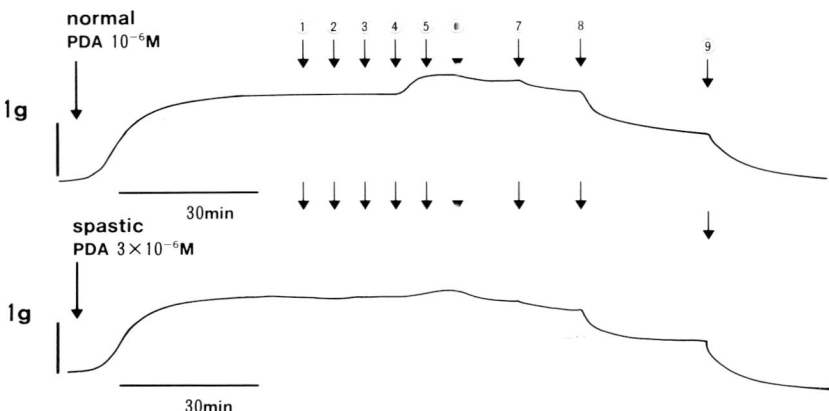

FIGURE 16. Tracings of actual recordings of the contraction induced by PDA. Above: the normal basilar artery; below: the spastic basilar artery. Each shows responses to PDA, and the cumulative addition of (1) atropine, (2) diphenhydramine, (3) phentolamine, (4) methysergide, (5) R24571 (a calmodulin inhibitor), (6) W-7 (a calmodulin inhibitor), (7) nicardipine (10^{-6} M), (8) H-7 (10^{-5} M), and (9) H-7 (10^{-4} M). Note the similar responses of the two arteries to addition of each agent.

The contraction developed by PDA was not significantly affected by cumulative addition of various antagonists to neurotransmitters, inhibitors of calmodulin, or a calcium antagonist, nicardipine. But, a relatively specific inhibitor of PKC, H-7,[29] markedly inhibited the PDA-induced contraction (Figure 16).[84] Although the data are not shown, pretreatment with staurospolin (a PKC inhibitor) at the concentration of 10^{-7} M, completely abolished the PDA-induced contraction. Both in the normal and the Ca^{2+}-free mediums, 8-bromo-cyclic GMP dose-dependently inhibited the PDA-induced contraction, whereas 8-bromo-cyclic AMP did not (Figure 17).[85]

Figure 18 shows the pattern of protein phosphorylation of the beagle basilar artery stimulated by PDA in the Ca^{2+}-free medium. While PDA caused a significant contraction of the artery in this condition, phosphorylation of 20 kDa protein (MLC) did not occur either in the control or the PDA-treated artery. But, proteins of 27- and 96 kDa were markedly phosphorylated in the PDA-stimulated artery.[85]

The above results suggest that PDA-induced contraction is mediated by activation of PKC, which phosphorylates proteins other than MLC even in the absence of external Ca^{2+}. This independence of PKC activity on external Ca^{2+} suggests that the canine basilar artery has the α-subtype of PKC,[48] which, recently was verified as such. Thus, the canine basilar artery is considered to have a powerful contractile system mediated by activation of PKC, in addition to that mediated by the Ca^{2+}/CaM/MLCK system.

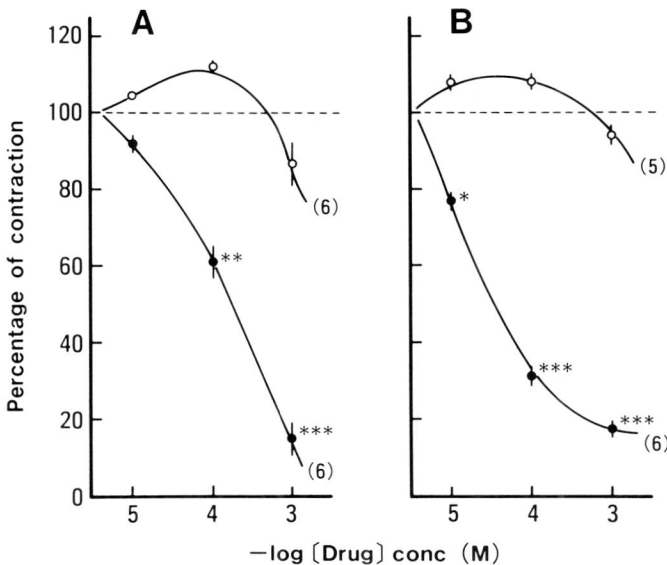

FIGURE 17. The effect of addition of 8-bromo-cyclic AMP and -GMP on the PDA-induced contraction of the canine basilar artery. (A) Effects in the normal Krebs-Henseleit solution; (B) effects in th Ca^{2+}-free solution. Note the dose-dependent inhibition of PDA-induced contraction by 8-bromo-cyclic GMP both in the normal and Ca^{2+}-free mediums.

VII. RELEVANCE OF THE PKC-SYSTEM TO LATE SPASM

Role of the PKC system in the occurrence of late spasm was evaluated *in vivo* as follows. Beagles were exposed to two-hemorrhage[73] and on day 7, the basilar artery undergoing late spasm was exposed via the transclival approach. The clot and the arachnoid membrane surrounding the basilar artery were meticulously removed, and the artery was superfused with the aerated Krebs-Henseleit solution (KHS) containing various pharmacological agents. The changes in the arterial diameter were recorded by repeated angiography. As shown in Figure 19, the constricted basilar artery did not respond to superfusion with KHS alone or with that containing each of the antagonists to neurotransmitters, inhibitors of calmodulin, and nicardipine. But, the basilar artery markedly and dose-dependently dilated on application of KHS containing H-7. This dilation induced by H-7 was reversible.[84]

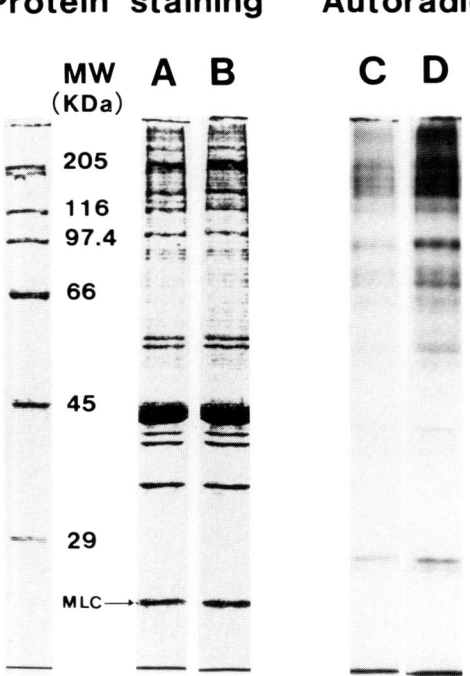

FIGURE 18. Protein phosphorylation pattern of the canine basilar artery induced by PDA. There was no difference in the amount of proteins obtained from the normal (A) and the PDA-treated (B) arteries. Addition of PDA in the Ca^{2+}-free medium caused marked increases in the phosphorylation of 27- and 96-kDa proteins (C) untreated, (D) PDA-treated. Note the absence of MLC phosphorylation in either the normal or the PDA-treated artery.

The basilar artery segment exposed to two-hemorrhage (day 7) was also subjected to an *in vitro* chamber study to examine its responsiveness to various agents. It would suffice to state here that the maximal tension developed by agonists such as $PGF_{2\alpha}$, serotonin, or PDA was slightly but significantly decreased in the spastic basilar artery as compared to that of the normal one, whereas the responsiveness to each agent in terms of ED_{50} values was unchanged.

Since the above result strongly points to the activation of PKC system, we proceeded to examine if the tissue level of the intrinsic activator of PKC, DAG, is increased in the basilar artery following SAH. Using the beagle two-hemorrhage model, the basilar artery, the diameter of which was measured by angiography just before sacrifice, was excised at 1 and 48 h after the first SAH and at 48 and 144 h after the second SAH. At the same time, samples of the CSF and the subarachnoid clot were obtained. The DAG content of each sample was assayed using the diacylglycerol kinase method.[52] The

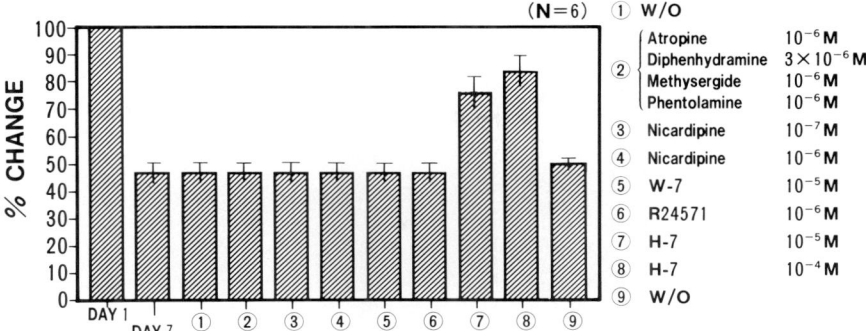

FIGURE 19. The change in the angiographical diameter of the spastic basilar artery in response to superfusion with various pharmacological agents. The figure shows the mean values of six dogs. Spasm of the basilar artery *in vivo* was not affected by wash with the normal Krebs-Henseleit solution (W/O) or other agents, except for H-7. Note the similarity in response between the *in vitro* (Figure 17) and *in vivo* conditions.

study disclosed a significant increase in the DAG content of the basilar artery from day 3 (48 h after the first SAH) to day 7 (Figure 20), whereas DAG was undetected in either the CSF or the clot.[84] This increase in the DAG content of the basilar artery was in the range of 150 to 190% of the normal basilar artery, which is sufficient to activate PKC.[69] Furthermore, a crude linear correlation was revealed between the DAG content and the angiographic diameter of each of the basilar arteries examined as shown in Figure 21.[84] Thus, our results show that the spastic artery *in vivo* exhibits the same responses to a battery of pharmacological agents as the normal artery stimulated by PKC *in vitro* does.

Noteworthy are the unresponsiveness of the spastic basilar artery to nicardipine, and its marked dilation by H-7. It may also be emphasized that the significant increase in the DAG content is consistent with the protracted time-course of late spasm and that it was roughly correlated with the severity of angiographical spasm.

Although much remains to be clarified, our data strongly suggest the involvement of PKC activation in the occurrence of late spasm. Since DAG was not detected in the CSF or the subarachnoid clot, the increase in the DAG content is presumably ascribed to some derangement in its synthesis and/or metabolism within the arterial wall. It is tempting to speculate that lipid peroxides and their metabolites, especially 12-HPETE and 12-HETE, trigger the whole train of events through inhibition of diacylglycerol kinase within the cerebral artery. Further investigation along these lines is in progress in our laboratory.

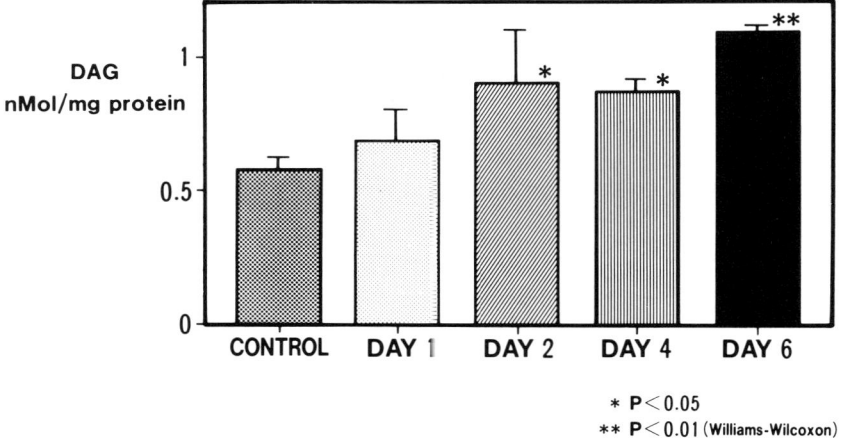

FIGURE 20. The 1,2-diacylglycerol content of the basilar artery exposed to two-hemorrhage. Note the significant increase from day 2 to day 6.

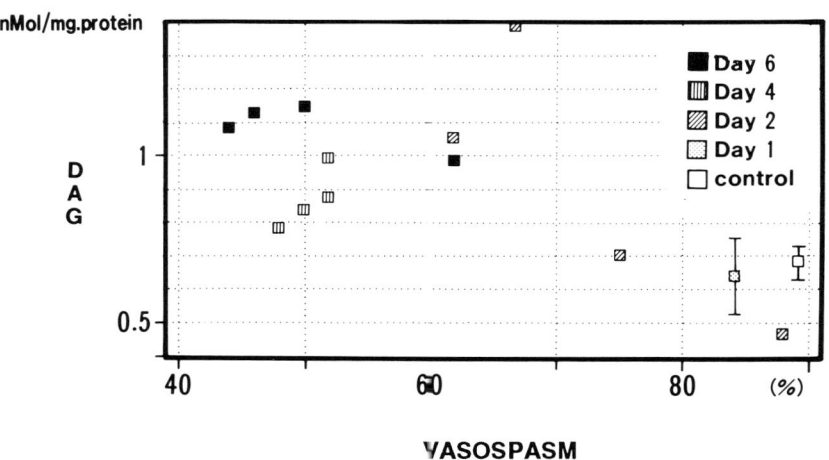

FIGURE 21. The scattergram of the angiographical diameter of the basilar artery following two hemorrhage (the ordinate) and the corresponding diacylglycerol content (DAG: the abscissa). The mean ± SD of values on day 1 is shown at the right, below. Note the existence of a crude correlation (r = 0.68, $p < 0.01$).

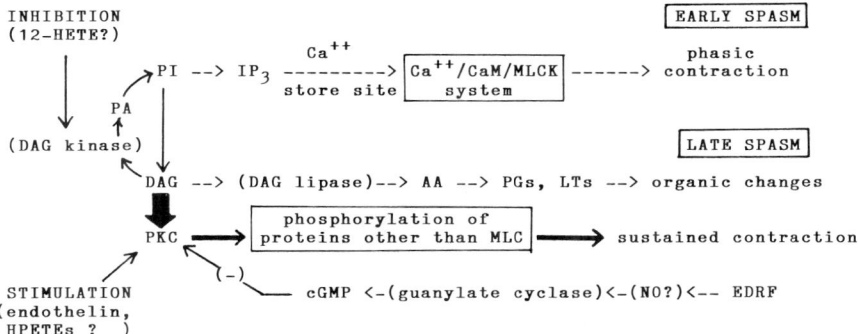

FIGURE 22. The suggested role of the protein kinase C (PKC) system activated by an increased tissue level of 1,2-diacylglycerol (DAG) in the occurrence of late spasm. Note that the Ca^{2+}/CaM/MLCK system participates only to manifestation of phasic contraction, to which the early spasm may be attributed. Protein kinase C is considered to be activated mainly through an increase in the tissue DAG level, but it may be directly activated by other agents such as HPETEs and endothelin. Endothelium-derived relaxing factor (EDRF) may thereby exert an inhibitory effect through generation of cyclic GMP (NO: nitrous oxide). Presently, inhibition of DAG kinase by 12-HETE is the most likely cause of the increase in the DAG level following SAH.

VIII. CONCLUSION

Our hypothesis concerning the pathogenetic mechanism of cerebral vasospasm is schematized in Figure 22. Conjectured in the scheme is that an active, sustained contraction of the arterial smooth muscle mediated by activation of the PKC system, not by the Ca^{2+}/CaM/MLCK system, plays a significant role in the occurrence of late spasm, together with the development of organic changes in the arterial wall. In this way, the enigma as to the role of smooth muscle contraction in the occurrence of vasospasm may be resolved. While the scheme focuses on the mechanism of sustained smooth muscle contraction, the presumed role of eicosanoids, particularly of the lipoxygenase products in the occurrence of inflammatory changes in the arterial wall, may be emphasized. Biochemically, indeed, the processes involved in activations of PKC and the lipoxygenase pathway are quite akin to each other and in some parts are overlapped. Inasmuch as our hypothesis needs further substantiation, it certainly provides a particularly interesting avenue to explore.

REFERENCES

1. **Aksoy, M. O., Murphy, R. A., and Kamm, K. E.,** *Am. J. Physiol.*, 242, C109–C116, 1982.
2. **Allen, G. S., Henderson, L. M., Chou, S. N., and French, L. A.,** *J. Neurosurg.*, 40, 433–441, 1974.
3. **Allen, G. S., Ahn, H. S., Preziosi, T. J., Battye, R., Boone, S. C., Chou, S. N., Kelly, D. L., Weir, B. K., Crabbe, R. A., Lavik, P. J., Rosenbloom, S. M., Dorsey, F. C., Ingram, C. R., Mellits, D. E., Bertsch, R. N., Boisvert, D. P. J., Hundley, M. B., Johnson, R. K., Strom, J. A. and Transou, C. R.,** *N. Engl. J. Med.*, 308, 619–624, 1983.
4. **Asano, T.,** in *Acute Aneurysm Surgery,* Sano, K., Asano, T., and Tamura, A., Eds., Springer-Verlag, Wien, 1987, 53–74.
5. **Asano, T., Tanishima, T., Sasaki, T., and Sano, K.,** in *Cerebral Arterial Spasm,* Wilkins, R. H., Ed., Williams & Wilkins, Baltimore, 1980, 190–201.
6. **Asano, T., Sasaki, T., Koide, T., Takakura, K., and Sano, K.,** *Neurol. Res.*, 6, 49–53, 1984.
7. **Asano, T., Watanabe, T., Takakura K., Sano, K., and Shimizu, T.,** in *Cerebral Vasospasm,* Wilkins, R. H., Ed., Raven Press, New York, 1988, 297–302.
8. **Auer, L. M.,** *Neurosurgery,* 15, 57–65, 1985.
9. **Baena, R. R., Gaetani, P., and Paoletti, P.,** *J. Neurol. Sci.*, 84, 329–335, 1988.
10. **Baraban, J. M., Gould, R. J., Peroutka, S. J., and Snyder, S. H.,** *Proc. Natl. Acad. Sci. U.S.A.*, 82, 604–607, 1985.
11. **Barber, A. A. and Bernheim, F.,** *Adv. Gerontol. Res.*, 2, 355–403, 1967.
12. **Berridge, D. M. J.,** *Annu. Rev. Biochem.*, 56, 159–193, 1987.
13. **Bevan, J. A. and Bevan, R. D.,** *Annu. Rev. Pharmacol. Toxicol.*, 28, 311–329, 1988.
14. **Brawley, B. M., Strandness, D. E., Jr., and Kelly, W. A.,** *J. Neurosurg.*, 28, 1–8, 1968.
15. **Chatterjee, M. and Tejada, M.,** *Am. J. Physiol.*, 251, C356–C361, 1986.
16. **Conway, L. W. and McDonald, L. W.,** *J. Neurosurg.*, 37, 715–723, 1972.
17. **Dale, M. M. and Obianime, W.,** *FEBS Lett.*, 190, 6–10, 1985.
18. **Danthurluri, N. R. and Deth, R. C.,** *Biochem. Biophys. Res. Commun.*, 125, 1103–1109, 1984.
19. **DeFeo, T. T. and Morgan, K. G.,** *J. Physiol.*, 369, 269–282, 1985.
20. **Espinosa, F., Weir, B., Overton, T., Castor, W., Grace, M., and Boisvert, D.,** *J. Neurosurg.*, 60, 1167–1175, 1984.
21. **Farooqui, A. A., Farooqui, T., Yates, A. J., and Horrocks, L. A.,** *Neurochem. Res.*, 13, 499–511, 1988.
22. **Fish, R. D., Sperti, G., Colucci, W. S., and Clapham, D. E.,** *Circ. Res.*, 62, 1049–1054, 1988.
23. **Fisher, C. M., Robertson, G. H., and Ojemann, R. G.,** *Neurosurgery,* 1, 245–248, 1977.
24. **Fisher, C. M., Kistler, J. P., and Davis, J. M.,** *Neurosurgery,* 6, 1–9, 1980.
25. **Forder, J., Scriabine, A., and Rasmussen, H.,** *J. Pharmacol. Exp. Ther.*, 235, 267–273, 1985.
26. **Gleason, M. M. and Flaim, S. F.,** *Biochem. Biophys. Res. Commun.*, 138, 1362–1369, 1986.
27. **Godfraind, T., Miller, R., and Wibo, M.,** *Pharmacol. Rev.*, 38, 321–416, 1986.
28. **Goetzl, E. J.,** *Med. Clin. North Am.*, 65, 809–828, 1981.
29. **Hidaka, H., Inagaki, M., Kawamoto, S., and Sasaki, Y.,** *Biochemistry,* 23, 5036–5041, 1984.
30. **Hughes, J. T. and Schianchi, P. M.,** *J. Neurosurg.*, 48, 515–525, 1978.
31. **Jackson, I. J.,** *Arch. Neurol. Psychiatry,* 65, 572–589, 1949.

32. **Kamm, K. E. and Stull, J. T.**, *Annu. Rev. Pharmacol. Toxicol.*, 25, 593–620, 1985.
33. **Kassel, N. F., Torner, J. C., and Drake, C. G.**, *Stroke*, 15, 566–570, 1984.
34. **Kellogg, E. W. and Fridovich, I.**, *J. Biol. Chem.*, 252, 6721–6728, 1977.
35. **Khalil, R. A. and van Breemen, C.**, *J. Pharmacol. Exp. Ther.*, 244, 537–542, 1987.
36. **Kikkawa, U. and Nishizuka, Y.**, *Annu. Rev. Cell. Biol.*, 2, 149–178, 1986.
37. **Kobayashi, T. and Levine, L.**, *J. Biol. Chem.*, 258, 9116–9121, 1983.
38. **Koide, T., Neichi, T., Takato, M., Matsushita, H., Sugioka, K., Nakano, M., and Hata, S.**, *J. Pharmacol. Exp. Ther.*, 221, 481–488, 1982.
39. **Lands, W. E. M., Kulmacz, R. J., and Marshall, P. J.**, in *Free Radical in Biology*, Vol. VI, Pryor, W. A., Ed., Academic Press, Orlando, FL, 1984, 39–63.
40. **Lewis, P. J., Weir, B. K. A., Nosko, M. G., Tanabe, T., and Grace, M. G.**, *Neurosurgery*, 22, 492–500, 1988.
41. **Liszczak, T. M., Varsos, V. G., Black, P. M., Kistler, J. P., and Zervas, N. T.**, *J. Neurosurg.*, 58, 18–26, 1983.
42. **Mee, E., Dorrance, D., Lowe, D., and Neil-Dwyer, G.**, *Neurosurgery*, 22, 484–491, 1988.
43. **Menkes, H., Baraban, J. M., and Snyder, S. H.**, *Eur. J. Pharmacol.*, 122, 19–27, 1986.
44. **Miller, J. R., Hawkins, D. J., and Wells, J. N.**, *J. Pharmacol. Exp. Ther.*, 239, 38–42, 1986.
45. **Minakuchi, R. Y., Takai, B., Yu, B., and Nishizuka, Y.**, *J. Biochem.*, 89, 1651–1654, 1981.
46. **Misra, H. P. and Fridovich, I.**, *J. Biol. Chem.*, 247, 6960–6962, 1972.
47. **Nishizuka, Y.**, *Science*, 225, 1365–1370, 1984.
48. **Nishizuka, Y.**, *Science*, 233, 305–312, 1986.
49. **Nosko, M., Weir, B., Krueger, C., Cook, D., Norris, S., Overton, T., and Boisvert, D.**, *Neurosurgery*, 16, 129–136, 1985.
50. **Park, S. and Rasmussen, H.**, *Proc. Natl. Acad. Sci. U.S.A.*, 82, 8835–8839, 1985.
51. **Pellettieri, L., Bolander, H., Carlsson, H., and Sjoelander, U.**, *Surg. Neurol.*, 30, 180–186, 1988.
52. **Preiss, J., Loomis, C. R., Bishop, W. R., Stein, R., Miedel, J. E., and Bell, R. M.**, *J. Biol. Chem.*, 261, 8597–8600, 1986.
53. **Rasmussen, H., Forder, J., Kojima, L., and Scriabine, A.**, *Biochem. Biophys. Res. Commun.*, 122, 776–784, 1984.
54. **Rasmussen, H., Takuwa, Y., and Park, S.**, *FASEB J.*, 1, 177–185, 1987.
55. **Saito, I., Ueda, Y., and Sano, K.**, *J. Neurosurg.*, 47, 412–429, 1977.
56. **Saito, I., Shigeno, T., Aritake, K., Tanishima, T., and Sano, K.**, *J. Neurosurg.*, 51, 466–475, 1979.
57. **Sakaki, S., Kuwabara, H., and Ohta, S.**, *Stroke*, 17, 196–202, 1986.
58. **Sakaki, S., Ohta, S., Nakamura, H., and Takeda, S.**, *J. Cereb. Blood Flow Metab.*, 8, 1–8, 1988.
59. **Samuelsson, B., Hammarstrom, S., Murphy, R. C., and Borgeat, P.**, *Allergy*, 35, 375–381, 1980.
60. **Sano, K.**, *Clin. Neurosurg.*, 30, 13–58, 1982.
61. **Sasaki, T., Wakai, S., Asano, T., Watanabe, T., Kirino, T., and Sano, K.**, *J. Neurosurg.*, 54, 357–365, 1981.
62. **Sasaki, T., Murota, S., Wakai, S., Asano, T., and Sano, K.**, *J. Neurosurg.*, 55, 771–778, 1982.
63. **Setty, B. N. Y., Graeber, J. E., and Stuart, M. J.**, *J. Biol. Chem.*, 262, 17613–17622, 1987.
64. **Shimizu, T., Watanabe, T., Asano, T., Seyama, Y., and Takakura, K.**, *J. Neurochem.*, 51, 1126–1131, 1988.
65. **Singer, H. A. and Baker, K. M.**, *J. Pharmacol. Exp. Ther.*, 243, 814–821, 1987.

66. Smith, R. R., Clower, B. R., Peeler, D. F., and Yoshioka, J., *Stroke*, 14, 240–245, 1983.
67. Smith, R. R., Clower, B. R., Grotendorst, G. M., Yabuno, N., and Cruse, J. M., *Neurosurgery*, 16, 171–176, 1985.
68. Sybertzt, E. J., Desiderio, D. M., Ttzloff, G., and Chiu, P. J. S., *J. Pharmacol. Exp. Ther.*, 239, 78–83, 1986.
69. Takuwa, Y., Takuwa, N., and Rasmussen, H., *J. Biol. Chem.*, 261, 14670–14675, 1986.
70. Takuwa, Y., Takuwa, N., and Rasmussen, H., *Am. J. Physiol.*, 253, C817–C827, 1987.
71. Takuwa, Y., Kelley, G., Takuwa, N., and Rasmussen, H., *Mol. Cell. Endocrinol.*, 60, 71–86, 1988.
72. Tappel, A. L., *Arch. Biochem. Biophys.* 44, 378–395, 1953.
73. Varsos, V. G., Liszczak, T. M., Dae, H. H., Kistler, J. P., Vielma, J., Black, P. M., Heros, R. C., and Zervas, N. T., *J. Neurosurg.*, 58, 11–17, 1983.
74. Vollmer, D. G., Kassel, N. F., Hongo, K., Ogawa, H., and Tsukahara, T., *Surg. Neurol.*, 311, 190–194, 1989.
75. Walker, V., Pickard, J. D., Smythe, P., Eastwood, S., and Perry, S., *J. Neurol. Neurosurg. Psychiatry*, 46, 119–125, 1983.
76. Watanabe, T., Asano, T., Shimizu, T., Seyama, Y., and Takakura, K., *J. Neurochem.*, 50, 1145–1150, 1988.
77. Watanabe, T., Sasaki, T., Asano, T., Takakura, K., Tuchinoue, T., Watanabe, K., Yoshimura, S., and Abe, K., *Neurol. Med. Chir. (Tokyo)*, 28, 645–649, 1988.
78. Weir, B., Grace, M., Hansen, J., and Rothberg, C., *J. Neurosurg.*, 48, 173–178, 1978.
79. Wilkins, R. H., *Neurosurgery*, 6, 198–2 0, 1980.
80. Willis, E. D., *Biochem. J.*, 99, 667–676, 1966.
81. Winterbourn, C. C., McGrath, B. M., and Carrell, R. W., *Biochem. J.*, 155, 493–502, 1976.
82. Wolfe, L. S., *J. Neurochem.*, 38, 1–14, 1982.
83. Yokota, M., Tani, E., Maeda, Y., and Kokubu, K., *Stroke*, 18, 512–518, 1987.
84. Matsui, T., Takuwa, Y., Johshita, H., Yamashita, K., and Asano, T., *J. Cerebr. Blood Flow Metabol.*, 11, 143-149, 1991.
85. Sugawa, M., Koide, T., Naitoh, S., Takato, M., Matsui, T., and Asano, T., *J. Cerebr. Blood Flow Metabol.*, 11, 135-142, 1991.

Chapter 10

STUDY OF FREE RADICAL GENERATION IN ISOLATED CELLS AND WHOLE TISSUES USING ELECTRON PARAMAGNETIC RESONANCE SPECTROSCOPY

Periannan Kuppusamy and Jay L. Zweier

TABLE OF CONTENTS

I. Introduction 242

II. Background 242
    A. Electron Paramagnetic Resonance Spectroscopy 242
    B. Oxygen Free Radicals 247

III. Applications 251
    A. Direct Measurements on Isolated Reperfused Hearts[13,15,42] 251
    B. Spin-Trapping Measurements on Isolated Reperfused Hearts[14,15,41] 254
    C. Spin Trap EPR Measurements on Cells[44] 257
    D. *In Vivo* EPR Methods 263

References 267

ISBN 0-8493-8091-X
© 1993 by CRC Press, Inc.

I. INTRODUCTION

Over the past two decades there has been increased interest in the possibility of cellular damage and dysfunction caused by oxidative biochemical reactions involving oxygen byproducts, particularly oxygen derived free radicals. Free radicals have been suggested to play an important role in a wide variety of clinical diseases including heart attack, stroke, respiratory distress syndrome, acute tubular necrosis of kidney, reperfusion injury of a wide variety of organs, and oncogenesis and tumor promotion.[1-7] Even the process of aging itself has been proposed to be due to the formation of reactive free radicals. In certain cases such as radiation injury and some chemical toxicities free radical injury has been considered to be the sole cause. Thus, free radicals have been proposed to mediate many of the most prevalent diseases causing morbidity and mortality.

Of the several methods available to study free radicals in biological tissues and cells, electron paramagnetic resonance, EPR, spectroscopy has been recognized to be the most important and direct technique. In this article we will describe the principles and techniques of EPR spectroscopy applied to study free radicals, with special emphasis on the application of EPR techniques to study oxygen mediated free radical injury in ischemic and reperfused myocardial cells and tissues.

II. BACKGROUND

A. ELECTRON PARAMAGNETIC RESONANCE SPECTROSCOPY

With the advances in microwave electronics and signal processing technology which occurred due to the development of radar during World War II it became possible to construct the Electron Paramagnetic Resonance (EPR) spectrometer. During 1945 and 1946 Zavoisky in the U.S.S.R. and Cummerow and Halliday in the U.S. performed the first EPR measurements.[8,9] By the 1950s and 1960s EPR was widely applied to study free radical and metal ion chemistry and it became the definitive techique for characterizing free radical intermediates in chemical reactions. During the 1960s and 1970s EPR was applied widely to biochemical systems and over the last decade EPR has been applied to measure free radicals in cells and tissues. Even more recently, over the last 5 years the first *in vivo* EPR experiments have been performed, and over just the last 2 years the first *in vivo* EPR imaging experiments of free radicals have been performed.[10-15]

EPR spectroscopy is concerned with the resonant absorption of microwave radiation by paramagnetic samples in the presence of an applied magnetic field. An electron, by virtue of its intrinsic spin, can exist in one of the two possible spin states. In the absence of any external magnetic field the two states are identical in energy and are thus said to be degenerate. In the presence of a magnetic field, however, interaction of the spin moments with the mag-

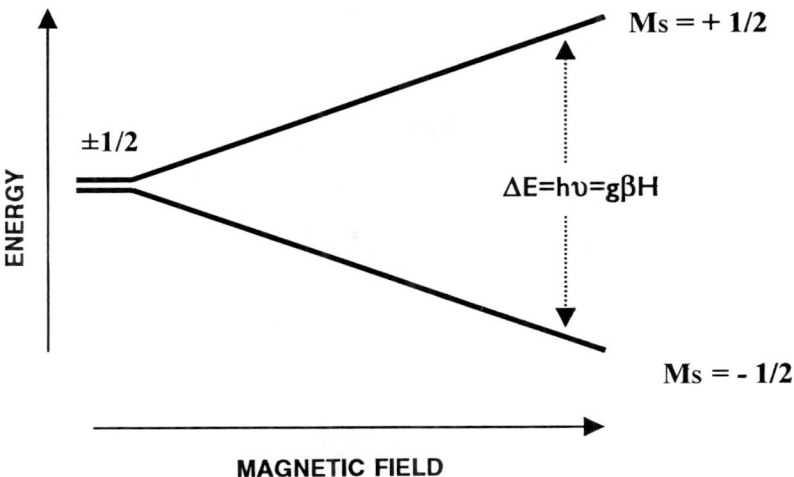

FIGURE 1. Zeeman splitting of a paramagnetic system with S = 1/2.

netic field results in a splitting of the energy levels of the spin states, thus causing the degeneracy to be lifted. This splitting is known as the Zeeman splitting (Figure 1). The magnitude of the splitting, ΔE, between the spin states is directly proportional to the strength of the applied magnetic field, H, as follows:

$$\Delta E = g\beta H$$

In the above expression g is a constant usually called *g-factor* which has a characteristic value for a given paramagnetic system and β is a constant called Bohr magneton.

Under conditions of EPR experiments the ground state ($m_s = -1/2$) will be excessively populated as per the Boltzmann distribution. Transition between the levels can be induced by irradiating the spin system with a radiation of energy equal to the splitting, ΔE. The frequency, υ of this radiation is given by

$$h\upsilon = g\beta H$$

where h is Planck's constant. For organic free radicals g will be approximately 2. Thus for a magnetic field strength of 3500 G the frequency υ is calculated to be around 9.5×10^9 Hz or 9.5 GHz. This frequency is in the microwave region of electromagnetic spectrum. When the condition given by the above expression is satisfied the free electron system will absorb the energy, $h\upsilon$, and go to the excited state ($m_s = +1/2$). The spin system is then said to be

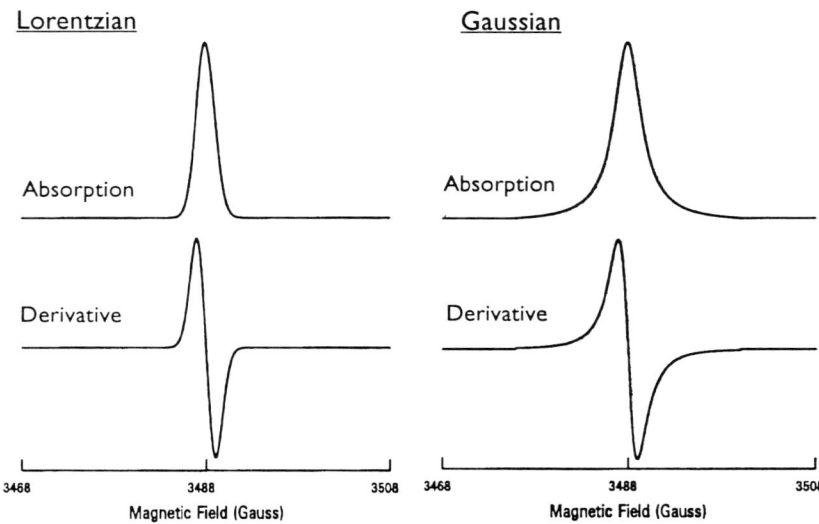

FIGURE 2. Typical EPR absorption shapes.

"in resonance". This resonant absorption of microwave radiation by paramagnetic systems placed in a magnetic field is termed as electron paramagnetic resonance.

In a typical continuous wave (CW) EPR experiment one irradiates the paramagnetic sample with a microwave radiation of constant frequency and sweeps the magnetic field to achieve the resonance condition. Typically an absorption function such as shown in Figure 2 is seen during the scan. In conventional EPR experiments, however, the derivative of absorption function is recorded. The correspondence between the two functions is shown in Figure 2.

The ~9 GHz region of the microwave energy spectrum is called X-band and hence EPR spectroscopy using this frequency is called X-band EPR spectroscopy. The EPR experiment, however, is not limited to X-band alone. Depending upon the sample conditions, sensitivity, information required, techniques available, etc. one can perform EPR measurements at a variety of microwave frequencies. Some other commonly used frequencies are Q-band (~35 GHz), K-band (~25 GHz), S-band (~3 GHz), L-band (~1 GHz), etc.

The area under absorption curve (single integral) or the derivative curve (double integral) is directly proportional to the amount of paramagentic material undergoing resonance in the cavity. Thus EPR can provide quantitative information. In addition, the splitting factor g is a tensor. It can provide a wealth of information regarding the symmetry and magnitude of the electron density distribution in the paramagnetic moiety. Based on the three principal values of the g-tensor, namely, g_{xx}, g_{yy} and g_{zz}, one can classify systems as

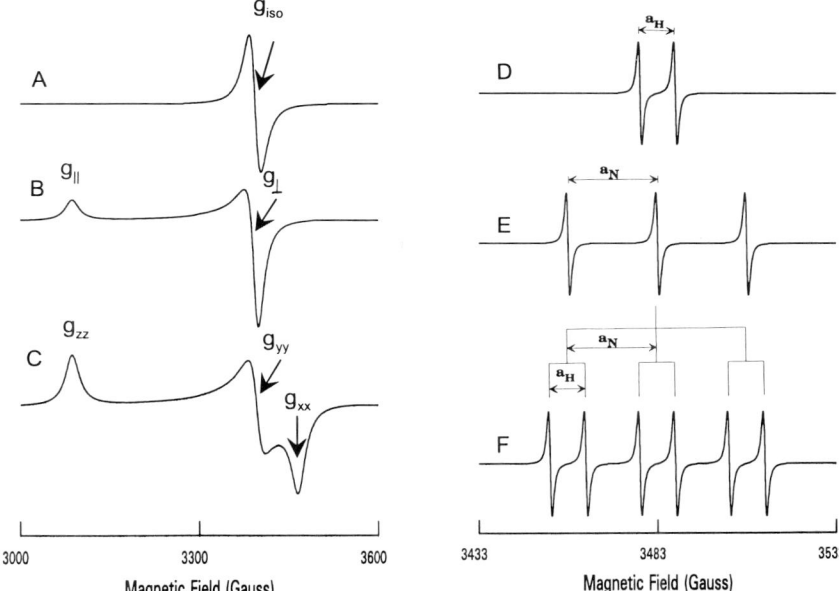

FIGURE 3. Typical EPR line shapes. Powder shapes with g anisotropy (A, isotropic; B, axially symmetric; and C, rhombic) and isotropic system with hyperfine splittings from a nitrogen, hydrogen, or both (D, E, and F). A: g_{iso} = 2.0; B: $g_{\parallel}$ = 2.2, $g_{\perp}$ = 2.0; C: g_{xx} = 1.96, g_{yy} = 2.0, g_{zz} = 2.2; D: a_H = 10 G; E: a_N = 25 G; F: a_N = 25 G, a_H = 10 G.

isotropic, axially symmetric, and rhombic as shown in Figure 3. Thus a careful analysis of the g-factor may often be useful in order to characterize the system.

In reality, however, one never has an isolated electron or literally "free" electron in the paramagnetic sample. The free electrons are part of atoms or ions and hence are associated with the nucleus. The nucleus may or may not have an associated magnetic moment, called nuclear magnetic moment. Nuclei with a non-zero magnetic moment, e.g., ^1H, ^{14}N, ^{19}F, etc., can interact with the electron moment and hence modulate the Zeeman levels. This will result in the multiplicity of splittings and hence multiple lines are observed in the EPR spectrum. The electron-nucleus interaction is called hyperfine coupling and the lines that arise due to this coupling are called hyperfine lines. The hyperfine interaction is measured as hyperfine coupling constant, HFC, the separation between the split lines and is normally expressed in field units, Gauss. Because of the extended delocalization of the unpaired electron spin density it is also possible to see super hyperfine interactions, that arise as a result of interaction with the nucleus of neighboring atoms in the molecule. Thus hyperfine coupling can also provide a wealth of information regarding the identity and environment of the free electron system.

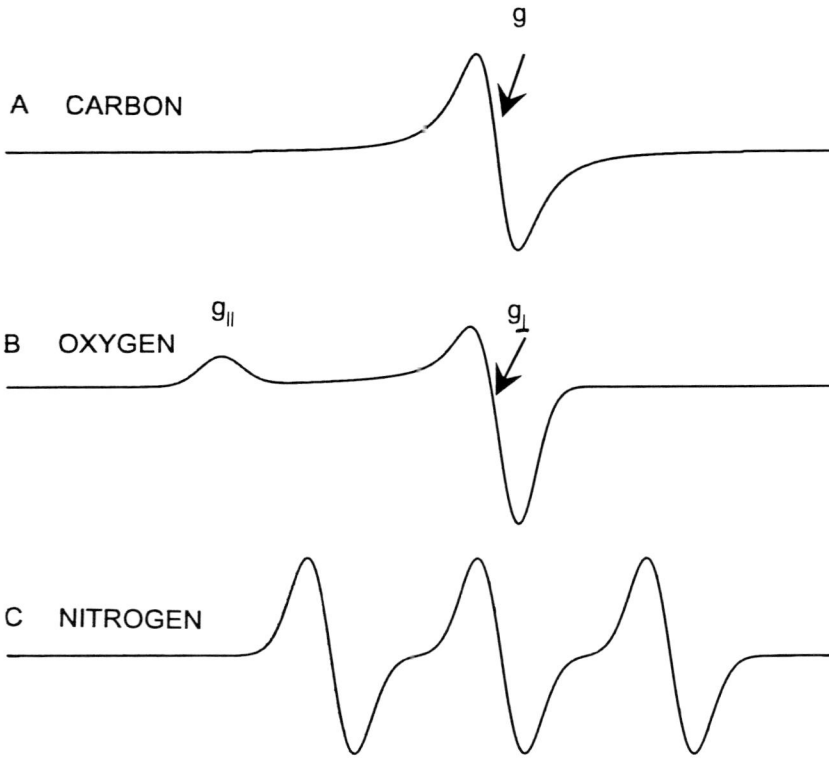

FIGURE 4. Typical EPR spectral shapes of (A) carbon, (B) oxygen, and (C) nitrogen centered free radicals.

Thus EPR can define the presence and concentration of a given free radical, the symmetry of electron environment, the presence of nuclei in the vicinity of the electron, and hence can serve to identify the chemical structure of a given free radical.

Structurally different free radicals give rise to different EPR spectra. An isotropic free radical, such as a one-electron reduced semiquinone, gives rise to a spectrum as shown in Figure 4A. Radicals with axial symmetry, such as an oxygen centered alkylperoxy radical, give rise to a spectrum as shown in Figure 4B. A nitrogen coupled free radical such as a nitroxide gives rise to a triplet spectrum, as shown in Figure 4C, based on the interaction of the nitrogen nuclear spin with the electron. The nitrogen nucleus has a spin of one, therefore it has three spin states resulting in a triplet hyperfine splitting, the magnitude of which is defined by the hyperfine coupling constant, a_N.

B. OXYGEN FREE RADICALS

In biological systems the oxygen molecule, O_2, normally undergoes a complete four-electron reduction to form water, H_2O. It can also undergo successive one-electron reductions to give, respectively, $\cdot O_2^-$, H_2O_2, $\cdot OH$, and H_2O. The energy derived from this conversion is utilized by the cells. However, the same reduction process when it is incomplete leads to free radical formation. For example, incomplete one-electron reduction of O_2 leads to the formation of superoxide anion, $\cdot O_2^-$, while the three-electron reduction produces hydroxyl radical, $\cdot OH$:

$$O_2 \longrightarrow \cdot O_2^- \longrightarrow H_2O_2 \longrightarrow \cdot OH \longrightarrow H_2O$$

$$O_2 + 1e^- \longrightarrow \cdot O_2^- \text{ (superoxide anion radical)}$$
$$O_2 + 2e^- + 2H^- \longrightarrow H_2O_2 \text{ (hydrogen peroxide)}$$
$$O_2 + 3e^- + 3H^+ \longrightarrow H_2O + \cdot OH \text{ (hydroxyl radical)}$$
$$O_2 + 4e^- + 4H^+ \longrightarrow 2H_2O \text{ (water)}$$

The superoxide and hydroxyl free radicals are highly reactive in aqueous solution and hence can potentially react with biological macromolecules thereby causing oxidative damage. The hydrogen peroxide is also a potent but sluggish oxidizing agent. In the presence of transition metal ions it generates $\cdot OH$ as follows:

$$H_2O_2 + Fe^{2+} \longrightarrow \cdot OH + OH^- + Fe^{3+}$$

The superoxide anion radical, $\cdot O_2^-$ has a life-time in the order of milli-seconds in aqueous solution at neutral pH. It undergoes spontaneous dismutation reaction to yield H_2O_2 and O_2 as follows:

$$2 \cdot O_2^- + 2H^+ \longrightarrow H_2O_2 + O_2 \quad (k = 2 \times 10^5 \text{ M}^{-1}\text{s}^{-1} \text{ at pH 7.4})$$

Because of this spontaneous and rapid dismutation reaction of $\cdot O_2^-$, generation of $\cdot O_2^-$ in biological systems is accompanied by H_2O_2 generation. The standard redox potential of $\cdot O_2^-$, -0.16 V,[17] makes it possible to reduce a variety of Fe^{3+} and Cu^{2+} complexes, which in turn may reduce H_2O_2 to more reactive $\cdot OH$ radicals. Further, in the presence of transition metal ions such as Fe^{3+} and Cu^{2+}, the combined presence of $\cdot O_2^-$ and H_2O_2 can lead to the generation of a more reactive $\cdot OH$ radical:

$$\cdot O_2^- + Fe^{3+} \longrightarrow O_2 + Fe^{2+}$$
$$H_2O_2 + Fe^{2+} \longrightarrow \cdot OH + OH^- + Fe^{3+}$$

Net reaction $\quad \cdot O_2^- + H_2O_2 \longrightarrow \cdot OH + OH^- + O_2$

The above reaction is usually referred to as the superoxide driven Fenton reaction or iron-catalyzed Haber-Weiss reaction. Since iron is ubiquitous in biological systems and low molecular weight chelates, hematin or protein-bound iron such as hemoglobin, myoglobin, or ferritin can potentially catalyze the formation of ˙OH.[18-20]

Though ˙O_2^- itself is sufficiently reactive to damage cells, the presence of a negative charge diminishes its strength as an oxidant. Neutralization of the charge with a proton or with metal cations, however, can make it a stronger oxidant. The ˙O_2^- oxidizes polyphenols, catecholamines, tocopherols, leucoflavins, ascorbate, and various thiols.[6,21] It inactivates catalases, peroxidases, dihydroxy acid dehydratase, and other enzymes.[22-25]

Of all the free radicals produced in biological systems, the ˙OH radical is the most reactive radical. It is extremely reactive ($k = 10^7$ to $10^{-1} M^{-1} s^{-1}$) and short-lived.[26] The ˙OH radical is highly electrophilic and can react with virtually any organic compound at diffusion-limited rates. The ˙OH radical has been demonstrated to induce lipid peroxidation of cell membranes. Polyunsaturated fatty acids are particularly very susceptible to peroxidation. The ˙OH radical can abstract a hydrogen atom from the carbon chain of the fatty acid resulting in an alkyl radical, L˙ with a subsequent molecular rearrangement to form a more stable conjugated diene moiety. Molecular oxygen is then taken up at the new carbon radical site with the formation of the lipid peroxyl radical LOO˙. This peroxyl radical can then extract a hydrogen atom from another fatty acid molecule, thus starting a self-perpetuating chain of lipid peroxidation as shown below:

LH + ˙OH $\longrightarrow$ L˙ + H_2O (akyl radical, initiation)
L˙ $\longrightarrow$ L˙   (dienyl radical, rearrangement)
L˙ + O_2 $\longrightarrow$ LOO˙   (alkoxy radical, peroxidation)
LOO˙ + LH $\longrightarrow$ LOOH + L˙   (peroxide, propagation)

The ˙OH radical can also add to aromatic compounds such as the aromatic amino acids resulting in the formation of hydroxylated derivatives. In addition, ˙OH oxidizes sulfhydryl groups (SH) on amino acids causing alterations in the protein structure or deactivation of the enzyme with SH groups at the catalytic sites. This radical also reacts with sugar moieties of DNA to yield a variety of degradation products, with eventual cleavage of DNA backbone. This process has been proposed to be the mechanism of action of the antineoplastic activity of Adriamycin and daunorubicin, both of which chelate iron and subsequently generate ˙OH.[27,28]

A wide range of techniques with varying specificity and sensitivity have been used to measure superoxide and hydroxyl free radicals. The chemical methods used for assaying superoxide anion radical take advantage of the redox reactions in which ˙O_2^- participates. Some of the chemical methods used to assay superoxide include adrenochrome formation,[29] cytochrome c reduction,[30] and nitroblue tetrazolium (NBT) reduction.[31] The methods used for hydroxyl radical assay follow the oxidative products of ˙OH radical scavengers. These include hydroxylation of aromatic compounds,[32] detection of dimethyl sulfoxide byproducts such as methane and acetaldehyde,[2,33,34] $^{14}CO_2$ production from ^{14}C-labeled benzoate,[34] ethylene formation from methional or other methionine derivatives.[2] All these methods are to some extent nonspecific. Oxygen radicals can also be assayed using chemiluminescence techniques.[35] The chemiluminescence techniques are more sensitive but still nonspecific.

All of these indirect techniques can yield false-positive results and hence it is often difficult to characterize or quantitate the radicals that are generated. Thus application of these techniques to cellular systems may not truly reflect the generation of free radicals.

EPR spectroscopy, on the other hand, is specific to free radicals and thus can provide a more definitive and direct evidence for the presence of free radicals. Using EPR technique one can perform detection, characterization, and quantitation of free radicals in biological systems. Oxygen free radicals in tissue samples can be measured by EPR spectroscopy in two different ways: (1) direct measurements at low temperatures and (2) indirect measurements at room or physiological temperatures. The direct method involves detection and measurement of radical concentration on the fast-frozen sample. These radicals are quite short-lived in aqueous medium or in biological tissues at room temperature or at physiological temperatures. Hence the samples are rapidly frozen to 77 K or below so as to increase the lifetime and the EPR measurements are then performed in the frozen state. Alternatively the short-lived radicals can be indirectly measured by trapping them with spin traps. The spin traps are a special class of compounds, usually nitrones or nitroso compounds. These spin traps themselves are not paramagnetic and hence do not show any EPR spectrum. The spin traps, however, form stable adducts with the short-lived oxygen radicals and the byproducts. The stability of the spin adducts over a period of a minute or longer makes it possible to perform EPR studies in solution at room temperatures. 5,5'-dimethyl-1-pyrroline-N-oxide (DMPO) is the most comomonly used spin trap molecule. It reacts with hydroxyl, superoxide, alkyl, or alkoxy radicals forming relatively stable EPR detectable nitroxides (Figure 5). It is the most versatile spin trap for biological studies and is the least toxic. Structurally different DMPO radical adducts have different hyperfine coupling constants as shown in Figure 5, therefore, one observes distinctly different EPR spectra for the superoxide, hydroxyl, and alkyl radical adducts.[10,36-41]

Each of these EPR techiques has inherent advantages and disadvantages. Advantages of rapid-freezing techniques are as follows:

1. This is the most conclusive technique in that one can perform direct measurements of the radicals of interest.

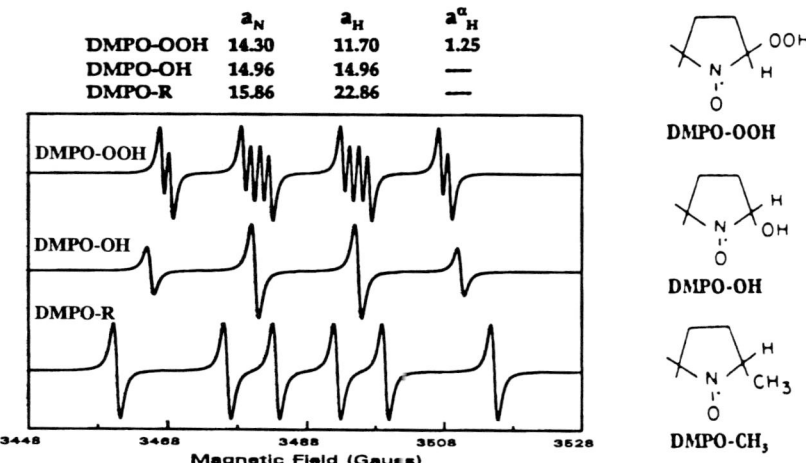

FIGURE 5. Molecular structure and sample EPR spectra of some DMPO spin adducts.

2. At cryogenic temperatures the radicals measured are more stable. Hence the study can be made for long periods of time; for example, one can employ EPR signal averaging techniques to enhance sensitivity of the signal.
3. One can study radical properties including power saturation and temperature stability (titration) to characterize the radicals formed.
4. Quantitative measurements are performed on the actual radicals formed.
5. No artifacts due to added spin traps, either to the function of interest or to the EPR spectra are recorded.

The disadvantages of the direct techniques are:

1. This method is destructive. One sample yields information at only one time point. Therefore one cannot simultaneously correlate radical concentration and subsequent physiological functions.
2. There are technical difficulties in fast-freezing as well as maintaining the samples at low temperature.
3. There are possible mechanical effects of tissue processing. Extreme care must be taken to avoid or control artifactual production of free radicals while tissue grinding, packing at low temperatures.
4. This technique is often less sensitive than spin-trapping techniques.
5. Highly labile radicals such as ˙OH are still not measurable.

The spin-trapping technique has the following advantages:

1. The technique is nondestructive thus allowing simultaneous measurements of radicals and organ function.

2. Measurements can be performed at physiological or room temperatures.
3. The enhanced stability of the adducts compared to the primary free radical makes integrative accumulation of the radical, thus increasing the sensitivity of measurement.
4. Very labile radicals like ˙OH can be trapped and studied.
5. Assignment and analysis of the radical signals are simple.

The disadvantages of the indirect techniques include:

1. Trapping is not 100% efficient.
2. Artifacts from impure spin trap reagent, light induced degradation of the trap, or nonspecific redox reactions of the trap may give rise to background signals. Therefore matched control experiments are always essential.
3. The spin trap can cause alterations in cells and tissues.

Thus both the direct and spin-trapping EPR methods for studying oxygen free radicals in biological tissues and cells have strengths and weaknesses. The two approaches are, however, complementary to each other and hence a combined use of both can provide more information than the use of either technique alone.

In the following sections we will describe a few typical examples to illustrate the use of both direct and indirect EPR methods for studying oxygen free radicals in biological samples. We have taken data from the work performed in our laboratories. Though EPR can be used to study a variety of biological cells and tissues our work is mainly focused on investigating the mechanism of oxidative injury to the myocardium during ischemia and reperfusion.

III. APPLICATIONS

A. DIRECT MEASUREMENTS ON ISOLATED REPERFUSED HEARTS[13,15,42]

Isolated Langendorff perfused rabbit hearts were used to measure free radical generation on post-ischemic reperfusion. The animals were anesthetized, the hearts removed, and the aorta cannulated and perfused with retrograde flow of Krebs bicarbonate buffered perfusate at a constant pressure of 80 mmHg. A balloon was placed within the left ventricle to measure contractile function. The balloon volume was adjusted to achieve a diastolic pressure in the range of 8 to 12 mmHg. The hearts were perfused for a period of at least 15 min prior to the start of the experiment to allow for stabilization. The intrinsic heart rate was about 150 bpm with left ventricular developed pressures of approximately 100 mmHg. The hearts were subjected to varying periods of ischemia and reperfusion. At the desired time the left ventricular balloon was removed and the hearts were freeze-clamped using Wollenberger tongs cooled to 77 K. The Freeze-clamped heart tissue was then ground or chopped to a granular powder under liquid nitrogen and the powder transferred to precision EPR tubes. Mechanical processing of the tissue sample was minimized by limiting the duration of grinding to less than 1.5 min with a resulting frozen particle size of 1.5 to 2.5 mm in diameter. In addition, care was taken to prevent exposure of the tissue to air. These steps minimized the generation of mechanically derived radicals. Further, a variety of techniques including grinding, chopping, core extrusion, and core biopsy were used to characterize the appearance or disappearance of artificial radicals. In order to make reliable quantitative comparisons the samples were filled in tubes to a height more than sufficient to achieve the optimum critical filling of the resonator. The measurements were performed under nonsaturating microwave power.

In general, three different radical signals were observed (Figure 6). These were (A) isotropic signal with $g = 2.004$ suggestive of a semiquinone radical; (B) signal with axial symmetry, $g_{\parallel} = 2.033$ and $g_{\perp} = 2.005$ identical to that of a peroxyl radical; (C) a triplet signal with $g = 2.000$ and $a = 24$ G suggestive of a radical with nitrogen coupling such as peroxylamine.[13] There is also an additional signal with g values of 2.027 and 1.936 attributable to an iron-sulfur protein with a Fe_2S_2 cluster.

The signals B and C are highly labile and disappear on warming. In normally perfused hearts A was observed with only small amounts of B and C. During ischemia, B and C increased, reaching a maximum after 45 min while A decreased. On reflow with oxygenated perfusate all three signals increased.

Experiments were performed to measure the time course of radical generation as a function of the duration of ischemia and reflow. Hearts were subjected to 10, 30, or 60 min of ischemia and then freeze-clamped after 2 to 30 min periods of reflow. A maximum increase in signals A and B was observed on reperfusion after 30 min of ischemia (Figure 7). Further, peak signal intensities were found to occur after 15 s of reflow following 30 min of ischemia as seen in Figure 7.

At shorter periods of ischemia, as shown by solid circles, of 10 min of ischemia there was less radical generation. With longer periods of ischemia, of 60 min, there was again less radical generation observed upon reperfusion. The decrease in radical generation which occurred after long durations of ischemia may be due to the occurrence of a no-reflow phenomenon, and in fact with long durations of ischemia one observes that coronary flows are markedly decreased. These data suggest that indeed there is a window of ischemic duration which results in maximum free radical generation upon post-ischemic reperfusion.

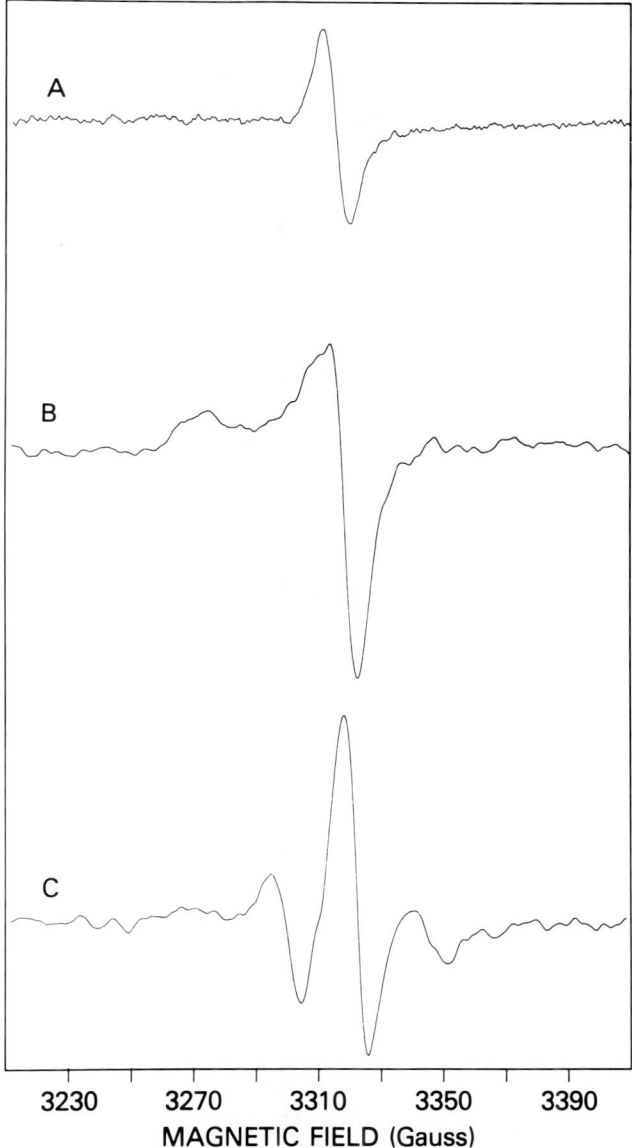

FIGURE 6. Component EPR spectra of frozen heart samples. (A) Isotropic signal with g = 2.004 suggestive of a semiquinone radical. (B) Signal with axial symmetry, $g_{\parallel}$ = 2.033 and $g_{\perp}$ = 2.005 identical to that of a peroxyl radical. (C) A triplet signal with g = 2.000 and a = 24 G, suggestive of a radical with nitrogen coupling such as peroxylamine.

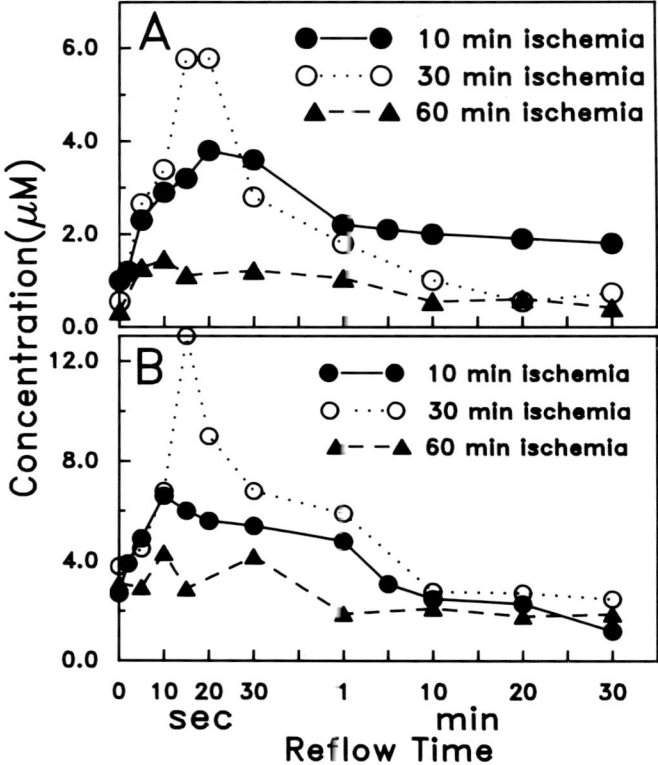

FIGURE 7. Graph of the concentration of signals A and B (reference Figure 6) as a function of the post-ischemic reflow time for hearts subjected to 10, 30, and 60 min of ischemia. (A) signal A. (B) Signal B.

B. SPIN-TRAPPING MEASUREMENTS ON ISOLATED REPERFUSED HEARTS[14,15,41]

In the spin-trapping method the Langendorff perfused hearts were subjected to varying duration of ischemia and reperfusion in the presence of 40 mM concentrations of DMPO. This concentration of DMPO was used based on the fact that concentrations of less than 50 mM induce no measurable toxicity on the heart, that is, cardiac contractile function remains unchanged with concentrations of less than 50 mM.[14] In these experiments the effluent was sampled, the EPR spectra were either immediately measured or frozen at 77 K and then subsequently measured. The EPR measurements were performed at room temperature in an EPR flat cell at X band.

We observed with a control heart perfused with the spin trap DMPO prior to ischemia that there was no measured radical signal. However, upon post-ischemic reperfusion, after 30 min of ischemia, a prominent signal was observed as early as the first 10 s of reperfusion.[14] The signal intensity peaked between 0 to 20 s and then gradually declined until at 5 min there was no longer any measurable radical signal (Figure 8).

The EPR spectrum of the DMPO perfused heart effluent consisted of two major components — a 1:2:2:1 quartet of DMPO-OH and a 6-peak spectrum due to the trapped alkyl radical, DMPO-R. This was confirmed by computer simulations. One can explain the experimental spectra as a combination of these two radical adduct signals.

In order to investigate the mechanisms of this radical generation we performed experiments reperfusing hearts in the presence of both denatured SOD and active SOD.[14,15] Figure 9A shows a spectrum obtained in which hearts were subjected to 30 min of global ischemia and then reperfused with SOD denatured by the procedure of Hodgson and Fridovic.[43] We see prominent radical signal and accompanying this radical generation we also see that after post-ischemic reperfusion the left ventricular developed pressure is markedly decreased with a high end diastolic pressure. In the matching experiment with active SOD we observed (Figure 9B) that there was approximately an 80% decrease in the observed radical signals and accompanying this there was a marked increase in the recovery of left ventricular developed pressure with a lower end diastolic pressure.

These experiments suggest that much of the radical generation which we observe is secondary to superoxide production as suggested by the incomplete scavenging by SOD. It was thought that there could be a competitive reaction of superoxide with iron resulting in the formation of the highly reactive hydroxyl radical which would not be scavenged by SOD. Hence we performed additional experiments in which the perfusate was chelexed. As shown in Figure 10, in the absence of the SOD again, we observed a prominent radical signal. With the chelexed perfusate in the presence of SOD, however, we observed that there was complete quenching of radical generation.

We then performed measurements to see if there was a correlation between measured radical generation and the impairment of cardiac contractile function as measured from the left ventricular developed pressure. We observed with denatured SOD that there was prominent radical generation and accompanying this radical generation there was a marked impairment in contractile function with a left ventricular developed pressure of only 24 mmHg observed at the end of reperfusion. With active SOD we observed that there was approximately once again an 80% decrease in radical generation with a marked increase in the recovery of left ventricular developed pressure. With chelexed perfusate in the presence of SOD there was complete quenching of radical generation and there was an even more enhanced recovery of left ventricular developed pressure with a value of 68 mmHg, and a much lower end diastolic pressure.

Thus, these spin trap studies demonstrate that superoxide derived hydroxyl and alkyl radicals are generated in the reperfused heart. Iron mediated Fenton

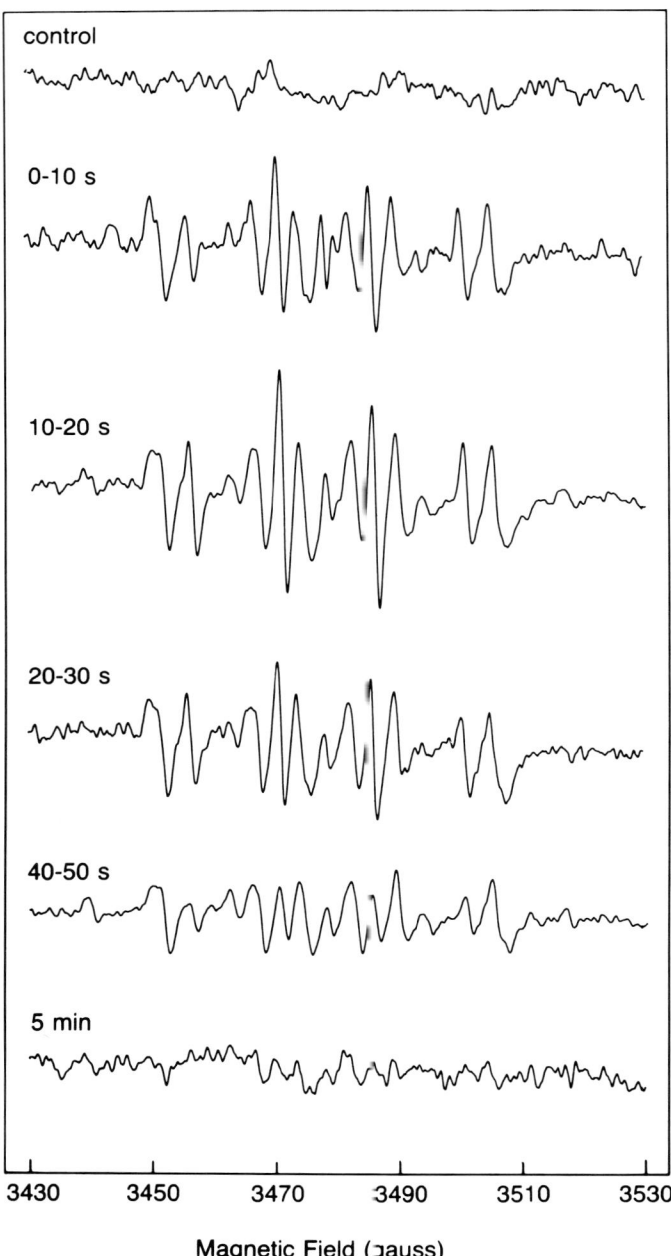

FIGURE 8. EPR spectra of the effluent perfusate of a heart perfused with 40 mM DMPO sampled over 10 s periods of reflow after 30 min of ischemia. The control sample was collected prior to ischemia.

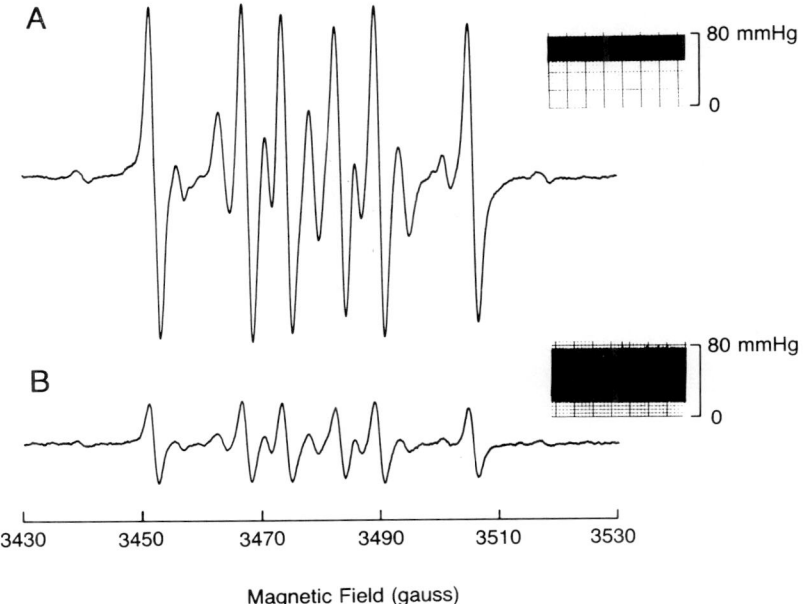

FIGURE 9. EPR spin trapping experiment of hearts perfused in the presence of enzymatically active recombinant human SOD (B) or identical concentrations of inactivated recombinant human SOD (A). The hearts were subjected to 30 min 37°C global ischemia and reflowed with active and inactive SOD. The spectra were recorded 10 to 20 s after the onset of reflow. The right insets show the recovered left ventricular function after 45 min of reperfusion.

chemistry appears to be an important mechanism of this radical generation. A burst of radical generation is observed early, peaking after only 10 to 20 s of reflow. There was a direct correlation between this measured radical generation and the impairment in cardiac contractile function. We observed that the efficacy of SOD at scavenging the reperfusion radical burst was modulated by the presence of adventitial iron. This reaction of superoxide with iron may potentially explain some of the conflicting results reported in the literature with *in vivo* infarct size models, where different groups have found differing efficacies of SOD in decreasing infarct size.

C. SPIN TRAP EPR MEASUREMENTS ON CELLS[44]

Though it is established that oxygen free radicals are generated in the post-ischemic tissues, the mechanism of this radical generation is not well understood. It is not known which cell types are responsible for the observed radical generation. It was hypothesized, based on the fact that intravascularly administered radical scavenging enzymes prevent reperfusion injury in a variety of tissues, that vascular endothelial cells could be the central source of the radical burst. We therefore performed experiments subjecting bovine aortic

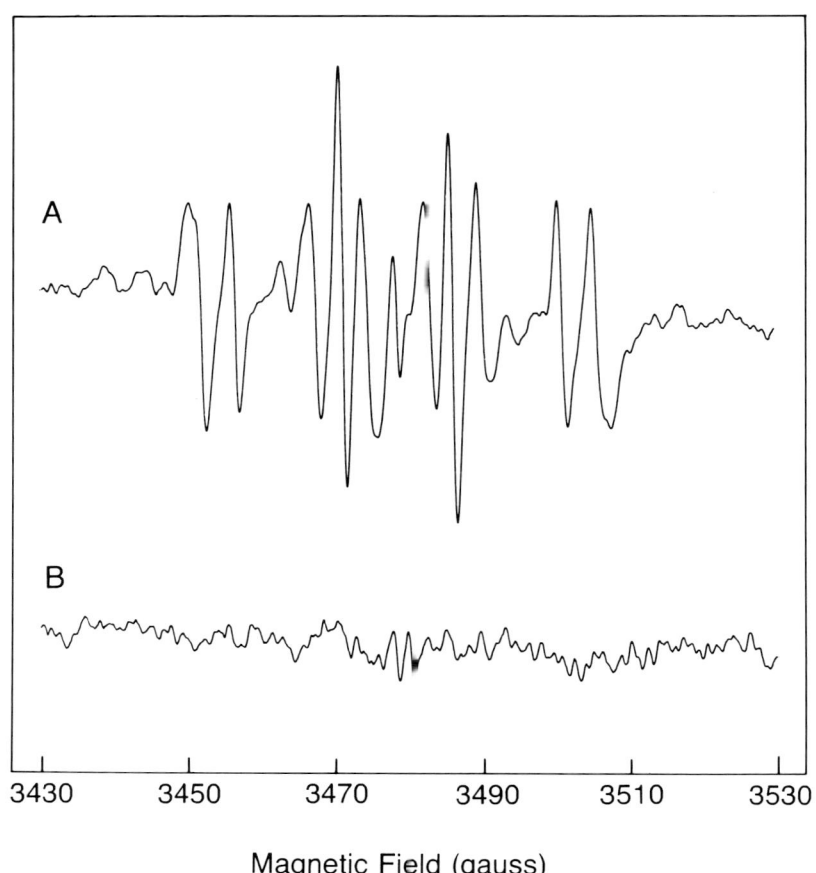

FIGURE 10. EPR spin trapping experiment of hearts perfused with chelexed perfusate after 30 min of global ischemia. (A) Reperfused without recombinant h-SOD. (B) Reperfused with active recombinant h-SOD.

endothelial cells to various periods of anoxia and reoxygenation in an effort to create an ischemia-reperfusion equivalent.[44] As shown in Figure 11 bovine endothelial cells, after reoxygenation, gave a prominent radical signal consisting of a 1:2:2:1 quartet of DMPO-OH; the matched cells in air, however, gave no signal. This radical signal was completely quenched by SOD and catalase; however, with the iron chelator deferoxamine there was little effect (Figure 12). In order to characterize the radicals actually generated by the reoxygenated endothelial cells we performed experiments in the presence of 1% ethanol. Hydroxyl radicals are highly reactive and in the presence of ethanol would extract hydrogen forming a hydroxyethyl radical, which when trapped with DMPO would give rise to a distinct six-line EPR spectrum. In

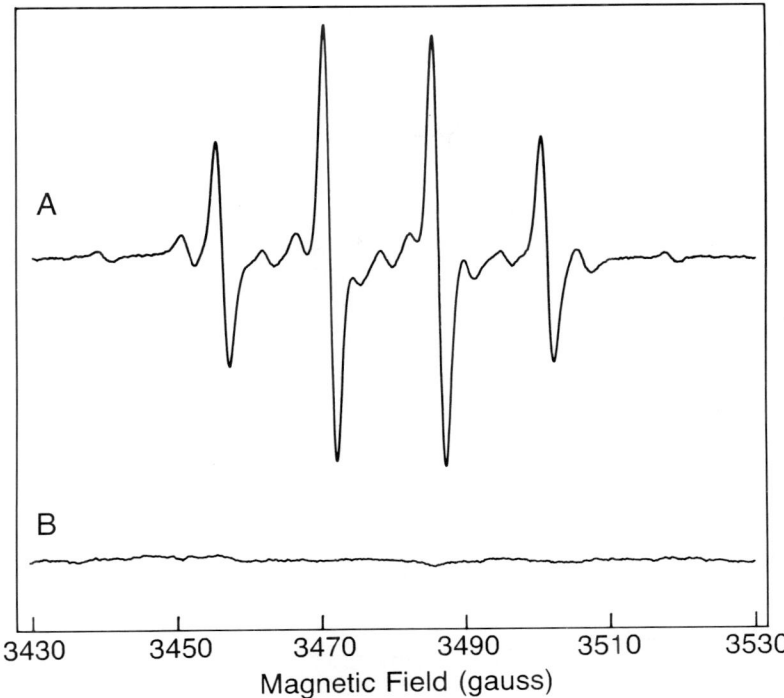

FIGURE 11. EPR spectrum of preparations of endothelial cells (1.6 × 10⁷ cells in 1 ml) in the presence of 100 mM DMPO. (A) Subjected to 45 min 37°C anoxia and then reoxygenated. (B) Identical preparation of cells not subjected to anoxia.

these experiments we observed that in addition to the 1:2:2:1 quartet of DMPO-OH, a clear six-line DMPO-hydroxyethyl radical adduct is formed as seen in Figure 13. The simulation shown in Figure 13B shows clearly that one can explain the experimental spectrum in Figure 13A as a combination of ethyl and hydroxyl adducts.

It was suggested by McCord that the enzyme xanthine oxidase may be an important source of radical generation in reperfused tissues.[45] Therefore we performed experiments both in the presence of allopurinol and the definitive xanthine oxidase blocker, oxypurinol. Experiments were performed incubating preparations of 8 million endothelial cells in 1 ml PBS under anaerobic conditions at 37°C for 45 min in the presence or absence of allopurinol or oxypurinol followed by reoxygenation, with a DMPO concentration of 50 mM (Figure 14). In the presence of allopurinol a 60% decrease in the DMPO-OH signal was observed compared to that observed in the absence of the drug from a concentration of 0.3 to 0.12 μM; however, a new alkyl signal was observed with a concentration of 0.08 μM (Figure 14B, Table 1). The observed alkyl signal could be due to the formation of a drug radical generated

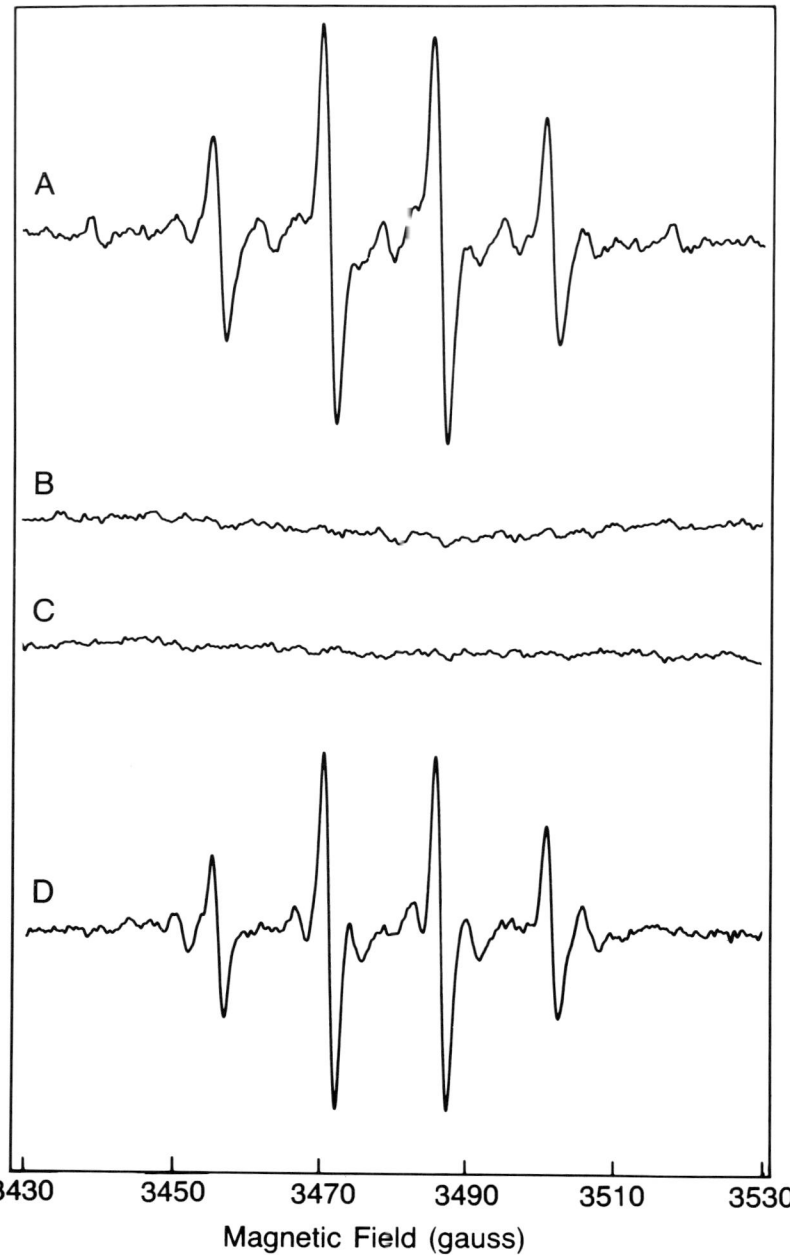

FIGURE 12. EPR spectra of preparations of endothelial cells (8×10^6 cells in one ml) subjected to 45 min 37°C anoxia then reoxygenated in the presence or absence of radicals scavengers. (A) No scavengers. (B) 5000 units/ml SOD. (C) 5000 units/ml catalase. (D) 1 mM deferoxamine.

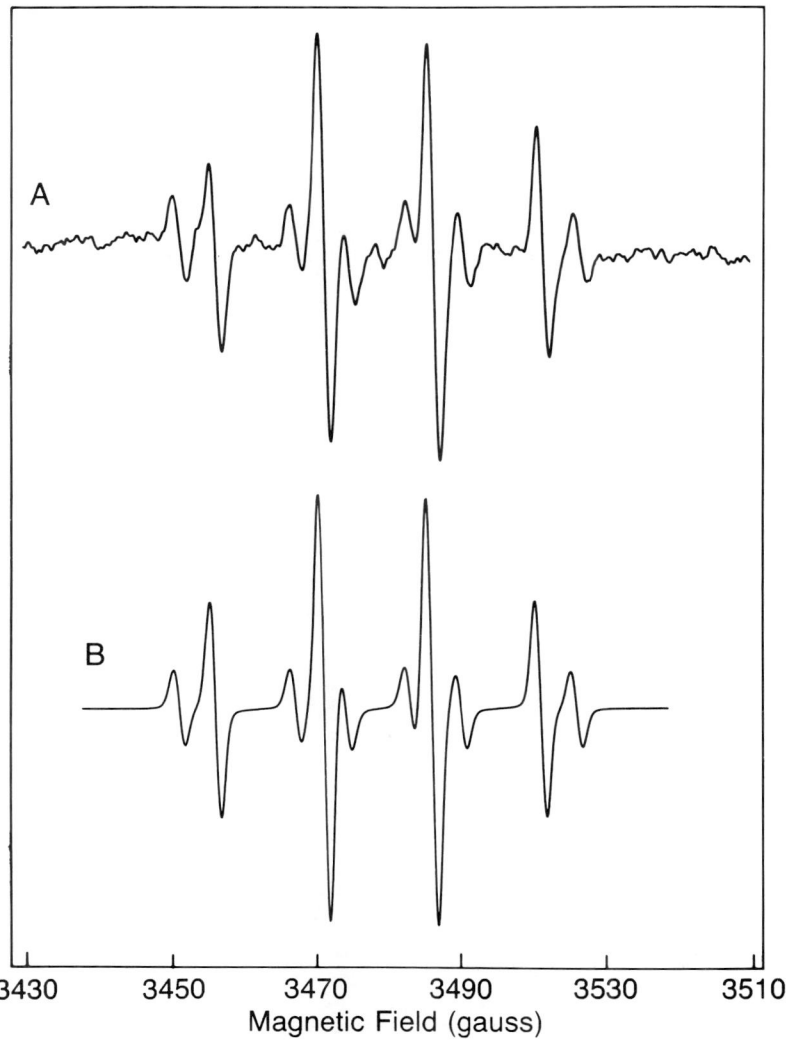

FIGURE 13. EPR spectrum of reoxygenated endothelial cells subjected to anoxia and reoxygenation in the presence of 1% ethanol. (A) Measured spectrum. (B) Computer simulation of A consisting of two components, a 1:2:2:1 quartet with $a_N = a_H = 14.9$ G, weight 0.7 and a 1:1:1:1:1:1 sextet $a_N = 15.8$ G and $a_H = 22.8$ G, weight 0.3.

by the scavenging of ˙OH by the drug. It has previously been suggested that allopurinol may prevent free radical injury by scavenging ˙OH rather than by the specific blocking of ˙O_2^- generation by xanthine oxidase.[46] The observed spectrum suggests that allopurinol is only partially effective as a blocker of ˙OH generation and partially effective as an ˙OH scavenger. Experiments

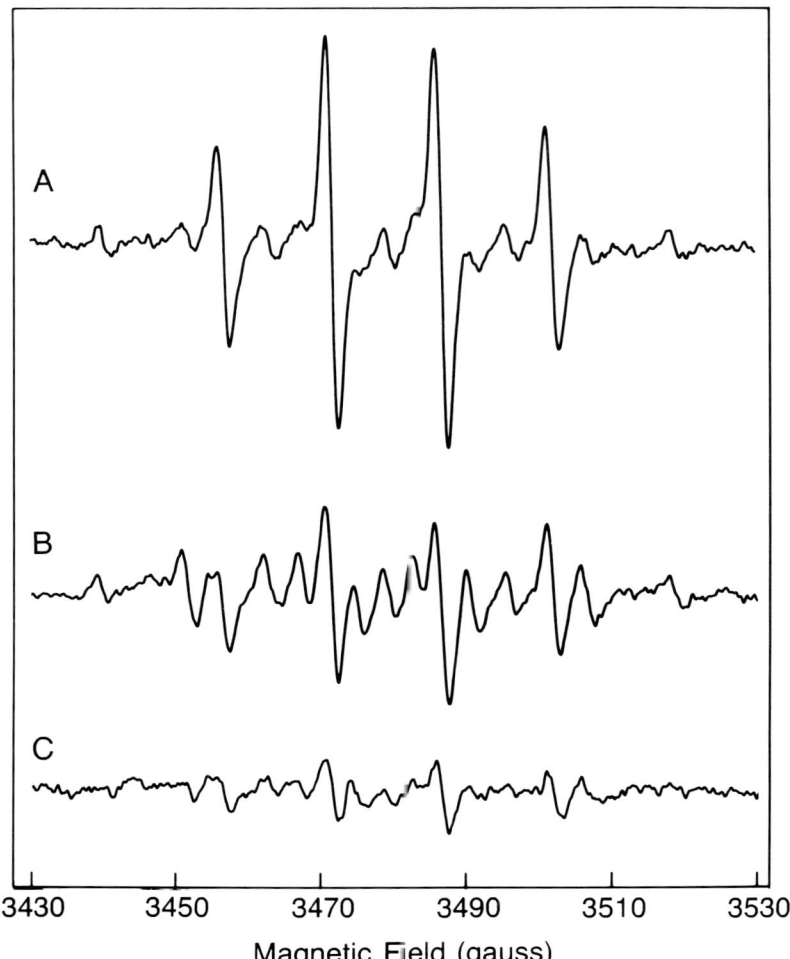

FIGURE 14. EPR spectra of reoxygenated endothelial cells in the presence and absence of xanthine oxidase inhibitors. (A) No added inhibitor. (B) With 4 mM allopurinol. (C) With 4 mM oxypurinol.

performed with the more potent xanthine oxidase inhibitor oxypurinol showed a marked 80 to 90% reduction in the DMPO-OH signal (Figure 14C). This marked reduction in ˙OH generation suggests that the enzyme xanthine oxidase is an important source of endothelial free radical generation.

In order to determine if endothelial free radical generation causes cell injury and cell death, cell viability was assessed in parallel with the above EPR measurements. Prior to transfer to the EPR flat cell small aliquots of cells were removed for measurement of trypan blue exclusion.

TABLE 1
Correlation of Radical Generation and Cell Injury

	Radical concentration (μM)		Cell viability
	DMPO-OH	DMPO-R	(% cells excluding trypan blue)
Control cells	0	0	99
Reoxygenated	0.3	0.03	10
Reoxyg. + SOD	0	0	96
Reoxyg. + catalase	0	0	92
Reoxyg. + allopurinol	0.12	0.08	70
Reoxyg. + oxypurinol	0.03	0.0	82

Note: Cells were counted 10 min after reoxygenation.

As shown in Table 1 cells not subjected to anoxia excluded trypan blue while cells subjected to anoxia and reoxygenation took up the dye. The time course of this reoxygenated endothelial cell death was studied by preparing slides of cells stained at different times after reoxygenation. Cell counts performed in the first 2 min after reoxygenation showed that only 25% of the cells took up trypan blue. After 4 min 50% of the cells took up the dye, and after 10 min >90% of the cells took up the dye. Cells not subjected to anoxia continued to exclude the dye with >95% of the cells excluding dye even after 30 min. These studies suggest that the ongoing generation of free radicals in reoxygenated endothelium induces gradually increasing cellular damage as radical production continues. In the presence of either SOD or catalase this cell damage was almost totally prevented in accordance with the abolishment of free radical generation (Table 1). Denatured SOD or denatured catalase, however, did not decrease the observed radical concentrations or prevent cell damage. The radical concentrations and the uptake of trypan blue were identical to preparations with no added enzyme. Allopurinol did decrease the proportion of the cells that took up the dye, but it was considerably less effective than SOD or catalase (Table 1). Oxypurinol, however, was considerably more effective than allopurinol. Thus, cell death appeared to correlate closely with measured radical concentrations suggesting that the observed free radical generation was sufficient to cause cell injury and cell death.

D. *IN VIVO* EPR METHODS

Conventional EPR spectrometers working at X- or Q-band frequencies set an upper limit of 1 or 0.2 mm, respectively, to the size of the nonfrozen aqueous sample to be studied. Hence EPR measurements cannot be performed on larger intact biological organs or tissues. We developed a specific type of resonator enabling us to accommodate much larger samples, up to 20 or 25 mm thickness of water. This spectrometer uses a recessed loop-gap resonator and a microwave source called an L-band bridge. Utilizing this type of instrumentation one can measure free radicals, for example, in a rat heart. As shown in Figure 15, strong EPR spectra are observed with concentrations down to the micromolar level in a rat heart infused with the nitroxide spin label TEMPO.[47,48]

With this technique we have performed measurements of free radical kinetics in the intact heart. One can look at the uptake of a radical label within the heart, and the kinetics of its clearance. One can similarly look at how kinetics are altered in the ischemic state. Figure 16 shows the kinetics of radical clearance when a heart is subjected to ischemia at time zero. In addition, this technique has the additional feature that oxygen concentrations can be determined within the myocardium. The basis of this is the fact that oxygen itself is intrinsically paramagnetic. Therefore, it gives rise to a broadening of the EPR signal of any given free radical label and from this broadening ΔH one can calculate the oxygen tension. As seen in Figure 16 the linewidth rapidly sharpens with ischemia, indicating that the oxygen tension decreases within a minute, dropping to that of air, and then within 5 to 10 min further decreasing to near zero.

The *in vivo* technique has some interesting features for both oximetry as well as for measuring radical kinetics. One can measure spectra very rapidly. Measurements of free radicals or oxygen can be performed with this technique in the millisecond time domain. In addition, EPR is a magnetic resonance technique, so magnetic field gradients can be incorporated and imaging experiments performed. This technique can be applied to image both simple or geometrically complicated objects. Recently we have applied this methodology to produce the first free radical images of the heart.

Thus, EPR spectroscopy is a powerful and versatile method for measuring free radical generation in cells and tissues. The recent development of *in vivo* EPR spectroscopy and imaging further opens the way to study a broad range of important questions in free radical biology.

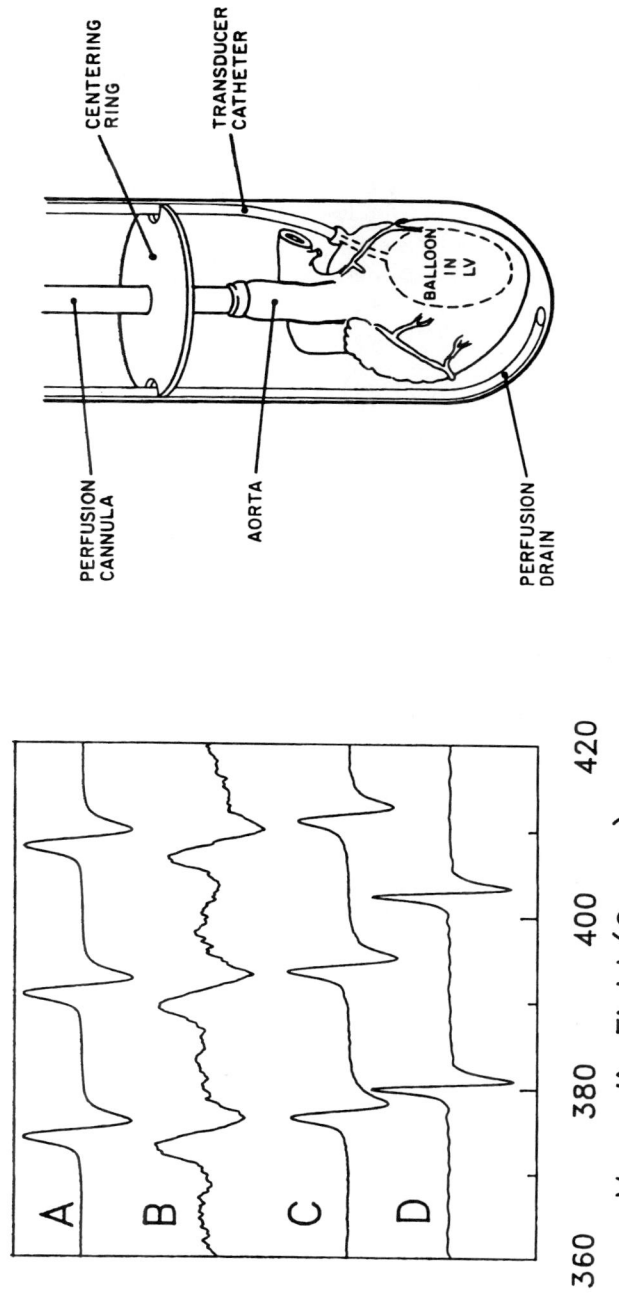

FIGURE 15. (Left) A, EPR spectrum of an aqueous 1.0 mM solution of TEMPO filling a 13 mm cylindrical tube. The resonator frequency was 1.085 GHz. B, With 2.0 μM TEMPO solution. C, EPR spectrum of rat heart perfused with 1.0 mM TEMPO. D, EPR spectrum of rat heart perfused with 1 mM ^{15}N-PDT. (Right) Diagram of perfused heart preparation for *in vivo* EPR. A balloon was inserted into the left ventricle to measure left ventricular developed pressure, enabling real time simultaneous measurements of free radicals within the heart along with cardiac contractile function.

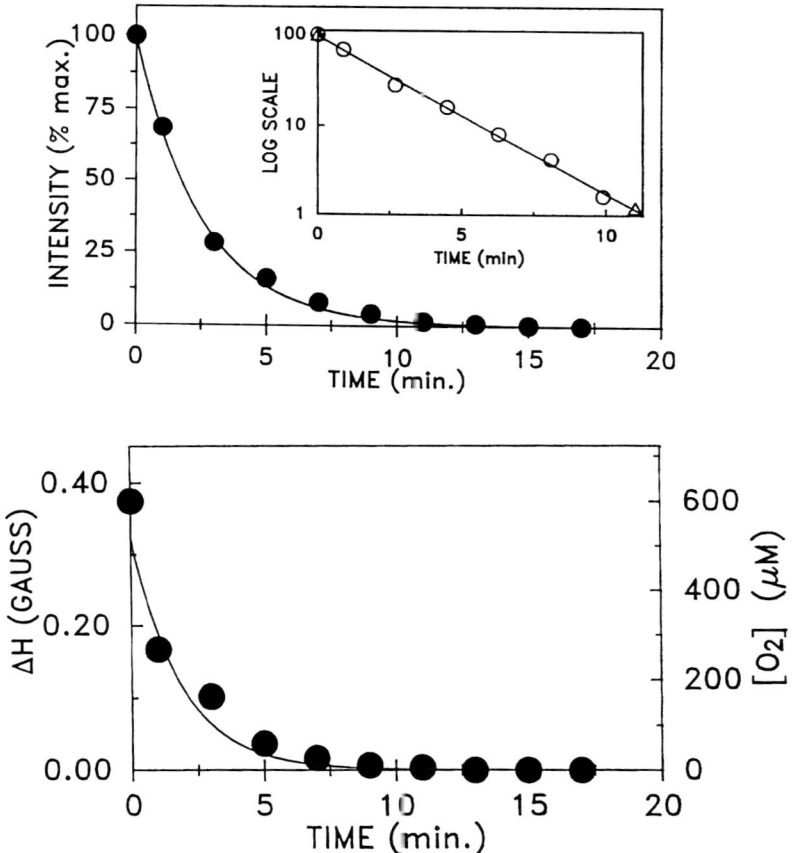

FIGURE 16. (Top) Graph of the kinetics of free radical metabolism in the ischemic heart. The heart was loaded with a 30 min infusion of 1 mM TEMPO followed by induction of ischemia at time 0. (Insert) Semilogarithmic plot of racical decay during ischemia; k = 0.40 min^{-1}. (Bottom) Graph of myocardial oxygen concentrations calculated from paramagnetic broadening by oxygen in the ischemic heart as a function cf duration of ischemia.

REFERENCES

1. **Korthius, R. J. and Granger, D. N.**, *Physiology of Oxygen Radicals*, Taylor, A. E., Matalon, S., and Ward, P. A., Eds., Williams & Wilkins, Baltimore, 1986, 217–249.
2. **Fantone, J. C. and Ward, P. A.**, Role of oxygen derived free radicals and metabolites in leukocyte-dependent inflammatory reactions, *Am. J. Pathol.*, 107, 395, 1982.
3. **Guarnieri, C., Flamigni, F., and Caldarera, C. M.**, Role of oxygen in the cellular damage induced by reoxygenation of hyperoxic hearts, *J. Mol. Cell. Cardiol.*, 12, 797, 1980.
4. **Hess, M. L. and Manson, N. H.**, Molecular oxygen: friend and foe. The role of the oxygen free radical system in the calcium paradox, the oxygen paradox and ischemia/reperfusion injury, *J. Mol. Cell. Cardiol.*, 16, 969, 1984.
5. **Burton, K. P., McCord, J. M., and Ghai, G.**, Myocardial alterations due to free radical generation, *Am. J. Physiol.*, 246, H776, 1984.
6. **Fridovich, I.**, Superoxide radical: an endogenous toxicant, *Annu. Rev. Pharmacol. Toxicol.*, 23, 239, 1983.
7. **Fridovich, I.**, Superoxide Dismutase: an adaptation to a paramagnetic gas, *J. Biol. Chem.*, 264, 7761, 1989.
8. **Zavoisky, E.**, *J. Phys. U.S.S.R.*, 9, 211, 1945.
9. **Cummerow, R. W. and Halliday, D.**, Paramagnetic losses in two manganous salts, *Phys. Rev.*, 70, 433, 1946.
10. **Janzen, E. G. and Blackburn, B. J.**, Detection and identification of short-lived free radicals by electron spin resonance trapping techniques (spin trapping), *J. Am. Chem. Soc.*, 91, 4481, 1969.
11. **Zweier, J. L.**, Reduction of oxygen by iron-adriamycin, *J. Biol. Chem.*, 259, 12759, 1983.
12. **Rosen, G. M. and Freeman, B. A.**, Detection of superoxide generated by endothelial cells, *Proc. Natl. Acad. Sci. U.S.A.*, 81, 7269, 1984.
13. **Zweier, J. L., Flaherty, J. T., and Weisfeldt, M. L.**, Direct measurement of free radical generation following reperfusion of ischemic myocardium, *Proc. Natl. Acad. Sci. U.S.A.*, 84, 1404, 1987.
14. **Zweier, J. L.**, Measurement of superoxide derived free radicals in the reperfused heart: evidence for a free radical mechanism of reperfusion injury, *J. Biol. Chem.*, 263, 1353–1357, 1988.
15. **Zweier, J. L., Kuppusamy, P., Williams, R., Rayburn, B. K., Smith, D., Weisfeldt, M. L., and Flaherty, J. T.**, Measurement and characterization of postischemic free radical generation in the isolated perfused heart, *J. Biol. Chem.*, 264, 18890–18895, 1989.
16. **Bielski, B. H. J. and Shieu, G. G.**, Reaction rates of superoxide radicals with essential amino acids, in *Oxygen Free Radicals and Tissue Damage*, Elsevier, New York, 1979, 43.
17. **Wood, P. M.**, *Biochem. J.*, 253, 287, 1988.
18. **Floyd, R. A.**, Direct demonstration that ferrous ion complexes of di- and triphosphate nucleotides catalyze hydroxyl free radicals formation from hydrogen peroxide, *Arch. Biochem. Biophys.*, 225, 263, 1983.
19. **Floyd, R. A. and Lewis, C. A.**, Hydroxyl free radical formation from hydrogen peroxide by ferrous iron-nucleotide complexes, *Biochemistry*, 22, 2645, 1983.
20. **Gutteridge, J. M.**, Reactivity of hydroxyl and hydroxyl-like radicals discriminated by release of thiobarbituric acid-reactive material from deoxy sugars, nucleosides and benzoate, *Biochem. J.*, 224, 761, 1984.
21. **Mashino, T. and Fridovich, I.**, *Arch. Biochem. Biophys.*, 254, 547–554, 1987.
22. **Kono, Y. and Fridovich, I.**, *J. Biol. Chem.*, 258, 13646–13648, 1983.
23. **Okajima, T. and Yamazaki, I.**, *Biochim. Biophys. Acta*, 284, 355–359, 1972.

24. **Kuo, C. F., Mashino, T., and Fridovich, I.,** *J Biol. Chem.*, 262, 4724–4727, 1987.
25. **Lin, W. S., Armstrong, D. A., and Lal, M.,** *Int. J. Radiat. Biol.*, 33, 231–243, 1978.
26. **Grisham, M. B. and McCord, J. M.,** Chemistry and cytotoxicity of reactive oxygen metabolites, in *Physiology of Oxygen Free Radicals,* Taylor, A. E., Matalon, S., and Ward, P. A., Eds., Williams & Wilkins, Baltimore, 1986, 1–18.
27. **Zweier, J. L., Gianni, L., Muindi, J., and Myers, C.,** Differences in oxygen reduction by the iron complexes of Adriamycin and duanomycin — the importance of the side chain hydroxyl group, *Biochim. Biophys. Acta*, 884, 326, 1986.
28. **Myers, C., Gianni, L., Zweier, J. L., Muindi, J., Sinha, B., and Eliot, H.,** The role of iron in Adriamycin biochemistry, *Fed. Proc.*, 45, 2792, 1986.
29. **Misra, H. P. and Fridovich, I.,** The role of superoxide anion in the auto oxidation of epinephrine and a simple assay for superoxide dismutase, *J. Biol. Chem.*, 247, 3170, 1972.
30. **Fridovich, I.,** Measurement of superoxide anion-cytochrome c, in *CRC Handbook of Methods for Oxygen Radical Research,* Greenwald, R. A., Ed., CRC Press, Boca Raton, FL, 1985, 121.
31. **Auclair, C. and Voisin, E.,** Nitroblue tetrazolium reduction, in *CRC Handbook of Methods for Oxygen Radical Research,* Greenwald, R. A., Ed., CRC Press, Boca Raton, FL, 1985, 123.
32. **Richmond, R., Halliwell, B., Chauhan, J., and Darbre, A.,** Superoxide dependent formation of hydroxylation of radicals: detection of hydroxyl radicals by the hydroxylation of aromatic compounds, *Anal. Biochem.*, 118, 328, 1981.
33. **Sbarra, A. J. and Karnovsky, M. L.,** The biochemical basis of phagocytosis. I. Metabolic changes during the ingestion of particles by polymorphonulcear leukocytes, *J. Biol. Chem.*, 234, 1355, 1959.
34. **Borg, D. C. and Schaich, K. M.,** Cytooxicity from coupled redox cycling of autooxidizing xenobiotics and metals, *Isr. J. Chem.*, 24, 38, 1984.
35. **Cadenas, E., Boveris, A., and Chance, B.,** Low-level chemiluminescence of biological systems, in *Free Radicals in Biology,* Pryor, W. A. Ed., Academic Press, San Diego, 1984, 211.
36. **Perkins, M. J.,** Spin trapping, *Adv. Phys. Org. Chem.*, 17, 1, 1980.
37. **Finkelstein, E., Rosen, G. M., and Rauckman, E. J.,** Spin trapping: kinetics of the reaction of superoxide and hydroxyl radicals with nitrones, *J. Am. Chem. Soc.*, 102, 4994, 1980.
38. **Brittigan, B. E., Cohen, M. S., and Rosen, G. M.,** Detection and production of oxygen centered free radicals by human neutrophils using spin trapping techniques: a critical perspective, *J. Leukocyte Biol.*, 41, 349–362, 1987.
39. **Buettner, G. R. and Oberly, L. W.,** Considerations in the spin trapping of superoxide and hydroxyl radical in aqueous systems using 5,5-dimethyl-1-pyrsoline-N-oxide, *Biochem. Biophys. Res. Commun.*, 83, 69–74, 1978.
40. **Buettner, G. R.,** Spin Trapping: ESR parameters of spin adducts, *Free Radical Biol. Med.*, 3, 259–303, 1987.
41. **Zweier, J. L., Rayburn, B. K., Flaherty, J. T., and Weisfeldt, M. L.,** Recombinant superoxide dismutase reduces oxygen free radical concentration in reperfused myocardium, *J. Clin. Invest.*, 80, 1728, 1987.
42. **Zweier, J. L., Flaherty, J. T., and Weisfeldt, M. L.,** Measurement of free radical generation in the post-ischemic heart, in *Oxy-Radicals in Biology and Pathology,* Alan R. Liss, New York, 1988, 365–383.
43. **Hodgson, E. K. and Fridovich, I.,** The interaction of bovine erythrocyte superoxide dismutase with hydrogen peroxide: inactivation of the enzyme, *Biochemistry,* 14, 5294, 1975.

44. **Zweier, J. L., Kuppusamy, P., and Lutty, G. A.,** Measurement of endothelial cell free radical generation: evidence for a central mechanism of free radical injury in postischemic tissues, *Proc. Natl. Acad. Sci. U.S.A.,* 85, 4046–4090, 1988.
45. **McCord, J. M.,** Oxygen derived free radicals in postischemic tissue injury, *N. Engl. J. Med.,* 312, 159–163, 1985.
46. **Das, D. K., Rao, P. S., and Engelman, R. M.,** *J. Mol. Cell. Cardiol.,* 19(Suppl. IV), S.78, 1987.
47. **Zweier, J. L. and Kuppusamy, P.,** Electron paramagnetic resonance measurements of free radicals in the intact beating heart: a technique for detection and characterization of free radicals in whole biological tissues, *Proc. Natl. Acad. Sci. U.S.A.,* 85, 5703–5707, 1988.
48. **Zweier, J. L. and Kuppusamy, P.,** Electron paramagnetic resonance studies of free radicals in the perfused heart, *Phys. Med.,* 2–4, 289, 1989.

Chapter 11

GASTRIC MUCOSAL INJURY

Toshikazu Yoshikawa and Motoharu Kondo

TABLE OF CONTENTS

I. Introduction 272

II. Gastric Mucosal Injury Produced by Oxygen Free Radicals 272

III. Gastric Mucosal Injury Induced by Platelet Activating Factor ... 272

IV. Gastric Mucosal Injury Induced by Stress 274

V. Gastric Mucosal Injury Induced by Ischemia-Reperfusion 275

VI. Gastric Mucosal Injury Induced by SOD Inhibitor 278

VII. Sources of Oxygen Radicals 278

VIII. Arachidonic Acid Cascade and Free Radicals 279

IX. Conclusion 281

References 282

I. INTRODUCTION

Gastric mucosal injury appears to result when the normal gastric mucosal defense mechanisms are overcome by factors that promote mucosal injury. This imbalance can occur either because the normal defenses including prostaglandins, mucous, and gastric mucosal blood flow become impaired, or because the injurious factors, such as gastric acid, pepsin, refluxed duodenal contents, and exogenous ulcerogens, exceed the capacity of defense mechanisms to protect. However, the cause of gastric mucosal injury remains unknown.

In recent years, it has been shown that active oxygens and free radicals are closely involved in various diseases.[1-4] These reactive species can attack and damage important biological molecules such as lipids, nucleic acids, enzymes, and proteins. In particular, polyunsaturated fatty acids located in the lipophilic section of cell membranes are prone to be attacked by free radicals, which produce toxic lipid peroxides through a chain reaction of lipid peroxidation (Figure 1). Lipid hydroperoxides produced by this process are thought to be the main factor in the damage to biological membranes caused by oxygen free radicals.[5]

Even in the digestive tract system, the gastric mucosa is exposed to foreign substances and toxins, and disturbances in the microcirculation are easily produced by several stresses. Thus, it is conceivable that mucosal damage can be easily produced by the generation of exogenous and endogenous active oxygens and free radicals. Since the report that experimental gastric mucosal injuries were suppressed by treatment with superoxide dismutase (SOD),[6] a selective scavenger of the superoxide anion radicals, various studies have been conducted. In this review, we outline the active oxygen and free radical generation system and the defensive system against oxidative stress in the stomach, and their significance in the development of gastric mucosal injury.

II. GASTRIC MUCOSAL INJURY PRODUCED BY OXYGEN FREE RADICALS

If ferrous iron and ascorbic acid are injected directly into the gastric submucosa, a distinct ulcerative lesion appears (unpublished data). With the injection of iron or ascorbic acid only, lesions do not appear, indicating that oxygen radicals produced by the iron-ascorbic acid system are the likely cause for the onset of these ulcers. In fact, these lesions can be significantly prevented by treatment with SOD (Table 1). Similarly, injecting the hypoxanthine-xanthine oxidase (HX-XO), well known as an *in vitro* generating system of superoxide anion radicals, via the celiac artery, increases the loss of ^{51}Cr-labeled erythrocytes across the gastric mucosa in stomachs.[7] Stein et al.[8] recently demonstrated that local infusion of HX and XO via the left gastric artery in rats caused marked gross injury in the stomach without producing

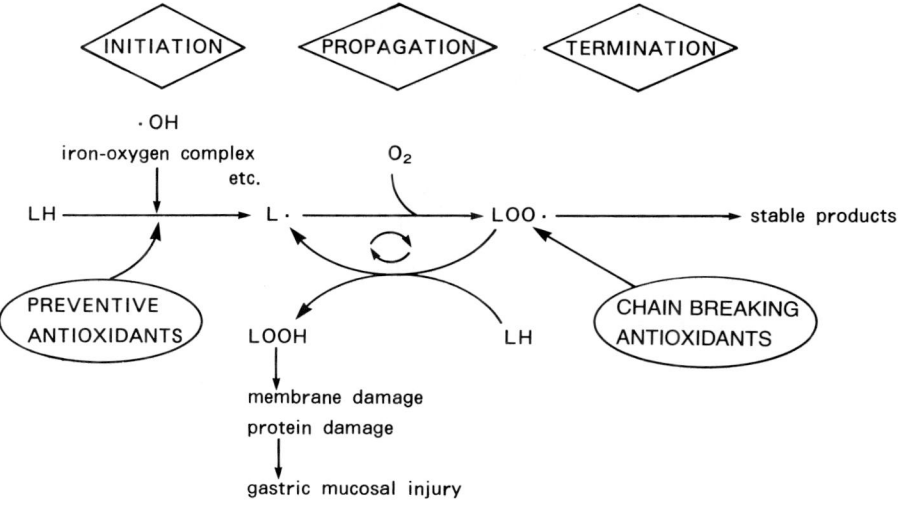

FIGURE 1. An outline mechanism of lipid peroxidation mediated by free radical chain reaction.

TABLE 1
Effects of Human-SOD on the Total Area of Gastric Ulcer and Lipid Peroxides in the Gastric Mucosa Following Local Injection of Ferrous-Ascorbate Mixture (FE + AA)

Group	(n)	Ulcerative area (mm²)	TBA-reactants (nmol/mg protein)
Physiological saline	(10)	0.0 ± 0.0	0.53 ± 0.16
FE + AA	(10)	17.4 ± 7.0[a]	0.99 ± 0.17[a]
FE + AA + human SOD	(10)	4.2 ± 4.5**	0.76 ± 0.17*
FE + AA + apo SOD	(5)	13.8 ± 6.3	0.80 ± 0.27
FE + AA + heated SOD	(5)	19.6 ± 8.8	1.05 ± 0.24

Note: Each value indicates the mean ± SD.

[a] $p < 0.001$ for difference in the values of control group injected with physiological saline. *$p < 0.05$ and **$p < 0.001$ for difference in the values of the group injected with ferrous-ascorbate mixture.

severe ischemic condition. This occurs even in the absence of exogenous luminal acid and is increased by luminal acidification. This injury is not dependent on luminal acidification and is not mediated by alteration in gastric mucosal blood flow. In addition, gross mucosal injury by HX-XO is largely prevented by SOD, supporting the assumption that this injury is caused by the release of oxygen radicals and not by the effects of the infusate. These findings give support to the concept that oxygen free radicals play an important role in the pathogenesis of gastric mucosal injury.

III. GASTRIC MUCOSAL INJURY INDUCED BY PLATELET ACTIVATING FACTOR

Rosam et al.[9] demonstrated that intravenous infusions of platelet activating factor(PAF)-acether at doses in the range of 10 to 100 ng/kg/min produced focal ulceration and hemorrhage in the rat stomach. Such damage was not produced when the precursor for PAF-acether, lyso-PAF, was administered in similar doses,[10] and was able to be prevented by pretreatment specific PAF-acether receptor antagonists.[11] Microcirculatory disturbance has been proposed to be important in the pathogenesis of the gastric mucosal injury induced by PAF-acether,[9] however, the precise mechanism of this injury is not yet known. Administration of PAF-acether to the rat at doses which cause gastric mucosal injury results in significant increases in tissue levels of myeloperoxidase, an enzyme found predominantly in neutrophils which is used as a biochemical marker for these cells.[12] Furthermore, Suematsu et al.[13] clearly demonstrated *in vivo* that myeloperoxdase-mediated oxidants from sticking granulocytes could actually attack the venular endothelial surface in PAF-induced microvascular changes, using a real-time visualization technique of luminol-dependent photoemission derived from granulocytes by using ultrasensitive video intensifier microscopy PAF can prime neutrophils to release superoxide anion radicals.[14] Yoshida et al.[15] recently demonstrated that the pretreatment with SOD and/or catalase could inhibit the aggravation of the gastric mucosal lesions induced by PAF-acether without altering the reduced gastric mucosal blood flow. In addition, in rats made neutropenic by treatment with methotrexate[16] or antineutrophil antibody,[15] PAF-induced gastric mucosal damage is significantly reduced. These findings are very supportive of the hypothesis that oxygen radicals generated by neutrophils play an important role in mediating the gastric mucosal damage induced by PAF.

IV. GASTRIC MUCOSAL INJURY INDUCED BY STRESS

It is well known that water immersion restraint or burn stress results in formation of multiple, hemorrhagic erosions at the corpus in the rat. We first reported that lipid peroxidation plays a significant role in the pathogenesis of gastric mucosal lesions induced by these stresses.[17,18] Especially in the burn stress model, there is a marked elevation of thiobarbituric acid (TBA)-reactive substances, an index of lipid peroxication, in the gastric mucosa before erosions of the mucosa could be detected with the naked eye, indicating that lipid peroxidation is etiologically significant, and is not the result of the gastric mucosal injury.[18] In addition, the formation of the gastric mucosal injury and the increase in lipid peroxides in the gastric mucosa induced by water-immersion restraint stress or burn stress are reportedly significantly inhibited by treatment with SOD and/or catalase (Table 2).[19] Reduced microcirculation of the gastric mucosa induced by stresses doesn't change following treatment

TABLE 2
Effects of SOD and Catalase on the Total Area of Erosions in the Gastric Mucosa 6 h after the Beginning of the Water-Immersion Restraint Stress or 3 h after the Burn Stress

Treatment	Water-immersion restraint		Burn	
	(mm^2)	(n)^a	(mm^2)	(n)^a
Saline (control)	55.4 ± 6.5	(10)	10.6 ± 3.5	(10)
SOD	36.2 ± 3.8	(6)***	4.3 ± 2.6	(8)***
Catalase	49.3 ± 8.7	(6)	4.9 ± 3.4	(8)**
SOD + catalase	26.1 ± 5.7	(6)***	5.3 ± 2.3	(8)**

Note: Each value indicates the mean ± SD. SOD at a dose of 20 mg/kg and/or catalase at a dose of 1 mg/kg were injected s.c. immediately prior to and at 3 h into the water-immersion restraint stress. SOD at a dose of 5 mg/kg and/or catalase at a dose of 2 mg/kg were injected i.v. 30 min before the burn stress.

^{a} Number of rats.

** $p < 0.01$ and *** $p < 0.001$ by Student's *t*-test for the difference between this value and that of rats treated with physiological saline as the control.

with scavengers, suggesting that the effectiveness of SOD and catalase on the gastric mucosal lesions induced by stress may occur via the scavenging of oxygen radicals, and not through an effect on gastric mucosal microcirculation.

V. GASTRIC MUCOSAL INJURY INDUCED BY ISCHEMIA-REPERFUSION

Ischemia in itself causes tissue damage and eventual death of cells. However, if oxygen is reintroduced by reperfusion after a certain period of ischemia, further tissue damage may occur. It has been reported that this reperfusion damage can be prevented by administering SOD, catalase, dimethylsulfoxide (DMSO), and other antioxidants in experimental models of the stomach.[6,20-23] Itoh and Guth[6] demonstrated that SOD could significantly protect against gastric mucosal injury induced by a withdrawal for 20 min and retransfusion in rats. Perry et al.,[20] using cat stomach, produced an ischemia-reperfusion model, and observed a suppression of gastric mucosal lesions by SOD and DMSO. In addition, Smith et al.,[22] using a model similar to the one used by Itoh and Guth, reported that leakage of ^{51}Cr-labeled erythrocytes into the gastric cavity was only accelerated during retransfusion, and this was significantly inhibited by SOD and the iron chelating agent deferoxamine. These studies suggest that active oxygen species derived from molecular oxygen such as superoxide radical, hydroxyl radical, and hydrogen peroxide, have an important role in the etiology of gastric mucosal injury induced by ischemia-reperfusion.

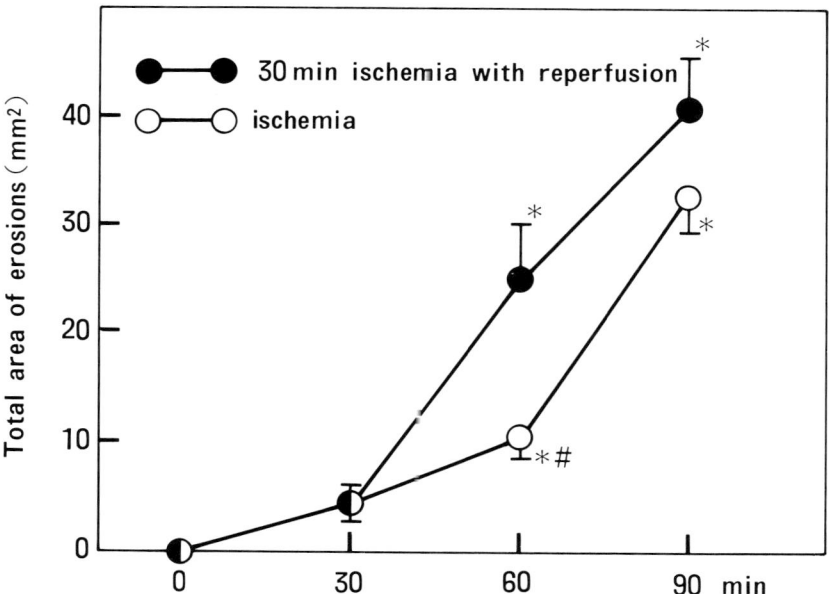

FIGURE 2. Changes in total area of gastric erosions after ischemia or ischemia-reperfusion. Each value indicates the mean ± SE of 5 to 12 rats. *p <0.001 for difference in the values of rats before clamping of the celiac artery. #p <0.05 for difference in the value of rats 30 min after reperfusion following 30 min of ischemia.

Currently, we are able to evaluate the relation between gastric mucosal damage and lipid peroxidation caused by ischemia-reperfusion using rats.[23] The animal model of ischemia-reperfusion damage was prepared by clamping the celiac artery for a certain period with a hemostat and then releasing the artery. By clamping the celiac artery the gastric mucosal blood flow decreased to 10% of that measured before clamping, that is the so-called low flow state but not complete ischemia, and recovered to the normal range by subsequent reperfusion.[23] The total area of erosions, a morphological index of gastric injury, didn't increase after 30 min of ischemia, and significantly increased after 60 min of ischemia, and 30 min of ischemia with 30 min reperfusion; however, the increase in the total area of erosions in the latter was significantly higher than that in the former (Figure 2).[23] Therefore, these results are consistent with the view that the injury produced by 30 min of reperfusion of a tissue subjected to 30 min of ischemia is more severe than that produced by 60 min of ischemia per se. Lipid peroxides in the gastric mucosa scarcely increased for 30 min of ischemia and accumulated remarkably and significantly following reperfusion. We previously indicated the possibility that hypoxia might be one of the factors predisposing the accumulation of lipid peroxide, and that peroxidation might occur in conditions of short supply of

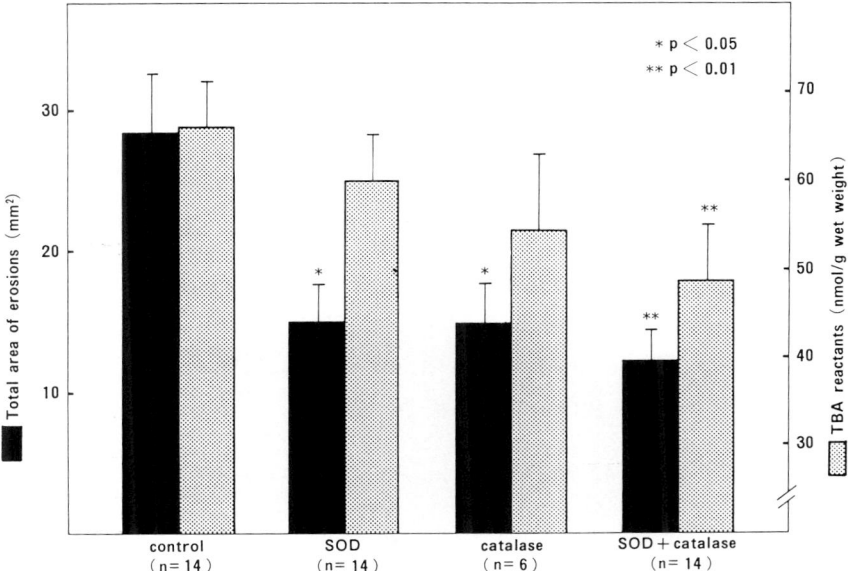

FIGURE 3. Effect of SOD and/or catalase on the total area of the erosions and the increase of TBA-reactive substances in the gastric mucosa induced by ischemia-reperfusion. Each value indicates the mean + SE. *p <0.05 and **p <0.01 for difference in the values of rats treated with physiological saline as the control.

oxygen.[24] α-Tocopherol in the gastric mucosa and serum decreased slightly after 30 min of ischemia, but decreased more significantly during reperfusion.[25] Vitamin E, which reacts with lipid peroxyradical and terminates free radical-mediated chain reaction, is speculated to be consumed in serum and gastric mucosa to prevent the development of tissue damage. Conversely, the increase of lipid peroxides and the decrease of vitamin E suggest the implication of free radicals in the gastric mucosal injury induced by ischemia-reperfusion. In addition, we have demonstrated that in vitamin E-deficient rats, gastric mucosal injury induced by ischemia-reperfusion was more severe than that in control rats fed an ordinary diet.[25]

To test the possibility that active oxygens may induce the lipid peroxidation-mediated gastric mucosal injury with ischemia-reperfusion, human SOD (50,000 U/kg, Nippon Kayaku Co., Ltd., Tokyo) and/or catalase (90,000 U/kg, Sigma, St. Louis, MO) were injected subcutaneously 1 h before ischemia, and 10,000 U/kg of SOD was intravenously injected just before reperfusion. The increase in the total area of the erosions was significantly inhibited by treatment with SOD, catalase, and SOD + catalase, and the increase in TBA-reactive substances was also significantly inhibited by SOD + catalase (Figure 3).[26] In addition, we have reported that Ebselen (Figure 4), which shows glutathione peroxidase-like activity, can protect against the gastric mucosal

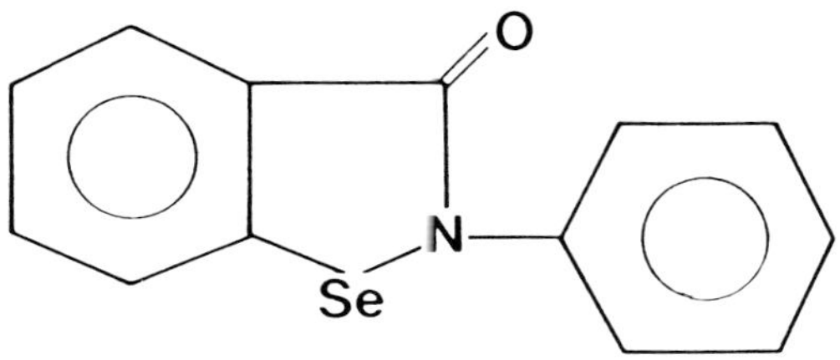

Ebselen

2-phenyl-1, 2-benzoisoselenazol-3(2H)-one

$C_{13}H_9NOSe$: Mol wt : 274.18

C 56.95% : H 3.31%: N 5.11%

O 5.83%   : Se 28.80%

FIGURE 4. Structure of seleno-organic compound, Ebselen.

injury induced by ischemia-reperfusion, and also inhibits the lipid peroxidation in the gastric mucosa.[27] These results indicate that scavenging hydrogen peroxides and lipid peroxides as well as superoxide anion radicals are important for protecting the gastric mucosa from ischemia-reperfusion injury.

VI. GASTRIC MUCOSAL INJURY INDUCED BY SOD INHIBITOR

The failure of the defensive mechanism against active oxygens and free radicals is one important factor in the development of gastric mucosal lesions. The SOD activity in the tissues is reportedly suppressed by diethyldithiocarbamate (DDC) in a dose-dependent manner and the mechanism of the inhibition is the chelation of Cu from the SOD protein.[28] Ogino et al.[29] demonstrated that DDC caused gastric mucosal lesions in the corpus and decreased SOD activity in the gastric mucosa. Ichiyama et al.[30] also reported that gastric ulcer with destruction of the muscular layer in the gastric antrum was produced by treatment with DDC and pylorus ligation, and Cu,Zn-SOD activity in the antrum decreased prior to the formation of ulcer without a change in Mn-SOD activity. These results show the possibility that in the gastric mucosa superoxide anion radical is an injurious offensive factor and Cu,Zn-SOD is a protecting defensive factor.

VII. SOURCES OF OXYGEN RADICALS

It has become clear that even among free radicals, the oxygen-derived radicals play an important role as a cause of various gastric mucosal injuries, but how are these oxygen radicals produced? At the present, as it is mainly being investigated in the small intestine,[31] the hypoxanthine(HX)-xanthine oxidase(XO) system induced by ischemia and the activated leukocytes are important as a source, and the view that the reciprocal action of both of these sources exacerbates the tissue damage is nearly true of the gastric mucosa also.[32] In normal conditions, the equilibrium of the synthesis and degradation of ATP (adenosine triphosphate) is maintained, but if ischemia continues, ATP is generated mainly by the action of adenylate kinase. Since this enzyme produces ATP and AMP (adenosine monophosphate) from two molecules of ADP, ischemic tissues tend to accumulate AMP, which results in an increase in hypoxanthine by way of inosine. At the same time, xanthine dehydrogenase is converted to xanthine oxidase accompanied by ischemia. With reperfusion (reoxygenation), xanthine oxidase catalyzes the reaction between hypoxanthine and molecular oxygen to form superoxide and hydrogen peroxide. Subsequently, these two active oxygens react in the presence of traditional metals or their chelates to form the highly reactive and cytotoxic hydroxyl radical. However, this hypothesis has many unknown elements which must be resolved in the future, including the time required to convert xanthine oxidase, the mechanism of the conversion, and the distribution of this enzyme within tissue. Furthermore, superoxide is generated from phagocytes including polymorphonuclear leukocytes stimulated by a variety of stimulants which activate NADPH oxidase present in the cell membrane.[33] In the gastric injury model induced by hemorrhagic shock, Itoh and Guth[6] found that gastric lesions are inhibited by treatment with allopurinol, an inhibitor of xanthine oxidase. Smith et al.[34] have reported that the depletion of neutrophils via antineutrophil serum resulted in a dramatic reduction in the area of gross lesions as well as a reduction in red blood cell flux into the lumen of the stomach after ischemia-reperfusion. We also carried out investigations using allopurinol and antineutrophil antibody, but suppressive effects against these gastric mucosal injuries varied depending on the experimental model (Table 3). Pretreatment with allopurinol significantly prevented the gastric mucosal injuries induced by burn shock,[19] by treatment with compound 48/80,[35] and by ischemia-reperfusion,[23] which indicates that xanthine oxidase is one of the major sources of oxygen radicals in these models. By the treatment with antineutrophil serum, the gastric mucosal injuries induced by PAF[15] and compound 48/80[35] were significantly inhibited, but burn shock[19] and ischemia-reperfusion injuries[23] were not inhibited. As compared with the xanthine oxidase system, neutrophils seem to play a relatively small part in the formation of gastric injuries induced by burn shock and ischemia-reperfusion.

VIII. ARACHIDONIC ACID CASCADE AND FREE RADICALS

Arachidonic acid cascade is one of the sources of free radicals in the biological system, and at the same time, free radicals suppress the arachidonic acid cascade.[36] As shown in Figure 5, in the conversion process of PGG_2 to

TABLE 3
Effects of Allopurinol or PMN-Depletion on Gastric Mucosal Injury in Rats

	Allopurinol	PMN-depletion
Gastric mucosal injury induced by		
Burn stress	O[a] (O)[b]	X[a] (X)[b]
Water-immersion restraint	X (X)	X (X)
Platelet activating factor	X (X)	O (O)
Compound 48/80	O (O)	O (O)
Ischemia-reperfusion	O (O)	X (X)

Note: 50 mg/kg/day of allopurinol dissolved in distilled water and pH-adjusted to 10.8 by 0.5 N NaOH was orally administered to rats 48 and 24 h before experiments; 18 h after the administration of anti-rat PMN antibody (10 ml/kg), PMN-depleted rats (below 500/mm^3) were used.

[a] Total area of gastric erosions.
[b] TBA-reactive substances in the gastric mucosa. O indicates effective, and X indicates ineffective.

PGH$_2$ in the cyclooxygenase pathway of arachidonic acid cascade, and when HPETE is converted to HETE in the lipoxygenase pathway, the hydroxyl radical is produced. Furthermore, the hydroxyl radical has also been reported to suppress PG synthesis due to inhibition of cyclooxygenase. In addition, HPETE, HETE, and leukotrienes are leukocyte chemotactic factors as well as stimulants which increase oxygen radical generation from leukocytes, which may contribute to tissue damage. Moreover, PGE$_1$ and PGE$_2$ inhibit superoxide generation by polymorphonuclear leukocytes,[37] which suggests that the production of superoxide radicals from leukocytes is regulated by the products from arachidonic acid cascade.

There are few reports pertaining to the relationship between the arachidonic acid cascade and free radicals in gastric mucosal damage, but it is interesting that Cu,Zn-SOD in the gastric mucosa has been shown to localize in the parietal cells in which prostaglandins localized according to immunohistochemical studies.[38] Furthermore, the administration of exogenous SOD causes an increase in PGE$_2$ of the gastric mucosa, and inhibits gastric mucosal damage induced by indomethacin. The involvement of leukotrienes has also been found to be important in gastric mucosal damage. By the treatment with 5-lipoxygenase inhibitor (AA-861, Takeda Pharmaceutical Co., Ltd., Tokyo), gastric mucosal injury and the increase in lipid peroxides in the gastric mucosa after ischemia-reperfusion were significantly inhibited.[39] A sulfide peptide antagonist (YM-638, Yamanouchi Pharmaceutical Co., Ltd., Tokyo) could show significant and stronger inhibition against gastric mucosal injury induced by ischemia-reperfusion. However, it is not clear if this protection is an effect

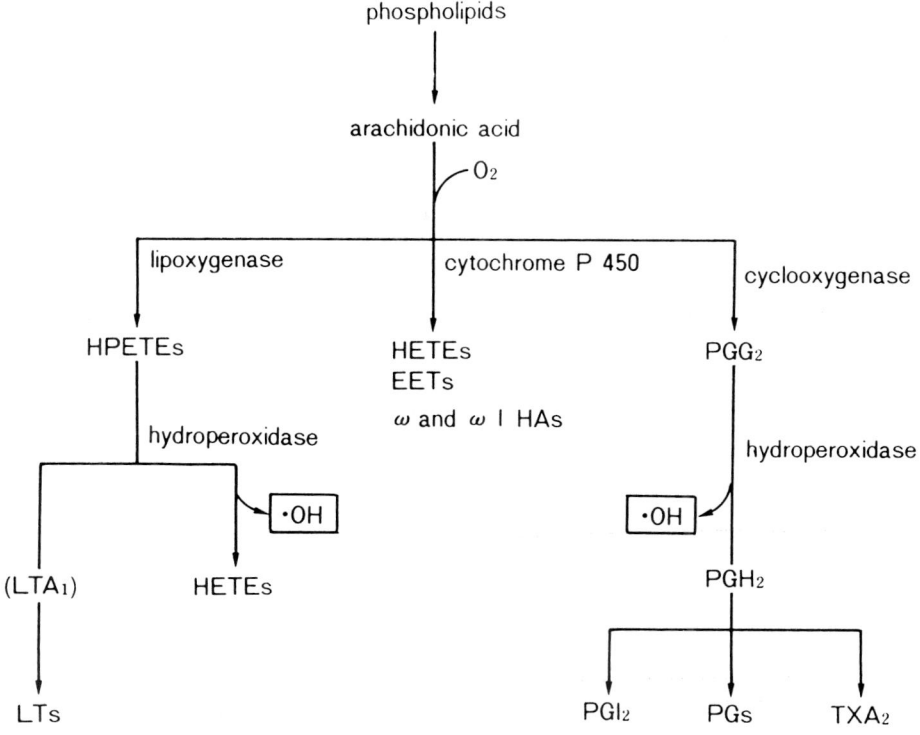

FIGURE 5. Metabolism of arachidonic acid cascade and production of hydroxyl radicals.

mediated through PGs. Thus, in the future, the relationship between free radicals and the arachidonic acid cascade in the development of gastric mucosal injury will be the subject of many studies.

IX. CONCLUSION

As stated above, it has become clear that lipid peroxidation of cell membranes mediated by active oxygens and free radicals plays an important role in the pathogenesis of acute gastric mucosal injury. Furthermore, the elimination mechanisms of these reactive species is an important function in the prevention or repair of lipid peroxidation-induced gastric mucosal injury. Recently, we also found that some antioxidants were effective in preventing experimental gastric mucosal damage induced by oxygen radicals, which was not prevented by histamine H_2 receptor antagonist. Clinically, when considering the prevention and treatment of gastric mucosal injury, it is understood that the inhibition of the radical production, enhancement of the elimination mechanism, and suppression of lipid peroxidation are important. Currently, among the anti-ulcer agents that are being and will be used clinically, there are many that have these types of effects,[40] and from this standpoint, we are anticipating the clinical application of newly developed agents.

REFERENCES

1. **Bulkley, G. B.**, The role of oxygen free radicals in human disease processes, *Surgery*, 94, 407–411, 1983.
2. **Halliwell, B.**, The expanding role of oxygen free radicals in clinical medicine, *West. J. Med.*, 144, 441–446, 1986.
3. **Fantone, J. C. and Ward, P. A.**, Role of oxygen-derived free radicals and metabolites in leukocyte-dependent inflammatory reactions, *A.J.P.*, 107, 397–418, 1982.
4. **Flaherty, J. T. and Weisfeldt, M. L.**, Reperfusion injury, *Free Radicals Biol. Med.*, 5, 409–419, 1988.
5. **Niki, E.**, Antioxidants in relation to lipid peroxidation, *Chem. Phys. Lipids*, 44, 227–253, 1987.
6. **Itoh, M. and Guth, P. H.**, Role of oxygen-derived free radicals in hemorrhagic shock-induced gastric lesions in the rat, *Gastroenterology*, 88, 1162–1167, 1985.
7. **Wadhwa, S. S. and Perry, M. A.**, Gastric injury induced by hemorrhage, local ischemia, and oxygen radical generation, *Am. J. Physiol.*, 253, G129–G133, 1987.
8. **Stein, H. J., Esplugues, J., Whittle. B. J. R., Bauerfeind, P., Hinder, R. A., and Blum, A. L.**, Direct cytotoxic effect of oxygen radicals on the gastric mucosa, *Surgery*, 106, 318–324, 1989.
9. **Rosam, A. C., Wallace, J. L., and Whittle, B. J. R.**, Potent ulcerogenic actions of platelet activating factor on the stomach, *Nature*, 319, 54–56, 1986.
10. **Dembinska-Kiec, A., Peskar, B. A., Muller, M. K., and Peskar, B. M.**, The effects of platelet activating factor on flow rate and eicosanoid release in the isolated perfused rat gastric vascular bed, *Prostaglandins*, 37, 69–91, 1989.
11. **Clostre, F., Etienne, A., Mencia-Huerta, J. M., and Braquet, P.**, Prevention of the platelet-activating factor-induced gastrointestinal damages by BN52021 and BN52063, in *Glikgolides — Chemistry, Biology, Pharmacology, and Clinical Perspectives*, Vol. 1., Braquet, P. and Prous, Jr., Eds., Prous Science Publishers, S. A., Barcelona, 1988, 541–551.
12. **Wallace, J. L. and MacNaughton, W. K.**, Gastrointestinal damage induced by platelet activating factor: role of leukotrienes *Eur. J. Pharmacol.*, 151, 43–50, 1988.
13. **Suematsu, M., Kurose, I., Asako, H., Miura, S., and Tsuchiya, M.**, In vivo visualization of oxyradical-dependent photoemission during endothelium-granulocytes interaction in microvascular beds treated with platelet-activating factor, *J. Biochem.*, 106, 355–360, 1989.
14. **Vercellotti, G. M., Yin, H. Q., Gustafsson, K. D., Nelson, R. D., and Jacob, H. S.**, Platelet-activating factor primes neutrophil responses to agonist: role in promoting neutrophil-mediated endothelial damage *Blood*, 71, 1100–1107, 1988.
15. **Yoshida, N., Yoshikawa, T., Ando, T., Naito, Y., Oyamada, H., Takemura, T., Tanigawa, T., Sugino, S., and Kondo, M.**, Pathogenesis of platelet-activating factor-induced gastric mucosal damage in rats, *Scand. J. Gastroenterol.*, 24(Suppl. 162), 210–214, 1989.
16. **Etienne, A., Thonier, F., Hecquet, F., and Braquet, P.**, Role of neutrophils in gastric damage induced by platelet activating factor, *Naunyn Schmiedebergs Arch. Pharmacol.*, 338, 422–425, 1988.

17. **Yoshikawa, T., Miyagawa, H., Yoshida, N., Sugino, S., and Kondo, M.**, Increase in lipid peroxidation in rat gastric mucosal lesions induced by water-immersion restraint stress, *J. Clin. Biochem. Nutr.*, 1, 271–277, 1986.
18. **Yoshikawa, T., Yoshida, N., Miyagawa, H., Takemura, T., Tanigawa, T., Sugino, S., and Kondo, M.**, Role of lipid peroxidation in gastric mucosal lesions induced by burn shock in rats, *J. Clin. Biochem. Nutr.*, 2, 163–170, 1987.
19. **Yoshikawa, T., Yoshida, N., Naito, Y., Takemura, T., Miyagawa, H., Tanigawa, T., and Kondo, M.**, Role of oxygen radicals in the pathogenesis of gastric mucosal lesions induced by water-immersion restraint stress and burn stress in rats, *J. Clin. Biochem. Nutr.*, 8, 227–234, 1990.
20. **Perry, M. A., Wadhwa, S., Parks, D. A., Pickard, W., and Granger, D. N.**, Role of oxygen radicals in ischemia-induced lesions in the cat stomach, *Gastroenterology*, 90, 362–367, 1986.
21. **von Ritter, C., Hinder, R. A., Oosthuizen, M. M. J., Svensson, L. G., Hunter, S. J. S., and Lambrecht, H.**, Gastric mucosal lesions induced by hemorrhagic shock in baboons; the role of oxygen derived free radicals, *Dig. Dis. Sci.*, 33, 857–864, 1988.
22. **Smith, S. M., Grisham, M. B., Manci, E. A., Granger, D. N., and Kvietys, P. R.**, Gastric mucosal injury in the rat. Role of iron and xanthine oxidase, *Gastroenterology*, 92, 950–956, 1987.
23. **Yoshikawa, T., Ueda, S., Naito, Y., Takahashi, S., Oyamada, H., Morita, Y., Yoneta, T., and Kondo, M.**, Role of oxygen-derived free radicals in gastric mucosal injury induced by ischemia or ischemia-reperfusion in rats, *Free Radicals Res. Commun.*, 7, 285–291, 1989.
24. **Yoshikawa, T., Furukawa, Y., Wakamatsu, Y., Takemura, S., Tanaka, H., and Kondo, M.**, Experimental hypoxia and lipid peroxide in rats, *Biochem. Med.*, 27, 207–213, 1982.
25. **Yoshikawa, T., Yasuda, M., Ueda, S., Naito, Y., Tanigawa, T., Oyamada, H., and Kondo, M.**, Vitamin E in gastric mucosal injury induced by ischemia-reperfusion, *Am. J. Clin. Nutr.*, 53, 2105–2145, 1991.
26. **Ueda, S., Yoshikawa, T., Takahashi, S., Ichikawa, H., Yasuda, M., Oyamada, H., Tanigawa, T., Sugino, S., and Kondo, M.**, Role of free radicals and lipid peroxidation in gastric mucosal injury induced by ischemia-reperfusion in rats, *Scand. J. Gastroenterol.*, 24(Suppl. 162), 55–58, 1989.
27. **Ueda, S., Yoshikawa, T., Takahashi, S., Naito, Y., Oyamada, H., Takemura, T., Morita, Y., Tanigawa, T., Sugino, S., and Kondo, M.**, Protection by seleno-organic compound, Ebselen, against acute gastric mucosal injury induced by ischemia-reperfusion in rats, in *Antioxidants in Therapy and Preventive Medicine,* Emerit, I. et al., Eds., Plenum Press, New York, 1990, 187–191.
28. **Heikkila, R. E., Cabbat, F. S., and Cohen, G.**, In vivo inhibition of superoxide dismutase in mice by diethyldithiocarbamate, *J. Biol. Chem.*, 251, 2182–2185, 1976.
29. **Ogino, K., Oka, S., Matsuura, S., Sakaida, I., Yoshimura, S., Matsuda, K., Sasaki, Y., Yamamoto, K., Yoshikawa, T., Okazaki, Y., and Takemoto, T.**, Ulcer formation in rat stomach with diethyldithiocarbamate, *J. Clin. Biochem. Nutr.*, 3, 189–193, 1987.
30. **Ishiyama, H., Yamasaki, K., Imaizumi, T., Kanbe, T., Ogino, K., Oka, S., Okazaki, Y., and Takemoto, T.**, Gastric antral ulcer produced by diethyldithiocarbamate in rats, *J. Clin. Biochem. Nutr.*, 5, 155–163, 1988.
31. **Granger, D. N., Rutili, G., and McCord, J. M.**, Superoxide radicals and feline intestinal ischemia, *Gastroenterology,* 81, 22–29, 1981.
32. **Grisham, M. B., Hernandez, L. A., and Granger, D. N.**, Xanthine oxidase and neutrophil infiltration in intestinal ischemia, *Am. J. Physiol.*, 251, G567–574, 1986.
33. **Babior, B. M., Kipnes, R. S., and Curnutte, J. T.**, Biological defense mechanisms. The production by leukocytes of superoxide, a potential bactericidal agent, *J. Clin. Invest.*, 52, 421–424, 1973.

34. **Smith, S. M., Holm-Rutili, L., Perry, M. A., Grisham, M. B., Arfors, K. E., Granger, D. N., and Kvietys, P. R.,** Role of neutrophils in hemorrhagic shock-induced gastric mucosal injury in the rat, *Gastroenterology,* 93, 468–471, 1987.
35. **Takemura, T., Yoshikawa, T., Yoshica, N., Takano, H., Tasaki, N., Naito, Y., Ueda, S., Sugino, S., and Kondo, M.,** Role of lipid peroxidation and oxygen radicals in compound 48/80-induced gastric mucosal injury in rats, *Scand. J. Gastroenterol.,* 24(Suppl.), 51–54, 1989.
36. **Egan, R. W., Paxton, J., and Kuehl, F. A.,** Mechanism for irreversible self-deactivation of prostaglandin synthetase, *J. Biol. Chem.,* 251, 7329, 1976.
37. **Yoshikawa, T., Ichikawa, H., Naio, Y., Oyamada, H., Ueda, S., and Kondo, M.,** Effects of prostaglandins on superoxide production from human polymorphonuclear leukocytes stimulated by various stimulants, *J. Clin. Exp. Med.,* 154, 259–260, 1990 (in Japanese).
38. **Makita, T., Ishida, T., and Ogino, K.,** Immunocytochemical localization of human Cu,Zn-SOD in the parietal cells of gastric gland of the baboon, in *Medical, Biochemical and Chemical Aspects of Free Radicals,* Hayaishi, O., Niki, E., Kondo, M., and Yoshikawa, T., Eds., Elsevier, Amsterdam, 1989, 683–684.
39. **Yoshikawa, T., Ichikawa, H., Naito, Y., and Kondo, M.,** Role of leukotrienes in gastric mucosal injury induced by ischemia-reperfusion in rats, *J. Clin. Gastroenterol.,* 14, 568–570, 1992.
40. **Yoshikawa, T., Naito, Y., Tanigawa, T., Yoneta, T., Oyamada, H., Ueda, S., Takemura, T., Sugino, S., and Kondo, M.,** Effect of zinc-carnosine chelate compound (Z-103) on burn-induced gastric mucosal injury in rats, *J. Clin. Biochem. Nutr.,* 7, 107–113, 1989.

Chapter 12

RENAL ISCHEMIA

Colin Green, Lisa Cotterill, and Jon Gower

TABLE OF CONTENTS

I.	Introduction 286	
	A. Anatomy and Physiology of the Kidneys 286	
II.	Renal Ischemia 289	
	A. Warm Ischemia 290	
	B. Cold Ischemia 291	
III.	Oxidative Stress 292	
	A. The Role of Iron 294	
IV.	The Role of Calcium 297	
V.	Second Messengers 301	
VI.	Conclusions 303	
References 305		

I. INTRODUCTION

A. ANATOMY AND PHYSIOLOGY OF THE KIDNEYS

The kidneys are paired organs which are vital in the homeostatic control of extracellular fluid, ensuring that its volume, osmolality, pH, and its content of salts and other solutes are held within narrow limits. They do this through a combination of several mechanisms. First, a proportion of blood plasma is continuously filtered through the glomeruli into the tubules but, as long as the salt and water balance of the body remains intact, the filtrate is reabsorbed by the tubular cells along with some of the solutes dissolved in it, and is returned to the extracellular compartment. In this way, the correct balance of water and electrolytes such as sodium, chloride, phosphate, and bicarbonate ions is maintained and the loss of metabolically essential solutes such as amino acids and glucose is prevented. Any excesses arising in the extracellular fluid are, at least if renal function is normal, excreted from the body in urine.

Extracellular fluids are continuously reprocessed by the kidneys. The paired organs normally filter a volume equal to the total body extracellular fluid every 2 h, a mass equivalent to about 20% of the body weight. If one of the kidneys has impaired function, the other rapidly makes up the deficit and a single organ can maintain normal fluid balance. Kidneys are relatively independent of direct nervous control but renal hormones interact with hormones from the adrenal cortex, parathyroid, and pituitary glands to fine-tune regulation of salt and water balance. Efficient function of these organs is therefore essential to health. However, because they are paired organs and because effective life-saving substitutes in the different modes of dialysis are available, renal ischemia need not be lethal and even acute renal failure can be reversed with time.

The unique functions of the kidneys are associated with their highly specialized anatomical organization. In sagittal section (Figure 1), an outer cortex is clearly demarcated from the outer and inner medullary zones or papillae. The smallest functional unit in the kidney is known as a nephron and each human kidney contains approximately 1.2 million. It consists of a glomerulus, a proximal convoluted tubule, loop of Henle, and a distal convoluted tubule which opens into a collecting duct (Figure 2). The glomerular tuft is composed of a coil of specialized capillaries fed by arterioles and lies within a space called Bowman's capsule (Figure 3). Ultrafiltration occurs across this capillary tuft and the fluid passes into the proximal convoluted tubule whose lining of epithelial cells is continuous with the glomerulus and has a prominent brush border. Glomeruli and proximal convoluted tubules are found only in the cortex and drain into the straight portions (partes rectae) of the proximal tubules which terminate in the outer strip of the outer zone of the medulla. The proximal tubule is the largest segment of the nephron. Between 70 to 90% of the sodium, chloride, potassium, and water are reabsorbed along its more proximal portions while more distally, secretion of weak acids and weak bases becomes more prominent.

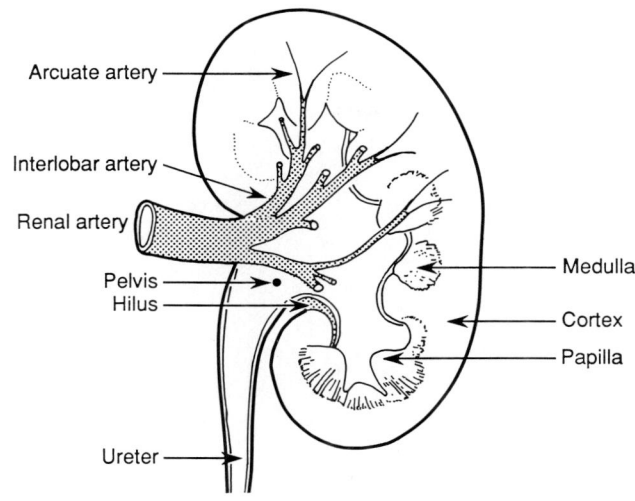

FIGURE 1. Structure of the cut surface of the kidney.

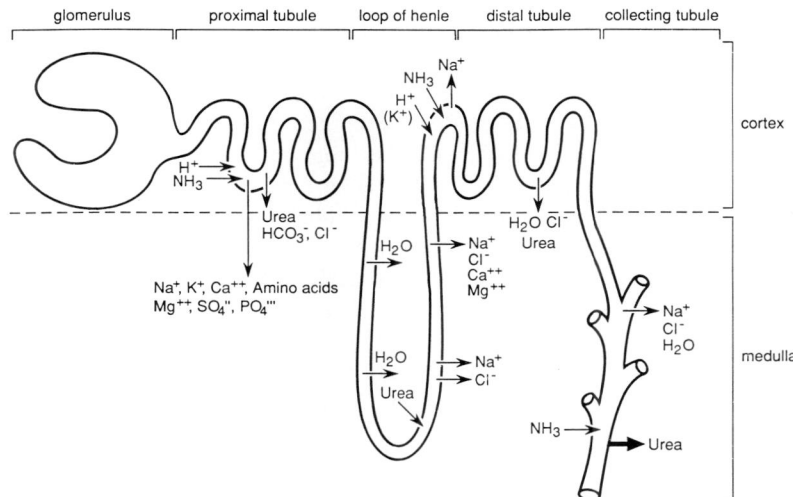

FIGURE 2. Diagramatic representation of a nephron, showing its location in the cortex and medulla.

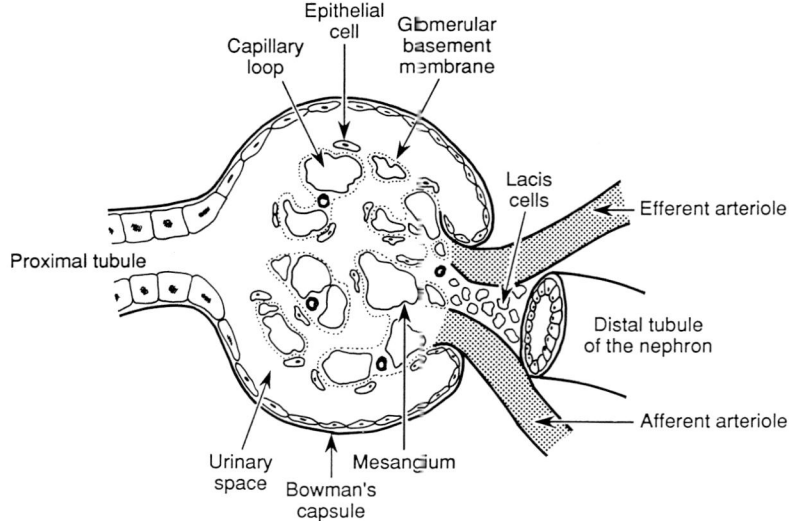

FIGURE 3. Diagram of the glomerulus and juxtaglomerular apparatus.

As the nephron passes into the medulla, it forms a U-shaped bend called the loop of Henle. This is surrounded by a dense capillary network, the vasa rectae, and functions as a counter-current multiplier. Fluid in the descending limb becomes progressively more concentrated during its passage from the cortico-medullary junction to the tip of the loop. In the ascending limb, sodium is reabsorbed more rapidly than water by an ion exchange mechanism. The fluid passing into the distal tubule is more dilute than that which entered the descending limb and so a concentration gradient is created between the two limbs. Water then diffuses out of the descending limb until the contents are concentrated to a level in equilibrium with interstitial fluid. After leaving the distal tubule, the urine then enters the collecting ducts and is concentrated as it passes through the medulla before leaving the kidney via the ureter.

The glomerular filtering membrane has three components: a layer of capillary endothelium, the basement membrane, and epithelial cells of the capsule. Glomerular filtration is a passive process driven by the sum of hydrostatic pressure and osmotic pressure gradients across the capillary membrane. Urea, glucose, and creatinine are as freely able to pass through the capillary wall as water and it is not until a molecular weight between 80,000 and 90,000 is reached that molecules are barred from passage. This corresponds to a pore size of 42 to 45 Å. A further barrier is presented by the glomerular basement membrane with channels ranging from 2 to 45 Å in diameter.

The glomerular filtration rate (GFR) in a normal adult man is about 120 ml/min and is independent of the rate of urine production. It is defined as the total volume of filtrate produced by all glomeruli per unit time. In clinical practice, the GFR is measured either by clearance of endogenous creatinine or by clearance of exogenously administered inulin (a freely filtered, nontoxic, polysaccharide which does not undergo protein binding and is neither absorbed nor secreted by tubular cells). In practice, a fall in GFR is usually associated with systemic hypotension or hypovolemia; with local or systemic disruption in hormonal controls (particularly those activating α-adrenoreceptors) of the afferent and efferent blood vessels supplying the glomerulae; or with physical slowing of the blood supply locally due to microthrombosis or vessel-wall stenosis.

In summary then, the nephron comprises many delicate and specialized structures which closely interact both physiologically and anatomically with the vascular bed. Renal impairment can result (a) directly from insult to the nephrons themselves; (b) indirectly from insult to the nephrons through vascular deviations; or (c) from a combination of direct and indirect insults. Ischemic damage falls into the latter category.

II. RENAL ISCHEMIA

Renal ischemia, in which the kidney is rendered anoxic, or partial ischemia, in which it is hypoxic, may occur naturally through falling blood pressure, vasospasm of the renal vessels and poor perfusion, or it may be iatrogenic and surgically induced. In the former case, ischemia will be normothermic (37°C), referred to hereafter as warm ischemia (WI). During iatrogenic ischemia, the kidney may be subjected to warm ischemia (30 to 37°C), cooling or rewarming ischemia (10 to 30°C) referred to as hypothermic ischemia (HI), and cold ischemia (0 to 10°C) hereafter referred to as CI. Iatrogenic WI may be induced (1) when the renal vessels are clamped *in situ* while the organ is explored for renoliths or tumors or when aberrant vessels are reconstructed by microsurgical techniques; (2) when the organ is removed from the donor, reconstructed *ex vivo* ("bench surgery") and then replanted orthotopically; and (3) for a time when a kidney is harvested for transplantation from a brain-dead cadaver. Iatrogenic HI ensues while the organ is cooling for storage and during the transplantation operation when it is rewarming slowly.

Cold ischemia (CI) results from deliberate cooling to 0 to 10°C for preservation of organs during storage and transport to a transplant recipient. Hence, a kidney may be subjected to WI, HI, and CI between the donor's death and restoration of blood after transplantation. Most kidneys are stored for 6 to 24 h using a simple initial vascular flush with a cold (4°C) asanguinous balanced salt solution followed by refrigeration at 0 to 4°C. A few centers worldwide use a continuous perfusion technique for the duration of the storage time and attempt to provide substrates for the limited metabolism occurring at 8 to 10°C. Many of these kidneys (50% in some reported series) fail to function immediately after engraftment if they have been stored for longer than 24 h; hence, long-term survival of the graft, inevitably under attack from the hosts' immune responses, is still further compromised by this ischemic insult. For bench surgery, 3 to 4 h of HI has to be tolerated by the kidney. In the future, longer periods may be needed. One can also predict that storage *ex vivo* for 10 to 14 days will be a requirement in transplantation to allow time for immunological assessment and manipulation of the scheduled host's immune repertoire before the organ is finally transplanted. There is therefore a considerable need to improve ways of preventing WI, HI, and CI damage. We may hope for serendipitous breakthroughs. Alternatively, we can explore the changes that occur at a molecular, biochemical, cell membrane, and cell biology level, and design therapeutic strategies in a logical way in response to that information. The rest of this review concentrates on WI and CI. The underlying mechanisms of damage may be similar if not identical in each situation, but the sequence of events is certainly different in temporal and possibly in spatial terms.

A. WARM ISCHEMIA

In simple normothermic ischemia (WI), many pathological changes only become evident after blood circulation has been restored. The events which have taken place *during* WI are therefore difficult to identify with precision. However, some aspects of cell function are known to change. Mitochondrial electron transport is inhibited and mitochondria swell with the net effect that ATP cannot be adequately regenerated. Adenine nucleotides are used up and are catabolized through AMP to hypoxanthine.[15] As ATP levels fall, membrane transport shuttles fail. The internal homeostasis in various intracellular compartments is lost. Ions diffuse down their respective electrochemical potential gradients and there is a net influx of water producing cell swelling or hydropic degeneration.[79] Potassium is lost to the extracellular compartment but the concentration of cytosolic calcium, sodium, and chloride rises as these ions enter the cell because energy-dependent pumps which normally extrude them fail. Reduced pyridine nucleotides (NADH, NADPH) accumulate because they cannot be processed via mitochondrial electron transfer and pH falls due to a switch to anaerobic type metabolism.[10] The altered intracellular environment induces lysozomal disruption with release of degradative enzymes into the cell.

At a more gross level, toxic metabolites such as lactic acid and denatured proteins accumulate.[102] In badly damaged kidneys, the end result is an outflow block after initial reflow of blood through the vascular bed when the organ is transplanted[92] and acute renal failure ensues (Figure 1), but what actually initiates the lesions has been the subject of much speculation. However, prevention or delay in the chain of events involving endothelial swelling, protein leaks to the extravascular space, edema, and loss of erythrocyte deformability have been claimed in several systems to which membrane stabilizers such as chlorpromazine,[12] relatively impermeant solutes such as mannitol,[84] and free radical scavengers such as superoxide dismutase,[42] α-tocopherol and co-enzyme Q_{10},[93] glutathione peroxidase,[64] ascorbic acid,[78] selenium salts,[35] or iron chelators[44] have been added. Hence there is now sufficient circumstantial evidence available to incriminate toxic free-radical products as the possible culprits in initiating this cascade.

In our experience, there are marked species differences in the length of time that kidneys can tolerate WI without irreversible and lethal damage. In rats, the maximum tolerated period is 60 min, in rabbits after 90 min of WI, 50% of the kidneys will survive but 120 min is always lethal; in dogs, the maximum tolerated time is 90 min. In man, it is generally accepted that WI at 37°C for 60 min severely compromises subsequent function and the kidney may take many days to recover.

B. COLD ISCHEMIA

Reduced temperature has been the cornerstone of all currently employed storage techniques ever since it was demonstrated that rapidly cooling organs resulted in a profound reduction in metabolic rate (Levy, 1959) and protected kidneys from ischemic damage albeit for relatively brief periods.[17] There is no clear CI time beyond which subsequent function can be predicted to fail but 72 h is generally regarded as the absolute limit for clinical transplantation. Occasional reports of 8 and 9-day storage in experimental animals usually refer to individuals within a series and it is certainly only in isolated instances.

The causes of deterioration during prolonged periods of *ex vivo* storage remain unresolved but several pathological changes have been documented. These include depletion of high energy adenine nucleotides,[16] accumulation of metabolites such as H^+ ions leading to a significant fall in intracellular pH,[90] release of lysosomal enzymes,[77] damage to cellular membranes with loss of constituent phospholipids,[91] impaired mitochondrial function,[3] damaged endoplasmic reticulum,[85] and vascular injury that results in edema, loss of circulating proteins, loss of erythrocyte deformability, rouleaux formation, and leakage of blood into the extravascular compartment.[31,98] Severely damaged kidneys are slow to reperfuse when revascularized, then swell and become cyanosed as a microcoagulopathy and outflow block develops within minutes. From this evidence, damage to the vascular bed, particularly the endothelial lining, seems to occur early in the damaging seqeunce. Whether damage to other species of renal cells such as the proximal tubular cells occurs directly through ischemia and cold itself or indirectly as secondary ischemia after reperfusion has blocked the vasculature has not been resolved.

Early work on preservation centered on the composition of perfusates and flush solutions after it had been shown that a short period of cold perfusion leads to rapid loss of intracellular cations.[57] This observation led to the formulation of so-called "intracellular" perfusates in which it was thought that high potassium and high magnesium concentrations were crucial[20,65] if pathologic ionic exchange across the cell membrane and cell swelling were to be prevented. The importance of conservation of intracellular potassium in the survival of cells at low temperature has been stressed by Willis[100] who, in studies comparing renal cells from non-hibernating and hibernating (hence relatively cold tolerant) species showed that loss of potassium and replacement by sodium had disastrous effects on ion transport, volume regulation, mitochondrial respiration and phosphorylation, calcium ion transport, and protein synthesis. However, the undoubted benefits of these intracellular formulations cannot be ascribed entirely to manipulating the concentrations of intracellular cations as cell swelling has also been demonstrated after high potassium solutions have been used.[1]

The role of osmotically active, relatively impermeant indiffusible anions such as phosphate and sulfate and neutral non-electrolytes such as glucose has been emphasized.[27,46] Maintenance of normal cell volume is an energy-requiring process to balance the osmotic effects of Donnan equilibrium produced by protein anions.[100] Thus cold itself may be damaging in the long term because it directly inhibits sodium pumps. It has even been suggested that the ATPase of vascular endothelium is peculiarly sensitive to cold and endothelial cell swelling is mainly responsible for poor reflow during reperfusion of the organ.[6]

As histological evidence of damage is generally only revealed in the kidneys after they have been reperfused with blood, the "storage damage syndrome" is now approached as combined ischemia and reperfusion injury. It is the mechanisms underlying this gross pathology which have yet to be elucidated. In our own approach to the problem, we have made two basic assumptions upon which to build our program. The first is that lowered adenine nucleotide levels are central to a number of events *during* ischemia; the second is that damage to endothelial cells and its subsequent interaction with incoming blood is inevitably linked to reperfusion injury. Just as in the WI damage discussed earlier, oxidative stress and the role of free radicals, transition metals, free fatty acids, and second messenger systems are perhaps prime movers in the membrane and intracellular changes observed.

III. OXIDATIVE STRESS

Most high energy adenine nucleotides (ATP) are generated by the reoxidation of reduced pyridine nucleotides (NADH, NADPH) via the mitochondrial electron transfer system and the tetravalent reduction of dioxygen (ground state oxygen O_2). Some dioxygen, however, undergoes univalent reduction with the production of free radical intermediates. Radicals are highly reactive species because of an unpaired electron in their outer orbital shells; this inherent instability is responsible for their reactivity as they seek to restore normal electron pairing in their orbitals. Single electron reduction of O_2 leads to the formation of superoxide radicals (O_2^-); further reduction yields H_2O_2

and the highly reactive hydroxyl radical (OH·). Under normal physiological circumstances, production of these reactive species is low because O_2 metabolism is carefully controlled by enzymes such as cytochrome oxidase which catalyzes the $4e^-$ reduction of O_2 directly to H_2O. Cells also contain a number of enzymes which protect against runaway of univalent reduction including superoxide dismutase which converts O_2^- to H_2O_2 and catalase and peroxidases which reduce H_2O_2 to H_2O. Ascorbic acid, glutathione, and lipid-soluble vitamin E also have important antioxidant activity. Although these systems ensure that oxidative damage is kept to a minimum under normal circumstances, there is mounting evidence that they are overwhelmed during reperfusion following periods of ischemia.

Oxygen-derived free radicals have been implicated in reperfusion damage to many tissues[13,67,88] including kidneys[61,76,80] and in the storage damage syndrome.[33,59] Although free radicals are known to attack proteins and DNA, as well as lipid molecules, most attention has been devoted to the peroxidation of polyunsaturated fatty acids,[99] a chain reaction which can result in extensive membrane damage (Figure 2).

In some of our earliest experiments with rabbit kidneys attempting to correlate increases in lipid peroxidation either with WI or CI per se or with reperfusion after transplanation, we noted significant increases in oxidative stress associated with reperfusion. Markers of lipid peroxidation such as diene conjugates, thiobarbituric acid-(TBA)-reactive products, and Schiffs bases were assayed in homogenates prepared from kidneys which had been subjected to ischemia. Both WI[44] and CI[45] were found to significantly increase the rate of formation of these indices during incubation at 37°C *in vitro*. In the CI experiments rabbit kidneys were either stored in a poor storage medium (isotonic saline) for 24 h or for periods up to 72 h in a relatively effective medium, the hypertonic citrate solution developed by Ross and co-workers.[82] The data showed that there was a good correlation between formation of markers of lipid peroxidation *in vitro* and the physiological dysfunction of the stored organs after transplantation *in vivo*. Reperfusion of the ischemic kidneys with oxygenated blood *in vivo* generally led to further rises in the extent of lipid peroxidation in these organs. Addition of free radical scavengers (mannitol and uric acid) or the iron-chelator deferoxamine to the flush and storage solutions significantly inhibited the rise in lipid peroxidation products following the ischemic period.[44]

From this circumstantial evidence, we concluded that the period of ischemia primed the kidneys for reperfusion injury. What might be happening to the kidneys during the ischemic period to render them so susceptible when re-exposed to oxygen? One possibility is that during ischemia the level of antioxidant defenses is reduced. Another is that enzymes alter during ischemia rendering them able to encourage free radical production upon reoxygenation. One such system has been much quoted since it was proposed — the conversion of the enzyme xanthine dehydrogenase to xanthine oxidase during ischemia,[83] possibly as a result of calcium-dependent proteolysis. It is thought that xanthine oxidase can generate O_2^- radicals from incoming O_2 and from the hypoxanthine which has accumulated during ischemia via the catabolism of ATP. Another probable reason for priming of organs to reperfusion damage is that mitochondrial injury allows reduced components of the electron transport chain to accumulate during ischemia allowing increased leakage of single electrons to O_2 and increased O_2^- production during reoxygenation. In addition, the release of chemotactic factors from damaged ischemic tissues may attract polymorphonucleocytes in sufficient numbers to damage endothelial cells by generating free radicals in a respiratory burst.[41] Finally, it seems likely that pathological alterations in metal ion fluxes during ischemia when energy-dependent membrane shuttles are disturbed and intracellualr pH falls have an important role in ischemia-reperfusion injury. Several metals have been incriminated including calcium, iron, aluminum, and copper. As most of our recent work has centered on the role of iron and calcium and the way these interact with polyunsaturated fatty acids and intracellular second messenger systems these will be discussed in some detail. The spatial and temporal sequence of events is most difficult, perhaps impossible, to resolve.

A. THE ROLE OF IRON

The importance of transition metals in catalyzing damaging free radical mediated sequences is well recognized. Particularly relevant to ischemia-reperfusion injury is the iron-catalyzed formation of OH˙ radicals from less reactive precursors (O_2^- and H_2O_2) via the Haber-Weiss reaction.[49] Highly reactive OH˙ radicals can damage all manner of biological macromolecules and initiate lipid peroxidation.[48] Perhaps just as important are iron-centered species which may themselves catalyze hydrogen abstraction from polyunsaturated fatty acids and directly initiate lipid peroxidation.[68] In addition, iron salts are known to decompose lipid hydroperoxides to reactive peroxy and alkoxy radicals which then attack further polyunsaturated fatty acids and propagate a chain reaction.[50] Some low molecular weight chelators of iron such as ATP and EDTA increase the reactivity of the metal[28] whereas high molecular weight chelators such as desferoxamine (DFX) bind iron with such high affinity (10^{31})[56] that the metal ion is prevented from catalyzing adverse reactions.[47]

In some experiments to investigate the role of iron, we found that i.v. administration of DFX to rabbits 15 min before reperfusion of kidneys which had been subjected to 60 or 120 min of WI *in situ* significantly inhibited rises in markers of lipid peroxidation (Green et al., 1986). DFX was also highly effective at reducing levels of oxidative membrane damage in kidneys which had been subjected to CI during storage in isotonic saline for 24 h[44] or for periods of up to 72 h in HCA.[38] The most effective regime proved to be i.v. administration of DFX both before the kidneys were harvested for storage and before reperfusion when they were autotransplanted. Analysis of biochemical markers of lipid peroxidation in homogenates from different regions of the kidney revealed that DFX was particularly effective in the cortex, whereas in the medulla, which contains relatively high levels of cyclooxygenase,[81] a less marked decrease in lipid peroxidation was effected by deferoxamine administration. Conversely, the cyclooxygenase inhibitor indomethacin inhibited oxidative stress in the medulla rather than cortex.[38] Concurrent administration of both drugs effectively inhibited peroxidation in both the cortex and medulla.

These results provided strong indirect evidence of a role for iron in storage damage. More direct evidence was now sought and the theoretical background was considered. Under normal physiological circumstances most iron in the body is stored in 'safe' sites which prevent transition from its redox state and hence catalysis of damaging reactions involving single electrons. These sites include hemoglobin and transferrin in the extracellular milieu and ferritin which is a predominantly intracellular protein capable of storing up to 4500 atoms of iron as ferric hydroxides.[2] There is also growing evidence for a small pool of intracellular iron chelated to low molecular weight (LMW) species such as ATP, citrate, and glycine.[4,71] These LMW chelates are thought most likely to be vehicles for iron in transit from extracellular transferrin to intracellular ferritin and perhaps also for providing iron for immediate synthetic purposes as required. Whatever their physiological role, it is likely that they will be important in oxidative stress.

We hypothesized that ischemia results in altered intracellular iron homeostasis leading to an increase in levels of catalytic forms of the metal (Figure 3). To gain evidence for this idea necessitated developing an accurate method for assaying the quantity of intracellular iron available for chelation by DFX.[39] Low speed supernatants of tissue homogenates were incubated with an excess of DFX for 60 min and the parent agent and its iron-bound form, feroxamine (FX), were extracted using solid phase extraction cartridges and analyzed by reversed phase high performance liquid chromatography (HPLC) with dual wavelength detection. Standard curves obtained after known amounts of iron were incubated with DFX demonstrated that the ratio of the area of the FX peak to the area of DFX peak provided an accurate determination of DFX-available (DFX-A) iron levels in the 1 to 25 μM range. This method was then used to determine DFX-A levels in kidneys subjected to periods of WI and CI and subsequently reperfused *ex vivo* with an oxygenated asanguinous medium on an isolated perfusion circuit at 37°C.

Measurable levels of DFX-A iron were present in both the cortex and medulla of fresh control rabbit kidneys prior to any ischemic insult. Following 1 h of WI or 24 h of cold storage in HCA, levels of DFX-A iron increased significantly by about twofold in both the cortex and medulla.[40] Storage in HCA for longer periods up to 72 h resulted in levels of DFX-A iron which were generally higher than after the less damaging 24 h period of CI. The differences between groups stored for 24 h and longer periods were much more marked after reperfusion.[1] After 24 h CI, the levels of DFX-A iron decreased immediately upon reperfusion and returned rapidly to control levels in both the cortex and medulla within 5 min. In contrast, following the physiologically more damaging 48 h of CI, DFX-A iron levels remained elevated in both regions of the kidney during the first 5 min of reperfusion and returned to control levels only after 30 min. Since there were no significant differences in total iron content in these kidneys when analyzed by atomic absorption spectroscopy, these data indicate that both WI and CI lead to a redistribution of intracellular iron to forms more available for chelation by DFX. We cannot be sure that DFX-A iron would be available to catalyze the initiation of lipid peroxidation but it must be a strong possibility that it is. In view of our earlier findings that DFX was more effective at inhibiting increased levels of lipid peroxidation in the cortex following ischemia,[38] it may be significant that increases in DFX-A iron during ischemia were more pronounced in this region than in the medulla.

It is likely, though not proven from our data, that increased levels of DFA-A were released from ferritin during ischemia. Although surprisingly little is known about the release of iron from ferritin, it is thought to involve the reduction of Fe^{3+} to Fe^{2+}.[34] This reduction could be facilitated by the low oxygen tension accompanying ischemia and the low pH which has been shown by NMR studies to be as low as 6.9 after only 6 h of simple storage.[32] This environment with increased levels of reducing species may release iron from ferritin. Exogenous redox-active quinones such as adriamycin have been shown to catalyze lipid peroxidation in the presence of ferritin under hypoxic conditions.[95] Experiments *in vitro* have demonstrated that lipid peroxidation in microsomes is stimulated in the presence of purified ferritin and flavin mononucleotide during aerobic incubation following a period of hypoxia.[37]

Further release of iron into the DFX-A pool on reperfusion of kidneys stored for longer periods may be the result of iron release from ferritin by reducing agents from the incoming oxygen. Superoxide anions produced in significant quantity upon reperfusion have been shown elsewhere to release iron from ferritin *in vitro*.[11,70] Alternatively, other reaction products of superoxide or oxygen itself with as yet unidentified cellular components may play an important role. It is also possible that the continued high levels of DFX-A iron during the reperfusion period observed in our studies were due to impairment of uptake of intracellular iron species into "safe" sites.

From this evidence, it appears that redistribution of iron to forms more accessible to DFX occurs as an early event during ischemia itself. It seems highly likely that it is also readily accessible to catalyze free radical generation and is one important underlying cause for the increased oxidative damage which occurs during reperfusion. Chelation of increased levels of intracellular LMW iron complexes by agents such as DFX inhibits the initiation and propagation of damaging events such as lipid peroxidation and may prove to be a useful therapeutic approach to post-ischemic renal failure.

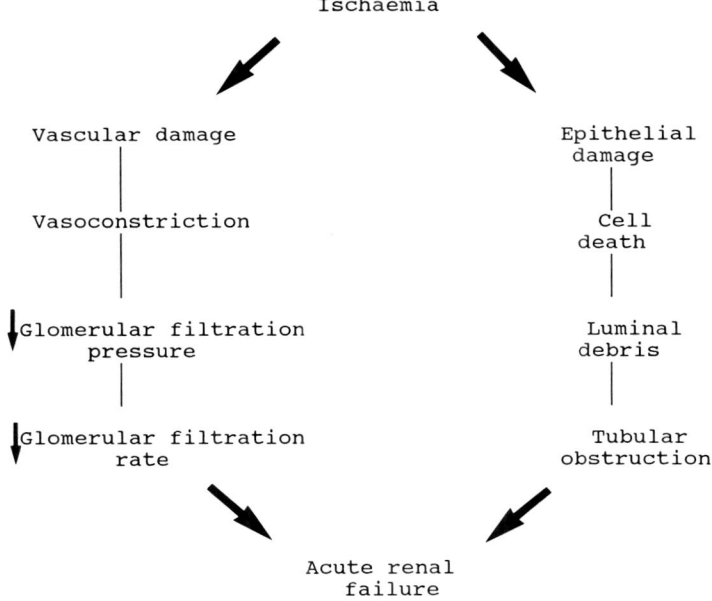

FIGURE 4. Vascular and epithelial components of ischemic renal cell injury which lead to acute renal failure.

IV. THE ROLE OF CALCIUM

Calcium ions are important in many biological functions.[29] These include regulation of cell metabolism through the activation of many enzymes and maintenance of normal blood flow via controlled production of a range of eicosanoids. Under normal physiological circumstances, a gradient between low (10^{-7} M) cytosolic levels and the high extracellular concentrations (10^{-3} M) is maintained by several mechanisms. Energy-dependent pumps remove calcium from cells and sequester excess calcium into intracellular organelles.[18] Calcium ions can only enter cells through specific voltage or receptor operated channels (Figure 4). Intracellular calcium levels are also controlled by the receptor-mediated turnover of phosphatidylinositols which involves a GTP-binding protein and a specific phospholipase C.[8]

Rapid depletion of ATP and ADP levels has been demonstrated in several studies. ^{31}P NMR spectroscopy has revealed that ATP and ADP peaks virtually disappear after 2 to 4 h CI even though the kidneys are flushed and stored with the most effective preservation solutions available.[32] As calcium pumps are energy dependent, a gain in cytosolic calcium is a consistent feature of tissues subjected to ischemia and reperfusion.[72] If the levels continue to rise, cell morphology and function may be irreversibly damaged.[30] The probability that calcium is intimately involved in ischemia-reperfusion damage has not escaped the notice of other workers.[19,74] Whether altered intracellular calcium homeostasis is an important initiator of damage or whether it is involved at a later stage in the pathology of ischemia-reperfusion, perhaps even resulting from it, will be difficult to resolve. We have carried out experiments in rabbit kidneys to try to address these questions.

We rendered rabbit kidneys cold ischemic in storage solutions containing various agents which either affect calcium movements or interfere with calcium-dependent enzymes.[22] Oxidative damage was assessed by measuring formation of markers of lipid peroxidation in tissue homogenates *in vitro*. Blockage of voltage-operated Ca-channels by verapamil reduced the extent of oxidative damage to low levels in both the cortex and medulla of kidneys after 24 h cold storage in isotonic saline but had no effect on oxidative damage after more prolonged (72 h) storage of organs in HCA solution. Elevation of extracellular calcium levels by addition of $CaCl_2$ to the storage medium increased oxidative damage significantly only when added to the isotonic saline and had no effect when added to HCA. It was concluded that influx of extracellular calium through voltage-operated channels was a significant mediator of oxidative damage to organs stored in saline. We believe that no effect was observed in kidneys flushed with HCA because the excess of citrate (55 mM) in that solution chelates excess calcium anyway and the superior ionic balance of HCA may have protected voltage operated channels during the ischemic period; the calcium chelating capacity of HCA may be one of the reasons why it is relatively so effective as a preservation solution.[22]

When A23187, an ionophore which permeabilizes both plasma and intracellular membranes to calcium, was added to the storage medium, postischemic rates of peroxidation were significantly raised above the high peroxidation levels already caused by the CI insult in both saline and HCA preserved kidneys.[22] Addition of ruthenium red, a polysaccharide dye which inhibits mitochondrial calcium transport, also potentiated oxidative damage to stored kidneys, regardless of which of the two media was used. These results suggested that even in the absence of extracellular calcium, redistribution of this metal takes place within cells during ischemia and contributes to increase peroxidation of cellular lipids upon reoxygenation.

Increased levels of cytosolic calcium could potentiate free radical mediated post-ischemic injury in several ways. Conversion of xanthine dehydrogenase to xanthine oxidase may be catalyzed by a calcium-dependent protease.[67] The evidence cited for the role of this enzyme conversion in ischemia-reperfusion injury is not entirely convincing as it has been difficult to demonstrate the presence of this enzyme in many of the tissues shown to undergo lipid peroxidation. We have provided some circumstantial evidence by showing in one set of experiments that allopurinol effectively inhibits lipid peroxidation of rabbit kidneys subjected to 120 min WI (Green et al., 1989). In our CI kidney model we have more recently demonstrated that addition of allopurinol to the saline storage medium partially prevented the increase in lipid peroxidation following storage in the presence of the calcium ionophore A23187.[23] Allopurinol is a potent xanthine oxidase inhibitor and it is perhaps most likely that it conferred benefit in our experiments as well as those of others (Vasko et al., 1972) during reperfusion rather than in preventing irreversible loss of purine nucleotides from hypoxic cells through inhibition of nucleotide dephosphorylating enzymes. Another probability is that calcium overload of mitochondria causes damage during reperfusion,[3] increasing leakage of single electrons from the electron transport chain onto O_2, thus increasing O_2^- production. Another link between raised intracellular calcium concentration and increased oxidative stress is the involvement of calcium-dependent phospholipases which hydrolyze membrane phospholipids releasing free fatty acids (FFA) and leaving behind lysophosphatide residues in the membrane. In experiments to test this possibility, we found that addition of dibucaine, a specific inhibitor of phospholipase A_2, to the storage solution resulted in significant protection against oxidative membrane damage following CI.[23]

Evidence for phospholipase activation during CI was obtained by analyzing FFAs in freeze-clamped renal tissue by gas liquid chromatography.[24] Levels of unsaturated FFAs ($C_{18:1}$, $C_{18:2}$, $C_{20:4}$) rose significantly in kidneys stored for 72 h after flush with HCA whereas the levels of saturated FFAs ($C_{16:0}$, $C_{18:0}$) did not change. The high concentration of free arachidonic acid ($C_{20:4}$) observed after 72 h CI is likely to be particularly important as release of this fatty acid from membranes is the rate-limiting step in the formation of prostaglandins.[54] This could, in turn, upset the balance in the production of these vasoactive eicosanoids with dire consequences for the endothelial lining when the organ is reperfused.[86] Evidence to support this idea has been provided in experiments which showed that after ischemia there is a decrease in the level of the potent vasodilating, platelet disaggregating agent prostacyclin and elevation in the formation of thromboxanes which cause vasoconstriction and aid platelet aggregation.[63,87] This may in turn contribute to the microcoagulopathy and poor perfusion which ensues if the kidney is badly damaged. The release of free arachidonic acid during IC is likely to be the main cause of the increased rate of indomethacin-inhibitable peroxidation via the cyclooxygenase pathway which we have observed in the medulla of stored kidneys.[38] Increased phospholipase activity may result in the formation of lipoxygenase products such as leukotrienes which are powerful mediators of vascular constriction (Figure 5) and have been implicated in ischemic injury.[62]

Storage of kidneys in the presence of dibucaine or A23187 revealed the existence of a good correlation between the extent of FFA accumulation and the rate of post-ischemic lipid peroxidation. It is still too early to state with confidence the spatial and temporal sequence of events involving calcium, phospholipase activity, free radical generation, and lipid peroxidation in ischemia-reperfusion injury. One possibility is shown in Figure 6. Early redis-

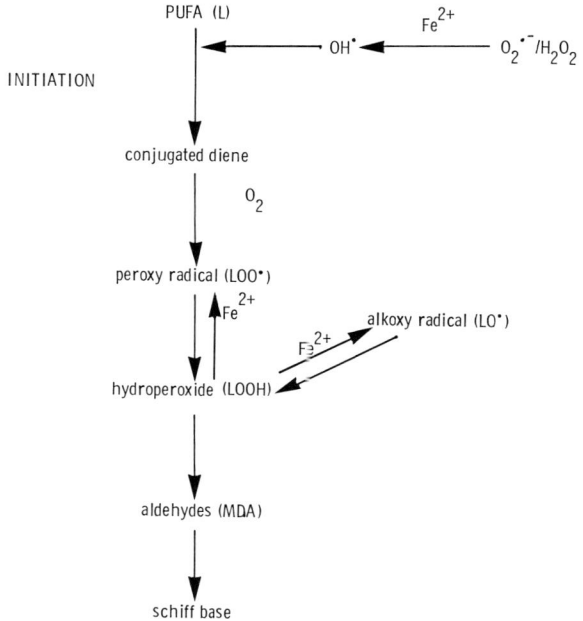

FIGURE 5. Lipid peroxidation.

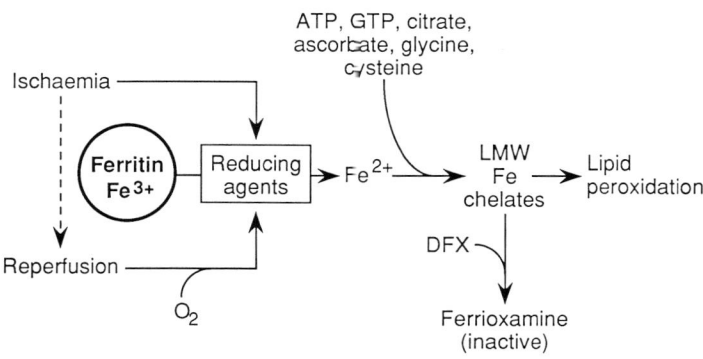

FIGURE 6.

tribution of calcium concentrations during ischemia leads to the activation of phospholipases. The released unsaturated FFAs, unprotected by the membrane-bound endogenous antioxidant vitamin E, provide excellent targets for free radical attack upon reoxygenation. The resulting peroxy radicals may then initiate peroxidation of membranes directly or may break down to relatively stable hydroperoxides which can diffuse to other sites inside the cell

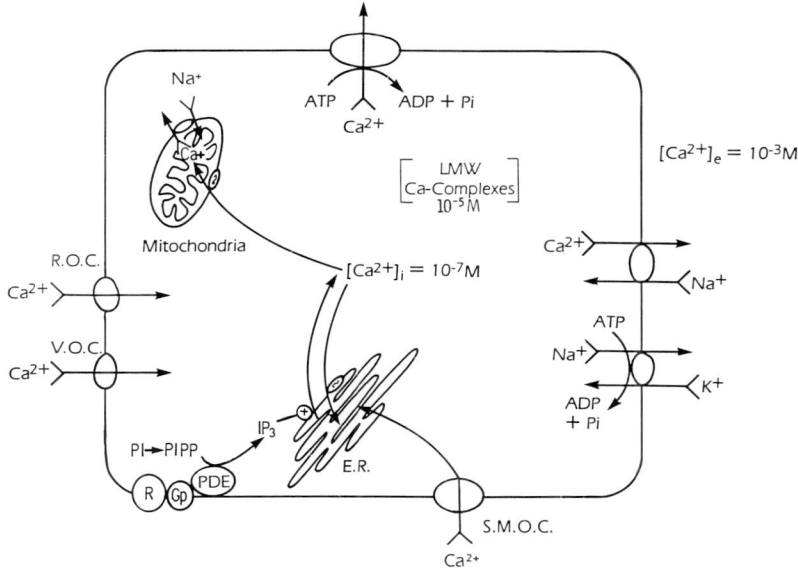

FIGURE 7. Regulation of calcium homeostasis in a normal mammalian cell.

and stimulate lipid peroxidation through interaction with catalytic iron complexes which regenerate reactive lipid radicals.[50] Meanwhile, the build-up of residual lysophosphatides in the membrane due to phospholipase activation alters fluidity and permeability[97] and possibly renders membranes more susceptible to free radical attack.[94] Peroxidation of membrane lipids results in further loss of fatty acids from the membranes and also increases lysophosphatide levels[94] and membrane rigidity.[25] As phospholipase A_2 activity is higher in rigid membranes[69] elevated rates of lipid peroxidation upon reoxygenation may lead to further increases in phospholipase activation. The damaging cycle of events may then escalate. This would lead to extensive damage to membranes which would become permeabilized to calcium so that cytosolic calcium rises still further. Oxidative damage has also been shown to inhibit plasma-membrane calcium-extruding systems[73] and to destroy the ability of intracellular organelles to sequester calcium.[5] Eventually this could result in irreversible cell injury (Figure 7).

V. SECOND MESSENGERS

We have recently started to investigate the possibility that ischemia followed by reoxygenation may affect the cleavage of membrane-bound phosphotidylinositols (PIP2) in the kidney. This secondary messenger system involves the formation of inositol triphosphate (IP3) and diacylglycerol (DAG).[8]

IP3 mobilizes calcium from intracellular stores and DAG stimulates phosphorylation of protein kinase C (PKc), a process which requires phospholipids and calcium for maximum activity.[8] In the kidney, PIP2 hydrolysis triggered by activation of α_1-adrenoreceptors evokes several responses including increased sodium reabsorption,[52] prostanoid production and vasoconstriction,[21] gluconeogenesis,[58] and inhibition of renin release.[66]

In our experiments, kidney cortical slices were incubated *in vitro* at 37°C either under an atmosphere of 95% O_2:5% CO_2 (control) or gassed with and incubated under N_2 (hypoxia). After 120 min all slices were then oxygenated and incubated aerobically for 30 min further. The formation of lipid peroxidation products increased during the first 60 min of incubation in the presence of O_2 and then leveled off over the remaining period. Lipid peroxidation was also evident in the hypoxic slices but proceeded at a slower rate than in the oxygenated samples and also leveled off after 60 min. Reoxygenation following 120 min of hypoxia resulted in a significant (p <0.0001) increase in the rate of lipid peroxidation which was not observed in the slices gassed with N_2 or when slices incubated in the presence of O_2 for 120 min were regassed. This *in vitro* system seemed therefore to closely mimic the increase in free radical-mediated oxidative membrane damage which we had observed in earlier experiments with whole rabbit kidneys subjected to ischemia-reperfusion.[38]

To determine the rate of PIP2 hydrolysis in these slices, radiolabel was incorporated into the membrane-bound phosphatidylinositol pool by incubating the slices in the presence of myo-(2-^{3}H)-inositol for 60 min at 37°C. The slices were then repeatedly washed and incubated under aerobic or hypoxic conditions for 120 min followed by oxygenation. Aliquots of slices were taken every 30 min and the hydrolysis products of PIP2 [IP3 and its subsequent metabolites inositol bisphosphate (IP2) and inositol monophosphate (IP1)] were analyzed by HPLC with scintillation counting.[53]

There was no change in the rate of PIP2 breakdown in the hypoxic slices during the 120 min incubation period compared with control slices incubated under aerobic conditions whether in the presence of calcium or in calcium-free medium containing EGTA. However, immediately upon reoxygenation of the hypoxic slices incubated in the presence of calcium, PIP2 breakdown increased rapidly. This was highly significant (p = 0.0002) and was maintained over the remaining 30 min of aerobic incubation. No increase in PIP2 breakdown was observed when slices incubated in the presence of calcium and oxygen for 120 min were regassed with 95% O_2:5% CO_2, nor did reoxygenation of hypoxic slices in calcium-free medium (+EGTA) alter the rate of PIP2 breakdown. No significant changes were seen in slices incubated in the presence of O_2 and EGTA.

These findings clearly demonstrated that hydrolysis of phosphatidylinositols to secondary messenger products is activated very rapidly upon reoxygenation of renal tissue after a period of hypoxia. Inhibition of this effect by

EGTA strongly suggested that calcium was involved at some stage in the sequence. Changes in PIP2 breakdown were not observed during hypoxia itself.

Clearly, much more work is needed before we can identify more closely the mechanisms responsible for and consequences of PIP2 cleavage in ischemia-reperfusion injury in kidneys. There are several possibilities. One of the products of lipid peroxidation, 4-hydroxynonenal, has been shown to stimulate adenylate cyclase, guanylate cyclase, and PIP2 breakdown *in vitro*.[26] It is possible then that increased levels of aldehydic products of lipid peroxidation produced during reperfusion stimulate increased PIP2 hydrolysis. In addition, lipid peroxidation and high calcium-dependent phospholipase A_2 activity alter membrane configuration[94] and this may affect the interaction of phospholipase C with membrane-bound regulatory components or make it more accessible to PIP2. Perhaps too, rapid changes in cytosolic calcium evoked an enhanced phospholipase C response. We suggest that rapid hydrolysis of PIP2 on reoxygenation following ischemia results in deregulation of receptor-mediated function via this intracellular secondary messenger system. The resultant loss of balance in eicosanoid production would then contribute to the vascular dysfunction described earlier (Figure 7).

VI. CONCLUSIONS

It is clear that no single biochemical event is responsible for the deterioration of kidneys subjected to WI or CI. It is therefore difficult to design pharmacological strategies which might protect these organs. The investigations reviewed in this chapter suggest that oxidative damage following renal ischemia can be significantly inhibited by many different agents which indicates that a complex interaction of a number of factors is responsible for post-ischemic damage. Some intracellular changes are summarized in Figure 8. It is not possible to say at what stage or in what combination these become irreversible. Thus, upon reperfusion ATP may be regenerated and provide energy for ionic pumps including those responsible for calcium, potassium, sodium, and magnesium homeostasis. Similarly, delocalized intracellular iron appears to be rapidly sequestered upon reoxygenation after a short ischemic period. However, these early events seem to be crucial in priming the kidney for subsequent reperfusion damage and the longer the period of ischemia the more important they become.

Upon reperfusion, a burst of O_2^- production from incoming O_2 would react with increased levels of catalytic iron to yield more reactive radical species. Damage to cellular components would follow including peroxidation of lipids in membranes already compromised by increased calcium-dependent phospholipase activity. The resulting loss of integrity of the plasma membrane and intracellular organelles would cause further imbalances in intracellular

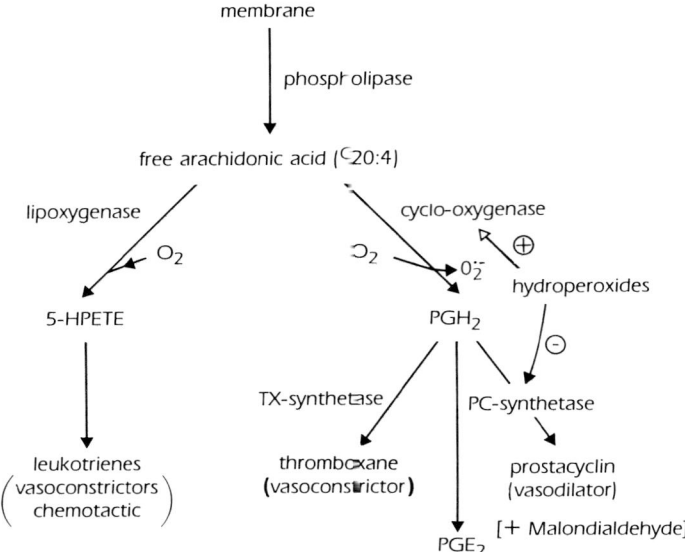

FIGURE 8. Cyclooxygenase- and lipoxygenase-catalyzed oxidation of arachidonic acid.

ionic homeostasis. A self-perpetuating cycle could then become established until cellular structures were so perturbed as to become irreversibly damaged.

Disturbances in the vasculature also contribute to post-ischemic organ failure. These can result from biochemical changes in the vessel wall and trapped ischemic blood, and through interactions between these components and incoming fresh blood during reperfusion. Imbalances in eicosanoid production due to calcium-dependent accumulation of free arachidonic acid, production of inflammatory mediators such as leukotrienes, release of chemotactic substances with subsequent adhesion, and activation of polymorphonucleocytes and derangement of receptor-mediated functions such as the phosphatidyl inositol secondary messenger system may each play a part in the vascular bed.

Evidence for the involvement of a number of biochemical mechanisms during ischemic-reperfusion injury is supported by the ability of many different pharmacological agents to afford at least some protection to kidneys subjected to ischemic insult. These include iron-chelators;[75] free radical scavengers including superoxide dismutase[14,29,76] and catalase;[14] allopurinol;[59] calcium antagonists[89] and prostacyclin analogs.[60] The very multiplicity of biochemical derangements makes it extremely unlikely that any one agent will be particularly effective. The most we can hope for, at least in the forseeable

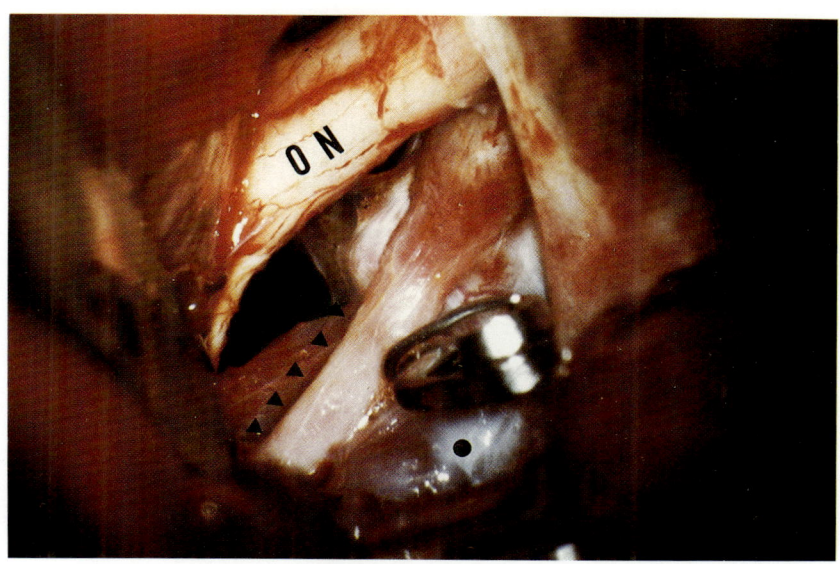

PLATE 1. An intraoperative photograph after aneurysmal clipping showing severe, segmental spasm of the internal carotid artery (Day 6). ON, the right optic nerve; ▲▲▲, narrowing of the internal carotid artery; ●, the aneurysmal sac.

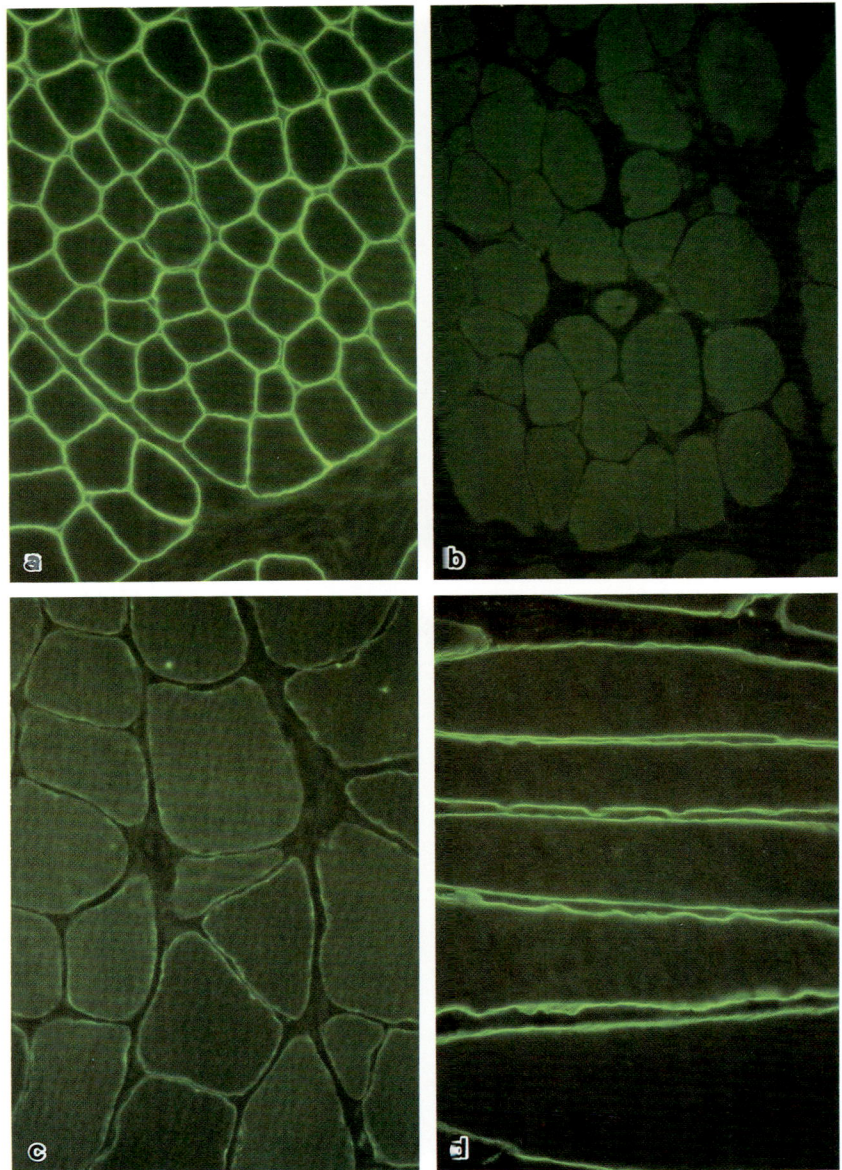

PLATE 2. Immunofluorescence staining patterns of dystrophin in skeletal muscle sections. Panel (a) shows continuous clear staining of the surface membrane in normal muscle, (b) shows the absence of staining in a muscle section from patient with DMD, (c) shows the discontinuous "patchy" membrane staining with reduced intensity in patient with Becker muscular dystrophy. Panel (d) shows a longitudinal section of normal muscle. (Original magnification for (a) × 210; for (b through d) × 420.)

future, is that administration of agents systemically to the patient or infused through the organ will delay the onset of irreversible damage. It is most likely that a combination of agents will prove successful.

Maintenance of the kidneys in optimum condition during *in situ* or bench surgery and the need for longer periods of cold storage of kidneys for transplantation are both increasingly important in the surgical management of patients in renal failure. Hypothermic storage in special solution with improved ionic composition has already yielded benefits.[7,20,82] Further advances will likely involve a combined pharmacological strategy including infusing kidneys with solutions rendered hyperosmolar (380 to 420 osm/l) with relatively inert and impermeable molecules such as mannitol, raffinose, trehalose, or sucrose; solutions containing impermeant anions such as sulfate or phosphate and buffering systems which are effective at low temperatures in static situations; solutions containing antioxidants (in the widest sense) such as deferoxamine, superoxide dismutase, catalase, allopurinol, or verapamil; and treating the patients with agents such as deferoxamine, verapamil, indomethacin, prostacyclin, mannitol, dextrans, antiplatelet aggregating agents, and vasodilating agents. It may prove necessary to give several agents together or it may be better to give them in a logical sequence. Retarding biochemical changes such as altered intracellular iron and calcium homeostasis and preventing the loss of antioxidant protection are most likely to make a significant contribution in this field.

REFERENCES

1. **Acquatella, H., Gonzales, M. P., Morales, J. M., and Whittembury, G.,** Ionic and histologic changes in the kidney after perfusion and storage for transplantation, *Transplantation*, 14, 480–489, 1972.
2. **Aisen, P. and Listowsky, I.,** Iron transport and storage proteins, *Annu. Rev. Biochem.*, 49, 357–393, 1980.
3. **Arnold, P. E., Lumlertgul, D., Burke, T. J., and Schrier, R. W.,** In vitro versus in vivo mitochondrial calcium loading in ischemic acute renal failure, *Am. J. Physiol.*, 248, F845–50, 1985.
4. **Bakkeren, D. L., Jeu-Jaspars, C. M. H., Van der Heul, C., and Van Eijk, H. G.,** Analysis of iron-binding components in the low molecular weight fraction of rat reticylocyte cytosol, *Int. J. Biochem.*, 17, 925–930, 1985.
5. **Bellomo, G., Richelmi, P., Mirabelli, F., Marioni, V., and Abbagnano, A.,** Inhibition of liver microsomal calcium ion sequestration by oxidative stress: role of protein sulphydryl groups, in *Free Radicals in Liver Injury*, Poli, G., Cheeseman, K. H., Dianzani, M. U., and Slater, T. F., Eds., IRL Press, Oxford, 1985, 139–142.
6. **Belzer, F. O., Hoffman, R., Huang, J., and Downs, G.,** Endothelial damage in perfused dog kidney and cold sensitivity of vascular Na-K-ATPase, *Cryobiology*, 9, 457–460, 1972.
7. **Belzer, F. O. and Southard, J. H.,** Principles of solid-organ preservation by cold storage, *Transplantation*, 45, 673–676, 1988.

8. **Berridge, M. J.,** Inositol triphosphate and diacylglycerol as second messengers, *Biochem. J.,* 220, 345–60, 1984.
9. **Berridge, M. J., Dawson, R. M. C., Downes, C. P., Heslop, J. P., and Irvine, R. F.,** Changes in the levels of inositol phosphates after agonist-dependent hydrolysis of membrane phosphoinositides, *Biochem. J.,* 2.2, 473–482, 1983.
10. **Berthon, B., Claret, M., Mazet, J. L., and Foggioli, J.,** Volume- and temperature-dependent permeabilities in isolated rat l ver cells, *J. Physiol. (London),* 305, 267–269, 1980.
11. **Biemond, P., Swaak, A. J. G., van Eijk, H. G., and Koster, J. F.,** Superoxide-dependent iron release from ferritin in inflammatory diseases, *Free Radical Biol. Med.,* 4, 185–198, 1988.
12. **Bilde, T. and Dahlager, J. I.,** The effect of chlorpromazine pretreatment on the vascular resistance in kidneys following warm ischaemia, *Scand. J. Urol. Nephrol.,* 11, 21–26, 1977.
13. **Bolli, R.,** Oxygen-derived free radicals and postischemic myocardial dysfunction ("stunned myocardium"), *J. Am. Coll. Cardiol.,* 12, 239–249, 1988.
14. **Bosco, P. J. and Schweizer, R. T.,** Use of oxygen radical scavengers on autografted pig kidneys after warm ischemia and 48-hour perfusion preservation, *Arch. Surg.,* 123, 601–604, 1988.
15. **Buhl, M. R. and Jorgensen, S.,** Breakdown of 5'-adenine nucleotides in ischaemic renal cortex estimated by oxypurine excretion during perfusion, *Scand. J. Clin. Lab. Invest.,* 35, 211–217, 1975.
16. **Calman, K. C., Quin, R. O., and Bell, P. R.,** Metabolic aspects of organ storage and the prediction of organ viability, in *Organ Preservation,* Pegg, D. E., Ed., Chruchill Press, London; 1973, 225–240.
17. **Calne, R. Y., Pegg, D. E., Pryse-Davis, J., and Leigh-Brown, F.,** Renal preservation by ice-cooling. An experimental study relating to kidney transplantation from cadavers, *Br. Med. J.,* 2, 651–655.
18. **Carofoli, E.,** Intracellular calcium homeostasis, *Annu. Rev. Biochem.,* 56, 395–433, 1987.
19. **Cheung, J. Y., Bonventure, J. V., Malis, C. D., and Leaf, A.,** Calcium and ischaemic injury, *N. Engl. J. Med.,* 314, 1670–1676, 1986.
20. **Collins, G. M., Bravo-Shugarman, M. B., and Terasaki, P. I.,** Kidney preservation for transplantation. Initial perfusion and 30 hour ice storage, *Lancet,* ii, 1219–1222, 1969.
21. **Cooper, C. L. and Malik, K. U.,** Prostaglandin synthesis and renal vasoconstriction elicited by adrenergic stimuli are linked to activation of alpha-1 adrenergic receptors in the isolated rat kidney, *J. Pharmacol. Exp. Ther.,* 233, 24–31, 1985.
22. **Cotterill, L. A., Gower, J. D., Fuller, B. J., and Green, C. J.,** Oxidative damage to kidney membranes during cold ischaemia: evidence of a role for calcium, *Transplantation,* 48, 745–751, 1989a.
23. **Cotterill, L. A., Gower, J. D., Fuller, B. J., and Green, C. J.,** Oxidative stress during hypothermic storage of rabbit kidneys possible mechanisms by which calcium mediates free radical damage, *CryoLett.,* 10, 119–126, 1989b.
24. **Cotterill, L. A., Gower, J. D., Fuller, B. J., and Green, C. J.,** Free fatty acid accumulation following cold ischaemia in rabbit kidneys and the involvement of a calcium dependent phospholipase A_2, *CryoLett.,* 11, 3–12, 1989c.
25. **Demopoulos, H. B., Flam, E. S., Pietronigro, D. D., and Seligman, M.,** The free radical pathology and the micro-circulation in the major central nervous system disorders, *Acta Physiol. Scand.,* 492, 91–119, 1980.
26. **Dianzani, M. U., Paradisi, L., Barrera, G., Rossi, M. A., and Parola, M.,** The action of 4-hydroxynonenal on the plasma membrane enzymes from rat hepatocytes, in *Free Radicals, Metal Ions and Biopolymers,* Beaumont, P. C., Deeble, D. J., Parsons, B. J., and Rice-Evans, C., Eds., Richelieu Press, London, 1989, 329–346.

27. **Downes, G., Hoffman, R., Huang, J., and Belzer, F. O.**, Mechanisms of action of washout solutions for kidney preservation, *Transplantation,* 16, 46–53, 1973.
28. **Dunford, H. B.**, Free radicals in iron-containing systems, *Free Radical Biol. Med.,* 3, 405–421, 1987.
29. **Evered, D. and Whelan, J.**, *Calcium and the Cell,* Evered, D. and Whelan, J., Eds., Ciba Foundation Symposium 122, John Wiley & Sons, Chichester, 1986.
30. **Farber, J. L.**, The role of calcium in cell death, *Life Sci.,* 29, 1289–1295, 1981.
31. **Flores, J., Di Bona, D. R., Frega, N., and Leaf, A.**, Cell volume regulation and ischaemic tissue damage, *J. Membr. Biol.,* 10, 331–343, 1972.
32. **Fuller, B. J., Busza, A. L., Proctor, E., Myles, M., Gadian, D., and Hobbs, K. E. F.**, Control of pH during hypothermic liver storage: role of the flushing solution, *Transplantation,* 45, 239–241, 1988.
33. **Fuller, B. J., Gower, J. D., and Green, C. J.**, Free radicals and organ preservation: fact or fiction?, *Cryobiology,* 25, 377–393, 1989.
34. **Funk, F., Lenders, J. P., Crichton, R. R., and Schneider, W.**, Reductive mobilisation of ferritin iron, *Eur. J. Biochem.,* 152, 167–172, 1985.
35. **Franconi, F., Manghi, N., Giotti, A., Martini, F., and Dini, M.**, Effect of selenium on the contractile force of isolated and perfused guinea-pig heart, *Acta Pharmacol. Toxicol. (Copenhagen),* 46, 98, 1980.
36. **Gingrich, G. A., Barker, G. R., Lui, P., and Stewart, S. C.**, Renal preservation following severe ischaemia and prophylactic calcium channel blockade, *J. Urol.,* 134, 408–410, 1985.
37. **Goddard, J. G., Serebin, S., Basford, D., and Sweeney, G. D.**, Microsomal chemiluminescence (lipid peroxidation) is stimulated by oxygenation in the presence of anaerobically released ferritin iron, *Fed. Proc.,* 45, 174, 1986.
38. **Gower, J. D., Healing, G., Fuller, B. J., Simpkin, S., and Green, C. J.**, Protection against oxidative damage in cold-stored rabbit kidneys by desferrioxamine and indomethacin, *Cryobiology,* 26, 309–317, 1989.
39. **Gower, J. D., Healing, G., and Green, C. J.**, Determination of desferrioxamine-available iron in biological tissues by high-pressure liquid chromatography, *Anal. Biochem.,* 180, 126–130, 1989a.
40. **Gower, J., Healing, G., and Green, C. J.**, Measurement by HPLC of desferrioxamine-available iron in rabbit kidneys to assess the effect of ischaemia on the distribution of iron within the total pool, *Free Radical Res. Commun.,* 5, 291–299, 1989b.
41. **Granger, D. N., Benoit, J. N., Suzuki, M., and Grisham, M. B.**, Leukocyte adherence to venular endothelium during ischaemia-reperfusion, *Am. J. Physiol.,* G683–688, 1989.
42. **Granger, D. N., Rutilo, G., and McCord, J. M.**, Superoxide radicals in intestinal ischaemia, *Gastroenterology,* 81, 22, 1981.
43. **Green, C. J., Healing, G., Lunec, J., Fuller, B. J., and Simpkin, S.**, Evidence of free radical-induced damage in rabbit kidneys after simple hypothermic preservation and autotransplantation, *Transplantation,* 41, 161–165, 1986a.
44. **Green, C. J., Healing, G., Simpkin, S., Fuller, B. J., and Lunec, J.**, Reduced susceptibility to lipid peroxidation in cold ischaemic rabbit kidneys after addition of desferrioxamine, mannitol or uric acid to the flush solution, *Cryobiology,* 23, 358–365, 1986b,.
45. **Green, C. J., Healing, G., Simpkin, S., Lunec, J., and Fuller, B. J.**, Desferrioxamine reduces susceptibility to lipid peroxidation in rabbit kidneys subjected to warm ischaemia and reperfusion, *Comp. Biochem. Physiol.,* 85B, 113–117, 1986c.
46. **Green, C. J. and Pegg, D. E.**, Mechanism of action of 'Intracellular' renal preservation solutions, *World J. Surg.,* 3, 115–120, 1979.
47. **Gutteridge, J. M. C., Richmond, R., and Halliwell, B.**, Inhibition of the iron-catalysed formation of hydroxyl radicals from superoxide and of lipid peroxidation by desferrioxamine, *Biochem. J.,* 184, 469–472, 1979.

48. **Gutteridge, J. M. C.,** Lipid peroxidation initiated by superoxide-dependent hydroxyl radicals using complexed iron and hydrogen peroxide, *FEBS Lett.,* 172, 245–249, 1984.
49. **Halliwell, B.,** Superoxide-dependent formation of hydroxyl radicals in the presence of iron salts, *FEBS Lett.,* 96, 238–242, 1978.
50. **Halliwell, B. and Gutteridge, J. M. C.,** Oxygen toxicity, oxygen radicals, transition metals and disease, *Biochem. J.,* 219, 1–14, 1984.
51. **Healing, G., Gower, J. D., Fuller, B. J., and Green, C. J.,** Intracellular iron redistribution: an important determinant of reperfusion damage to rabbit kidneys, *Biochem. Pharmacol.,* 39, 1239–1245, 1990.
52. **Hesse, I. F. A. and Johns, E. J.,** The subtype of α-adrenoceptor involved in the neural control of renal tubular sodium reabsorption in the rabbit, *J. Physiol.,* 328, 527–538, 1984.
53. **Irvine, R. F., Anggard, E. E., Letcher, A. J., and Downes, C. P.,** Metabolism of inositol 1,4,5-trisphosphate and inositol 1,3,4-trisphosphate in rat parotid glands, *Biochem. J.,* 229, 505–511, 1985.
54. **Isakson, P. C., Raz, A., Hsueh, W., and Needleman, P.,** Lipases and prostaglandin biosynthesis, in *Advances in Prostaglandin and Thromboxane Research,* Vol. 3, Galli, C., Ed., Raven Press, New York, 1978, 13–119.
55. **Jewell, S. A., Bellomo, G., Thor, H., Orrenius, S., and Smith, M. T.,** Bleb formation in hepatocytes during drug metabolism is caused by disturbances in thiol and calcium ion homeostasis, *Science,* 217, 1257–1259, 1982.
56. **Keberle, H.,** The biochemistry of desferrioxamine and its relation to iron metabolism, *Ann. N.Y. Acad. Sci.,* 119, 758–768, 1964.
57. **Keeler, R., Swinney, J., Taylor, R. M. R., and Uldall, P. R.,** The problem of renal preservation, *Br. J. Urol.,* 38, 653–656, 1966.
58. **Kessar, P. and Saggerson, E. D.,** Evidence that catecholamines stimulate renal gluconeogenisis through α-type of adrenoceptor, *Biochem. J.,* 190, 119–123, 1980.
59. **Koyama, I., Bulkley, G. B., Williams, G. M., and Im, M. J.,** The role of oxygen free radicals in mediating the reperfusion injury of cold-preserved ischaemic kidneys, *Transplantation,* 40, 590–595, 1985.
60. **Langkopf, B., Rebmann, U., Schabel, J., Pauer, H. -D., Heynemann, H., and Forster, W.,** Improvement in the preservation of ischemically impaired renal transplants of pigs by iloprost (ZK 36,374), *Prostaglandins Leukotrienes Med.,* 21, 23–28, 1986.
61. **Laurent, B. and Ardaillou, R.,** Reactive oxygen species: production and role in the kidney, *Am. J. Physiol.,* 251, F765–776, 1986.
62. **Lefer, A. M.,** Eicosanoids as mediators of ischaemia and shock, *Fed. Proc.,* 44, 275–280, 1985.
63. **Lelcuk, S., Alexander, F., Kobzik, L., Valeri, C. R., Shepro, D., and Hechtman, H. B.,** Prostaglandins and thromboxane A2 moderate post-ischaemic renal failure, *Surgery,* 98, 207–212, 1985.
64. **McKay, P. B., Gibson, D. B., Fong, K. L., and Hornbrook, K. R.,** Effect of glutathione peroxidase activity on lipid peroxidation in biological membranes, *Biochim. Biophys. Acta,* 431, 459, 1976.
65. **Martin, D. C., Smith, G., and Fareed, D. O.,** Experimental renal preservation, *J. Urol.,* 103, 681–685, 1970.
66. **Matsumura, Y., Miyawaki, N., Sasaki, Y., and Morimoto, S.,** Inhibitory effects of norepinephrine, methoxamine and phenylephrine on renin release from rat kidney cortical slices, *J. Pharmacol. Exp. Ther.,* 233, 782–787, 1985.
67. **McCord, J. M.,** Oxygen-derived free radicals in post-ischaemic tissue injury, *N. Engl. J. Med.,* 312, 159–163, 1985.
68. **Minotti, G. and Aust, S. D.,** The role of iron in the initiation of lipid peroxidation, *Chem. Phys. Lipids,* 44, 191–208, 1987.

69. **Momchilova, A., Petkova, D., and Koumanov, K.**, Rat liver microsomal phospholipase A2 and membrane fluidity, *Int. J. Biochem.*, 18, 659–663, 1986.
70. **Monteiro, H. P. and Winterbourne, C. C.**, The superoxide-dependent transfer of iron from ferritin to transferrin and lactoferrin, *Biochem. J.*, 256, 923–928, 1988.
71. **Mulligan, M., Althaus, B., and Linder, M. C.**, Non-ferritin, non-heme iron pools in rat tissues, *Int. J. Biochem.*, 18, 791–798, 1986.
72. **Naylor, W. G., Panagiotopoulos, S., Elz, J. S., and Daly, M. J.**, Calcium-mediated damage during post-ischaemic reperfusion, *J. Mol. Cell. Cardiol.*, 20 (Suppl. 2), 41–54, 1988.
73. **Nicotera, P. L., Moore, M., Mirabelli, F., Bellomo, G., and Orrenius, S.**, Inhibition of hepatocyte plasma membrane Ca^{2+} ATPase activity by menadione metabolism and its restoration by thiols, *FEBS Lett.*, 181, 149–153, 1985.
74. **Opie, L. H.**, Proposed role of calcium in reperfusion injury, *Int. J. Cardiol.*, 23, 159–164, 1989.
75. **Paller, M. S., Hedlund, B. E., Sikora, J. J., Faassen, A., and Waterfield, R.**, Role of iron in postischaemic renal injury in the rat, *Kidney Int.*, 34, 474–480, 1988.
76. **Paller, M. S., Hoidal, J. R., and Ferris, T. F.**, Oxygen free radicals in ischaemic acute renal failure in the rat, *J. Clin. Invest.*, 74, 1156–1164, 1984.
77. **Pavlock, G. S., Southard, J. H., Starling, J. R., and Belzer, F. O.**, Lysosomal enzyme release in hypothermically perfused dog kidneys, *Cryobiology*, 21, 521–528, 1984.
78. **Pryor, W. A.**, The role of free radical reactions in biological systems, in *Free Radicals in Biology*, Vol. 1, Pryor, W. A., Ed., Academic Press, New York, 1976.
79. **Quinn, P. J.**, A lipid-phase separation model of low temperature damage to biological membranes, *Cryobiology*, 22, 128–146, 1985.
80. **Ratych, R. E. and Bulkley, G. B.**, Free-radical-mediated postischemic reperfusion injury in the kidney, *J. Free Radicals Biol. Med.*, 2, 311–319, 1986.
81. **Robak, J. and Sobanska, B.**, Relationship between lipid peroxidation and prostaglandin generation in rabbit tissues, *Biochem. Pharmacol.*, 25, 2233–2236, 1976.
82. **Ross, H., Marshall, V. C., and Escott, M. L.**, 72-hr canine kidney preservation without continuous perfusion, *Transplantation*, 21, 498–501, 1976.
83. **Roy, R. S. and McCord, J. M.**, Superoxide and ischaemia: conversion of xanthine dehydrogenase to xanthine oxidase, in *Oxyradicals and their Scavenging Systems*, Vol. 2, Greenwald, R. and Cohen, G., Eds., Elsevier, New York, 1983, 145–153.
84. **Sacks, S. A., Petritsch, P. H., and Kaufman, J. J.**, Canine kidney preservation using a new perfusate, *Lancet*, i, 1024–1028, 1973.
85. **Schieppati, A., Wilson, P. D., Burke, T. J., and Schrier, R. W.**, Effect of renal ischaemia on cortical microsomal calcium accumulation, *Am. J. Physiol.*, 249, C476–483, 1985.
86. **Schlondorff, D. and Ardaillon, R.**, Prostaglandins and other arachidonic acid metabolites in the kidney, *Kidney Int.*, 29, 108–119, 1986.
87. **Schmitz, J. M., Apprill, P. G., Buja, L. M., Willerson, J. T., and Campbell, W. B.**, Vascular prostaglandin and thromboxane production in a canine model of myocardial ischaemia, *Circ. Res.*, 57, 223–231, 1985.
88. **Schoenberg, M., Younes, M., Muhl, E., Sellin, D., Fredholm, B., and Schildberg, F. W.**, Free radical involvement in ischaemic damage to the small intestine, in *Oxyradicals and their Scavenger Systems*, Vol. 2, Greenwald, R. and Cohen, G., Eds., Elsevier, New York, 1983, 154–157.
89. **Schrier, R. W., Arnold, P. E., vanPutten, V. J., and Burke, T. J.**, Cellular calcium in ischemic acute renal failure: role of calcium entry blockers, *Kidney Int.*, 32, 313–321, 1987.

90. **Sehr, P. A., Bore, P. J., Papatheofanis, J., and Radda, G. K.**, Non-destructive measurement of metabolites and tissue pH in the kidney by ^{31}P nuclear magnetic resonance, *Br. J. Exp. Pathol.*, 60, 632–641, 1979.
91. **Southard, J. H., Ametani, M. S., Lutz M. F., and Belzer, F. O.**, Effects of hypothermic perfusion of kidneys on tissue and mitochondrial phospholipids, *Cryobiology*, 21, 20–24, 1984.
92. **Summers, W. K., and Jamison, R. L.**, The no reflow phenomenon in renal ischaemia, *Lab. Invest.*, 25, 635, 1975.
93. **Takenaka, M., Tatsukuwa, Y., Dohi, K., Ezaki, H., Matsukawa, K., and Kawasaki, T.**, Protective effects of α-tocopherol and co-enzyme Q_{10} on warm ischaemic damage of the rat kidney, *Transplantation*, 32, 137, 1981.
94. **Ungemach, F. R.**, Plasma membrane damage of hepatocytes following lipid peroxidation: involvement of phospholipase A2, in *Free Radicals in Liver Injury*, Poli, G., Cheeseman, K. H., Dianzani, M. U., and Slater, T. F, Eds., IRL Press, Oxford, 1985, 127–134.
95. **Vile, G. F. and Winterbourne, C. C.**, Adriamycin-dependent peroxidation of rat liver and heart microsomes catalysed by iron chelates and ferritin, *Biochem. Pharmacol.*, 37, 2893–2897, 1988.
96. **Vogt, M. T. and Farber, E.**, On the molecular pathology of ischaemic renal cell death. Reversible and irreversible cellular mitochondrial metabolic alteration, *Am. J. Pathol.*, 53, 1–24, 1968.
97. **Weltzem, H. U.**, Cytolytic and membrane-perturbing properties of lysophosphatidylcholine, *Biochim. Biophys. Acta*, 559, 259–287, 1979.
98. **Weed, R. I., La Celle, P. I., and Merrill, E. W.**, Metabolic dependence of red cell deformability, *J. Clin. Invest.*, 48, 795–809, 1969.
99. **Wills, E. D.**, Lipid peroxide formation in microsomes: the role of non-heme iron, *Biochem. J.*, 113, 325–332, 1969.
100. **Willis, J. S.**, The possible role of cellular K for survival of cells at low temperature, *Cryobiology*, 9, 351–366, 1972.

Part IV
*Membrane Alterations
and Membrane-Acting Drugs*

Chapter 13

CANCER AND CELL MEMBRANES

Tommaso Galeotti and Lanfranco Masotti

TABLE OF CONTENTS

I.	Introduction	314
II.	Plasma Membranes	314
	A. Defective Regulatory Mechanisms of Cell Growth	315
	1. Transduction of Mitogenic Signals	315
	2. Gene Expression of Growth Factors and Growth Factor Receptors	318
	B. Tumor Cell Spreading to Secondary Sites	320
	1. Local Invasion	323
	2. Migration to Secondary Sites	326
III.	Endoplasmic Reticulum	328
IV.	Mitochondrial Membranes	331
Acknowledgments		334
References		334

ISBN 0-8493-8091-X
© 1993 by CRC Press, Inc.

I. INTRODUCTION

Cancer is a term currently used almost exclusively to indicate a malignant neoplasm. However, in the recent literature the term malignant is often used as a synonym for neoplastic and tumorigenic. We think it useful, before going into the discussion of the role of cellular membranes in cancer, to define the main features of malignant neoplasms: they are noncapsulated and often characterized by numerous and abnormal mitoses, as well as by rapid growth; they are invasive, poorly differentiated and anaplastic to varying degrees; finally, they have the potential to metastasize. The tendency to metastasize, like the other mentioned biological functions, is a probability function and not an absolute quality.[1] However, this last characteristic constitutes the critical difference between benign and malignant neoplasms because, while the majority of the differences between the former and the latter are relative, by definition benign neoplasms do not exhibit metastatic growth.

A further concept that we think useful to define as a tool for understanding the contents of subsequent sections is that of transformation. Its definition cannot certainly be expressed in simple terms; however, for our purposes a transformed cell is a genotypically altered, latent tumor cell. This "initiated" cell can be converted into a phenotypically altered cell whose promotion results in its clonal expansion.

II. PLASMA MEMBRANES

Isolated plasma membranes and plasmalemmas of whole tumor cells exhibit a wide variety of changes as compared to the homologous normal cells. These changes concern both physical and functional characteristics of the various membrane components, e.g., the molecular order of lipids, lateral and rotational diffusion of lipids and proteins, enzyme activities, antigen and receptor distribution, as well as more general membrane properties such as transport, relationship with the cytoskeleton, cell-to-cell communication. Although none of these modified properties has to date been proven to be specifically associated with malignant transformation, nonetheless the essential changes in the tumor cell phenotype, i.e., the loss of growth and positional control and the escape from the immune surveillance, do involve the cell surface. We shall, therefore, examine which of these changes is thought to be significant in defining the malignancy of the cell.

Membrane subcellular components, such as mitochondria or endoplasmic reticulum, may be considered logistically and functionally as autonomous entities. By contrast, the plasma membrane is intimately interconnected with the cytoskeleton and the extracellular matrix. Therefore, any plasma membrane alteration hardly occurs without the involvement of these important structures. Among all, a significant example is given by the events, described in detail in the following sections, which take place during spreading and growth at secondary sites, i.e., invasion and metastases.

Due to these complex structural and functional relationships, the section on plasma membrane will be developed as follows. We will consider first the regulatory mechanisms of cell growth; second, the loss of positional control (local invasion); third, the migration to secondary sites (metastatization). An important feature of the malignant cell and of the role played by the plasma membrane, i.e., the escape from the immune control, will not be treated owing to the vastness of the field which would require a specific, separate review article.

A. DEFECTIVE REGULATORY MECHANISMS OF CELL GROWTH

1. Transduction of Mitogenic Signals

The progression of the cell through the different phases of the mitotic cycle is controlled by polypeptide growth factors, regulatory peptides, and a variety of other activators. The binding of these factors to specific receptors, located in a relatively small number in the plasma membrane, triggers an amplification pathway, involving one or more second messengers. This results in the activation of metabolic responses leading to new DNA synthesis and cell proliferation. In mammalian cells three distinct mitogenic pathways have been outlined: (1) the cyclic AMP (cAMP) system, (2) the phosphatidylinositol (PI)-Ca^{2+}-diacylglycerol (DAG) cascade, and (3) the growth factor protein-tyrosine kinase pathway. Figure 1 presents some details of these three pathways. The simultaneous operation of two or all three of them depends on the cell type; they can be mitogenic or, on the contrary, exert an inhibitory action on cell division. In the former case three different levels of interplay between these pathways can be considered: they can operate independently at the phase of signal transduction (receptor binding and tyrosine kinase activation), overlap at the level of gene expression and protein synthesis, and finally converge on DNA replication and cell division.

cAMP has been the first described and therefore the best known second messenger involved in the regulation of cell proliferation. However, such a role has been controversial: an antimitogenic effect was demonstrated in early studies, whereas more recently a positive control has been shown in other cell types. A possible explanation of these conflicting reports lies in the fact that the regulatory function of the cAMP system may be associated with the differentiation state of the cell. In malignant cells that are generally poorly differentiated, the control is often negative; in normal cells and in tumors with a high degree of differentiation the control is positive (see Reference 2 for review). Alternatively, this double effect of the cAMP system has been ascribed to the requirement of an optimal intracellular concentration of the nucleotide which varies in the different cell types.[3]

The second mitogenic pathway, i.e., the PI cascade, is triggered by growth factors that are peptides or extracellular small signal proteins. Their action is mediated by specific, high affinity receptors located on the plasma membrane.

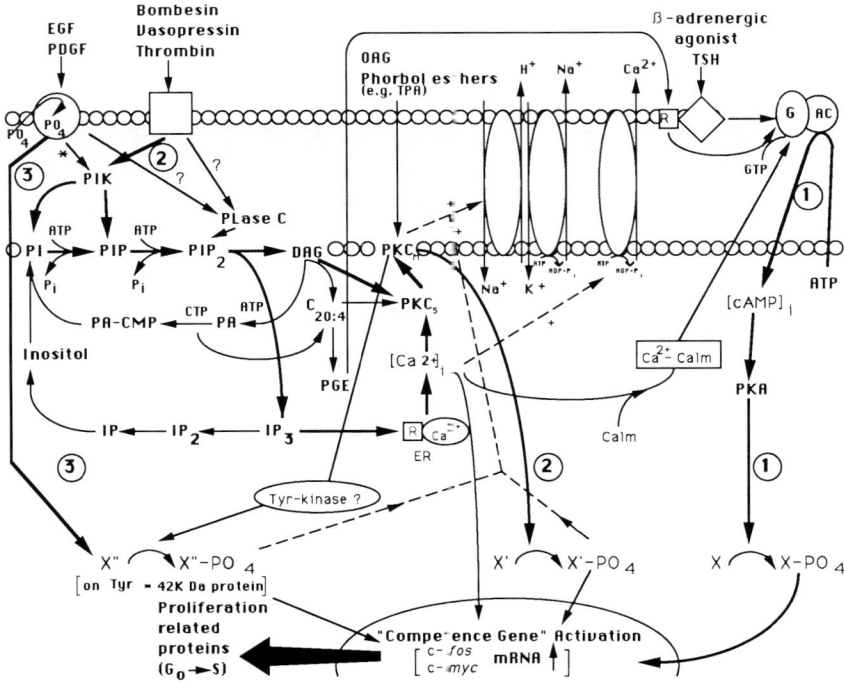

FIGURE 1. Schematic representation of mitogenic pathways in mammalian cells.

These receptors share a common three-domain structure (Figure 2): the first domain is the amino terminal, extracellular binding site, specific for the appropriate growth factor; the second is a hydrophobic transmembrane domain; the third is the carboxy terminal, intracellular functional domain that, when endowed with a tyrosine kinase activity, defines the protein-tyrosine kinase family of growth factor receptors. To this family belong the receptors for EGF, PDGF, CSF-1, and insulin (IGF). The cytosolic domain of these receptors has the largest degree of similarity and its enzymatic activity appears to be essential for the propagation of the proliferation response.[5] Alternatively, the PI cascade can be activated by mitogens such as bombesin, vasopressin, and thrombin which bind to receptors that do not have tyrosine kinase activity. For the protein-tyrosine kinase family of receptors, in spite of extensive work done by several laboratories, the targets of the tyrosine kinase activity have not yet been identified except for the receptor itself. This might be explained in several ways. One possible explanation is that autophosphorylation could induce a conformational change that affects a noncovalent interaction with a target GTP-binding protein (G protein), thus suggesting a model similar to that elucidated for the β-adrenergic receptor (see below). This G protein would stimulate the membrane-associated phospholipase C.[6] Following tyrosine ki-

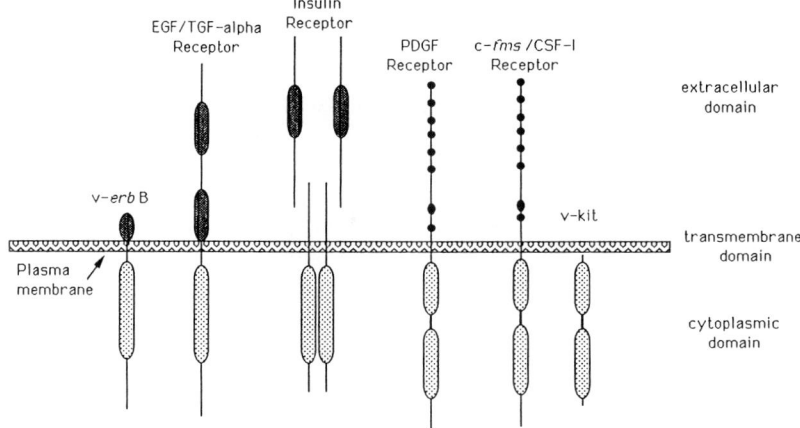

FIGURE 2. Schematic representation of the three domain structure of cell surface receptors and oncogene products. (Adapted from Yarden, Y. et al., *Nature,* 323, 226–232, 1986.)

nase activation, stimulation of PI turnover in the plasma membrane occurs. The two second messengers, DAG and inositol 1,4,5-triphosphate (IP3), regulate different metabolic pathways. DAG promotes the translocation of the cytosolic protein kinase C (PKC) to the plasma membrane. DAG action can be mimicked by phorbol esters and synthetic diacylglycerol (OAG). The membrane bound PKC subsequently phosphorylates a number of intracellular unidentified proteins involved in DNA duplication. IP3 stimulates the release of Ca^{2+} from the intracellular stores, thus increasing the concentration of intracellular free Ca^{2+}. The latter operates, in turn, as an activating signal for PKC, for several enzymatic reactions as well as for genes encoding proliferation-related proteins. Activation of PKC leads either directly or indirectly to stimulation of Na^+, K^+, H^+, and Ca^{2+} fluxes across the plasma membrane.[7] In fact, the increased intracellular Na^+ concentration, brought about by the stimulated Na^+/H^+ antiport system, causes cytoplasmic alkalinization and, as a secondary effect, the stimulation of the Na^+/K^+ pump. Finally, the increase of cytosolic free Ca^{2+} induces Ca^{2+} efflux by activation of plasma membrane Ca^{2+}-ATPase. Cytoplasmic alkalinization and ATP utilization (activation of plasma membrane ATPases) could be partially responsible for an increase in glycolytic activity in the context of growth factor-induced acceleration of cellular metabolism. As shown in Figure 1, a correlation exists between the PI cascade and the cAMP mediated metabolic pathway. The Ca^{2+}-calmodulin complex and the E-type prostaglandins (PGs, PGE), derived by arachidonate (released from DAG and other sources), stimulate, through their membrane receptors, the G protein that is a component of the adenylate cyclase complex.

The third mitogenic pathway is the growth factor protein-tyrosine kinase pathway (Figure 1). For most cell types the signal elicits exclusively the phosphorylation of 42 kDa proteins (X''-PO_4) on tyrosine, bypassing the PI cascade. Only in some systems the PI turnover and PKC are activated. A comparison between pathways 2 and 3 shows that they are distinct at the receptor level, the receptors of the former lacking the protein tyrosin kinase activity. Furthermore, in pathway 2, PKC activates an unknown tyrosine kinase that phosphorylates the 42 kDa proteins, as well as phosphorylates a different class of proteins (X'-PO_4). Both pathways stimulate the Na^+/H^+ antiport, which is contrariwise unaffected by protein phosphorylation occurring in the cAMP-dependent pathway. Finally, all three pathways control c-*fos* and c-*myc* proto-oncogene expression. c-*fos* and c-*myc* are growth-regulated genes (also called "early genes" or "competence genes"), i.e., genes that are not expressed in resting (G_0) cells, but become expressed when cells transit to the growing stage by the addition of appropriate mitogens.[8] Since these cellular oncogenes (that are the cellular equivalents of retroviral transforming genes) encode nuclear proteins, they could be involved in the communication of second messenger response to nuclear targets.

2. Gene Expression of Growth Factors and Growth Factor Receptors

Cancer genes, i.e., genes whose mutation or deregulation can cause loss of contact inhibition of cells in culture and induction of tumors in animals,[9,10] are called *transforming genes* or *oncogenes*. Two families have been recognized, the v-oncogenes in retroviruses and the homologous normal cellular genes (c-oncogenes), named proto-oncogenes. According to a widely shared hypothesis,[11] some forms of neoplasia are consequent to the accidental activation or altered expression of an oncogene, induced by a number of mechanisms including insertion of a provirus or point mutation caused by chemical or physical mutagens (Figure 3). Table 1 reports a list of genes encoding proteins located on or interacting with plasma membrane, that confer growth autonomy to the cell. These oncogenes operate generally by three different mechanisms. The first is the *autocrine stimulation* in which the gene encodes or indirectly induces the synthesis and the release of a growth factor whose receptor is present in the same cell type. Such a mechanism renders the cell independent from the control normally exercised by external regulators. As a consequence, an excessive growth of the tumor cells as well as of the adjacent nontumor cells occurs, thus allowing the tumor mass to develop. An example is given by the simian sarcoma virus oncogene v-*sis* that encodes a PDGF-like protein (Figure 4a).

The second mechanism is that by which a mutated oncogene codes for a structurally altered receptor that is continuously activated in the absence of the mitogen. One example of this class is the *erb*B oncogene that encodes a truncated EGF receptor lacking the amino terminal, extracellular binding site. This results in constitutive activation of the intracellular tyrosine kinase domain (Figure 4b).

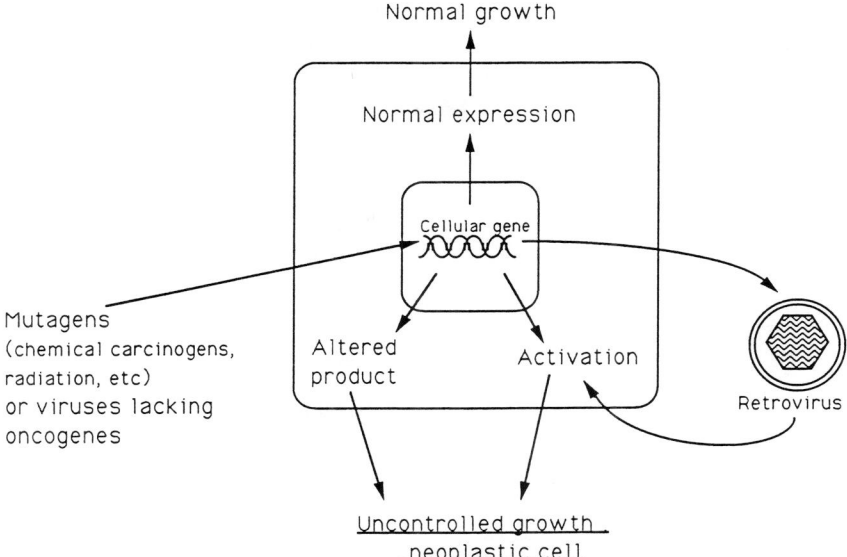

FIGURE 3. The cellular oncogene hypothesis. (Adapted from Taussig, M. J., Ed., *Processes in Pathology and Microbiology,* Blackwell Scientific, Oxford, 1984, 805–809.

By the third mechanism the gene codes for a protein that acts as a transducer of signals from the receptor to cytosolic second messengers. For example, the transforming protein encoded by v-*abl* is a tyrosine kinase whose normal homolog is not endowed with such enzymatic activity. The protein, which lacks the transmembrane and extracellular domains, is at least partly bound to the inner face of the plasma membrane, probably by covalently linked fatty acids (Figure 4c). Finally, the *ras* family represents another group of oncogenes that act through the third mechanism. These genes encode small p21 proteins with GTPase activity and have sequence homology to the family of GTP-binding proteins that transduce vision and hormone responses (Figure 4d).

We think it useful to stress the point that the oncogenes families reported in Table 1 are only those whose products are associated with the plasma membrane. There are, however, other oncogenes whose products are found in the nucleus or in the cytoplasm. Oncogenes are demonstrably associated with only 10 to 20% of human cancers. While it cannot be excluded that several more oncogenes will be discovered in the future, it is also useful to underline that, beside the genetic mechanism of tumorigenesis described so far, tumors may also be determined as the result of an abnormal pathway of differentiation. Inheritable epigenetic changes might involve cellular membranes and special plasma membranes in terms of modified cell adhesion and cell-to-cell communication.

TABLE 1
Oncogenes Conferring Growth Autonomy to a Cell whose Products are Located on or Interact with the Plasma Membrane

Class	Oncogen	Subcellular location of protein	Nature of encoded protein	Mechanism of action
I	v-*sis*	Secreted	PDGF β subunit-like molecule	Autocrine stimulation
II	*erb*B	Plasma membrane	Truncated EGF receptor	
	v-*ros*	Plasma membrane(?) Cytoplasmic membranes	Receptor (uncharacterized GF)	Continuously activated tyrosine kinase
	neu	Plasma membrane	EGF receptor homologous protein	
III	v-*src* v-*abl* v-*fps*	Plasma membrane	Signal transducer	Tyrosine protein kinase
	ras family	Plasma membrane	Guanine nucleotide-binding protein	GTPase

B. TUMOR CELL SPREADING TO SECONDARY SITES

The invasion of cells of one tissue into neighboring or distant tissues is a process comprising physiological as well as pathological aspects. Among the former one can comprehend, for example, angiogenesis (endothelial cell invasion), cell migration during embryonic development, monocyte migration to various tissues and organs of the body (to constitute the mononuclear phagocyte system); the spread of malignant tumors represents a classical example of the second type of process. One of the first steps of neoplastic invasion is the detachment of cells from the "primary tumor" due to loss of cohesiveness. Subsequently, the cells migrate into the adjacent tissue, possibly driven by chemotactic factors. This stage constitutes the so-called *local invasion* and is the result of the loss of positional control. To this local spreading a *migration to secondary sites* ensues. This process involves a series of sequential steps. In most cases the invading cells reach the blood or lymphatic vessels to whose walls they adhere. Thereafter, they penetrate into vessels (intravasation) and disseminate throughout the circulatory system as homotypic or heterotypic emboli. Then extravasation occurs and multiplication at distant locations ("secondary colonies") follows. This cascade of events is illustrated in Figure 5. In several of these events the plasma membrane plays a pivotal role in unison with the connected structures, i.e., the cytoskeleton and the extracellular matrix.

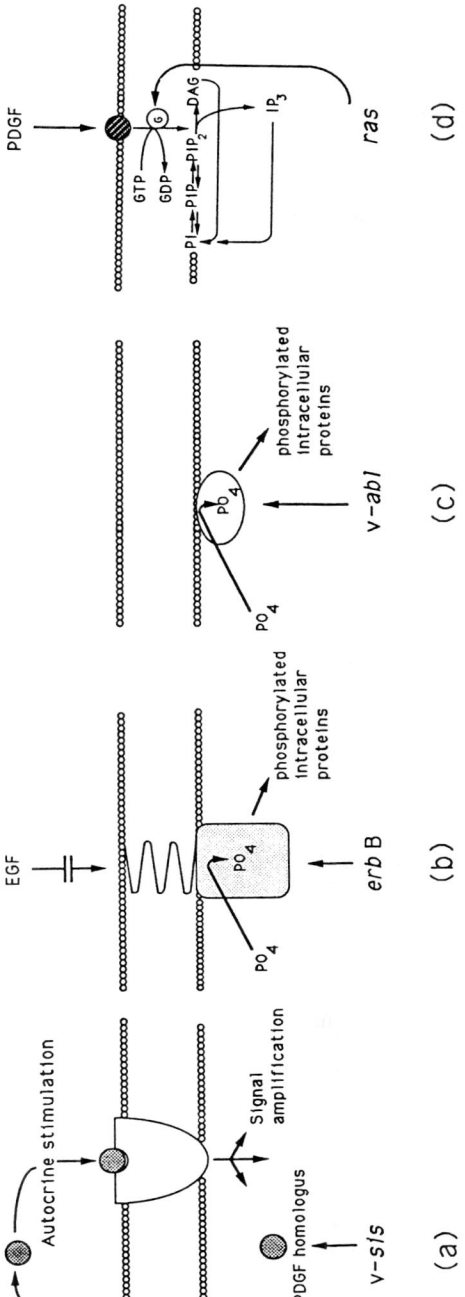

FIGURE 4. Postulated mechanisms of action of oncogenes interacting with the plasma membrane.

The metastatic cascade	Main biological events
LOCAL INVASION	
Detachment from the "primary tumor"	Reduced cell adhesiveness; loss of positional control
Migration to the adjacent tissue	Release of host- and tumor cell-derived chemotactic factors (e.g., matrix molecules degradation products, autocrine motility factors)
MIGRATION TO SECONDARY SITES	
Adhesion to the wall of lymphatic and blood vessels	Altered interaction with basement membrane and endothelial cells; cellular binding to laminin
Entry into the vessels (intravasation)	Crossing basement membrane and the endothelial layer; release of proteolytic enzymes; formation of podosomes
Dissemination through the circulatory system	Homotypic or heterotypic emboli formation
Arrest in a specific tissue	Adherence to or mechanical trapping in the vessels; organ-specific chemoattractants
Leaving the circulation (extravasation)	Crossing the endothelial and basement membrane barrier; cellular binding to laminin; release of proteolytic enzymes
Multiplication at a distant location (secondary colonies formation)	Release of growth and angiogenetic factors; escape from immunological surveillance.

FIGURE 5. The several distinct steps of the metastatic process.

The cytoskeleton is an intracellular fibrous network that contains three major classes of fibers: 7-nm-diameter actin microfilaments, 24-nm-diameter microtubules, and 10-nm-diameter tonofilaments (intermediate filaments). Many plasma membrane proteins are immobilized by contacts with the cytoskeleton. The network is involved in the maintenance of the cell shape and in the cell movement. Other important functions of these fibers are in the attachment of cells to the substratum, in the division as well as in endo- and exocytosis.

The extracellular matrix is composed of the basement membrane (BM) and the interstitial connective tissue. BM main components are the fibrous proteins type IV collagen, laminin, entactin, and the proteoglycans heparan sulfate and chondroitin sulfate. The structural skeleton of the BM is constituted by an open network of type IV collagen to which fibronectin, laminin, and heparan sulfate proteoglycans are bound. The last two components are also bound to each other. The interstitial connective tissue is made up of three major collagen types (I, II, and III), fibronectin, elastin, proteoglycans, and other components. They form a matrix in which cells such as fibroblasts, osteoblasts, macrophages, and chondrocytes are embedded. In tissues the extracellular matrix mediates the contact between adjacent cells. It provides the cells with a frame that, besides architectural functions, regulates the cell shape, metabolism, degree of differentiation, and proliferation.

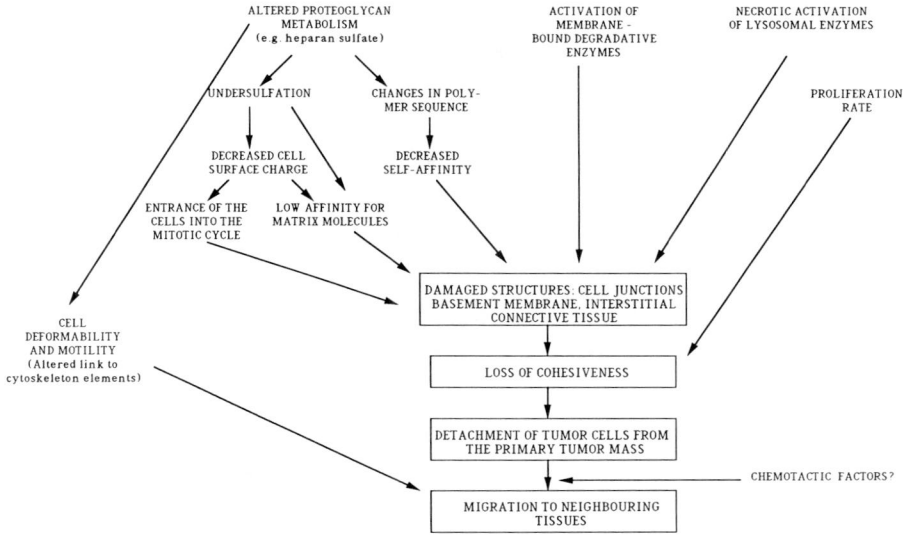

FIGURE 6. Major steps for the detachment of tumor cells from the primary tumor and their migration to adjacent tissues.

1. Local Invasion

The detachment of cancer cells from the tumor mass is indicative of a property, adhesiveness, which seems to be less pronounced in these cells with respect to those of a normal tissue. Such property is influenced by a variety of factors, including elevated activity of some proteinases, lysosomal activation by necrosis, growth rate, cell deformability, and locomotion. Chemical and structural changes of tumor plasma membrane may be implicated in many of these functional characteristics. The suggested mechanisms of detachment of cells from the primary tumor are shown in Figure 6.

Cell surface proteoglycans, particularly those bearing heparan sulfate side chains, are complex macromolecules endowed with a variety of functions, which allow the plasma membrane to communicate properly with the intracellular and the surrounding environment. Their multifunctional role is related to the ability to bind to and interact with both cytoskeletal (spectrin, actin, vimentin) and matrix (heparan sulfate, collagen I, II, and V, fibronectin, and laminin) constituents. Hence one can understand why these molecules have been implicated in the transduction of environmental signals associated with the maintenance of cellular polarity, shape, and position, and in other processes such as proliferation, recognition, and adhesion. Qualitative and quantitative changes in heparan sulfate proteoglycans at both the cell surface and extracellular level, i.e., a reduced degree of sulfation, abnormal polymer sequence, and overproduction, have been reported as a consequence of malignant transformation. Some of these alterations are considered partially responsible for the tumor local invasion.

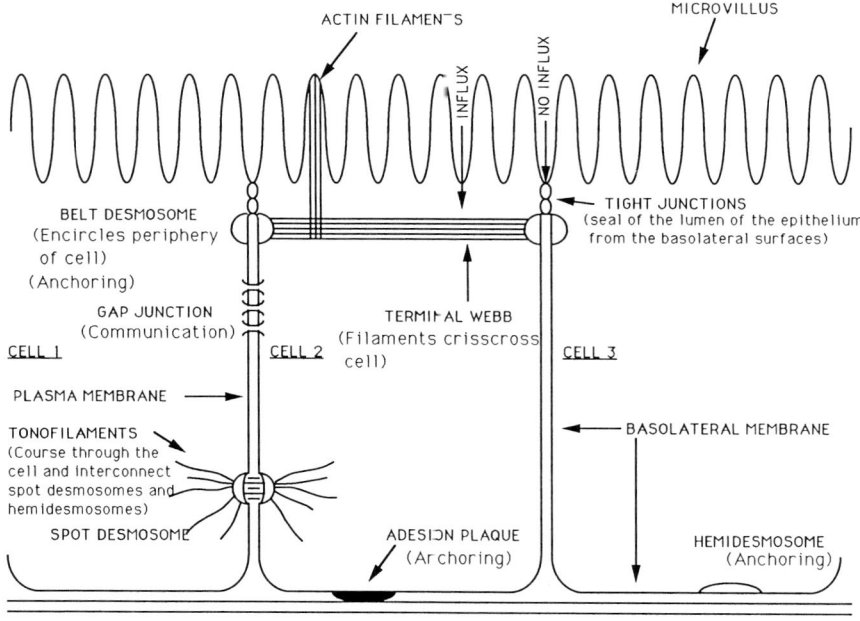

FIGURE 7. Schematic diagram of the principal types of cell-to-cell interconnections.

Disruption of the stromal connective tissue and cell-cell junctions (Figure 7) is another important event that allows the spreading of tumor cells. It is widely recognized that an increased release and/or activation of matrix-degrading proteinases is associated with the malignant phenotype. These enzymes are produced either by the tumor cells themselves or by cells of the host tissue and their activities often correlate directly with the invasiveness of tumor cells (see References 13 and 14 for reviews). Under physiological conditions the proteolysis of the matrix is regulated by a proteinase activation/inhibition balance, through an intracellular mechanism of biosynthesis and secretion and an extracellular mechanism of activation of latent enzymes and action of proteinase inhibitors. The neoplastic state is clearly characterized by a shift in the balance in favor of the proteinase activation. The degradative enzymes most frequently associated with the malignant phenotype include plasminogen activators (t-PA, tissue-type; u-PA, urokinase-like), collagenases (interstitial; type IV), proteoglycanases, cathepsins, and others, such as elastase or gelatinase. Recent studies have suggested the involvement of the plasma membrane in the proteolytic removal of several matrix components. Tumor-promoting phorbol esters stimulate the synthesis and the secretion of PAs and collagenases in endothelial cells and fibroblasts in culture. The secretion of the same enzymes is also stimulated by EGF and PDGF. More-

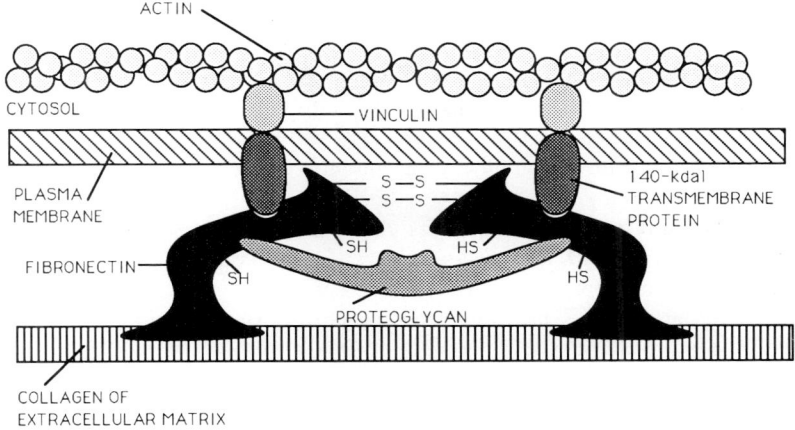

FIGURE 8. Model of the interaction between the fibronectin receptor with its substrate and with actin.

over, it has been shown that a number of degradative enzymes must bind to and interact with the cell surface to elicit their catabolic effect. A high affinity cell-surface receptor for u-PA has been recently identified.[15]

As already pointed out, plasma membrane plays a key role in the establishment of structural and functional interactions between the extracellular matrix and the cytoskeleton as well as of cell-cell contacts. Cell adhesion is a selective process that involves receptors located in the plasma membrane and a variety of adhesion proteins. The receptors comprehend a family of cell surface glycoproteins termed *integrins* and the laminin receptor. The adhesion proteins include matrix molecules, such as fibronectin, type I collagen, vitronectin, laminin, and plasma proteins like fibrin, von Willebrand factor, and thrombospondin. Integrins are transmembrane proteins constituted of α (140 kDa) and β (95 kDa) subunits. Their binding site seems to recognize preferentially proteins that carry specific ligand sequences adjacent to the Arg-Gly-Asp (RGD) sequence also found in fibronectin. For this reason integrins are also named RGD recognition receptors. As an example of integrin we shall briefly outline the fibronectin receptor (Figure 8). This three-domain receptor, mainly located in the adhesion plaque (see Figure 7), on the extracytoplasmic phase is glycosylated on both subunits and also contains in the β subunit an SH-rich region. This type of structure is frequently found in plasma membrane surface receptors: it serves to generate a globular structure that contributes to the formation of binding sites. The intracellular domain binds the actin fibers through an accessory protein, vinculin. It also carries a short consensus sequence containing a tyrosine residue. This can be phosphorylated by tyrosine kinases with the consequence of structural modification of the adhesion plaque and the likely disanchoring of the cytoskeleton. Whereas in normal cells this is a transient process, in many neoplastic cells it becomes a permanent characteristic. Furthermore, many tumor cells either totally lack or have greatly reduced amounts of fibronectin.[13] In several instances these cells synthesize a normal amount of fibronectin but appear unable to bind it to their surface due to still unknown reasons. Some tumors, instead, produce an altered fibronectin and in other cases proteolytic enzymes are released that degrade the molecule.

The laminin receptor is a 65 kDa protein that mediates adhesion of epithelial cells to the BM. Many types of carcinoma cells contain laminin receptors whose number, distribution on the cell surface, and degree of occupancy may be altered. The laminin receptors of normal epithelia seem to be located at the basal site and bound by laminin in the BM, while those of invading carcinoma cells seem to be uniformly distributed over the cell surface.[16] Moreover, laminin binding to its receptor results in an increased secretion of type IV collagenase in metastatic tumor cells.[17] Thus the metastatic potential might be related to the ability of the cells to use this adhesion protein to attach themselves to the subepithelial BM prior to the penetration of the matrix and to the subendothelial BM prior to intra- and extravasation.

After the two steps of attachment to the matrix and its lysis, invasion proceeds through the migration of tumor cells into the region of the degraded matrix. Chemotactic factors, released by the host cells and by the tumor cells, drive the invading cell locomotion.

2. Migration to Secondary Sites

To enter the bloodstream (this is not needed for lymphatics which lack the BM) the tumor cells have to penetrate the endothelial BM. Again the process requires adhesion to and hydrolytic disruption of the BM, and the plasma membrane contributes to it not only with attachment and lysis but also with the formation of podosomes (Figure 9). In fact a role of membrane lipids in the genesis of such protrusions, that allow cell mobility, has been hypothesized. Cytokines, such as the autocrine motility factor (AMF), activate a transducer system involving phospholipases C and A_2. It is, therefore, conceivable that the phosphatidylinositol cascade is activated, thus inducing, through a release of Ca^{2+} from intracellular stores, a rapid cytoskeleton reassembly. The pseudopodal protrusions intercalate between the endothelial cells and drag the whole cell inside the vessel (intravasation).

Once in the bloodstream, tumor cells tend to agglutinate generating homotypic emboli and aggregate with host leukocytes, fibrin, or platelets, forming heterotypic emboli. Lectins, cell surface-localized binding proteins, contribute to this process.[18] Lectins are proteins that have two or more specific binding sites for a particular sugar. Since most plasma membrane glycoproteins contain multiple exposed sugar residues, lectins cross-link those that have the appropriate carbohydrate substituents. They can also bind to serum glycoproteins, thus generating cross-linking intercelular bridges. In cells transformed by

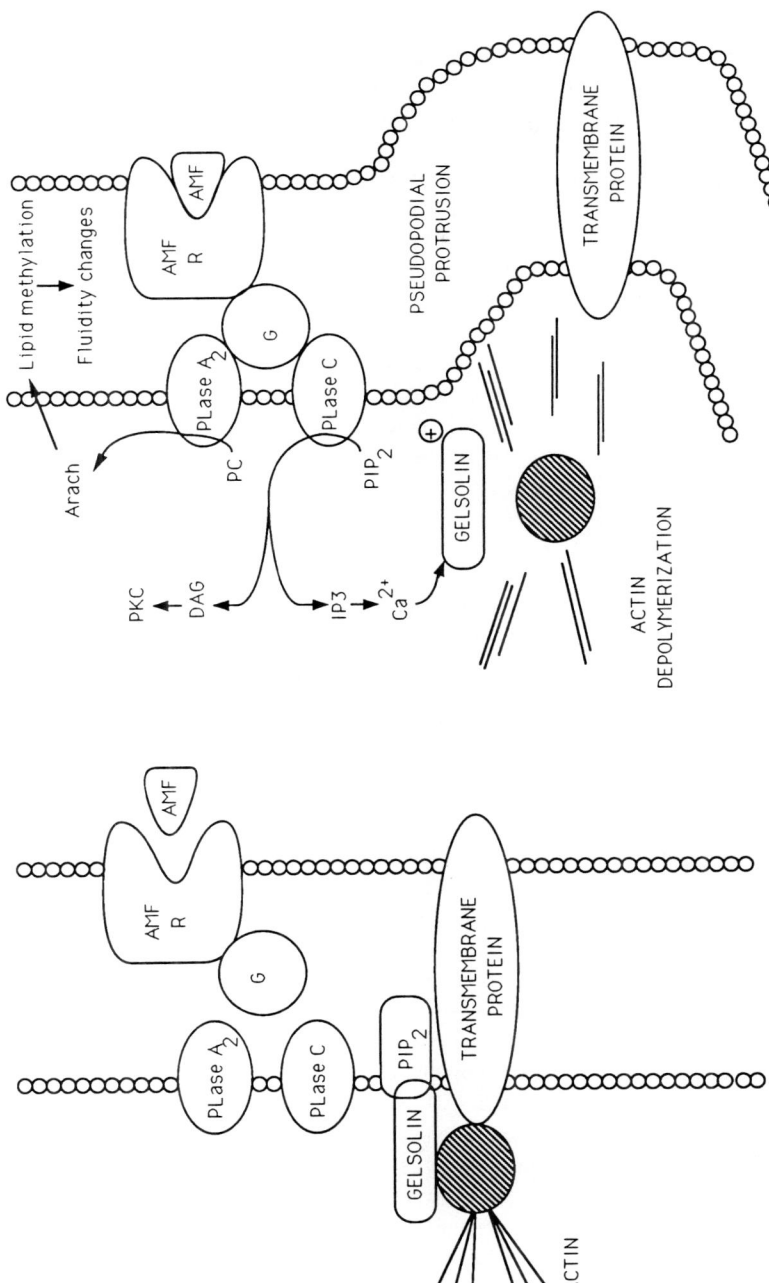

FIGURE 9. Tumor cell motility: hypothetical role of cytokines (such as AMF) in modulating rapid cytoskeleton remodeling and lipid fluidity that results in pseudopodial protrusion. The phosphoinositol cascade has been described in Figure 1. The turnover of PIP_2, that is known to form a complex with the cytoplasmic domain of transmembrane proteins, and the release of Ca^{2+}, could contribute to actin depolimerization through gelsolin, a calcium dependent actin binding protein. PC: phosphatidylcholine. Arach: arachidonic acid.

oncogenes or retroviruses, lectins are always increased compared to normal cells. In tumor cells they are similar in molecular size, antigenicity and sugar-binding specificity to those found in normal cells. They are exposed on the exterior of all tumor cells so far examined and higher levels are correlated with higher metastatic potential.[18]

Circulating clusters of tumor cells arrest in the vessels of the target organ, by adherence or mechanical trapping. To exit the circulation (extravasation) tumor cells again interact by means of adhesion proteins with the RGD-type receptors of the endothelial cells and induce their active retraction. Subendothelial BM becomes exposed and adhesion (laminin and thrombospondin receptors) of the tumor cell follows. The BM is then dissolved by degradative proteinases, type IV collagenase, heparanase, and cathepsins. The micrometastatic process is thus completed. Growth of the secondary colonies which allows the metastasis to become clinically detectable is dependent on the local availability of growth and angiogenesis factors as well as on the host immunological defenses.

III. ENDOPLASMIC RETICULUM

In the frame of cell metabolism compartmentation, proteins and lipids are synthesized within a membrane network, the rough and smooth endoplasmic reticulum (ER). Furthermore, only liver ER and, although somewhat different, adrenal cortex mitochondria are endowed with a mixed-function oxidase (MFO) system which catalyzes hydroxylation of a variety of xenobiotics and endogenous substrates. Several enzymes such as N-demethylases, epoxide hydrases, UDP-glucuronyl transferase, sulfotransferases, and others, are also implicated in the processing of such substances. In the MFO system, cytochrome P-450 operates the reduction and activation of molecular oxygen by a two-electron transfer mechanism. The first requires the reduction of a cytochrome P-450-substrate complex via a reduced flavoprotein (NADPH-cytochrome c reductase), involving a one-electron transfer. Subsequently, the complex interacts with molecular oxygen giving rise to a ternary reduced cytochrome P-450-substrate-oxygen complex. The latter accepts a second electron with the formation of an active oxygen-cytochrome P-450-substrate complex that, after the transfer of one oxygen atom and the uptake of a proton, dissociates into oxidized cytochrome P-450, H_2O, and a hydroxylated product. The electron and proton donor to the cytochrome is NADPH which is provided by the hexose monophosphate shunt. Glycolysis can also furnish reducing equivalents in the form of NADH, that instead reduce cytochrome b_5 through a specific reductase. Thus, microsomal electron transport is involved both in the exergonic oxidoreductions and in the endergonic syntheses (see Figure 10). In tumors, microsomal electron transport has been shown to be curtailed, and the balance shifted in favor of synthetic pathways.

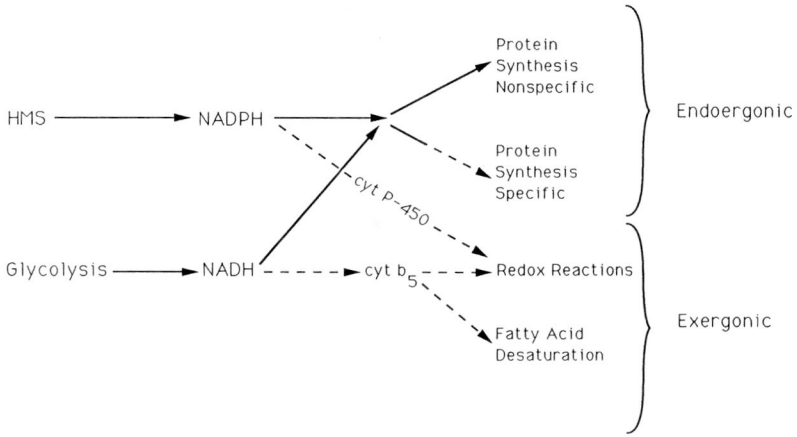

FIGURE 10. Microsomal electron transport, curtailed in neoplastic cells (stippled arrows). HMS: hexose monophosphate shunt. (Adapted from Apffel, L. A., in *Membranes in Tumor Growth*, Galeotti, T., Cittadini, A., Neri, G., and Papa, S., Eds., Elsevier, Amsterdam, 1982, 471–479.

Due to its central role in anabolic processes and catabolism of xenobiotics, particularly carcinogens, ER has been extensively studied both in normal and tumor cells. The studies performed before, during, and after malignant transformation point to this membrane system as the site of initial events, the most important of which appear to be the following: (1) significant changes in the ultrastructural, biophysical, and biochemical features; (2) a functional impairment of the reducing equivalent flux from the catabolic pathways of glucose to molecular oxygen via the microsomal cytochromes and the coenzymes NAD and NADP; (3) the defective attachment of ribosomes to the membrane; and (4) the irreversible decline of the cytochrome P-450 content by restriction of its synthesis and/or by peroxidative degradation.[19]

Lipid composition of ER in tumors has been investigated in detail. Hepatomas with varying growth rates and degrees of malignancy have been the models mostly employed. Microsomal membranes isolated from normal liver and hepatomas of the Morris line[20] contain the following phospholipids: phosphatidylethanolamine (PE), phosphatidylcholine (PC), sphingomyelin (SM), phosphatidylinositol (PI), and phosphatidylserine (PS). Cardiolipin was found to be virtually absent and confined to mitochondria both in normal and tumorigenic tissues. The relative amount of PE is variable in the hepatomas, increased in many, and normal in others. PC is generally normal or slightly reduced. SM from tumors of rapid or intermediate growth rate is significantly increased with respect to the liver, whereas it is normal in slow-growing hepatomas. PI is normal or reduced, PS is increased in some hepatomas.

TABLE 2
Biochemical and Physical Parameters of Endoplasmic Reticulum (ER) Membranes from Hepatomas with Varying Growth Rates[a]

	Total microsomes				Smooth ER			Rough ER		
	Liver	9618A	44	3924A	Liver	9618A	3924A	Liver	9618A	3924A
Protein content (mg/g wet tissue)	18.7	5.7	NA[i]	3.6	7.6	1.5	1.4	12.1	3.8	2.1
Phospholipid/protein (w/w)	0.52	0.46	0.43	0.21	0.56	0.56	0.33	0.54	0.34	0.21
Cholesterol/protein (nmol/mg)	91	105	NA	145	110	160	304	84	66	113
Cholesterol/phospholipid (mol/mol)	0.14	0.18	0.24	0.53	0.15	0.31	0.74	0.12	0.15	0.42
DBI[b]	174	152	NA	120	149	136	95	160	147	110
SFA[c]	46	44.7	NA	39.4	50.8	45.2	45.3	47.6	43.5	42.1
MFA[d]	8.5	19.3	NA	33.2	9.5	23.9	36.8	10.2	23.9	34.4
PUFA[e]	45.5	36.0	NA	27.4	39.7	30.9	17.9	42.2	32.6	23.5
MFA/PUFA	0.18	0.54	NA	1.21	0.24	0.77	2.06	0.24	0.73	1.46
$\langle P_2 \rangle$[f]	0.47	0.51	NA	0.65	0.5	0.57	0.49	0.47	0.62	0.71
T_R[g]	3.3	4.5	NA	4.3	2.4	3.2	4.3	3.3	4.1	4.9
NADPH-cytochrome c(P-450) reductase (nmol mg protein^{-1} min^{-1})	83.1	ND[h]	42.4	17.8						
Cytochrome b_5 (nmol mg protein^{-1})	0.32	ND	0.15	0.08						
Cytochrome P-450 (nmol mg protein^{-1})	1.03	0.16	0.14	Nil	0.95	0.15	Nil	0.73	0.15	Nil

[a] Tumor growth rate (cm/month) 9618A: 0.45; 44: 0.89; 3924A: 2.45.
[b] Double bond index.
[c] Saturated fatty acids.
[d] Monoenoic fatty acids.
[e] Polyunsaturated fatty acids.
[f] Second-rank order parameter of 1,6-diphenyl-1,3,5-hexatriene (DPH).
[g] Rotational correlation time of DPH (ns).
[h] Not determined.
[i] Not available (data are not available because in the course of experimentation the hepatomae line was lost).

In contrast to the relatively minor changes in the phospholipid composition, expressed as percentage of total lipid phosphorus, the content of phospholipid relative to membrane protein is dramatically altered in hepatomas when compared to that found in normal liver. In all hepatomas, except the 9618A (very slow growing), the content is decreased. The decrease, ranging from 28 to 44%, is directly proportional to the growth rate of the tumor.

Studies have also been performed on cholesterol and fatty acid content and composition, on redox protein content, and on the molecular order and fluidity of the membrane of both smooth and rough ER from Morris hepatomas with different growth rate (Table 2).[2] In terms of lipid composition the general trend of a marked decrease in PUFA content with increased deviation of the tumor appears to be well documented. Tumor membranes are less fluid and more ordered than those of hepatocytes. The value of the order parameter rises consistently with the gradual increase of the cholesterol/phospholipid ratio in the two ER fractions, owing to either the increase in cholesterol or the decrease in phospholipid content or both. The value of the correlation time gradually increases, and fluidity therefore decreases, in the smooth and rough fractions, going from the control liver to the more deviated tumors.

An abnormal fatty acid composition is exhibited by the majority of tumor cells. This phenomenon is likely to be partially dependent on changes in the activity of the fatty acid desaturase enzymes. Indeed, microsomal fatty acid desaturation was reported to be affected in experimental tumors or tissue culture models. Particularly, in a human lung, fast growing tumor, transplantated in nude mice, microsomes were able to desaturate 16:0, α-18:3 and 20:3 acids by $\Delta 9$, $\Delta 6$, and $\Delta 5$ desaturase, respectively. However, $\Delta 6$ desaturase activity for linoleic acid was not detectable. The microsomal elongation system was active in all fatty acid series tested, except for linoleic acid.[22]

The content in electron carriers of microsomal membranes decreases in hepatomas proportionally to the increase of the tumor growth rate. It is noteworthy that cytochrome P-450 is virtually absent in the maximal deviation Morris hepatoma 3924A.

Irreversible degranulation of the ER has been observed in premalignant cells, during carcinogenesis and in neoplastic cells of established tumors. From the pathogenetic point of view, Apffel[19] has proposed that degranulation begins with the interaction of a carcinogenic agent with the ER membrane and continues with the activation of the MFO system, leading to an overproduction of free radicals and to lipid peroxidation. The consequent membrane alterations, which comprehend increased cholesterol to phospholipid and saturated to unsaturated fatty acids ratios, impair the anchoring to the lattice of the ER membrane of the attachment proteins, ribophorin I and II, which normally mediate the binding of polysomes to membranes. A further effect of the radical attack is the destruction of cytochrome P-450 that cannot be adequately replaced, owing to the limited capacity of neoplastic cells to synthesize the microsomal cytochromes. In fact, cytochromes P-450 and b_5 are heme glycoproteins, whose synthesis can only occur on attached polysomes. The major consequence of degranulation is the shift of the microsomal electron transport from exergonic redox reactions to syntheses that require energy and from the synthesis of specific to that of nonspecific proteins. In this model, therefore, the ER would play a key role in the carcinogenic process.

IV. MITOCHONDRIAL MEMBRANES

Mitochondria occupy 15 to 50% of the cytoplasmic volume of most animal cells and participate in more metabolisms than any other subcellular organelle. They are the compartment where electron transport is coupled to ATP synthesis; they contain a genetic apparatus that codes for several mitochondrial hydrophobic proteins that are synthesized on mitochondrial ribosomes; they play a key role in both catabolic and anabolic metabolism. During the catabolic phase of metabolism, mitochondria, via shuttle systems, participate in the regeneration of NAD^+ for the glycolytic pathway. During prolonged starvation, liver mitochondria produce ketone bodies which are then used by brain and muscle mitochondria to provide energy in these tissues. Liver mitochondria participate also in the synthesis of urea. During the anabolic phase of metabolism they supply the carbon skeleton for carbohydrate, fatty acid, and cholesterol synthesis. Significantly, mitochondria can also use ATP to form NADH by the process of reverse electron flow, or to transhydrogenate, i.e., to form NADPH from NADH, or to support calcium uptake. They contain adenylate kinase that tends to buffer the adenine nucleotide pool in the cell, thus playing a pivotal role in metabolic regulation by adjusting the cellular energy charge. Finally there are indications that viral replication and/or viral production in some animal cell lines may require mitochondrial participation.

Despite this essential role of mitochondria in cell metabolism, the extent to which they contribute to promoting and maintaining neoplastic transformation is largely unknown. Most of the research work on tumor mitochondria was carried out in the 1960s and 1970s. While these studies did not show any real significant difference in mitochondrial energy metabolism between tumor cells when compared to normal cells from the same tissue of origin, an interesting series of observations was made that indicated that *in vitro* and *in vivo* many cancer cell lines exhibit a high rate of glycolysis to supply their energy requirements.

The increased glycolysis in tumors is correlated with a number of changes in metabolic processes. Among these the more significant are (1) an increase in plasma membrane glucose carriers; (2) an increase of the level of several key glycolytic enzymes, such as phosphofructokinase, pyruvate kinase, and hexokinase; (3) the prevalent presence of isozymes of the fetal type; and (4) a change in the compartmental localization of hexokinase. Pedersen and co-workers[23] demonstrated that hexokinase is specifically and preferentially bound to the outer mitochondrial membrane in highly glycolytic tumors. The receptor site for the enzyme is part of a pore-forming protein, named porin. The main characteristics of the mitochondrially bound hexokinase are the following: (1) high affinity for glucose; (2) decreased sensitivity to feedback inhibition by glucose-6-phosphate; (3) preferential access to oxidative phosphorylation rather than to cytosolic ATP; and (4) similarity to type II and type III hexokinase.

The preferential utilization of mitochondrially generated ATP by hexokinase leads to an increased availability of glucose-6-phosphate. As a result, higher levels of cellular ATP are achieved, as well as of glycolytic intermediates that bring about a high utilization of the hexose monophosphate shunt, of the nucleic acids and lipid biosynthetic pathways, and in general a high rate of cell division and growth. Figure 11 shows a schematic model of this metabolic situation.

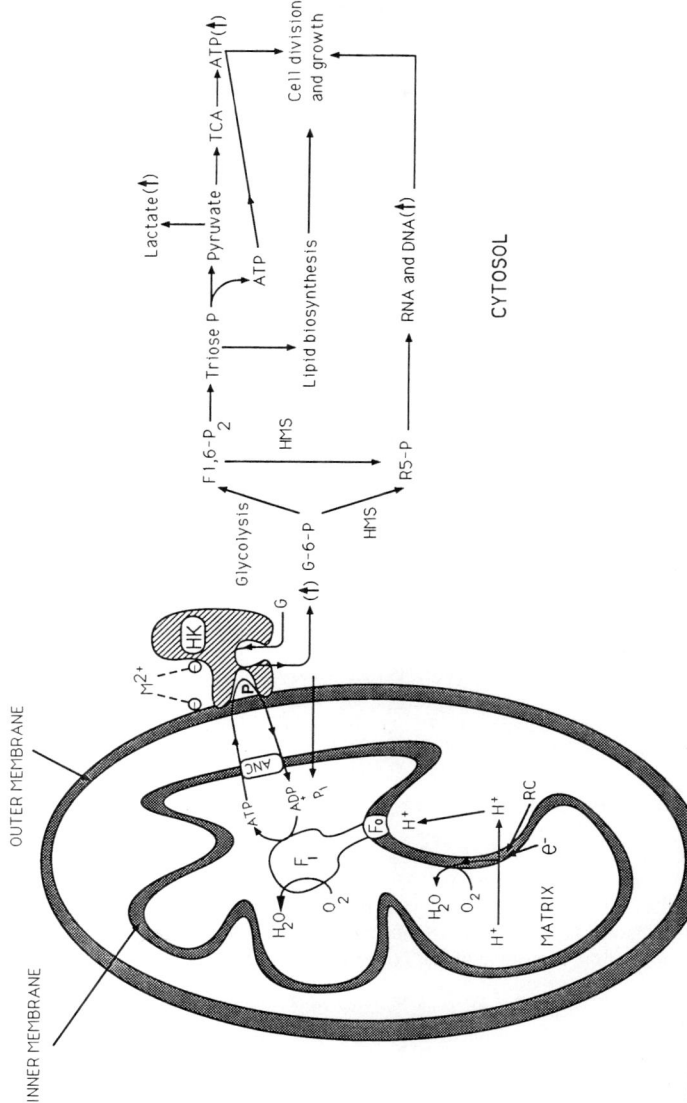

FIGURE 11. A scheme proposing the functional integration of mitochondrially associated Hk within tumor cell glucose metabolism. Hk: hexokinase; P: porin; G: glucose; G6P: glucose-6-phosphate; R5-P: ribose-5-phosphate; HMS: shunt of the hexose monophosphate; RC: respiratory chain; F-1,6-P$_2$: fructose-1,6-diphosphate. ANC: carrier of adenine nucleotides (Redrawn from Arora, K. K. and Pedersen, P. L., *J. Biol. Chem.*, 263, 17,422–17,428, 1988. With permission.)

Finally, other recent studies[24] on the H^+-ATP synthase have shown that in certain tumors there is a deficiency in the ATPase activity that can be attributed to causes not yet well defined, that comprise a lower content in the mitochondrial ATPase, an enhanced content of the ATPase inhibitor protein, or a defect in the coupling function of the enzyme.

ACKNOWLEDGMENTS

We thank Dr. Emanuela Bartoccioni for drawing the diagrams, Professor Giovanni Neri for helpful criticism of the manuscript, and Mrs. Clotilde Castellani for secretarial work.

This work was supported by grants from AIRC 1988, C.N.R. Special Project "Oncology", Grant N. 87.01295.44 (T.G.) and Grant No. 87.01346.44 (L.M.) and MURST 40 to 60% (T.G. L.M.).

REFERENCES

1. **Prehn, R. T.**, Neoplasia, in *Principles of Pathobiology*, La Via, M. F., and Hill, R. B., Eds., Oxford University Press, London, 1971, 191–241.
2. **Dumont, J. E., Janniaux, J. -C., and Roger, P. P.**, The cyclic AMP-mediated stimulation of cell proliferation, *Trends Biochem. Sci.*, 14, 67–71, 1989.
3. **Pawelek, J.**, Evidence suggesting that a cyclic AMP-dependent protein kinase is a positive regulator of proliferation in Cloudman S91 melanoma cells, *J. Cell. Physiol.*, 98, 619–626, 1979.
4. **Yarden, Y., Escobedo, J. A., Kuang, W. -J., Yang-Feng, T. L., Daniel, T. O., Tremble, P. M., Chen, E. Y., Ando, M. E., Harkins, R. N., Francke, U., Fried, V. A., Ullrich, A., and Williams, L. T.**, Structure of the receptor for platelet-derived growth factor helps define a family of closely related growth factor receptors, *Nature*, 323, 226–232, 1986.
5. **Hunter, T. and Cooper, J. A.**, Protein-tyrosine kinases, *Annu. Rev. Biochem.*, 54, 897–930, 1985.
6. **Serunian, L. A. and Cantley, L. C.**, Growth factor and oncogene influences on cell growth regulation, *Ann. N.Y. Acad. Sc.*, 551, 309–319, 1988.
7. **Rozengurt, E.**, Early signals in the mitogenic response, *Science*, 234, 161–166, 1986.
8. **Baserga, R., Calabretta, B., Travali, S., Jaskulski, D., Lipson, K. E., and De-Riel, J. K.**, Regulation of the expression of cell cycle genes, *Ann. N.Y. Acad. Sci.*, 551, 283–289, 1988.
9. **Bishop, J. M.**, Viral oncogenes, *Cell*, 42, 23–38, 1985.
10. **Weinberg, R. A.**, The action of oncogenes in the cytoplasm and the nucleus, *Science*, 230, 770–776, 1985.
11. **Druker, B. J., Mamon, H. J., and Roberts, T. M.**, Oncogenes, growth factors, and signal transduction, *N. Engl. J. Med.*, 321, 1383–1391, 1989.
12. **Taussig, M. J., Ed.**, *Processes in Pathology and Microbiology*, Blackwell Scientific, Oxford, 1984, 805–809.
13. **Tryggvason, K., Höyhtyä, M., and Salo, T.**, Proteolytic degradation of extracellular matrix in tumor invasion, *Biochim. Biophys. Acta*, 907, 191–217, 1987.

14. **Moscatelli, D. and Rifkin, D. B.**, Membrane and matrix localization of proteinases: a common theme in tumor cell invasion and angiogenesis, *Biochim. Biophys. Acta,* 948, 67–85, 1988.
15. **Appella, E., Robinson, E. A., Ullrich, S. J., Stoppelli, M. P., Corti, A., Cassani, G., and Blasi, F.,** The receptor-binding sequence of urokinase. A biological function for the growth-factor module of proteases, *J. Biol. Chem.,* 262, 4437–4440, 1987.
16. **Liotta, L. A. and Stetler-Stevenson, W. G.**, Principles of molecular cell biology of cancer: cancer metastasis, in *Cancer: Principles and Practice of Oncology,* De Vita, V. T., Hellman, S., and Rosenberg, S. A., Eds., J. B. Lippincott, Philadelphia, 1989, 98–115.
17. **Terranova, V. P., Rao, C. N., Kalebic, T., Margulies, I. M., and Liotta, L. A.,** Laminin receptor on human breast carcinoma cells, *Proc. Natl. Acad. Sci. U.S.A.,* 80, 444–448, 1983.
18. **Lotan, R. and Raz, A.,** Lectins in cancer cells, *Ann. N.Y. Acad. Sci.,* 551, 385–398, 1988.
19. **Apffel, C. A.,** The increased irreversible detachment of polysomes and its bearing on the neoplastic state, in *Membranes in Tumour Growth,* Galeotti, T., Cittadini, A., Neri, G., and Papa, S., Eds., Elsevier, Amsterdam, 1982, 471–479.
20. **Morris, H. P. and Wagner, B. P.,** Induction and transplantation of rat hepatomas with different growth rate (including ''minimal deviation'' hepatomas), in *Methods in Cancer Research,* Vol. IV, Busch, H., Ed., Academic Press, New York, 1968, 125–152.
21. **Masotti, L., Casali, E., Gesmundo, N., Sartor, G., Galeotti, T., Borrello, S., Piretti, M. V., and Pagliuca, G.,** Lipid peroxidation in cancer cells: chemical and physical studies, *Ann. N.Y. Acad. Sci.,* 551, 47–58, 1988.
22. **De Antueno, R. J., Niedfeld, G., De Tomas, M. E., Mercuri, O. F., and Montoro, L.,** Microsomal fatty acid desaturation and elongation in a human lung carcinoma grown in nude mice, *Biochem. Int.,* 16, 413–420, 1988.
23. **Arora, K. K. and Pedersen, P. L.,** Functional significance of mitochondrial bound hexokinase in tumor cell metabolism. Evidence for preferential phosphorylation of glucose by intramitochondrially generated ATP, *J. Biol. Chem.,* 263, 17422–17428, 1988.
24. **Papa, S. and Capuano, F.,** The H^+-ATP synthase of mitochondria in tissue regeneration and neoplasia, *Ann. N.Y. Acad. Sci.,* 551, 168–178, 1988.

Chapter 14

ANTI-HIV COMPOUNDS WITH MEMBRANE ORIENTED SPECIFICITY—EARLY RESULTS*

C. A. Stein and Ranajit Pal

TABLE OF CONTENTS

I.	Introduction	338
II.	Inhibition of the Maturation of HIV-1 Envelope Glycoproteins	339
III.	Inhibition of Myristylation of *gag* Gene Products	340
IV.	sCD4 as Therapy for HIV-1 Disease	341
V.	Polyanionic Compounds Active at the Cell Surface	344
	A. Dextran Sulfate and Related Molecules	344
	B. Other Anionic Compounds	346
	C. Sulfonic Acids	347
	D. Phosphorothioate Oligodeoxynucleotides	347
VI.	Conclusions	348
References		349

* This review was written in the spring of 1990 and is current only to that point.

ISBN 0-8493-8091-X
© 1993 by CRC Press, Inc.

I. INTRODUCTION

Nearly 8 years ago, a new lethal disease was observed in the U.S. which was characterized by irreversible immune dysfunction associated with opportunistic infections and rare malignancies, such as Kaposi's sarcoma. This disease, commonly referred to as acquired immune deficiency syndrome (AIDS), was first found to be prevalent among young male homosexuals, but subsequent epidemiological studies have demonstrated that the disease is also common among intravenous drug users and hemophiliacs. Infection with a retrovirus called human immunodeficiency virus (HIV) has been shown to be the cause of the disease AIDS.[1-4] It is not yet clear how the virus causes immunodeficiency in the infected host The simplest explanation presently offered is that infection of CD4+ T cells with HIV eliminates the T helper cells. Indeed activated CD4+ cells appear to be highly susceptible and undergo lytic infection *in vitro* when exposed to HIV.[5] Along with cell death, HIV may interfere with T cell function in a number of other ways. Binding of viral envelope glycoprotein to the CD4 molecule may perturb the structure that normally interacts with the antigen, thereby disturbing the post-receptor signal transduction pathway that is essential for normal function of T helper cells.[6,7]

A number of isolates of HIV have been cloned and sequenced in different laboratories.[8-10] The HIV genome is approximately 10 Kb long and contains two flanking long terminal repeat (LTR) elements. The complete sequence contains seven open reading frames of which three are identified as the *gag*, *pol*, and *env* genes. The *gag* gene encodes the viral capsid proteins and the *pol* gene encodes nonstructural proteins involved in virus replication. These include a virus specific protease, the viral reverse transcriptase (RT), and an endonuclease. The *env* gene encodes two envelope proteins, gp120 and transmembrane gp41. The other genes are referred to as *vif*, which encodes a 23,000 Da protein, important for virus infectivity; *nef*, which encodes a 27,000 Da protein most likely involved in down-regulation of virus replication; *tat*, which encodes a 14,000 Da protein required for virus transactivation; *rev*, whose product is required for synthesis and transport of envelope proteins; and *vpu* and *vpr* which are important for virus assembly.

The urgent need for chemotherapy of AIDS has directed considerable research interest toward effective anti-HIV agents. There are multiple steps in the virus replication cycle which may be relevant for therapeutic strategies. The replication cycle of HIV can be classified into a few distinct steps. The first step in infection is the binding of the virion to the host cell. Binding is mediated by the specific interaction of viral envelope glycoprotein and the CD4 on the surface of the infected cell. The virus then enters the host cell by direct fusion of the viral envelope with the host cell membrane. Upon entry, the viral genome is ejected into the cytoplasm where transcription from RNA to DNA is catalyzed by the viral reverse transcriptase. In the subsequent step, viral RNA in the RNA-DNA hybrid is degraded by the RNAse H activity of the RT and the newly synthesized double-stranded DNA then integrates into the host genome. The expression of the structural proteins of the virus takes place mainly under the control of *tat, rev,* and *nef* gene products. After synthesis of the components, the assembly of the virus takes place partly in the cytoplasm and partly at the plasma membrane of the host cell. Mature virions eventually bud from the infected cells. A number of antiviral agents inhibiting a variety of steps in the virus life are being tested for possible use in the treatment of AIDS and related diseases. Many of these drugs are directed towards the viral reverse transcriptase and protease enzyme. Several excellent reviews have been written on strategies for antiviral therapy in AIDS and HIV infection.[11,12] In this review we offer a comprehensive description of anti-HIV compounds with specificity towards membranes. The first part of the review describes antivirals which affect the maturation of envelope and *gag* gene products. Compounds interfering with the binding and fusion of HIV with host cell membranes are described in the second portion of the review.

II. INHIBITION OF THE MATURATION OF HIV-1 ENVELOPE GLYCOPROTEINS

The envelope glycoprotein gp120 of HIV-1 is the surface component of the virus.[4] It plays a key role in viral adhesion and initiation of infection through interaction with the CD4 receptor of T lymphocytes.[13-15] The glycoprotein is richly glycosylated with approximately half of the molecular mass consisting of carbohydrate distributed on nearly 20 N-glycosylation sites.[8] The envelope glycoprotein is initially synthesized as gp160 which upon proteolytic processing gives rise to gp120 and gp41.[16,17] The functional significance of the extensive glycosylation on the gp120 molecule is not fully understood. It has been demonstrated that gp120 oligosaccharides are necessary for binding of the glycoprotein to the host cell CD4, as deglycosylation of gp120 abolishes[18] or impairs[19] its binding to the receptor.

The major glycosylation pathway utilized for both viral and cellular membrane glycoprotein begins with the transfer of the precursor oligosaccharide Glc_3-Man_9-$glcNAc_2$ from the carrier lipid dolichol phosphate to the nascent polypeptide chain of the glycoprotein.[20] Complex type oligosacchrides are generated via the trimming pathway. The first step of this processing is the removal of three glucose residues by the glucosidase I and II enzymes. Subsequently, the action of the mannosidase I and II enzymes in the Golgi complex completes the removal of four mannose residues. The addition of peripheral sugar residues such as galactose, N-acetyl glucosamine and sialic acid by specific glycosyl transferases results in the synthesis of a complex type oligosaccharide moiety in the glycoprotein molecule. Several inhibitors of the trimming enzymes are now available.[21] The amino sugar 1-deoxynojirimycin and castanospermine have been shown to inhibit glucosidase I activity in the endoplasmic reticulum and 1-deoxymannojirimycin and swainsonine inhibit mannosidase I and II, respectively, in the Golgi complex.

Inhibitors of the trimming enzymes have been shown to reduce or abolish the infectivity and cell fusion activity of HIV-1.[22-26] Both castanospermine and deoxynojirimycin were shown to inhibit syncytium formation induced by the envelope glycoprotein of HIV, and to inhibit viral replication.[22-24] The decrease in syncytium formation in the presence of glucosidase I inhibitors can be attributed to the inhibition of the cleavage of the envelope glycoprotein precursor gp160 resulting in decreased cell surface expression of the mature glycoprotein, gp120. In addition, these compounds can cause defects in membrane fusion events occurring after binding of the viral glycoprotein with the CD4 receptor. Although processing of gp160 to gp120 was observed in the presence of deoxymannojirimycin, an inhibitor of mannosidase I in the Golgi complex, the infectivity of the virus released from chronically infected cells was sharply attenuated.[25,26] In contrast, treatment with swainsonine, an inhibitor of mannosidase II in the Golgi complex, had no effect on the infectivity of the progeny virus. These results suggest that inhibition of trimming of glucose and primary trimming of mannose residues from HIV glycoprotein attenuates viral infectivity.

III. INHIBITION OF MYRISTYLATION OF *gag* GENE PRODUCTS

The primary *gag* gene product of HIV-1 is synthesized utilizing an initiation site near the 5' end of the viral genomic RNA. This precursor polyprotein Pr53^{gag} is proteolytically processed into proteins of molecular weight 17, 24, 7, and 6 kDa.[4] The *gag* protein p17 is modified by the addition of a myristic acid residue through an acyl linkage to the amino terminal glycine.[4] This myristate modification of p17 appears to be critical for the assembly of HIV-1. Thus, mutagenesis of the myristoylation site in p17 from glycine to alanine abolishes the assembly of HIV-1 particles.[27,28] It is not clear how myristate linked to the *gag* protein p17 affects virus assembly. It is possible that the fatty acyl residue on p17 could provide a hydrophobic domain at the N-terminus, thus facilitating its interaction with the lipid bilayer of the plasma membrane. Such interaction may be necessary for the assembly of the virus. Several inhibitors of *N*-myristoylation have been described recently. *N*-myristoyl glycinal diethylacetal and similar derivatives inhibit myristoylation of p17 and production of mature HIV-1 particles in infected cells.[29] Several sulfur- and oxygen-substituted analogs of myristate that are similar in length were shown to inhibit HIV-1 replication markedly.[30] The antiviral effect of these compounds is not accompanied by any apparent toxicity to the growth of CD4+ cells.

IV. sCD4 AS THERAPY FOR HIV-1 DISEASE

This section will be devoted to those therapeutic strategies which have attempted to take advantage of the fact that the CD4 molecule is used by HIV-1 as its mode of cellular entry.

The immune abnormality in HIV disease is presumably caused by a depletion of CD4+ lymphocytes.[31] CD4 and CD8 are cell surface molecules which define distinct subsets of mature peripheral T-cells whose ability to interact with antigen is MHC class II and I restricted, respectively. The CD4 molecule[32] consists of an N-terminal hydrophobic sequence, four extracellular domains with homology to the immunoglobulin variable and joining regions, a hydrophobic transmembrane domain, and a polar cytoplasmic portion. It has recently been shown[33] that the two N-terminal domains of CD4 can inhibit HIV infection of target cells.

The affinity constant of CD4 for gp120 is approximately 10^{-9}. Smith et al.[31] produced CD4 analogs with comparable affinity for gp120 by eliminating the transmembrane and cytoplasmic domains of the protein. This truncated, secreted, soluble species bound to gp120 and abolished the growth of HIV-1 when incubated with virus infected cells, as measured by indirect immunofluorescence. Shortly after this report, a number of other authors[32-35] reported their results with soluble CD4 (sCD4). For example, as measured by p24 antigen production,[34] recombinant sCD4 blocked viral reproduction at 10 µg/ml. HIV-induced syncytia formation was also inhibited at this concentration. Others[36] reported inhibition of syncytia formation between chronically infected H9/HTLV-IIIB cells and CD4+ SupT1 cells at 2 µg/ml, and inhibition of p24 viral protein expression at 0.2 µg/ml. Significantly, sCD4 did not inhibit the interaction of cell surface CD4 with its natural ligand, MHC II, and thus failed to inhibit cytotoxic T-lymphocyte effector function, as measured by clonal proliferation in response to tetanus toxoid.[35] Furthermore,[36] the *in vitro* effect of soluble gp120, which has itself been shown to inhibit antigen-driven proliferation of T-cells, may in fact be overcome by treatment with sCD4.

The ability of sCD4 to block the binding of gp120 to cell surface CD4 may be seen even after adsorption of virions to CD4+ cells.[37] However, even though sCD4 can block HIV-1 and SIV binding, it is less effective with HIV-2. In fact, HIV-2 isolates require a 25-fold higher concentration to achieve an identical effect: the IC_{80} for HIV-1 (HTLVIII/RF) = 0.3 µg/ml while for LAV 2 = 5 µg/ml.

Watanabe et al.[38] treated SIV-infected rhesus monkeys with sCD4. These animals, when infected with the simian immunodeficiency virus (SIV_{MAC}), develop an AIDS-like syndrome. The sCD4 persisted in plasma for 8 h, with a $t^1/_2$ of about 6 h. In two animals, both peripheral blood lymphocytes (PBL) and bone marrow cultures became virus negative on treatment and remained so until 1 month after completion of treatment. Furthermore, bone marrow cultures of these treated animals could not be reinfected with SIV_{MAC} during the treatment period. In addition, BFU-E and CFU-GM bone marrow cells increased during the course of treatment.

On the basis of the ability of sCD4 to inhibit the replication of HIV-1 *in vitro* and in animal experimental models, and on the anticipation of lack of any serious systemic toxicity, a phase I trial of sCD4 was instituted in humans. Note that the phase I designation defines the trial as one interested in toxicity data only, with therapeutic efficacy only incidentally explored. The results of these trials were recently published. Initially,[39] the study involved 29 patients with AIDS and 13 with AIDS-related complex (ARC). Six subjects each were enrolled at doses of 1, 10, 30, 100, and 300 µg/kg IV; six were given 300 µg/kg subcutaneously; and six were treated with the latter dose IM. The first dose was given on day 1, then daily × 10 days, and finally three times per week × 8 weeks. After day 72, maintenance sCD4 was given. No intolerable symptoms were noted. Two patients had fever (max. 38.6°C), three patients reported asthenia, and one each reported nausea, twitching, diarrhea, paresthesias, or depression. There was no change in performance status or body weight. In ten patients, an approximately 20 to 25% decrease in systolic blood pressure after sCD4 administration was noted. This was symptomatic and did not recur. Leukopenia (<1000 WBC/mm^3) was seen in two subjects, thrombocytopenia in one, and increased liver function tests (non-sustained) in four. A mild transient coagulopathy was also observed. No effect was seen on CD4 or CD8 populations, and there was no consistently significant change in HIV serum antigen concentrations. Interestingly, two patients developed detectable anti-sCD4 antibodies, but these had no detectable clinical significance.

Pharmacokinetic data was also reported in this study. The initial volume of distribution was equal to the plasma volume and the clearance of the protein was constant over the 300-fold dosing range. The half-life associated with the major phase of drug elimination was, however, only 1 h. A 4 to 5 h terminal half-life was noted for the 100 to 300 µg/kg doses, but this contributed a minor amount of the total drug concentration × time product.

A second phase I clinical trial of recombinant sCD4 has also been published recently.[40] Twenty-five male patients (ages 28 to 61) were entered, and the maximum drug dose received was 30 mg/day × 28 d via IM administration. The drug was well tolerated. Three patients developed opportunistic infections or neoplasms within 4 weeks after cessation of therapy, but no changes were seen in hepatic, renal, or hematologic measurements. One patient with ARC who received drug at the 9 mg/day dose level developed a maculopapular rash which resolved when the drug was discontinued. One patient developed anti-CD4 antibodies which were associated with Coombs positivity but not with hemolysis. These antibodies did not recognize native CD4 expressed by PBL. Anti-CD4 antibodies were also found in four other patients. Once again, as in the previous trial, no changes in the number of CD4 or CD8 cells were seen. However, in the group of patients receiving the highest dose (30 mg/day), the average p24 level fell from 1341 pg/ml to 789 pg/ml, although virus could still be isolated from the PBL of these patients. Significantly, the serum sCD4 levels achieved in these patients (steady state level of 300 pg/ml) were felt to be in the range required to inhibit HIV-1 *in vitro*. This trial also confirmed the extremely short serum half-life of sCD4 (45 min) after IV injection. After IM injection, however, peak serum levels were reached in 4 to 6 h.

In an attempt to increase the serum half-life of the recombinant sCD4, a CD4-Fc fusion protein has been created.[41] The properties of this construct take advantage of a number of factors, including the fact that the Fc fragment of immunoglobulin heavy and light chains has a longer serum half-life than the Fab fragment. Furthermore, this new species was hypothesized to incorporate such functions as Fc receptor binding, complement fixation, and placental transfer, all of which are properties of Fc. Fusion proteins were constructed using the first two domains of CD4 or all four domains. Both heavy and light chain Fc fragments were also utilized. The fusion proteins, termed "immunoadhesins" had similar affinity for gp120 as did normal CD4. A CD4-Fc immunoadhesin blocked viral killing of ATH8 cells with the same potency as sCD4 (0.05 μM) and viral production was also inhibited at day 7 after infection. The compound was also active in a monocyte line. Pharmacokinetic data in rabbits revealed a marked increase in the serum half-life (from 15 min for sCD4 to about 5 h for the fusion protein). However, although the immunoadhesins bound to the Fc receptor, they did not bind the Clq component of complement. In order to circumvent this problem, Traunecker et al.[42] designed a fusion protein in which the V_H and C_{H1} domains of either mouse η2a or μ heavy chains were replaced with the first two N-terminal domains of CD. This construct differs from the one previously described insofar as the former authors[41] retained the C_{H1} domain. The pentameric IgM-CD4 chimera was found to be 1000-fold more active than dimeric IgG-CD4 in syncytium inhibition assays (IC_{50} = 10 ng/ml). Furthermore, the properties of both Fc receptor binding and Clq binding were retained. No clinical trials with this extremely interesting material have yet been reported.

Other investigators[43] have found that small, derivatized CD4 fragments also have the ability to block syncytia formation. The most active fragments appeared to be a dibenzylated 12-residue CD4 peptide (amino acids 83–94). Molecules of this type, as they suggest, might permit the design of anti-HIV agents that are more accessible to tissue compartments than the larger, full-length CD4 molecule.

Another novel approach to the therapy of HIV disease attempted to take advantage of the fact that the HIV envelope glycoprotein gp120 is expressed on the surface of many HIV infected cells. Till et al.[44] constructed a sCD4 conjugate with the potent inhibitor of protein synthesis, ricin A chain. The IC_{50} of this covalent conjugate in infected H9 cells was $1.5 \pm 0.53 \times 10^{-10}$ M, and uninfected cells were not killed at $<5 \times 10^{-8}$ M. The cytotoxic effect of the conjugate could be blocked by monoclonal antibodies directed against the gp120 binding site (e.g., Leu3a) or by sCD4 itself. Interestingly, Daudi cells expressing MHC II, which, as mentioned above is the natural ligand for CD4, were not killed by the conjugate. Chaudary et al.[45] used the first 178 amino acids of CD4 and amino acids 1–3 and 253–613 of Pseudomonas exotoxin A to produce a fusion protein. The toxin segment lacked domain I but retained domains II and III, which are responsible for translocation and ADP-ribosylation of ribosomal elongation factor 2. When the CD4+ cell line 8E5 was treated with this immunotoxin, a marked decrease in protein synthesis was noted, with an ID_{50} of 100 ng/ml. Thus, production of immunotoxins of the type described here may allow for the possibility of the selective killing of HIV-infected cells via interaction with virally derived cell surface molecules.

The use of sCD4 in combination therapy for HIV-1 disease has also recently been suggested.[46] sCD4 and azidodeoxythymidine (AZT), a potent inhibitor of the viral RT, have been examined in a variety of *in vitro* systems. In peripheral blood mononuclear cells, for example, the combination was synergistic at blocking viral replication at concentrations of sCD4 >0.16 µg/ml, and AZT >0.01/µM. In the monocyte line BT4 similar results were seen and concentrations as low as 0.001 µg/ml of sCD4 and 0.01 µM of AZT were synergistic. These results were confirmed by Hayachi et al.[47] who also noted synergy between sCD4 and dideoxynucleotides, which also inhibit HIV replication at the level of RT. The combination of sCD4 and dideoxyinosine appeared to be the most potent. Combinations of this type may, at some point, lead to interesting clinical trials.

V. POLYANIONIC COMPOUNDS ACTIVE AT THE CELL SURFACE

A. DEXTRAN SULFATE AND RELATED MOLECULES

Nakashima et al.[48] published the first report of anti-HIV activity of the polysulfated carbohydrates dextran sulfate (mol. wt. 34,000) xylofuranan sulfate (pentosan polysulfate, mol. wt. 3000), and ribofuranan sulfate (mol. wt. 10,000). They note inhibition of syncytia when Molt4 cells were cocultured with Molt4 cells chronically infected with HIV/HTLV-IIIB. Maximal inhibition was observed at a concentration of 100 µg/ml. In a subsequent report,[49] lentinan sulfate, (mol. wt. about 20,000), which also contains a repeating β-1,3-glycan unit, was also able to inhibit giant cell formation.

Mitsuya et al.[50] noted that dextran sulfate (DS), at a concentration of 6.75 µM, completely inhibited syncytia formation in cocultures of ATH8/H9 (HIV-1) cells. At a concentration of 1 µM, dextran sulfate inhibited the binding of radiolabeled virions to H9 cells. Similar conclusions were reached by Baba et al.,[51,52] who examined the binding of ^{32}P or 5-^{3}H-uridine labeled virus after incubation with MT4 cells. In the presence of heparin or dextran sulfate (mol.

wt. 5000, 25 μg/ml), no cellular uptake of radioactivity was noted, indicating that the drugs prevented the adsorption of HIV-1 to the cell surface. Furthermore, full protective activity was achieved when the compounds were present only during the 2 h virus adsorption period. Schols et al.[53] examined HIV-1 binding to MT4 cells after addition of a high titer polyclonal antibody derived from an ARC patient. They then stained the antibody with a fluorescein-labeled rabbit anti-human Ig F(ab)2 and measured cellular fluorescence by cell sorting. Heparin, pentosan polysulfate, and dextran sulfate caused a decrease in the fluorescence of MT4 cells exposed to the virus (25 μg/ml). Dermatan sulfate, which does not inhibit HIV replication, did not block virus binding. Suramin, at 125 μM, caused incomplete inhibition of HIV-1 binding. The concentration of inhibitors suppressing virus binding was similar to that necessary to overcome the viral cytopathic effect.

While dextran sulfate seemed capable of blocking the binding of virus to cells, it was, at the same time, incapable of blocking the binding of anti-CD4 antibodies (Leu3a, OKT4, OKT4a) to the surface of target CD4+ cells.[51,54] Accordingly, there have been several attempts made to elucidate the precise site of action of dextran sulfate. This question becomes particularly difficult to answer because polysulfated compounds (pentosan, fucoidan, dextran sulfate, heparin) are also potent HIV reverse transcriptase (RT) inhibitors.[52] However, it is unclear whether these compounds penetrate cells, and the inhibition of RT is achieved at much higher concentrations than that required to block either virus adsorption or virus replication.[55] In one set of experiments, Lederman et al.[56] immobilized gp120 on polystyrene surfaces. The binding of sCD4 was measured by an ELISA assay. Both low and higher molecular weight dextran sulfate, heparin, and fucoidin inhibited binding of the sCD4 to the immobilized gp120. If the inhibitors were removed and the plate washed before sCD4 binding, no inhibition of binding was observed. Similarly, if the sCD4 was bound to the plate first, pretreatment with the inhibitors blocked gp120 binding. Thus, the evidence, all together could be interpreted as indicating that dextran sulfate binds to CD4, but not to the HIV-1 binding epitope. However, another set of experiments performed by Schols et al.[57] leads to a different site of inhibitory action. These authors examined a persistently HIV-1 infected gp120+, CD4+ cell line (HUT78), and incubated it with an anti-gp120 antibody recognizing the gp120 epitope responsible for syncytia formation. Both dextran sulfate and pentosan polysulfate inhibited binding of this anti-gp120 antibody (IC_{50} = 0.16 to 0.8 μg/ml). Furthermore, if the cells were pretreated with dextran sulfate, the binding of the antibody was also blocked. Supporting evidence for this mechanism[54] includes experiments which have shown that preincubation of HIV-1 virions in dextran sulfate (mol. wt. 5000) for 14 h caused a 97% reduction in p24 antigen released into the supernatant of *de novo* infected SupT1 cells after 20 to 22 days.

Thus, on the basis of all the available evidence, the precise site of action of dextran sulfate in blocking HIV-1 replication cannot be localized with certainty. Nevertheless, because of its undeniably potent properties, a clinical trial of dextran sulfate in HIV-1 disease has recently been undertaken.[58] Unfortunately, this trial suffered from two serious problems:

1. The drug was given exclusively by the oral route, but the oral bioavailability of dextran sulfate is quite low (<1%).[59]
2. At the time no reproducible assay for serum dextran sulfate concentration had been developed, thus making determination of serum levels impossible.

Thirty-four patients were given dextran sulfate in the molecular weight range 7000 to 8000. The first group received 300 mg TID × 8 weeks, and a second group received escalating doses to a maximum of 5400 mg qd. The oral formulation was, in general, well tolerated, but four patients developed toxicity requiring withdrawal from the study. Mental excitability and insomnia were frequently described. One patient had headache, one became confused and lethargic, but a decrease in the drug dose led to symptom resolution. Leukopenia was seen, as well as increased hepatic transaminases in ten patients. However, this too returned to normal with drug dose reduction. Diarrhea and other gastrointestinal symptoms were seen in 12 individuals. There was no consistent change in CD4 counts or β-microglobulin levels. Furthermore, it has subsequently been determined[59] that plasma concentrations of dextran sulfate of 0.3 to 16 μg/ml, the approximate IC_{50} for HIV-1 replication, will cause significant coagulopathy, with activated partial thromboplastin times (APTT) of 25 to 80 s. Thus, the therapeutic index of dextran sulfate may indeed be very small. A similar problem may not occur for pentosan, which might be more active and less anticoagulating. Trials of intravenously administered dextran sulfate are currently in progress.

Combinations of dextran sulfate with other anti-HIV agents have also been examined *in vitro*.[37] Dextran sulfate is synergistic in combination with either ddC, ddI, or AZT.[60] On the other hand, dextran sulfate and sCD4 are antagonistic at lower DS concentrations (0.125 μM), but weakly synergistic at 0.625 μM. Significantly, none of the combinations were toxic for bone marrow cells.

B. OTHER ANIONIC COMPOUNDS

Glycyrrhizn sulfate,[61] a discrete, sulfated glycosylated steroid, was also found to block syncytia formation in the Molt4/Molt4/HIV/HTLV$_{IIIB}$ system, while glycyrrhizn itself was only about 25% as effective. No mechanistic studies with this compound have as yet been performed. The anti-HIV effect of polyanionic compounds is not, however, confined to sulfated species. Aurintricarboxylic acid (ATA)[62] is a commercially available calcium chelator previously thought to be a triphenylmethane derivative. Recent research by workers at Purdue University indicates that the substance may be a polymeric polycarboxylic acid. Schols et al.[62] have recently shown that this material, in a dose dependent and reversible fashion, blocked the binding of OKT4a to cell surface CD4 (90% inhibition at 10 μM). OKT4 binding was affected to a lesser extent. ATA also blocked the binding of labeled virion to MT4 cells, again presumably by interacting with CD4 and the HIV binding site. The effect was vitiated by addition of 10% fetal calf serum.

C. SULFONIC ACIDS

The anti-HIV activity of compounds of the bis-naphthalene sulfonic acid type has been recognized for some years.[63] For example, the tetra anion Evans Blue (EB) at 100 $\mu g/ml$ was protective of ATH8 cells against the cytopathogenicity of HTLV-III. The compound Direct Yellow 50 was also active, but to a lesser degree. Suramin was about equal in potency to Evans Blue.

Pal et al.[64] have studied the effect of these compounds on the process of formation of multinucleated giant cells, with CEM cells being used as targets for HIV-1 and HIV-2 infected Molt3 cells. Syncytium formation induced by Molt3/HIV-1 was markedly inhibited in the presence of EB, but much less so in the presence of its structural isomer Trypan Blue. EB was shown to have no effect on the synthesis, processing, or secretion of viral glycoprotein. Syncytia induced by HIV-2 infected cells were far less susceptible to inhibition by these compounds. Furthermore, while EB was capable of blocking the infection of CEM cells by cell-free HIV-1 virions, the effect was less pronounced for HIV-2. This occurred in spite of the fact that the RT activity of both viruses were equally inhibited by EB, with nearly 90% inhibition occurring at a concentration of 10 μM.

Eb was incapable of blocking the binding of OKT4 or Leu3a to its cell surface receptor, but at 10 $\mu g/ml$ entirely inhibited virus binding to target cells.[53] Using a polystyrene plate assay similar to that employed by Lederman et al.[56] it could be shown that EB could inhibit the binding of gp120 to CD4 in a dose dependent manner. However, even at a concentration of 80 μM, this effect was only partial. By contrast, Schols et al.[57] noted that EB blocked the binding of an anti-gp120 antibody to gp120. Thus, the precise location of action of EB, as is also true of DS, is at present uncertain. However, the relative lack of toxicity of this compound at therapeutically effective levels *in vitro* makes it an attractive candidate for further development.

Finally, an unusual glycolipid sulfonic acid was recently[65] isolated from cultured cyanobacteria (blue-green algae), and was shown to inhibit HIV-1 induced syncytium formation. This may represent the first of a novel series of anti-HIV compounds.

D. PHOSPHOROTHIOATE OLIGODEOXYNUCLEOTIDES

Phosphorothioate oligodeoxynucleotides are a class of oligodeoxynucleotide in which one of the nonbridging oxygen atoms is replaced by a sulfur. This substitution retains the original aqueous solubility properties of the compound and renders it relatively nuclease resistant.[66] Recently, Matsukura et al.[67] have shown that these materials can be potent sequence-specific inhibitors of HIV replication in chronically infected H9 cells. However, phosphorothioate oligomers are also potent non-sequence specific inhibitors of the cytopathic effect of HIV-1 in *de novo* infected ATH8 cells. The cytoprotective effect was highly dependent on the length of the oligomer, but independent of the oligomer sequence. The most effective compound was a 28-mer homopolymer of cytidine, SdC28. A 15-mer cytidine homopolymer was about as effective as the 28-mer, while a 5-mer was ineffective. The oxygen congener OdC28 also was ineffective. Phosphorothioate oligodeoxynucleotides are potent inhibitors of the viral RT,[68] as are many other polyanions, with the inhibition being competitive with respect to template-primer binding. It is possible that this mechanism is responsible for the observed cytoprotective effect.

Stein et al.,[69] however, incubated sCD4 with SdC28 and found that the protein retarded the mobility of the oligomer on a polyacrylamide gel. SdC28 was also shown to be a potent inhibitor of HIV-1 as well as HIV-2[64] induced syncytia formation. SdC5, on the other hand, was a very weak inhibitor of syncytia formation and did not form a detectable complex with CD4 on a gel. Pal et al.[64] examined HIV-1 induced syncytia formation using the 8E5 cell line. These CD4− cells carry a single copy of the entire HIV-1 genome but produce noninfectious virus particles because of a point mutation in the gene encoding reverse transcriptase. SdC28 (2 μM) was a potent inhibitor of syncytia formation when these cells were cocultured with the target CD4+ SupT1 cells. Incubation with 20 μM OdC28 was not inhibitory. SdC28 also partially inhibited the binding of gp120 to CD4 in the solid phase ELISA binding assay as described above, while OdC28 was not an effective inhibitor in this system. Furthermore, SdC15 and SdC28 reduced FITC-gp120 binding to CD4+ target cells by 80 and 90%, respectively, whereas SdC5 and OdC28 were ineffective. Leu3a binding was also inhibited to a similar extent, while the binding of OKT4, an anti-CD4 antibody that does not interfere with gp120 binding, was reduced by 30 to 50%. The binding of the pan-T cell antibody Leu 9 was only minimally affected by the phosphorothioate oligomer. Thus, phosphorothioate oligomers represent another class of polyanions that appear to inhibit HIV replication by, at least in part, interacting with molecules in the cell membrane.

VI. CONCLUSIONS

Over the past several years, a variety of anti-HIV agents have been developed that appear to act predominantly at the cell membrane of the target cell, and are capable of blocking viral binding and infection. Intuitively, and there is now some supporting laboratory data, such compounds should achieve their maximum therapeutic potential in combination with agents that act at the level of the viral RT, such as AZT or the dideoxynucleosides. Furthermore, the clinical use of such agents interdicts the virus at a point in its life cycle before integration into the host genome occurs. The ultimate utility of this approach will most certainly be tested in those clinical trials already or soon to be underway.

REFERENCES

1. **Popovic, M., Sarngadharan, M. G., Read, E., and Gallo, R. C.**, Detection, isolation and continuous production of cytopathic retroviruses (HTLV-III) from patients with AIDS and pre-AIDS, *Science,* 224, 497–500, 1984.
2. **Gallo, R. C., Salahuddin, S. Z., Popovic, M., Shearer, G. M., Kaplan, M., Haynes, B. F., Palker, T. J., Redfield, R., Oleske, J., Safai, B., White, G., Foster, P., and Markham, P. D.**, Frequent detection and isolation of cytopathic retroviruses (HTLV-III) from patients with AIDS and at risk for AIDS, *Science,* 224, 500–503, 1984.
3. **Barre-Sinoussi, F., Chermann, J.-C., Rey, F., Nugeyre, M. T., Chamaret, S., Gruest, J., Dauguet, C., Axler-Blin, C., Brun-Vezinet, F., Rouzioux, C., Rozenbaum, W., and Montagnier, L.**, Isolation of a T-lymphotropic retrovirus from a patient at risk for acquired immune deficiency syndrom (AIDS), *Science,* 220, 868–871, 1983.
4. **Sarngadharan, M. G. and Markham, P. D.**, in *Acquired Immunodeficiency Syndrome and Other Manifestations of HIV-Infection,* Wormser, Stahl, Bottone, Eds., Noyes, Park Ridge, NJ, 1987, 186.
5. **Klatzmann, D., Barre-Sinoussi, F., Nugeyre, M. T., Dauquet, C., Vilmer, E., Griscelli, C., Brun-Vezinet, F., Rouzioux, C., Gluckman, J., Cherman, J., and Montagnier, L.**, Selective tropism of lymphadenopathy-associated virus (LAV) for helper-inducer T lymphocytes, *Science,* 225, 59–63, 1984.
6. **Lane, H. C., Depper, J. M., Green, W. C., Whalen, G., Waldmann, T. A., and Fauci, A. S.**, Qualitative analysis of immune function in patients with the acquired immunodeficiency syndrome: evidence for a selective defect in soluble antigen recognition, *N. Engl. J. Med.,* 313, 79–84, 1985.
7. **Chirmule, N., Kalyanaraman, V. S., Oyaizu, N., Slade, H. B, and Pahwa, S.**, Inhibition of functional properties of tetanus antigen-specific T-cell clones by enveloped glycoprotein gp120 of Human Immunodeficiency virus, *Blood,* 75, 152–159, 1990.
8. **Ratner, L., Haseltine, W., Patarca, R., Livak, K. J., Starcich, B., Josephs, S. F., Doran, E. R., Rafalski, J. A., Whitehorn, E. A., Baumeister, K., Ivanoff, L., Petteway, S. R., Jr., Pearson, M. L., Lautenberger, J. A., Papas, T. S., Ghrayeb, J., Chang, N. T., Gallo, R. C., and Wong-Staal, F.**, Complete nucleotide sequence of the AIDS virus, HTLV-III, *Nature,* 313, 277–284, 1985.
9. **Sanchez-Pescador, R., Power, M. D., Barr, P. J., Steimer, K. S., Stempien, M. M., Brown-Shimer, S. L., Gee, W. W., Renard, A., Randolph, A., Levy, J. A., Dina, D., and Luciew, P. A.**, Nucleotide sequence and expression of an AIDS-associated retrovirus (ARV-2), *Science,* 227, 484–492, 1985.
10. **Srinivasan, A., Anand, R., York, D., Ranganathan, P., Feorino, P., Schochetman, G., Curran, J., Kalyanaraman, V. S., Luciew, P. A., and Sanchez-Pescador, R.**, Molecular characterization of human immunodeficiency virus from Zaire: nucleotide sequence analysis identifies conserved and variable domains in the enveloped gene, *Gene,* 52, 71–82, 1987.
11. **Mitsuya, H. and Broder, S.**, Strategies for antiviral therapy in AIDS, *Nature,* 325, 773–778, 1987.

12. **Yarchoan, R., Mitsuya, H., Myers, C., and Broder, S.**, Chemical pharmacology of 3'-azido-2', 3'-dideoxy thimidine (Zidovudine) and related dideoxynucleotides, *N. Engl. J. Med.*, 321, 726–738, 1990.
13. **Klatzmann, D., Champagne, E., Chamaret, S., Gruest, J., Guetard, D., Hercend, T., Gluckman, J. C., and Montagnier, L.**, T-lymphocyte T4 molecule behaves as the receptor for human retrovirus LAV, *Nature (London)*, 312, 767–768, 1985.
14. **Dalgleish, A. G., Beverley, P. C. L., Clapham, P. R., Crawford, D. H., Greaves, M. F., and Weiss, R. A.**, The CD4(T4) antigen is an essential component of the receptor for the AIDS retrovirus, *Nature (London)*, 312, 763–767, 1985.
15. **McDougal, J. S., Kennedy, M. S., Sligh, J. M., Cort, S. P., Mawle, A., and Nicholson, J. K. A.**, Binding of HTLV-III/LAV to T4 + T cells by a complex of the 110k viral protein and the T4 molecule, *Science*, 231, 382–385, 1986.
16. **Allan, J. S., Coligan, J. E., Barin, F., Mclane, M. F., Sodroski, J. G., Rosen, C. A., Haseltine, W. A., Lee, T. H., and Essex, M.**, Major glycoprotein antigens that induce antibodies in AIDS patients are encoded by HTLV-III, *Science*, 228, 1091–1094, 1985.
17. **Veronese, F. D., DeVico, A. L., Copeland, T. D., Oroszlan, S., Gallo, R. C., and Sarngadharan, M. G.**, Characterization of gp41 as the transmembrane protein coded by the HTLV-II/LAV envelope gene, *Science*, 229, 1402–1405, 1985.
18. **Mathews, T. J., Weinhold, K. J., Lyerly, H. K., Langlois, A. J., Wigzell, H., and Bolognesi, D. P.**, Interaction between the human T cell lymphotropic virus IIIB envelope glycoprotein gp120 and the surface antigen CD4: role of carbohydrates in binding and cell fusion, *Proc. Natl. Acad. Sci. U.S.A*, 84, 5424–5428, 1987.
19. **Fenouillet, E. B., Clerget-Raslain, B., Gluckman, J. C., Guetard, D., Montagnier, L., and Bahraoui, E.**, Role of N-linked glycans in the interaction between the envelope glycoprotein of human immunodeficiency virus and its CD4 cellular receptor, *J. Exp. Med.*, 169, 807–822, 1989.
20. **Kornfeld, R. and Kornfeld, S.**, Assembly of asparagine-linked oligosaccharides, *Annu. Rev. Biochem.*, 54, 631–664, 1985.
21. **Elbein, A. D.**, Glycosylation inhibitors for N-linked glycoproteins, *Methods Enzymol.*, 138, 661–709, 1987.
22. **Walker, B. D., Kowalski, M., Goh, W. C., Kozarsky, K., Krieger, M., Rosen, C., Rohrschneider, L., Haseltine, W. A., and Sodroski, J.**, Inhibitors of human immunodeficiency virus syncytium formation and virus replication by castanospermine, *Proc. Natl. Acad. Sci. U.S.A.*, 84, 8120–8124 1987.
23. **Gruters, R. A., Neefjes, J. J., Tersmette, M., de Goede, R. E., Tulp, A., Huisman, H. G., Miedema, F., and Ploegh, H. L.**, Interference with HIV-induced syncytium formation and viral infectivity by inhibitors of trimming glucosidase, *Nature (London)*, 330, 74–77, 1987.
24. **Pal, R., Kalyanaraman, V., Hoke, G., and Sarngadharan, M. G.**, Processing and secretion of evelope glycoproteins of human immunodeficiency virus type 1 in the presence of trimming glucosidase inhibitor deoxynojirimycin, *Intervirology*, 30, 27–35, 1989.
25. **Pal, R., Hoke, G. M., and Sarngadharan, M. G.**, Role of oligosaccharides in the processing and maturation of envelope glycoprotein of human immunodeficiency virus type 1 (HIV 1), *Proc. Natl. Acad. Sci. U.S.A.*, 86, 3384–3388, 1985.
26. **Montefiori, D. C., Robinson, W. E., Jr., and Mitchell, H. M.**, Role of protein N-glycosylation in the pathogenesis of human immunodeficiency virus type 1, *Proc. Natl. Acad. Sci. U.S.A.*, 85, 9248–9252, 1983.
27. **Pal, R., Reitz, M. S., Jr., Tschachler, E., Gallo, R. C., Sarngadharan, M. G., and Veronese, F. D.**, Myristoylation of gag proteins of HIV-1 plays an important role in virus assembly, *AIDS Res. Hum. Retroviruses*, 6, 721–730, 1990.
28. **Gottlinger, H. G., Sodroski, J. G., and Haseltine, W. A.**, Role of capsid precurrence processing and infectivity of human immunodeficiency virus type 1, *Proc. Natl. Acad. Sci. U.S.A.*, 86, 5781–5786.

29. **Tashiro, A., Shoji, S., and Kubota, Y.**, Antimyristoylation of the *gag* proteins in the human immunodeficiency virus infected cells with N-myristoyl glycinal diethylacetal results in inhibition of virus production, *Biochem. Biophys. Res. Commun.*, 165, 1145–1154, 1989.
30. **Bryant, M. L., Heuckeroth, R. O., Kimata, J. T., Ratner, L., and Gordon, J. I.**, Replication of human immunodeficiency virus 1 and moloney murine leukemia virus is inhibited by different heteroatom-containing analogs of myristic acid, *Proc. Natl. Acad. Sci. U.S.A.*, 86, 8655–8659, 1989.
31. **Smith, D. H., Byrn, R., Marsters, S., Gregory, T., Groopman, J., and Capon, D.**, Blocking of HIV-1 infectivity by a soluble, secreted form of the CD4 antigen, *Science*, 238, 1704–1707, 1987.
32. **Deen, K., McDougal, J. S., Inacker, R., Folena-Wasserman, G., Arthos, J., Rosenberg, J., Maddon, P. J., Axel, R., and Sweet, R.**, A soluble molecule of CD4 (T4) inhibits AIDS virus infection, *Nature*, 331, 82–84, 1988.
33. **Traunecker, A., Luke, W., and Karjaiainen, K.**, Soluble CD4 molecules neutralized human immunodeficiency virus type 1, *Nature*, 331, 84–86, 1988.
34. **Fisher, R., Bertonsi, J., Meier, W., Johnson, V., Costopoulos, D., Liu, T., Tizare, R., Walder, B., Hirsch, M., Schooley, R., and Flavell, R.**, HIV infection is blocked *in vitro* by recombinant soluble CD4, *Nature*, 331, 76–78, 1988.
35. **Hussey, R., Richardson, N., Kowalski, M., Brown, N., Chang, H. C., Siliciano, R., Dorfman, T., Walker, B., Sodrowski, J., and Reinherz, E.**, A soluble CD4 protein selectively inhibits HIV replication and syncytium formation, *Nature*, 331, 78–81, 1988.
36. **Manca, F., Habeshaw, J. A., and Dalgleish, A.**, HIV envelope glycoprotein, antigen specific T-cell responses, and soluble CD4, *Lancet*, 335, 811–815, 1990.
37. **Clapham, P. R., Weber, J., Whitby, D., McIntosh, K., Dalgleish, A., Maddon, P., Deen, K., Sweet, R., and Weiss, R.**, Soluble CD4 blocks the infectivity of diverse strains of HIV and SIV for T cells and monocytes but not for brain and muscle cells, *Nature*, 337, 368–370, 1989.
38. **Watanabe, M., Reimann, K., DeLong, P., Liu, T., Fisher, R., and Letvin, N.**, Effect of recombinant soluble CD4 in rhesus monkeys infected with simian immunodeficiency virus of macaques, *Nature*, 337, 267–270, 1989.
39. **Kahn, J. O., Allan, J. D., Hodges, T., Kaplan, L., Arri, C., Fitch, H., Izu, A., Mordenti, J., Sherwin, S., Groopman, J., and Volberding, P.**, The safety and pharmacokinetics of recombinant soluble CD4 (rCD4) in subjects with the acquired immunodeficiency syndrome (AIDS) and AIDS-related complex A phase I study, *Ann. Intern. Med.*, 112, 254–261, 1990.
40. **Schooley, R., Merigan, T., Gaut, P., Hirsch, M., Holodniy, M., Flynn, T., Liu, S., Byington, R., Henochowicz, S., Gubish, E., Spriggs, D., Kufe, D., Schindler, J., Dawson, A., Thomas, D., Hanson, D., Letwin, B., Liu, T., Gulinello, H., Kennedy, S., Fisher, R., and Ho, D.**, Recombinant soluble CD4 therapy in patients with the acquired immunodeficiency syndrome (AIDS) and AIDS-related complex: a phase I–II escalating dosage trail, *Ann. Intern. Med.*, 112, 247–253, 1990.
41. **Capon, D., Chamow, S., Mordenti, J., Master, S., Gregory, T., Mitsuya, H., Byrn, R., Lucas, C., Wurm, F., Groopman, J., Broder, S., and Smith, D.**, Designing CD4 immunoadhesions for AIDS therapy, *Nature*, 337, 525–531, 1989.
42. **Traunecker, A. Schneider, J., Kiefer, H., and Karjalainen, K.**, Highly efficient neutralization of HIV with recombinant CD4-immunoglobulin molecules, *Nature*, 339, 68–70, 1989.
43. **Lifson, J., Hwang, K. M., Nara, P., Fraser, B., Padget, M., Dunlop, N., and Eiden, L.**, Synthetic CD4 peptide derivatives that inhibit HIV infection and cytopathicity, *Science*, 241, 712–715, 1988.
44. **Till, M., Ghetie, V., Gregory, T., Patzer, E., Porter, J., Uhr, J., Capon, D., and Vitetta, E.**, HIV-infected cells are killed by rCD4-Ricin A chain, *Science*, 242, 1166–1168, 1988.

45. Chaudhary, V., Mizukami, T., Fuerst, T., FitzGerald, D., Moss, B., Pastan, I., and Berger, E., Selective killing of HIV-infected cells by recombinant human CD4-Pseudomonas exotoxin hybrid protein, *Nature*, 335, 369–372, 1988.
46. Johnson, V., Barlow, M., Chou, T. -C., Risher, R., Walder, B., Hirsch, M., and Schooley, R., Synergistic inhibition of human immunodeficiency virus type I (HIV-1) replication *in vitro* by recombinant soluble CD4 and 3-azido-3-deoxythymidine, *J. Infect. Dis.*, 159, 837–844, 1989.
47. Hayashi, S., Fine, R., Chou, T. -C., Currens, K. M., Border, S., and Mitsuya, H., In vitro inhibition of the infectivity and replication of human immunodeficiency virus type 1 by combination of antiretroviral 2′ 3′-dideoxynucleosides and virus-binding inhibitors, *Antimicrob. Agents Chemother.*, 34, 82–88, 1990.
48. Nakashima, H., Yoshida, O., Tochikura, T., Yoshida, T., Mimura, T., Kido, Y., Motoki, Y., Kaneko, Y., Uryu, T., and Yamamoto, N., Sulfation of polysaccharides generates potent and selective inhibitors of human immunodeficiency virus infection and replication, *in vitro, Jpn. J. Cancer Res.*, 78, 1164–1168, 1987.
49. Yoshida, O., Nakashima, H., Yoshida, T., Kaneko, Y., Yamamoto, I., Matsuzaki, K., Uryu, T., and Yamamoto, N., Sulfation of the immunomodulating polysaccharide lentinan: a novel strategy for antivirals to human immunodeficiency virus (HIV), *Biochem. Pharmacol.*, 37, 2887–2891, 1988.
50. Mitsuya, H., Looney, D., Kuno, S., Veno, R., Wong-Staal, F., and Broder, S., Dextran sulfate suppression of viruses in the HIV family: inhibition of virion binding to CD4+ cells, *Science*, 240, 646–648, 1988.
51. Baba, M., Pauwels, R., Balzarini, J., Arnout, J., Desmyter, J., and De Clercq, E., Mechanism of inhibitory effect of dextran sulfate and heparin on replication of human immunodeficiency virus *in vitro, Proc. Nat . Acad. Sci. U.S.A.*, 85, 6132–6136, 1988.
52. Baba, M., Nakajima, M., Schols, D., Pauwels, R., Balzarini, J., and De Clercq, E., Pentosan polysulfate, a sulfated oligosaccharide is a potent and selective anti-HIV agent *in vitro, Antiviral Res.*, 8, 335–343, 1988.
53. Schols, D., Baba, M., Pauwels, R., and De Clercq, E., Flow cytometric method to demonstrate whether anti-HIV-1 agents inhibit virion binding to T4+ cells, *J. AIDS*, 2, 10–15, 1989.
54. Bagasra, O. and Lischner, H., Activity of dextran sulfate and other polyanioic polysaccharides against human immunodeficiency virus, *J. AIDS*, 2, 10–15, 1988.
55. Nakashima, H., Yoshida, O., Baba, M., De Clercq, E., and Yamamoto, N., Anti-HIV activity of dextran sulphate as determined under different experimental conditions, *Antiviral Res.*, 11, 233–246, 1989.
56. Lederman, S., Gulick, R., and Chess, L., Dextran sulfate and heparin interact with CD4 molecules to inhibit the binding of coat protein (gp 120) of HIV, *J. Immunol.*, 143, 1149–1154, 1989.
57. Schols, D., Pauwels, R., Desmyter, J., and De Clercq, E., Dextran sulfate and other polyanioic anti-HIV compounds specifically interact with the viral gp 120 glycoprotein expressed by T-cells persistently infected with HIV-1, *Virology*, 175, 556–561, 1990.
58. Abrams, D., Kuno, S., Wong, R., Jefford, R., Nash, M., Lolaghan, J. B., Gorter, R., and Veno, R., Oral dextran sulfate (UA001) in the treatment of the acquired immunodeficiency syndrome (AIDS) and AIDS-related complex, *Ann Intern. Med.*, 110, 183–188, 1989.
59. Lorentsen, K., Hendrix, C., Collins, J., Kornhauser, D., Petty, B., Klecker, R., Flexner, C., Eckel, R., and Lietman, P., Dextran sulfate is poorly absorbed after oral administration, *Ann. Intern. Med.*, 111, 561–566, 1989.
60. Veno, R. and Kuno, S., Dextran sulphate, a potent anti-HIV agent *in vitro* having synergism with zidovudine, *Lancet*, ii, 1379 1987.
61. Nakashima, H., Matsui, T., Yoshida, O., Isowa, Y., Kido, Y., Motoki, Y., Ito, M., Shigeta, S., Mori, T., and Yamamoto, N., A new anti-human immunodeficiency virus substance, glycyrrhizin sulfate, endowment of glycyrrhizin with reverse transcriptase-inhibitory activity by chemical modification, *Jpn. J. Cancer Res.*, 78, 767–771, 1987.

62. **Schols, D., Baba, M., Pauwels, R., Desmyter, J., and De Clercq, E.,** Specific interaction of aurintricarboxylic acid with the human immunodeficiency virus/CD4 cell receptor, *Proc. Natl. Acad. Sci. U.S.A.,* 86, 3322–3326, 1989.
63. **Balzarini, J., Mitsuya, M., De Clercq, E., and Broder, S.,** Comparative inhibitor effects of suramin and other selected compounds on the infectivity and replication of human T-cell lymphotrophic virus (HTLV-III)/lymphadenopathy-associated virus (LAV), *Int. J. Cancer.,* 37, 451–457, 1986.
64. **Pal, R., Mumbauer, S., Hoke, G., LaRocca, R., Myers, C., Sarngadharan, M., and Stein, C. A.,** Comparative effects of polyanionic compounds on syncytia formation and infectivity of human immunodeficiency virus type I and type II *in vitro,* submitted.
65. **Gustafuson, K. R., Cardellina, J., Fuller, R., Weislow, O., Kiser, R., Snader, K., Patterson, G., and Boyd, M.,** AIDS-antiviral sulfolipids from cyanobacteria (blue-green algae), *J. Natl. Cancer Inst.,* 81, 1254–1258, 1989.
66. **Stein, C. A., Subasinghe, C., Shinozuka, K., and Cohen, J.,** Physicochemical properties of phosphorothioate oligodeoxynucleotides, *Nucleic Acids. Res.,* 16, 3209–3221, 1988.
67. **Matsukura, M., Zon, G., Shinozuka, K., Robert-Guroff, M., Shimada, T., Stein, C. A., Mitsuya, H., Wong-Staal, R., Cohen, J., and Broder, S.,** Regulation of viral expression of human immunodeficiency virus *in vitro* by an antisense phosphorothioate oligodeoxynucleotide against rev (art/trs) in chronically infected cells, *Proc. Natl. Acad. Sci. U.S.A.,* 86, 4244–4249, 1989.
68. **Majumdar, C., Stein, C. A., Cohen, J., Broder, S., and Wilson, S.,** Stepwise mechanism of HIV reverse transcriptase: primer function of phosphorothioate oligodeoxynucleotide, *Biochemistry,* 28, 1340–1346, 1989.
69. **Stein, C. A., Pal, R., Nair, B. C., Mumbauer, S., Hoke, G., and Neckers, L.,** Phosphorothioate oligodeoxycytidine inhibits binding of HIV-1 gp 120 to CD4, submitted.

Chapter 15

PROSTAGLANDIN DERIVATIVES AS CHEMOTHERAPEUTIC AGENTS IN CANCER, AIDS, AND MALARIA

S. Tsuyoshi Ohnishi

TABLE OF CONTENTS

I.	Introduction	356
II.	Compounds	356
III.	Cancer Chemotherapy	356
	A. Anti-Cancer Activity in Mouse Models	356
	B. Reduction of Toxic Side Effects of 5-FU by OC-5186	358
	C. Anti-Cancer Activity of PGOs in a Chemically Induced Cancer	359
IV.	Chemotherapy in Retroviral Diseases	359
	A. Efficacy Against HIV	359
	B. Effect of OC-5186 Against Mouse Leukemia Virus	360
	C. Protection Against AZT Toxicity	361
V.	Malaria	362
VI.	Discussion	363
References		364

ISBN 0-8493-8091-X
© 1993 by CRC Press, Inc.

I. INTRODUCTION

Prostaglandin oligomeric derivatives (PGOs) have been known to have various physiological activities.[1-12] It has been known that PGOs do not have the hormonal activities which monomeric prostaglandin compounds have.[1-3] The activities of PGOs in protecting tissues and organs against ischemia and trauma were the focus of the original studies of these compounds.[1-10] The mechanism of this protective action may be related to their antioxidant activity.[11,12] We synthesized several ester compounds. As compared with free-acid form oligomers, their esterified derivatives were found to have greater efficacy. Some of the ester compounds have greater affinity with membranes,[5-9,12] while others contain the ester moiety which could provide additional antioxidant activities.[11,12]

Besides protective effects in ischemia, PGOs were found to have other beneficial effects; they could be used as chemotherapeutic agents in cancer,[13-14] AIDS, or malaria.[9,15-17] This article will review these new aspects of the uses of the compounds.

II. COMPOUNDS

PGOs were synthesized from prostaglandin B_2 (PGB_2) and prostaglandin E1 (PGE_1) as the starting monomers. In PGB_2 oligomers, the free acid was called OC-2186. The acetoxymethyl ester-form and ascorbyl ester-form were called OC-3186 and OC-5186, respectively.[11,12] In PGE_1 oligomers, the free acid-form was called MR-256 and acetoxymethyl ester-form was called MR-356[6,7] (Figure 1).

III. CANCER CHEMOTHERAPY

A. ANTI-CANCER ACTIVITY IN MOUSE MODELS

We used two mouse models developed by Tomomatsu et al.:[18] Ehrlich's ascites tumor (EAT) model and a solid tumor model in which EAT cells were innoculated subcutaneously. In the EAT model 10^7 tumor cells were injected IP into each mouse (average body weight of 20 g). In the solid tumor model the same amount of tumor cells were injected subcutaneously into mice of similar weight. After 21 days, we measured several parameters: (1) the change of body weight; (2) average time to their death, from which we measured extension of life span, and (3) survival rate.[14]

In the ascites tumor model, an increase of body weight is related to the increase in ascites. When the animal dies, we collect the ascites fluid and measure the number of tumor cells found in the ascites. In the solid tumor model, we measured the extension of the life span and the survival rate. When the animal died, we measured the weight of the solid tumor (Table 1). In experiment 1, from day 4 until day 13 (for 10 days), anti-cancer drugs were

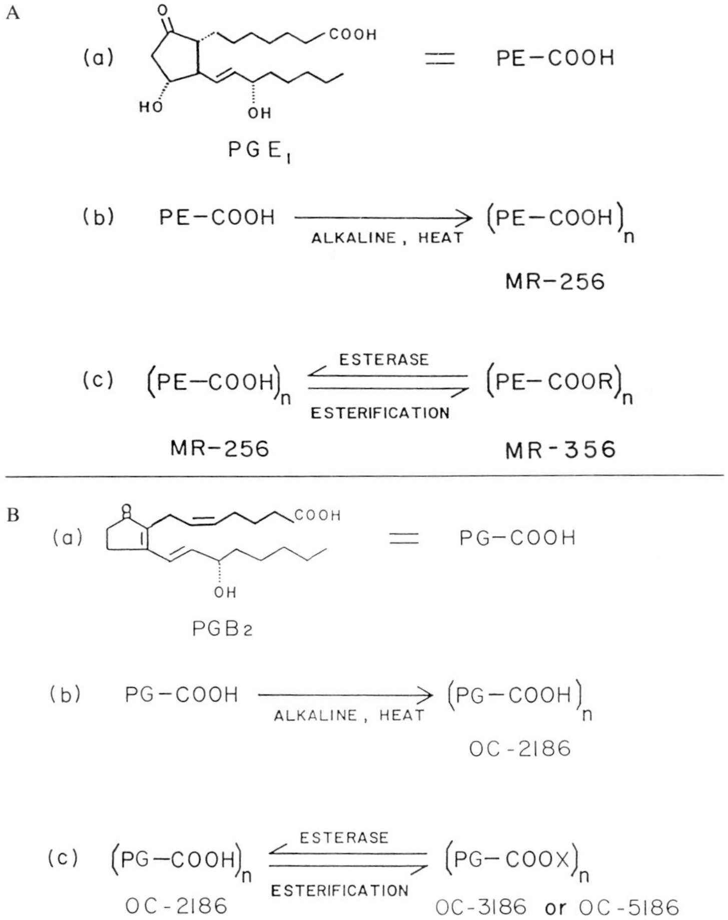

FIGURE 1. (A) Synthesis of MR-356 from PGE_1. Numbers in the structure (a) indicate the carbon position. (B) Synthesis of OC-3186 and OC-5186 from PGB_2.

administered. We used OC-5186 at a dose of 10 mg/kg/day and 5-fluorouracil (5-FU: a widely used anti-cancer agent) at a dose of 20 mg/kg/day. In the combination therapy, both drugs were administered simultaneously at their respective doses. In the EAT model, untreated control animals died in an average of 17 days, whereas animals with OC-5186 (10 mg/kg/day) died in an average of 19.7 days. With the 5-FU, they died in an average of 20.3 days. The survival rate of untreated control mice was 25%, that for OC-5186 was 33.3%, and that for 5-FU was 40%. When animals were treated with both OC-5186 and 5-FU, the average life span was extended to 20.8 days, and the survival rate was improved to 75% (Table 1). In the tumor model, a

TABLE 1
Results of Experiment 1 on Ehrlich's Ascites Tumor (EAT) Model and Solid Tumor Model

	EAT model		Solid tumor model
	Average survival (days)	21-day survival (%)	Solid tumor weight (g) (on 21 days)
Control	17.0	25	1.2
OC-5186 (10 mg/kg/day)	19.7	33.3	1.0
5-FU (20 mg/kg/day)	20.3	40	0.70
OC-5186 and 5-FU	20.8	75	0.31

TABLE 2
Results of Experiment 2 on Ehrlich's Ascites Tumor (EAT) Model and Solid Tumor Model

	EAT model		Solid tumor model	
	Average survival (days)	21-day survival (%)	Average survival (days)	21-day survival (%)
Control	15.3	0	23.2	25
OC-5186 (1 mg/kg/day)	17.0	0	23.9	30
5-FU (20 mg/kg/day)	20.3	40	26	40
OC-5186 and 5-FU	25.5	50	30.5	75

similar tendency was observed. It is also noted that the growth of the solid tumor was also suppressed with this regimen (Table 1).

In experiment 2, the amount of the oligomer was cut down to 1 mg/kg/day to see whether we could still observe anti-cancer activity. OC-5186 and/or 5-FU (20 mg/kg/day) were administered for 23 days. Even at this low dose, OC-5186 still demonstrated its efficacy. Especially in the combined therapy (OC-5186 plus 5-FU), the average life span in the EAT model was extended to 25.5 days and the survival rate was again improved from 0 to 50% (Table 2). In the solid tumor model the combination therapy was also effective (Table 2). These results indicated a synergism between the two agents in the combination therapy.

B. REDUCTION OF TOXIC SIDE EFFECTS OF 5-FU BY OC-5186

It should be emphasized that in the combination therapy in experiment 2, both OC-5186 and 5-FU were administered for 23 days, but most of the animals survived the chemotherapy. Without OC-5186, 5-FU cannot be administered for longer than 10 days because of the toxicity of 5-FU; all animals die if it is administered for 23 days. This suggests that OC-5186 protects animals against the lethal toxicity of 5-FU.

C. ANTI-CANCER ACTIVITY OF PGOS IN A CHEMICALLY INDUCED CANCER

A tongue cancer was produced in the hamster by painting 9,10-dimethyl-1,2-benzanthracene (DMBA) three times a week at the lateral border of the middle third of the tongue by the method of Fujita et al.[19] After 10 weeks, grossly carcinomatous lesions grew (of the size larger than 10 mm). After 17 weeks, the treatment was stopped for 10 days, but the carcinoma did not recede. Then the animals were given 5 mg/kg of OC-5186 IP once a day for 4 days. The size of the carcionma decreased by more than 60% in 4 days.[13]

D. EFFECTS OF MULTI-DRUG RESISTANCE

Multi-drug resistance is a serious problem in cancer chemotherapy. It has been known that when the cells became resistant to one type of chemotherapeutic drug, they also acquire resistance to other drugs, which are even structurally unrelated. Several mechanisms have been proposed for this multi-drug resistance. It was found that a membrane glycoprotein with a molecular weight of 170,000 Da (P170) is expressed, and that the protein is able to transport out drugs using the energy of ATP.[20-22] However, this is not the only mechanism. It was shown that the membranes of resistance cells are less permeable to drugs.[23] Therefore, if the drug permeability could be restored, the drug might regain effect. In collaboration with Dr. Leonard Warren of the Wistar Institute, we have examined the question of whether or not PGOs could be used to solve this problem.

We used human lymphoblastic leukemia cells (CEM) and measured the influx of radioactive H-3 vinblastin during a 10 min exposure. In the drug-sensitive cells, the influx was 5200 cpm, while in the drug-resistant cells it was 510 cpm. However, when 0.2 and 1.0 mg/ml of PGOs were used in conjunction with vinblastin, the influxes were restored to 1580 and 5300 cpm, respectively. In the drug-sensitive cells, PGOs did not change the permeability. This suggests the possibility that PGOs might be used in chemotherapy as an adjunctive agent to reverse multi-drug resistance. It is interesting to note in this connection that PGOs were found to reverse drug resistance in mouse malaria (see Section V).

IV. CHEMOTHERAPY IN RETROVIRAL DISEASES

A. EFFICACY AGAINST HIV

We measured the effect of OC-5186 against HIV-1 using an *in vitro* tissue culture model described by Nakashima et al.[24] MT-4 cells, a human T4-positive cell line carrying HTLV-I, were exposed to HIV at a multiplicity of infection (MOI) of 0.002 and incubated for 60 min at 37°C. After virus

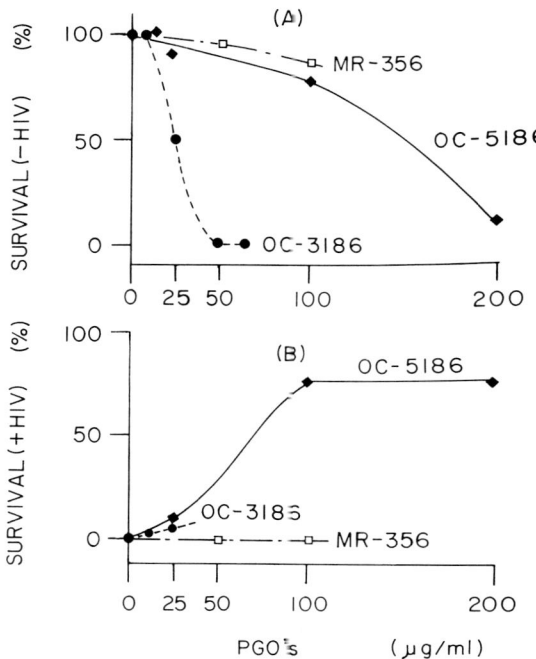

FIGURE 2. (A) Cytotoxicity of PGOs. The experiments were done in the absence of HIV-1. (B) Protection of cells against the infection of HIV-1. Measurements were done on the 6th day of incubation. See text for details.

adsorption, the cells were washed and adjusted to 6×10^5 cells/ml with RPMI 1640 medium supplemented with 10% fetal calf serum and antibiotics. The infected cell suspension was mixed with the same medium containing various concentrations of OC-5186 with a 1:1 ratio. It was then cultured in a CO_2 incubator. On the third day, half of the medium was exchanged with a medium containing the same amount of drug. On the sixth day, the number of viable cells was determined by the trypan blue dye exclusion method. The toxicity of the drug was measured by the same incubation method using uninfected MT-4 cells.

The dose-response relationship of the effects of MR-356, OC-3186, and OC-5186 on non-infected MT-4 cells is shown in Figure 2A, and their effect on HIV-infected MT-4 cells is demonstrated in Figure 2B. As shown in these figures, OC-5186 had a beneficial effect. At 100 µg/ml, the toxicity was still small. However, it improved the survival of infected-cells by six times.

B. EFFECT OF OC-5186 AGAINST MOUSE LEUKEMIA VIRUS

We employed a mouse model infected by RLV (retrovirus similar to HIV), since Ruprecht et al. demonstrated the usefulness of this model in the eval-

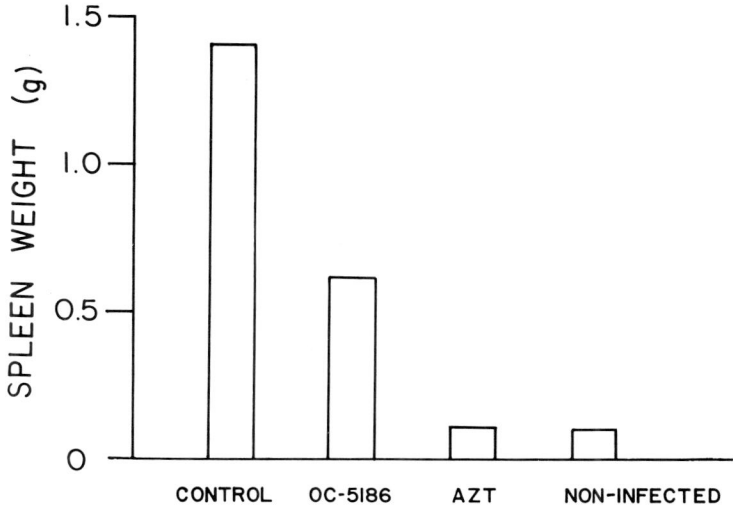

FIGURE 3. Effects of OC-5186 (20 mg/kg/day) and AZT (33 mg/kg/day) on the spleen weight of mice infected by RLV. The column on the extreme right shows the spleen weight of non-infected mice. Averages of 4 to 5 mice.

uation of anti-AIDS chemotherapy.[25] Currently, zidovudine (3′-azido-3′-deoxythymidine or abbreviated as AZT) is the most effective anti-AIDS agent. Therefore, we compared the efficacy of OC-5186 with that of AZT.

RLV derived from the strain RVB3 was used. Female Balb/C mice (10 to 12 g) were purchased from Charles River Laboratories (Wilmington, MA). Saline or 10,000 plaque-forming units of RLV were given on day 0. Four hours after inoculation, AZT was made available through drinking water at a concentration of 0.1 mg/ml. Mice had free access to drinking water and food. They drank an average of 4 ml daily which indicates that the approximate AZT intake was 33 mg/kg/day.

Since RLV is known to cause splenomegaly, and the spleen weight on day 20 has been used to evaluate the viral titer, we sacrificed animals on day 20 to measure spleen and body weights. As shown in Figure 3, OC-5186 suppressed the splenomegaly by 60%, while AZT almost completely inhibited it. Similar to the case with AIDS, OC-5186 has a beneficial effect in RLV.

C. PROTECTION AGAINST AZT TOXICITY

It has been recognized that, in long term administration, AZT causes profound hematologic effects, especially anemia and leukopenia.[25-27] Therefore, a drug which could prevent AZT toxicity has been desired. It was found that OC-5186 prevented AZT toxicity. Figure 4 indicates an experiment in which three groups of virus-infected mice were observed. In the no-drug group, the weight increase was 3.6 g in 20 days. In contrast, the increase in

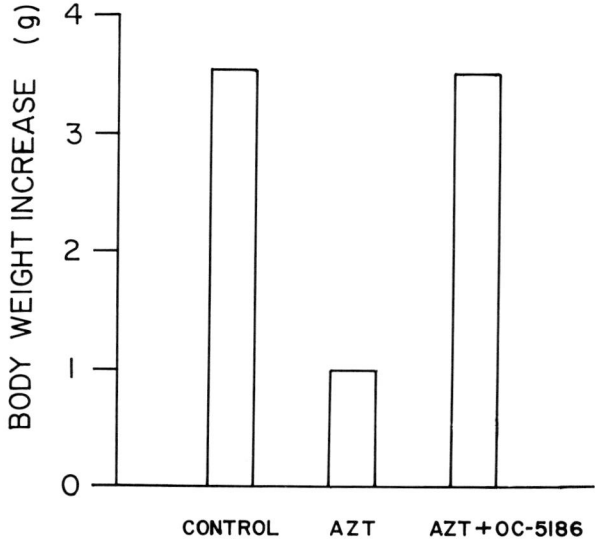

FIGURE 4. Effects of AZT (33 mg/kg/day) and simultaneous addition of AZT (the same dose) plus OC-5186 (20 mg/kg/day) on the increase of body weight of mice. Average of four mice were used per group.

the AZT-administered group was inhibited to 1.0 g. However, with simultaneous administration of OC-5186, the increase was restored to 3.5 g. OC-5186 almost completely eliminated the toxicity of AZT (Figure 4).

Thus, OC-5186 should be considered for use as an adjunct in anti-HIV chemotherapy. We expect that the simultaneous use of this drug with AZT would produce synergistic effects, namely, more efficacy and less toxic side effects.

V. MALARIA

Malaria has become an increasingly serious health problem in tropical developing countries. When the massive use of the insecticide, DDT, started in the late 1950s, and when chloroquine was developed as an effective antimalarial drug, it was thought for a while that malaria had been eradicated. In Sri Lanka, only 17 cases were reported in 1963. However, an epidemic erupted affecting millions of people in 1968.[28] The causes were that the mosquito which carries the malarial parasites became resistant to insecticides, and the parasites acquired resistance against chloroquine. The drug resistance in malaria is similar to that in cancer. Once a strain becomes resistant to a particular drug, it becomes resistant to other drugs, which even have unrelated structures. Today, the situation for malaria is even worse than it was in the 1950s.

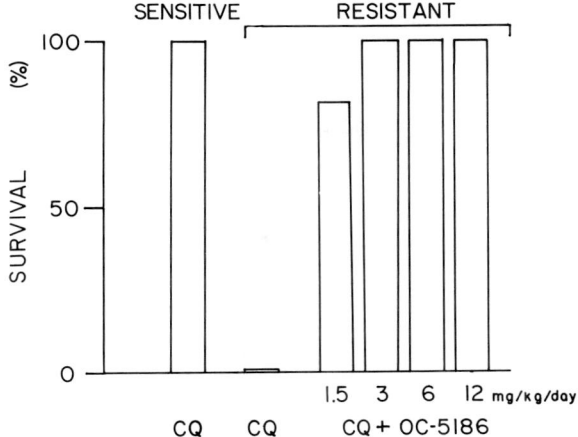

FIGURE 5. Reversal of chloroquine resistance in mice malaria. In each group, five mice were infected on day 0 by either chloroquine-sensitive or -resistant strain of *P. berghei*. Survival of mice was measured on the 13th day of infection. If mice were infected with the sensitive strain, 15 mg/kg/day of chloroquine for four days completely protected the mice. If mice were infected with chloroquine-resistant parasites, chloroquine (the same dose) did not show any efficacy. However, in this case, if 3 to 12 mg/kg/day of OC-5186 for four days was added together with chloroquine (15 mg/kg/day), it completely reversed the drug resistance.

In an attempt to develop a new approach to this problem, we started to study the efficacy of PGOs against malarial parasites. It was found that PGOs could prevent parasitemia in mouse malaria.[9] However, the inhibitory activity is not strong enough to allow the use of a PGO as a single chemotherapeutic agent. Recently, with the collaboration of Dr. B.N. Dhawan (Lucknow, India), we discovered that OC-5186 can reverse the drug resistance of parasites at a concentration so low that it has no effect by itself.[17] Figure 5 shows an experiment in which chloroquine-sensitive and -resistant strains of *Plasmodium berghei* were used. On day 0, mice were infected with 10^6 parasitized erythrocytes. When no chemotherapy was given, both strains killed all mice within 13 days. When the mice were infected with a sensitive strain, chloroquine (15 mg/kg/day) resulted in 100% survival.

At the same dose, chloroquine had no beneficial effect when mice were infected with a resistant strain. However, in this case, if 3 to 12 mg/kg/day of OC-5186 was administered in conjunction with chloroquine, the mice were completely protected (Figure 5).

VI. DISCUSSION

Originally we investigated PGOs as membrane protecting agents against ischemia and trauma. However, recent studies revealed that PGOs are also beneficial as adjuncts in the chemotherapy in cancer, AIDS, and malaria.

Although the precise mechanism of action is not known at this time, several of the possibilities could be the following:

1. Inhibition of proteases and phospholipiases.[5]
2. Chelation of calcium ions and other bivalent metal ions.[4,29] It has been known that chelation of some metal ions might have beneficial effects in viral diseases.[30]
3. Antioxidant activity.[11,12] Since oxygen free radicals are known to react with membrane proteins and lipids, and destroy their physiological activity, the antioxidant activity of PGOs could play important roles in protecting cells and cell membranes.
4. Intrinsic free radicals. PGOs were shown to have intrinsic free radicals.[1,12] This may be related to some of their pharmacologic activities.
5. Modification of physicochemical properties of membranes. Any compounds, which have a strong affinity with lipids, could modify membrane properties. Thus, PGOs may be expected to do the same.

In summary, it seems important to further pursue the roles of PGOs as adjunctive chemotherapeutic agents in various diseases.

REFERENCES

1. **Polis, B. D., Polis, E., and Kwong, S.**, Protection and reactivation of oxydative phosphorylation in mitochondria by a stable free-radical prostaglandin polymer (PGBx), *Proc. Natl. Acad. Sci. U.S.A.*, 76, 1598–1602, 1979.
2. **Angelakos, E. T., Riley, R. L., and Polis, B. D.**, Recovery of monkeys after myocardial infarction with ventricular fibrillation: effects of PGBx, *Physiol. Chem. Phys.*, 12, 81–96, 1980.
3. **Kolata, R. J. and Polis, B. D.**, Facilitation of recovery from ischemic brain damage in rabbits by polymeric prostaglandin PGBx, a mitochondrial protective agent, *Physiol. Chem. Phys.*, 12, 545–550, 1980.
4. **Ohnishi, S. T. and Devlin, T. M.**, Calcium ionophore activity of a prostaglandin B1 derivative (PGBx), *Biochem. Biophys. Res. Commun.*, 89, 240–245, (1979).
5. **Ohnishi, S. T., Katsuoka, M., and Hidaka, S.**, A prostaglandin oligomeric derivative inhibits activities of phospholipase and protease: a possible mechanism of membrane protection during ischemia, *Cell Biochem. Function*, 7, 51–55, 1989.
6. **Ohnishi, S. T., Barr, J. K., Katagi, C., and Katsuoka, M.**, Protection of rat spinal cord against contusion injury by new prostaglandin derivatives, *Arzneimittel-Forschung Drug Res.*, 39, 236–239, 1989.
7. **Ohnishi, S. T. and Katsuoka, M.**, Protection of the heart from ischemia by a new oligomeric derivative of prostaglandin E_1, *Prostagl. Leukotr. Essential Fatty Acids*, 36, 159–163, 1989.
8. **Ohnishi, S. T., Tominaga, T., and Katsuoka, M.**, Inhibition of ischemic brain edema formation by post-ischemic administration of a prostaglandin oligomer, *Prostagl. Leukotr. Essential Fatty Acids*, 37: 107–111 (1989).

9. **Ohnishi, S. T., Sadanaga, K., Katsuoka, M., and Weidanz, W. P.**, Effect of membrane-acting drugs on plasmodium species and sickle cell erythrocytes, *Mol. Cell. Biochem.*, 91, 159–165, 1989.
10. **Kurata, M., Okuda, M., Muneyuki, M. ,and Ohnishi, S. T.**, ^{31}P-MRS study of the protective effects of prostaglandin oligomers on forebrain ischemia in rats, *Brain Res.*, 545, 315–318, 1991.
11. **Sakamoto, A., Ohnishi, S. T., Ohnishi, T., and Ogawa, R.**, Protective effect of a new anti-oxidant on the rat brain exposed to ischemia-reperfusion injury: inhibition of free radical formation and lipid peroxidation, *Free Radical Biol. Med.*, 11, 385–391, 1991.
12. **Ohnishi, S. T. and Ohnishi, T.**, Biologically significant physico-chemical properties of antioxidative prostaglandin derivatives, *Arzneimittel-Forschung Drug Res.*, 41, 1201–1205, 1991.
13. **Hidaka, S., Morita, N., Ohnishi, S. T., Oga, Y., and Abe, K.**, Inhibitory effect of prostaglandin oligomeric derivatives on 9,10-dimethyl-1,2-benzanthracene-induced hamster lingual carcinomas, *Cancer Lett.*, 61, 171–176, 1992.
14. **Hidaka, S., Morita, N., Ohnishi, S. T., Abe, K., and Inoue, M.**, Studies on antitumor activity of prostaglandin derivatives (in Japanese), *J. Jpn. Soc. Cancer Ther.*, 26, 121, 1991 (meeting abstr.).
15. **Ohnishi, S. T., Ohnishi, N., Oda, Y., and Katsuoka, M.**, Prostaglandin derivatives inhibit the growth of malarial parasites in mice, *Cell Biochem. Function*, 7, 105–109, 1989.
16. **Brooks, A., Ohnishi, S. T., and Weidanz, W. P.**, Antimalarial activities of prostaglandin derivatives *in vitro*, to be published.
17. **Chandra, S., and Ohnishi, S. T., and Dhawan, B. N.**, Reversal of chloroquine resistance in murine malarial parasite by prostaglandin derivatives, *Am. J. Trop. Med. Hyg.*, to be published.
18. **Tomomatsu, Morita, Nagai, Shiwaku, Ichiki and Yunoki**, *Acta Med. Univ. Kagoshima*, 17, 99, 1975.
19. **Fujita, Kaku, Sasaki, and Onoe**, *J. Dent. Res.*, 52, 327–332, 1973.
20. **Gottesman, M. M. and Pastan, I.**, The multidrug-transporter, a double-edged sword, *J. Biol. Chem.*, 263, 12163–12166, 1988.
21. **Endicott, J. A. and Ling, V.**, The biochemistry of p-glycoprotein-mediated multidrug resistance, *Annu. Rev. Biochem.*, 58, 137–171, 1989.
22. **Pastan, I., Willingham, M. C., and Gottesman, M.**, Molecular manipulations of the multidrug transporter: a new role for transgenic mice, *FASEB. J.*, 5, 2523–2528, 1991.
23. **Warren, L., Jardillier, J. C., and Ordentlich, P.**, Secretion of lysosomal enzymes by drug-sensitive and multiple drug-resistant cells, *Cancer Res.*, 51, 1996–2001, 1991.
24. **Nakashima et al.**, *Jpn. J. Cancer Res.*, (GANN), 78, 1164–1168, 1987.
25. **Ruprecht, R. M., O'Brien, L. G., Rossoni, L. D., and Nusinoff-Lehrman, S.**, Suppression of mouse viraemia and retroviral disease by 3'-azido-3'deoxythymidine, *Nature*, 323, 467–469, 1986.
26. **Dournon, E., Matheron, S., Rozenbaum, W., and Gharakhanian, S.**, Effect of zidovudine in 365 consecutive patients with AIDS or AIDS-related complex, *Lancet*, ii, 1297–1302, 1988.
27. **Walker, R. E., Parker, R. I., and Kovacs, J. A.**, Anemia and erythropoiesis in patients with the acquired immunodeficiency syndrome (AIDS) and Kaposi's sarcoma treated with azidovudine, *Ann. Intern. Med.*, 108, 372–376, 1988.
28. **Marshal, E.**, Malaria research — What next?, *Science*, 247, 399–402, 1990.
29. **Kometani, T., Devlin, T. M., and Ohnishi, S. T.**, Studies of the cation binding properties of an oligomeric derivative of prostaglandin B1, *Prostagl. Leukotr. Essential Fatty Acids*, 37, 37–43, 1989.
30. **Hutchinson, D. W.**, Metal chelators as potential antiviral agents, *Antiviral Res.*, 5, 193–205, 1985.

Chapter 16

DYSTROPHIN DEFECT IN DUCHENNE AND BECKER MUSCULAR DYSTROPHY

Kiichi Arahata and Hideo Sugita

TABLE OF CONTENTS

I.	Introduction	368
II.	Clinical Features of Duchenne (DMD) and Becker (BMD) Muscular Dystrophies	368
III.	Localization of a DMD/BMD Gene to Xp21	369
IV.	Cloning of the DMD/BMD Gene	369
V.	Discovery of Dystrophin, the Missing DMD Product	370
VI.	Membrane Localization of Dystrophin	371
VII.	Membrane Hypothesis of DMD/BMD	376
VIII.	Current Topics of DMD/BMD Research	378
	A. Tissue Specific Isoforms of Dystrophin	378
	B. Dystrophin Family	379
	C. Treatment of Muscular Dystrophy and MDX Mouse	379
References		380

I. INTRODUCTION

There has been growing evidence during the past few years showing the primary features of the pathogenesis of Duchenne muscular dystrophy (DMD), which is the most common fatal X-linked recessive disease of muscle in children.[1,2] Complete cloning of the coding sequence for the DMD gene, and discovery of the gene product, dystrophin, is truly an epoch-making success in the history of DMD research.[3-12] Dystrophin is a previously unknown protein which is now thought to be a membrane associated cytoskeletal protein (underlying the plasma membrane known in muscle as the sarcolemma) of muscle fiber.[13-19] In DMD patients, the gene coding for dystrophin is defective, and hence the crucial protein is absent. Thus the genetic, biochemical, and morphological basis of DMD has been elucidated greatly. In this review article, our interest is particularly directed toward the understanding of membrane hypothesis of DMD. To begin with, we outline the clinical features of DMD, localization and cloning of the DMD gene, and molecular model of dystrophin. The membrane localization of dystrophin in muscles from normal and various neuromuscular diseases including DMD, BMD (Becker muscular dystrophy which is allelically related to DMD but has less severe clinical phenotype), and DMD carrier will then be discussed. These results may augment membrane hypothesis of Duchenne muscular dystrophy. Several current topics in DMD/BMD research will also be discussed.

II. CLINICAL FEATURES OF DUCHENNE (DMD) AND BECKER (BMD) MUSCULAR DYSTROPHIES

This fatal X-linked muscular disease, DMD, occurs with an incidence of approximately 1 in 3300 live male births, and one third of the patients are considered as new mutations.[1,2] Although walking delay is observed in about 50% of patients, most of the affected boys can walk normally until age 3 to 4, when toe- and clumsy walking is found. Progressive weakness and atrophy of skeletal muscles then appears in proximal limbs which produce Gowers' sign and waddling gait. Enlargement of calf muscles is frequently observed. Subsequently, most patients become wheelchair-bound by about 11 years of age and die in their early twenties mainly from respiratory failure. Involvement of cardiac muscle is also very common, and abnormal electrocardiogram appears in about 90% of the patients.[1,2]

Tissues other than skeletal and cardiac muscles are also involved in DMD patients. Of these, much attention has been paid to the central nervous system and smooth muscle. Mental impairment is more common in DMD patients than in the general population.[2] Although no specific pathological and electroencephalographical abnormalities have been detected, about 20% of the patients show IQs (intelligence quotient) below 70. But the mental impairment is neither progressive nor related to physical handicaps. Gastric hypomotility and dilatation are also well known clinical manifestations of DMD patients. Although the causes of these abnormalities in brain and smooth muscle are still unknown, deficiency of dystrophin in these tissues may play an important role.

Becker muscular dystrophy (BMD) is a benign Xp21 muscular dystrophy (allelic to DMD) with a similar clinical distribution of skeletal muscle involvement in proximal limbs and cardiac muscles to those of DMD. BMD has a later onset and a slower progression of clinical symptoms in most cases, and average age of wheelchair-bound patients in BMD is around 30 years, whereas in DMD it is 9 years.[1] Although BMD patients can survive much longer than DMD, death in the earlier ages may occur due to heart failure caused by cardiomyopathy. Mental impairment in BMD patients is less common than in DMD.[2] Incidence of BMD is about 1/10 to 1/5 that of DMD.

III. LOCALIZATION OF A DMD/BMD GENE TO Xp21

The localization of the DMD/BMD gene to the short arm of the human X-chromosome, Xp21, has been indicated by cytogenetic analysis of rare females with the DMD/BMD phenotype.[6,10] Each of them had an X; autosome translocation, and the break always occurred at the constant place "Xp21", but no apparent preference for the autosomal region involved suggested the presence of DMD/BMD gene in the Xp21 region. Usually the clinical findings of DMD patients in females with an X; autosome translocation are less severe than male DMD. This can be explained by non-random inactivation of the normal X-chromosome.[6] If 100% of the normal (non-translocated) X-chromosomes were inactivated during early embryogenesis, the patient becomes most severely affected because the translocated (abnormal) X-chromosome is always inactive.

Other approaches that mapped the DMD gene to Xp21 were made by analysis of DMD males with multiple X-linked diseases which produced microscopically visible deletions of Xp21,[3-6] or by RFLPs (restriction fragment length polymorphisms) analysis with RC-8 and L1.28 probes which places the p21 band in between.[20,21] For the earlier studies of linkage analysis, RC-8 was used as the first marker of the DMD gene (mapped to Xp21–22), and L1.28 as the second marker (mapped to Xp11). But both markers were still not close enough to the DMD gene.

IV. CLONING OF THE DMD/BMD GENE

During the next few years, the studies with DNA probes that detect RFLPs were developed in several laboratories. The first breakthrough came in 1985, when Kunkel and associates cloned a pERT87 clone from the DNA of a young boy (B.B.) who was unfortunately affected with five different genetic disorders linked to the X-chromosome, i.e., DMD, X-linked retinitis pigmentosa, chronic granulomatous disease, acanthocytosis of red blood cells (McLeod syndrome), and mental retardation.[3] The deletion in patient B.B. was microscopically visible around Xp21. The pERT87 fragment was then enlarged by chromosome walking method.[4-6] Finally, using the identified exon, a 14 kb (kilobase) cytoplasmic mRNA transcript was found in the skeletal muscle of the human fetus,[8,22] which encodes at least 65 exons distributed over 2500 kb in size, the largest known for any human gene occupying approximately 0.1% of the whole gene. It is estimated to be 1500 times larger than the -globin gene. Subsequently, mean intron size should become very large, and the estimated exon/intron ratio of the DMD gene becomes extremely small (0.007).[23] The deletion hot-spot region (exons 45 to 55) is known to start between exons 44 and 45 where a very large P-20 intron (150 kb) exists. This fact may explain the unusually high mutation rate of DMD/BMD. So far it has been reported that 60 to 65% of patients affected with DMD/BMD exhibit deletion mutations for some of the exons.[24-27]

Another type of the cloning strategy was the translocation cloning which was conducted by Worton and co-workers, using X(p21); autosome (21p12) translocation breakpoint (XJ1.1 clone (10,28). Today, both XJ and pERT genomic clones are employed for the genetic diagnosis of DMD.

V. DISCOVERY OF DYSTROPHIN, THE MISSING DMD GENE PRODUCT

Hoffman and co-workers have found a new protein of approximately 400 kDa in the extracts of normal human and mice skeletal muscles. It is encoded by the DMD locus and has been named "dystrophin" because of the selective absence in DMD and mdx muscles.[11,12] Antigens used to raise the antibodies for the Western blot analysis were made in *E. coli* using the fusion peptides of DMD cDNA and bacterial trp E gene. Dystrophin was also detected in normal cardiac and visceral smooth muscles and in brain but was completely absent in DMA patients. The abundance of this protein was extremely low and accounted only for 0.002% of total muscle protein, which was consistent with its putative cardiac and skeletal muscle mRNA (0.01 to 0.001% of total muscle protein). Interestingly, the protein was abnormal in molecular weight in BMD patients according to the internal deletion of the gene.[22,29]

The entire DMD cDNA sequence and the deduced primary structure of dystrophin have been identified by Koenig and co-workers by 1988 (Figure 1).[8,9] It consists of an uninterrupted translational reading frame starting from nucleotide 200 (ATG initiation codon) and extends over 11 kb followed by a noncoding 2.7 kb segment. The predicted 427 kDa dystrophin protein of human skeletal muscle contains 3685 amino acids which can be separated into four distinct domains: (a) the 240-amino acid N-terminal domain that is highly homologous to cytoskeletal-actinin, (b) the second largest rod-like domain composed of 26 repeats of 109 amino acids each of which resembles

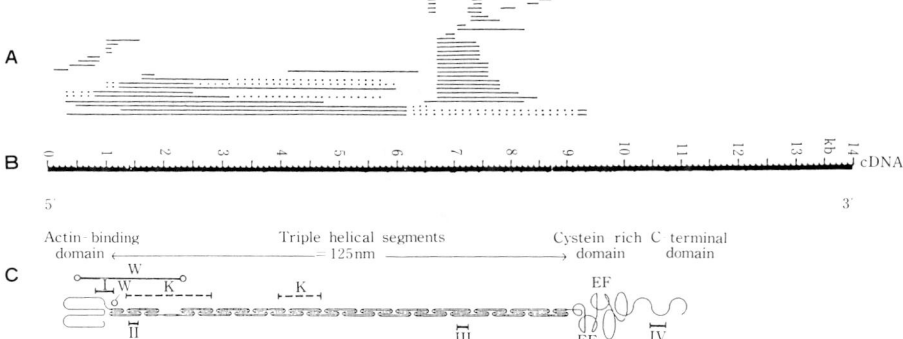

FIGURE 1. Structure of the dystrophin molecule (C) and the regions of the dystrophin protein that were used to generate anti-dystrophin antisera (K: by Kunkel's group, W: by Worton's group, I, II, III, and IV are authors'). In 104 patients with DMD, a spectrum of 53 deletions was found; broken lines indicate uncertainty about the exact length of the deletion (A). (Figure 1A from Koenig, M. et al., *Cell,* 50, 509–517, 1987; Figure 1C from Koenig, M. et al., *Cell,* 53, 219–228, 1988. With permission.)

the repeat domains of spectrin and -actinin, (c) the third represents cysteine-rich domain and is related to the carboxy-terminus of -actinin, (d) the C-terminal domain consisting of 420 amino acids that has no resemblance to any other previously characterized protein but is highly conserved between chicken and man.[30] Importance of the fourth domain will be discussed later. From the sequence analyses, dystrophin is thought to be a member of a family of cytoskeletal proteins.

Dystrophin is a highly conserved protein during evolution and is observed in human, mouse, rat, dog, cat, swine, rabbit, and chicken. Of these, dystrophin-defective animal was first found in mouse. Gene abnormality of mdx mouse, an X-linked mutant which is considered an animal model of DMD, lies within the dystrophin gene.[31] Recently, a point mutation (a single base substitution with replacement of a cytosine by thymine at nucleotide position 3185) was detected:[32] thus, in mdx mice, truncated N-terminal dystrophin is calculated from its gene analysis. Both immunoblot and immunofluorescence examination, however, do not recognize any dystrophin in the muscle, suggesting the rapid degradation of the truncated dystrophin with no C-terminal region.

VI. MEMBRANE LOCALIZATION OF DYSTROPHIN

The surface membrane localization of dystrophin in normal human and mouse skeletal and cardiac muscles, but not that of DMD patients and mdx mice, has been identified immunohistochemically using antisera raised against peptide fragments prepared as predicted from the distinctive regions of DMD

TABLE 1
Immunohistochemical Analysis of DMDP Expression in the Surface Membrane of Human and Mouse Skeletal and Cardiac Muscles

Clinical diagnosis (mouse strain)	No. of cases	DMDP expression Skeletal muscle	DMDP expression Cardiac muscle
Duchenne dystrophy	22	−	−
Pre-clinical DMD	5	−	ND
Female DMD (X;5 translocation)	1	−	ND
Becker dystrophy	9	−/+	ND
Emery-Dreifuss dystrophy	1	+	ND
Congenital dystrophy (Fukuyama)	5	+	ND
Limb-girdle dystrophy	5	+	ND
FSH dystrophy	8	+	ND
Myotonic dystrophy	5	+	ND
Dermatomyositis (child)	6	+	ND
Polymyositis	5	+	ND
Spinal muscular atrophy	6	+	ND
Other diseases[a]	7	+	ND
Normal human muscle[b]	9	+	+
mdx mice (C57BL/10 ScSn-mdx)	5	−	−
Control mice (C57BL/10 ScSn)	5	+	+
Total	104		

Note: Tissues were prepared and examined as described in Figure 1. ND = not determined; −, negative; +, positive; −/+, immunostaining of the surface membrane was largely negative, but some muscle fibers show limited or diffuse membrane staining of the fiber as shown in Figure 1.

[a] These include nemaline myopathy, central core disease, scleroderma, and hypothyroid myopathy.
[b] Obtained from the autopsy cases which were free of muscle disease.

Adapted from Arahata, K. et al., *Nature*, 333, 466–469, 1988.

cDNA,[13-17] or against the fusion proteins.[17-19] On cross sections of normal human and mouse skeletal muscle fibers, all the surface membranes are stained clearly with anti-dystrophin antiserum without interruption, and no focal staining is observed inside the muscle fiber (Plate 2* and Table 1). On longitudinal sections of normal and DMD muscles, there is no periodic staining corresponding to the repeat of sarcomere (Figure 2). In accordance with the results obtained from skeletal muscle, the surface membrane of cardiac muscles from both normal human and control (B10) mice stained equally for dystrophin, whereas those from DMD patients and mdx mice did not.[13,14]

* Plate 2 follows page 304.

By immunogold electron microscopy of ultrathin cryosections, dystrophin was mainly detected in the inner face of the plasma membrane of muscle fibers, and possibly some at the T-tubules which are the invaginated surface membrane.[19] Gold particles inside the plasma membrane have shown periodical labeling approximately 125 nm, which is well correlated with the molecular model of dystrophin deduced from its nucleotide sequence. Thus, dystrophin is strongly suggested to be a membrane-associated cytoskeletal component of muscle fiber.

From the clinical point of view, the model of dystrophin expressions in muscle specimens is very important and has been analyzed together with immunofluorescence and immunoblotting methods in a wide range of neuromuscular diseases.[33] Among them, symptomatic carriers of DMD can be identified by a distinct mosaic pattern in the immunohistochemical staining of the surface membrane,[15,34] and this would have very useful implications for genetic counseling when there is no deletion detected by Southern blot analysis.[35,36] In all three symptomatic carriers of DMD we examined, most individual muscle fibers reacted either strongly or not at all to the antiserum for dystrophin, and only 2 to 8% of the fibers showed partial immunostaining (Figure 2).[15] Although the mosaic pattern was seen in all histochemical fiber types, more than 80% of type 2B and 2C fibers did not show any dystrophin. This suggests a primary role of dystrophin in the disappearance of type 2 fibers — and/or dystrophin-deficient fibers may fall more frequently into cycles of muscle fiber degeneration and regeneration. However, we are still uncertain about the difference of turnover rate of these dystrophin-deficient and dystropin-positive muscle fibers. We have examined four additional asymptomatic obligatory carriers of DMD whose serum CK levels were slightly elevated (up to four times the upper limit of normal) and all patients were found to have a similar mosaic pattern of dystrophin immunostaining.[33]

The underlying biochemical defects responsible for both DMD and BMD are abnormalities of dystrophin. Although dystrophin is absent from DMD muscle, patients with BMD have been shown to have dystrophin of abnormal molecular weight and/or lower relative cellular abundance compared to normal muscle.[29] In fact, dystrophin can easily be identified by both immunoblotting (abnormal molecular weight, variable quantities of dystrophin) and immunofluorescence (faint, patchy, and discontinuous immunostaining) techniques in BMD muscle[33,37] (Plate 2, Figure 3). These findings suggest a partial and low-level expression of dystrophin at the surface membrane of BMD muscle, and might explain the relatively benign clinical course in comparison with DMD. Recently, Hoffman and co-workers found that the clinical severity of BMD patients is attributed to the content of dystrophin in the skeletal muscle rather than the size, i.e., severe BMD patients have only 3 to 10% of the normal level of dystrophin, whereas moderate to mild patients have over 20%.

Much attention has been paid to the relationship between clinical phenotypes and the deletion mutation pattern of dystrophin gene in DMD and BMD.

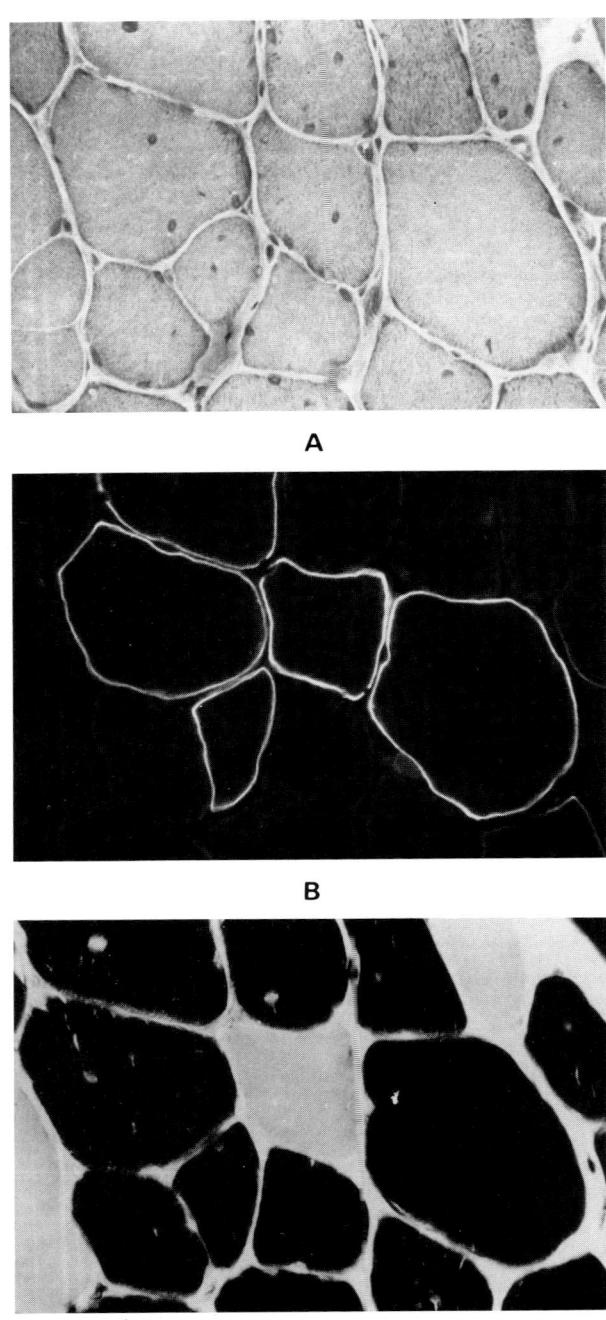

A

B

C

FIGURE 2

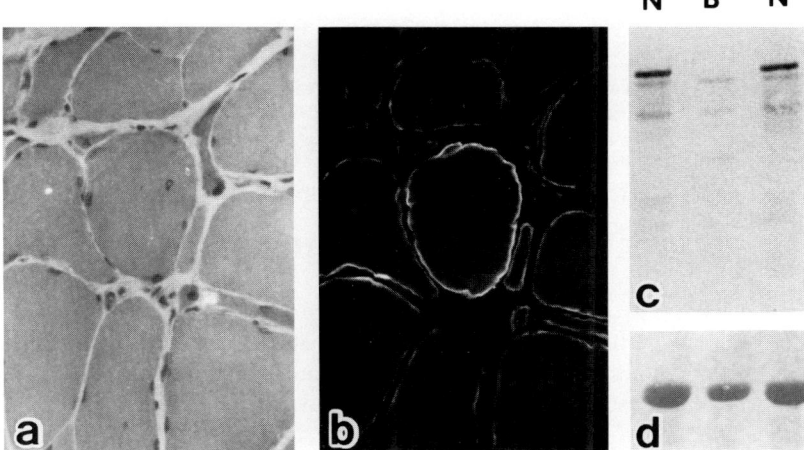

FIGURE 3. Immunofluorescent staining pattern with antidystrophin antiserum in Becker muscular dystrophy showing the discontinuous "patchy" fluorescence at the surface membrane (b), together with the section stained by hematoxylin and eosin (H&E) (a), immunoblotting (c) and Coomassie blue staining of myosin heavy chain after blot transfer (d). Dystrophin is evident in normal muscle (N) as the expected protein of 400 kDa, but exhibits low molecular mass (380 kDa) in Becker muscular dystrophy (B). (Immunoblotting performed by Hoffman, E. P.)

Interestingly, deletion size itself did not predict DMD and BMD;[38-40] instead, from detailed DNA sequence analysis, the "reading frame" hypothesis was proposed by Monaco and co-workers to explain this paradox.[41] BMD patients have shown internal deletion of the reading frame of mRNA (in-frame deletion). By contrast, in the case of DMD, a shift in the reading frame occurs (out-of-frame deletion), resulting in the production of severely truncated dystrophin that would be unstable and may rapidly be degraded.[22,41] The exception to this hypothesis is accounted for only in a small percentage of patients. Recently, both Kunkel's and Worton's groups have studied large numbers of DMD/BMD cases (258 and 80, respectively) independently, and the agreement with the reading frame hypothesis was found in more than 90% of the patients.[26,27] Small numbers of the exceptional cases have been explained by the re-initiation of the mRNA reading frame with alternative splicing mechanisms such as exon skipping or use of cryptic splice site, and/or occurrence of additional mutations as indicated by Malhotra[42] and Baumbach.[40]

FIGURE 2. Dystrophin expression pattern in the patient with symptomatic carrier of DMD. Serial cryostat sections stained trichromatically (A), immunohistochemically for dystrophin (B), and reacted for ATPase (pH 4.3) (C). In Panel E, an immunohistologic mosaic pattern of the surface membrane, with positive and negative patches, is observed, which is not corresponding with the histochemical fiber types as shown in Panel C.

TABLE 2
Membrane Involvement in DMD/BMD

Leakage of cytoplasmic enzymes[4-45] (1950s)
Discontinuities in the sarcolemma[6-48] (1970s)
Elevated cytoplasmic Ca^{2+} concentration[49,50] (1970s)
Depletion of intramembranous particles[51] (1970s)
Absence of dystrophin at the plasmamembrane[13-19] (1980s)

VII. MEMBRANE HYPOTHESIS OF DMD

Selective lack of dystrophin from the surface membrane of skeletal muscle obtained from DMD patients certainly substantiated the membrane hypothesis of DMD at a molecular level. The idea that some defect must exist in plasma membrane in muscle fibers of DMD is relatively old (Table 2). In 1949, Sibley and Lehninger noted elevated serum aldolase in two patients with muscular dystrophy.[43] Ebashi and co-workers found, in 1959, high serum creatine kinase (CK) activity in muscular dystrophy, especially DMD, but not neurogenic atrophy.[45] In 1966, Sugita and co-workers examined the calcium binding activity of the microsome fraction of muscle from patients with

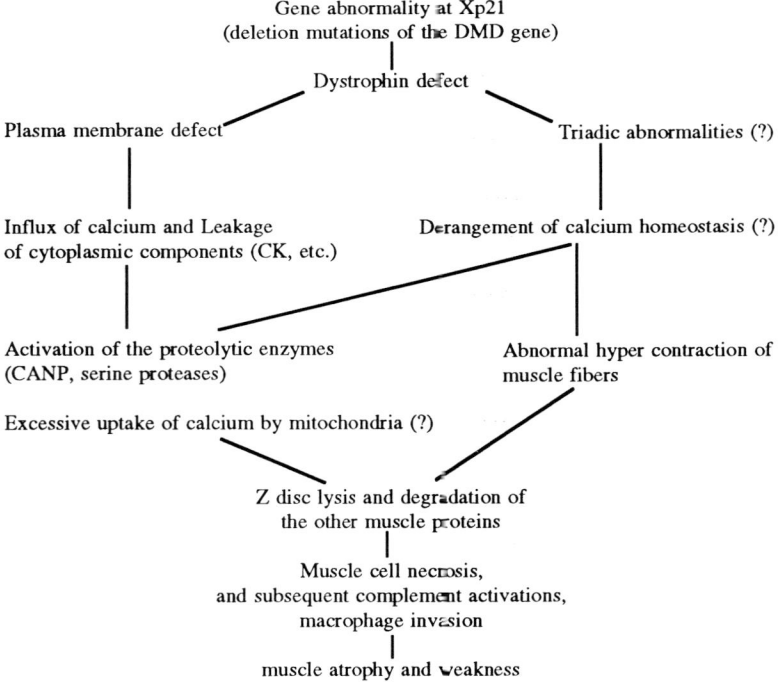

Scheme I. Schematic illustration of the pathogenesis of DMD.

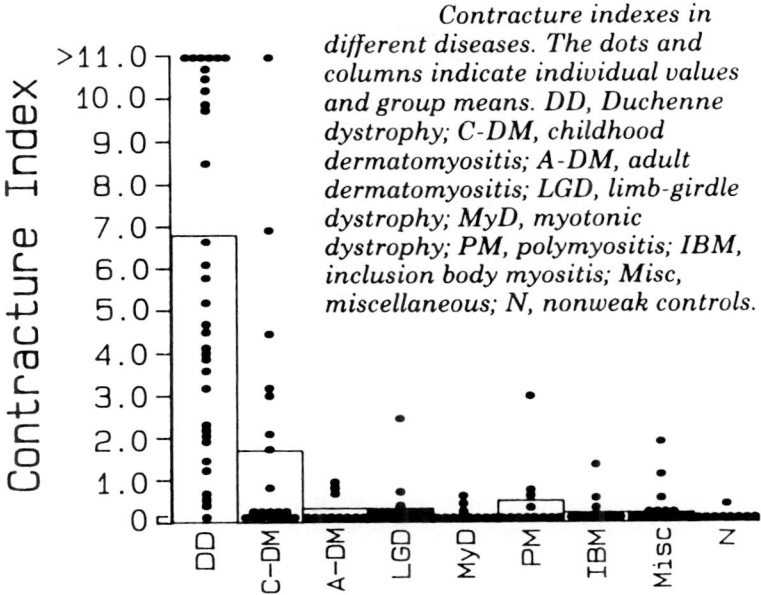

FIGURE 4. The incidence of hypercontracted fibers in DMD is significantly higher than in any other myopathies. (From Lotz, B. P. and Engel, A. G., Neurology, 37, 1466–1475, 1987. With permission.)

progressive muscular dystrophy.[52,53] They found a decrease in Ca^{2+} and an increase in Mg^{2+}, Ca^{2+}-dependent ATPase activity of the muscle microsome and suggested the presence of some degenerative alterations in sarcoplasmic reticulum (SR), which is the closely related structure with transverse tubules (T-tubules; an invaginated plasma membrane). Thus, they speculated the T-tubules as the possible site of leakage of CK in the serum. By elegant electron microscopy Engel and co-workers, in 1975, clearly demonstrated the plasma membrane defect in non-necrotic muscle fibers, which made small gaps in the membrane with preservation of the basal lamina.[46] Other investigators have also found similar defects in the plasma membrane.[47,48] Freeze fracture electron micron microscopy also revealed a membrane abnormality with the depletion of intramembranous particles in the dystrophic muscle.[51] But the density of the particles of cultured DMD myotube was not significantly different from normal controls.[54]

The membrane defect would facilitate the leakage of cytoplasmic components including aldolase and CK. Such a defect in the plasma membrane also allows excessive influx of calcium into the fiber[49,50] (Table 2) and may

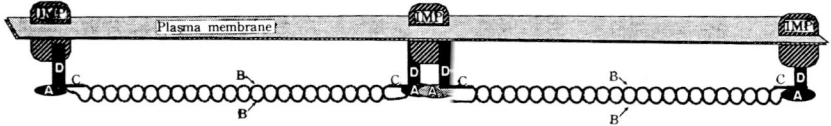

FIGURE 5. A hypothetical model of dystrophin structure and subcellular organization. (From Hoffman, E. P. and Kunkel, L. M., *Neuron*, 2, 1019–1029, 1989. With permission.)

cause hypercontraction (contracture) of muscle. Actually, hypercontracted opaque fibers appear more frequently in DMD muscle than the other neuromuscular diseases[55] (Figure 4). Probably the overloaded calcium would directly[56,57] or through the functional and structural damage of mitochondria due to excessive uptake of calcium by mitochondria[58] activate the intracellular calcium-activated proteases. Thus, the proteolytic degradation of myofibrillar proteins may occur, which would be followed by the necrotic changes of the fiber with the deposition of complement components.[59,60] Among the proteolytic enzymes contained in the skeletal muscle, calcium-activated neutral protease (CANP) might play an initial and important role in the disappearance of the fiber.[57] SDS gel electrophoretic patterns of both myosin and troponin in DMD muscle have shown a similar pattern which was observed in the trypsin-digested monkey myofibrils that was completely inhibited by CANP inhibitors such as leupeptin and E-64. In addition, proteases other than CANP, i.e., lysosomal cathepsins B and D, might also contribute to the degradation and repair of myofibrillar proteins.[61,62] Interestingly, normal skeletal muscle contains a small amount of cathepsin D.

Today, dystrophin has been shown to bind to integral membrane glycoprotein(s) through its C-terminal domain.[63] From this result, it has been speculated that dystrophin may have a similar role to ankyrin of the red blood cell (RBC) that binds to the transmembrane Band 3 protein (anion channel of RBC) and/or to spectrin which binds to actin and ankyrin. Thus, although the precise physiological function of dystrophin still remains to be solved, dystrophin could possibly be associated with the ion channel in the membrane. It may also play an important role in stabilizing the plasma membrane against the stress associated with rigorous contraction and relaxation of the muscle fiber as already indicated. As dystrophin has a similar structure to spectrin,[9] it is postulated to form an antiparallel homodimer under the plasma membrane of muscle fiber[22] (Figure 5). More recently, together with detailed immunoelectron microscopy, a flexible honeycomb network model of dystrophin at the inner face of the plasma membrane is proposed (Koenig, personal communication). Our immunohistochemical observations that myotendinous junctions express higher amounts of dystrophin (unpublished data) imply the existence of physiological compensation of the skeletal muscle against the mechanical stress.

The patchy localization of dystrophin at the surface membrane is a characteristic feature of BMD muscle which suggests the abnormal arrangement of dystrophin to the inner surface of plasma membrane, probably because of the abnormal molecular size or localized protein instability in the muscle.[33]

If dystrophin is missing from the triadic junction of the skeletal muscle,[11,19] the derangement of calcium homeostasis of the fiber could appear and this again might possibly explain the cause of abnormal muscle fiber contraction which may accelerate pathological changes of the fiber. We still, however, do not know precisely how the absence of dystrophin from the surface membrane and possibly triads could cause damage of muscle fibers. The relationship between the abnormalities of the intramembranous particles and orthogonal arrays and defect of dystrophin is also a very important question to be clarified in the future.

VIII. CURRENT TOPICS OF DMD/BMD RESEARCH

A. TISSUE SPECIFIC ISOFORMS OF DYSTROPHIN

Tissue distribution of dystrophin mRNA was analyzed by a highly sensitive assay method based on PCR (polymerase chain reaction) technique,[64] and the dystrophin transcript was detected not only in skeletal, cardiac, and visceral smooth muscles, but also in the other organs such as brain, lung, and kidney. However, because the amount of the transcript in the latter tissues was only 1% or less than that of skeletal muscle, immunohistochemically, we are still uncertain about the cellular localization of dystrophin in the brain. By using elegant Western blot technique, Hoffman and co-workers demonstrated the neuronal localization of dystrophin in the cultured embryonic rat brain.[65] A 400 kDa dystrophin band was recognized only in the cultures enriched for neurons but not in the glial cells of normal rat. It was totally absent from mdx brain cultures. From these results, the mental impairment of DMD patients is speculated to be caused by a defect of dystrophin. But whether the mental impairment can simply be attributed to the defect of dystrophin in the brain has not yet been confirmed, because mental retardation occurs only in about 30% of DMD patients in spite of the complete absence of dystrophin in the muscle of all the patients examined.

Tissue specific isoforms of dystrophin have been shown using sophisticated PCR techique.[66] Several possible isoforms are produced through the alternative splicing of amino and/or carboxyl terminals of dystrophin. The first exon of dystrophin transcript in the brain is different from that of the skeletal muscle, suggesting the different types of promotors for brain and muscle. In smooth muscle, alternative splicing produces a very small dystrophin (405 kDa) isoform. From these evidences, conceivably, dystrophin molecule interacts with different proteins in different tissues. In this sense, dystrophin anchoring membrane protein(s), which bind to the C-terminal domain of dystrophin, may possibly differ between each tissue. Thus it becomes an intriguing part of DMD research to address the biological characteristics of these proteins that interact with dystrophin.

B. DYSTROPHIN FAMILY

One of the most important findings of dystrophin study is that the DMD gene has a close relative on human chromosome 6.[67] A dystrophin-related autosomal transcript of 13 kb mRNA was found in fetal and adult human skeletal muscle including DMD. The predicted amino acid sequence of the autosomal transcript was highly homologous (65 to 80%) to the C-terminal region of dystrophin. This unknown protein may be included in a so-called 'dystrophin family' together with -actinin and spectrin. Although the biochemical and histopathological characteristics and the clinical significance of this protein are largely unknown, it will be an important protein for the understanding of autosomal muscular dystrophies.

C. TREATMENT OF MUSCULAR DYSTROPHY AND MDX MOUSE

An experimental myoblast transfer therapy for muscular dystrophies is truly one of the most important studies for DMD research. In 1988, Partridge and co-workers injected normal mouse myoblasts into dystrophin defective mdx mice, and successfully fused the injected myoblasts with regenerating muscle fibers in 39 of 70 (55.7%) mdx mice, which was detected by glucose-6-phosphate isomerase analysis.[68] A higher percentage of the myoblasts fusion was observed in the mdx/nude hosts group (79%) than in the mdx hosts (47%). More importantly, the injected muscles expressed normal dystrophin as much as 30 to 40% of levels found in normal muscle, which would probably be enough to improve clinical symptoms.[37] After that, Karpati and co-workers used clonal cultures of normal human myoblasts as the donor cells, and injected them into mdx leg muscles. The human myonuclei, which were radiolabeled before the injection, were incorporated into some of the regenerated mdx muscle fiber, and have expressed dystrophin at the surface membrane of the fiber with approximately 2 to 7% of all fibers per muscle of transverse section.[69] Of course, we need to overcome several side effects including immune rejection of implanted myogenic cells as indicated by Partridge et al. The most effective injection site, timing, i.e., age of the patient, and numbers of myoblast, also should be explored.

To our surprise, although dystrophin-lacking dog (CXMD) shows very severe clinical symptoms as observed in DMD patients,[70-72] dystrophin-lacking mdx mice have no obvious clinical symptoms. As recently reported, gene abnormality of mdx mouse is a single point mutation of the Dmd gene. The premature termination of translation still possibly produces truncated (115 kDa, and 27% in length) N-terminal dystrophin. The abnormal dystrophin may then help the stability of the plasma membrane, and thus could produce the benign clinical course of mdx mouse. But this is unlikely, because the truncated dystrophin protein was not detected in mdx muscle when we used antibodies against the N-terminal regions. Therefore, in mdx mice, dystrophin defect can be compensated for by the other protein(s) from the dystrophin family, or by the better regenerating capacity and/or lower amount of fibrosis in muscle. Thus, we could probably find a key for the therapy from mdx mice to improve the poor clinical course of DMD.

REFERENCES

1. **Engel, A. G.**, Duchenne dystrophy, in *Myology*, Vol. 2, Engel, A. G. and Banker, B. Q., Eds., McGraw-Hill, New York, 1986, 1185–1250.
2. **Emery, A. H. E.**, *Duchenne Muscular Dystrophy*, Emery, A. H. E., Ed., Oxford Medical Publishers, Oxford, 1987.
3. **Kunkel, L. M., Monaco, A. P., Middlesworth, W., Ochs, H. D., and Latt, S. A.**, Specific cloning of DNA fragments absent from the DNA of a male patient with an X chromosome deletion, *Proc. Natl. Acad. Sci. U.S.A.*, 82, 4778–4782, 1985.
4. **Monaco, A. P., Bertelson, C. J., Middlesworth, W., Colletti, C. A., Aldridge, J., Fishbeck, K. H., Bartlett, R., Pericak-Vance, M. A., Roses, A. D., and Kunkel, L. M.**, Detection of deletions spanning the Duchenne muscular dystrophy locus using a tightly linked DNA segment, *Nature*, 316, 842–845, 1985.
5. **Monaco, A. P., Neve, R., Colletti-Feener, C., Bertelson, C. J., Kurnit, D. M., and Kunkel, L. M.**, Isolation of candidate cDNAs for portions of the Duchenne muscular dystrophy gene, *Nature*, 323, 646–650, 1986.
6. **Monaco, A. P. and Kunkel, L. M.**, Cloning of the Duchenne/Becker muscular dystrophy locus, in *Advances in Human Genetics*, Vol. 17, Harris, H. and Hirschorn, K., Eds., Plenum Press, New York, 1988, 61–98.
7. **Hoffman, E. P., Monaco, A. P., Feener, C. C., and Kunkel, L. M.**, Conservation of the Duchenne muscular dystrophy gene in mice and humans, *Science*, 238, 347–350, 1987.
8. **Koenig, M., Hoffman, E. P., Bertelson, C. J., Monaco, A. P., Feener, C., and Kunkel, L. M.**, Complete cloning of the Duchenne muscular dystrophy (DMD) cDNA and preliminary genomic organization of the DMD gene in normal and affected individuals, *Cell*, 50, 509–517, 1987.
9. **Koenig, M., Monaco, A. P., and Kunkel, L. M.**, The complete sequence of dystrophin predicts a rod-shaped cytoskeletal protein, *Cell*, 53, 219–228, 1988.
10. **Worton, R. G. and Thompson, M. W.**, Genetics of Duchenne muscular dystrophy, in *Annual Review of Genetics*, Vol. 22, Annual Reviews, 1988, 601–629.
11. **Hoffman, E. P., Kundson, C. M., Campbell, K. P., and Kunkel, L. M.**, Subcellular fraction of dystrophin to the triads of skeletal muscle, *Nature*, 330, 754–758, 1987.
12. **Hoffman, E. P., Brown, R. H., Jr., and Kunkel, L. M.**, Dystrophin: the protein product of the Duchenne muscular dystrophy locus, *Cell*, 51, 919–928, 1987.
13. **Sugita, H., Arahata, K., Ishiguro, T., Suhara, Y., Tsukahara, T., Ishiura, S., Eguchi, C., Nonaka, I., and Ozawa, E.**, Negative immunostaining of Duchenne muscular dystrophy (DMD) and mdx mouse muscle surface membrane with antibody against synthetic peptide fragment predicted from DMD cDNA, *Proc. Jpn. Acad.*, 64, 210–212, 1988.

14. **Arahata, K., Ishiura, S., Ishiguro, T., Tsukahara, T., Suhara, Y., Eguchi, C., Ishihara, T., Nonaka, I., Ozawa, E., and Sugita, H.,** Immunostaining of skeletal and cardiac muscle surface membrane with antibody against Duchenne muscular dystrophy peptide, *Nature,* 333, 466–469, 1988.
15. **Arahata, K., Ishihara, T., Kamakura, K., Tsukahara, T., Ishiura, S., Baba, C., Matsumoto, T., Nonaka, I., and Sugita, H.,** Mosaic expression of dystrophin in symptomatic carriers of Duchenne's muscular dystrophy, *N. Engl. J. Med.,* 320, 138–142, 1989.
16. **Arahata, K., Ishiura, S., Tsukahara, T., and Sugita, H.,** Dystrophin digest, *Nature,* 337, 606, 1989.
17. **Zubrzycka-Gaarn, E., Bulman, D. E., et al.,** The Duchenne muscular dystrophy gene product is localized in sarcolemma of human skeletal muscle, *Nature,* 333, 466–469, 1988.
18. **Bonilla, E., Samitt, C. E., et al.,** Duchenne muscular dystrophy: deficiency of dystrophin at the muscle cell surface, *Cell,* 54, 447–452, 1988.
19. **Watkins, S. C., Hoffman, E. P., et al.,** Immunoelectromicroscopic localization of dystrophin in myofibers, *Nature,* 333, 863–866, 1988.
20. **Murray, J. M., Davies, K. E., Harper, P. S., Meredith, L., Mueller, C. R., and Williamson, R.,** Linkage relationship of a cloned DNA sequence on the short arm of the X chromosome to Duchenne muscular dystrophy, *Nature,* 300, 69–71, 1982.
21. **Davies, K. E., Pearson, P. L., Harper, P. S., Murray, J. M., O'Brien, T., et al.,** Linkage analysis of two cloned DNA sequences flanking the Duchenne muscular dystrophy locus on the short arm of the human X chromosome, *Nucleic Acids Res.,* 11, 2303–2312, 1983.
22. **Hoffman, E. P., and Kunkel, L. M.,** Dystrophin abnormalities in Duchenne/Becker muscular dystrophy, *Neuron,* 2, 1019–1029, 1989.
23. **Infante, J. P. and Huzagh, V. A.,** On the nature of the Duchenne muscular dystrophy locus: a portion of a complex of related gene clusters of recent pseudoautosomal origin?, *Mol. Cell. Biochem.,* 81, 103–119, 1988.
24. **Darras, B. T., Koenig, M., et al.,** Direct method for prenatal diagnosis and carrier detection in Duchenne/Becker muscular dystrophy using the entire dystrophin cDNA, *Am. J. Med. Genet.,* 29, 713–726, 1988.
25. **Forrest, S. M., Cross, G. S., Flint, T., Speer, A., Robson, K. J. H., and Davis, K. E.,** Further studies of gene deletions that cause Duchenne and Becker muscular dystrophies, *Genomics,* 2, 109–114, 1988.
26. **Koenig, M., Beggs, A. H., Moyer, M., Scherpf, S., Heidrich, K., et al.,** The molecular basis for Duchenne versus Becker muscular dystrophy: correlation of severity with type of deletion, *Am. J. Hum. Genet.,* 45, 498–506, 1989.
27. **Gillard, E. F., Chamberlain, J. S., Murphy, E. G., Duff, C. L., Smith, B., et al.,** Molecular and phenotypic analysis of patients with deletions within the deletion-rich region of the Duchenne muscular dystrophy (DMD) gene, *Am. J. Hum. Genet.,* 45, 507–520, 1989.
28. **Ray, P. N., Belfall, B., Duff, C., Logan, C., Kean, V., Thompson, M. W., Sylvester, J. E., Gorski, J. L., Schmickel, R. D., and Worton, E. G.,** Cloning of the breakpoint of an X;21 translocation associated with Duchenne muscular dystrophy, *Nature,* 318, 672–675, 1985.
29. **Hoffman, E. P., Fishbeck, K. H., Brown, R. H., Johnson, M., Medori, R., Loike, J. D., Harris, J. B., Waterston, R., Brooke, M., Specht, L., Kupsky, W., Chamberlain, J., Caskey, T., Shapiro, F., and Kunkel, L. M.,** Characterization of dystrophin in muscle biopsy specimens from patients with Duchenne's and Becker's muscular dystrophy, *N. Engl. J. Med.,* 318, 1363–1368 1988.
30. **Lemaire, C., Heilig, R., and Mandel, J. L.,** The chicken dystrophin cDNA: striking conservation of the C-terminal coding and 3' untranslated regions between human and chicken, *EMBO J.,* 7, 4157–4162, 1988.

31. Ryder-Cook, A. S., Sicinski, P., Thomas, K., Davies, K. E., Worton, R. G., Barnard, E. A., Darlison, M. G., and Barnard, P. J., Localization of the mdx mutation within the mouse dystrophin gene, *EMBO J.*, 7, 3017–3021, 1988.
32. Sicinski, P., Geng, Y., Ryder-Cook, A. S., and Barnard, P. J., The molecular basis of muscular dystrophy in the mdx mouse: a point mutation, *Science*, 244, 1578–1580, 1989.
33. Arahata, K., Hoffman, E. P., Kunkel, L. M., Ishiura, S., Tsukahara, T., Ishihara, T., Sunohara, N., Nonaka, I., Ozawa, E., and Sugita, H., Dystrophin diagnosis: comparison of dystrophin abnormalities by immunofluorescent and immunoblot analysis, *Proc. Natl. Acad. Sci. U.S.A.*, 86, 7154–7158, 1989.
34. Bonilla, E., Schmidt, B., Samitt, C., Miranda, A. M., Hays, A. P., DeOriveira, A., Chang, H. W., Servidei, S., Ricci, E., Younger, D. S., and DiMauro, S., Normal and dystrophin-deficient muscle fibers in carriers of the gene for Duchenne muscular dystrophy, *Am. J. Pathol.*, 133, 440–445, 1988.
35. Miranda, A. F., Francke, U., Bonilla, E., Martucci, G., Schmidt, B., et. al., Dystrophin immunohistochemistry in muscle culture: detection of a carrier of Duchenne muscular dystrophy, *Am. J. Med. Genet.*, 32, 268–273, 1989.
36. Hurko, O., Hoffman, E. P., McKee, L., Johns, D. R., and Kunkel, L. M., Dystrophin analysis in clonal myoblast derived from a Duchenne muscular dystrophy carrier, *Am. J. Hum. Genet.*, 44, 820–826, 1989.
37. Hoffman, E. P., Kunkel, L. M., Angelini, C., Clarke, A., Johnson, M., and Harris, J. B., Improved diagnosis of Becker muscular dystrophy via dystrophin testing, *Neurology*, 39, 1011–1017, 1989.
38. Forrest, S. M., Cross, G. S., Flint, T., Robson, K. J. H., and Davies, K. E., Further studies of gene deletions that cause Duchenne and Becker muscular dystrophies, *Genomics*, 2, 109–114, 1988.
39. Lindlof, M., Kaarianen, H., van Ommen, G. J. B., and de la Chapell, A., Microdeletions in patients with X-linked muscular dystrophy: molecular-clinical correlations, *Clin. Genet.*, 33, 131–139, 1988.
40. Baumbach, L. L., Chamberlain, J. S., Ward, P. A., Farwell, N. J., and Caskey, C. T., Molecular and clinical correlations of deletions leading to Duchenne and Becker muscular dystrophies, *Neurology*, 39, 465–474, 1989.
41. Monaco, A. P., Bertelson, C. J., Liechti-Gallati, S., Moser, H., and Kunkel, L. M., An explanation for the phenotypic differences between patients bearing partial deletions of the DMD locus, *Genomics*, 2, 90–95, 1988.
42. Malhotra, S. B., Hart, K. A., Klamut, H. J., Thomas, N. S. T., Bodrug, S. E., Burghes, H. M., Bobrow, M., Harper, P. S., Thompson, M. W., Ray, P. N., and Worton, R. G., Frame-shift deletions in patients with Duchenne and Becker muscular dystrophy, *Science*, 242, 755–759, 1988.
43. Sibley, J. A. and Lehninger, A. L., Aldolase in the serum and tissues of tumor-bearing animals, *J. Natl. Cancer Inst.*, 9, 303–309, 1949.
44. Schapira, G. and Dreyfus, J. C., Biochemistry of progressive muscular dystrophy, in *Muscular Dystrophy in Man and Animals*, Bourne, G. H. and Golarz, M. N., Eds., Hafner, New York, 1963, 48–87.
45. Ebashi, S., Toyokura, Y., Momoi, H., and Sugita, H., High creatine phosphokinase activity of sera of progressive muscular dystrophy, *J. Biochem. Tokyo*, 46, 103–104, 1959.
46. Mokri, B. and Engel, A. G., Duchenne dystrophy. Electronmicroscopic findings pointing to a basic or early abnormality in the plasma membrane of the muscle fiber, *Neurology*, 25, 1111–1120, 1975.
47. Schmalbruch, H., Segmental fiber breakdown and defects of the plasmalemma in diseased human muscle, *Acta Neuropathol. (Berlin)*, 33, 129–141, 1975.

48. **Carpenter, S. and Karpati, G.**, Duchenne muscular dystrophy plasma membrane loss initiates muscle cell necrosis unless it is repaired, *Brain*, 102, 147–161, 1979.
49. **Bodensteiner, J. B. and Engel, A. G.**, Intracellular calcium accumulation in Duchenne dystrophy and other myopathies: a study of 567,000 muscle fibers in 114 biopsies, *Neurology*, 28, 439–446, 1978.
50. **Emery, A. E. H. and Burt, D.**, Intracellular calcium and pathogenesis and antenatal diagnosis of Duchenne muscular dystrophy, *Br. Med. J.*, 280, 355–357, 1980.
51. **Schotland, D. L., Bonilla, E., and Van Meter, M.**, Duchenne dystrophy: alteration in muscle plasma membrane structure, *Science*, 196, 1005–1007, 1977.
52. **Sugita, H., Okimoto, K., and Ebashi, S.**, Some observations on the microsome fraction of biopsied muscle from patients with progressive muscular dystrophy, *Proc. Jpn. Acad.*, 42, 295–298, 1966.
53. **Sugita, H., Okimoto, K., Ebashi, S., and Okinaka, S.**, Biochemical alterations in progressive muscular dystrophy with special reference to the sarcoplasmic reticulum, in *Exploratory Concepts in Muscular Dystrophy and Related Disorders*, Harriman Co., New York, 1966, 321–326.
54. **Osame, M., Engel, A. G., Reboushe, C. L., and Scott, R. E.**, Freezefracture electron-microscopic analysis of plasma membranes of cultured muscle cells in Duchenne dystrophy, *Neurology*, 31, 972–979, 1981.
55. **Lotz, B. P. and Engel, A. G.**, Are hypercontracted muscle fibers artefacts and do they cause rupture of the plasma membrane?, *Neurology*, 37, 1466–1475, 1987.
56. **Sugita, H. and Toyokura, Y.**, Alteration of troponin subunits in progressive muscular dystrophy (PMD). II. Mechanism of the alteration of troponin subunit in PMD, *Proc. Jpn. Acad.*, 52, 260–263, 1976.
57. **Ebashi, E. and Sugita, H.**, The role of calcium in physiological and pathological process of skeletal muscle, in *Current Topics in Nerve and Muscle Research*, selected papers of the symposia held at the IV Int. Congr. *Neuromuscular Diseases*, Montreal, Canada, 1978, 73–84.
58. **Wrogemann, K. and Pena, S. D.**, Mitochondrial calcium load: a general mechanism for cell-necrosis in muscle disease, *Lancet*, I, 672–673, 1976.
59. **Engel, A. G. and Biesecker, G.**, Complement activation in muscle fiber necrosis: demonstration of the membrane attack complex of complement in necrotic fibers, *Ann. Neurol.*, 12, 289–296, 1982.
60. **Cornelio, R. and Dones, I.**, Muscle fiber degeneration and necrosis in muscular dystrophy and other muscle diseases: cytochemical and immunocytochemical data, *Ann. Neurol.*, 16, 694–701, 1984.
61. **Bird, J. W. C., Schwartz, W. N., and Spanier, A. M.**, Degradation of myofibrillar proteins by cathepsins B and D, *Acta Biol. Med. Ger.*, 36, 1587–1604, 1977.
62. **Whitaker, J. N., Bertorini, T. E., and Mendell, J. R.**, Immunocytochemical studies of cathepsin D in human skeletal muscle, *Ann. Neurol.*, 13, 133–142, 1983.
63. **Campbell, K. P. and Kahl, S. D.**, Association of dystrophin and an integral membrane glycoprotein, *Nature*, 338, 259–262, 1989.
64. **Chelly, J., Kaplan, J.-C., Marie, P., Gautron, S., and Kahn, A.**, Transcription of the dystrophin gene in human muscle and non-muscle tissues, *Nature*, 333, 858–860, 1988.
65. **Hoffman, E. P., Hudecki, M. S., Rosenberg, P. A., Pollina, C. M., and Kunkel, L. M.**, Cell and fiber type distribution of dystrophin, *Neuron*, 1, 411–420, 1988.
66. **Feener, C. A., Koenig, M., and Kunkel, L. M.**, Alternative splicing of human dystrophin mRNA generate isoforms at the carboxy terminus, *Nature*, 338, 509–511, 1989.
67. **Love, D. R., Hill, D. F., Dickson, G., Spurr, N. K., Byth, B. C., Marsden, R. F., Walsh, F. S., Edwards, Y. H., and Davies, K. E.**, An autosomal transcript in skeletal muscle with homology to dystrophin, *Nature*, 339, 55–58, 1989.

68. **Partridge, T. A., Morgan, J. E., Coulton, G. R., Hoffman, E. P., and Kunkel, L. M.,** Conversion of mdx myofibers from dystrophin-negative to -positive by injection of normal myoblasts, *Nature,* 337, 176–179, 1989.
69. **Karpati, G., Poulito, Y., Zubrzycka-Gaarn, E., Carpenter, S., Ray, P. N., et al.,** Dystrophin is expressed in mdx skeletal muscle fibers after normal myoblast implantation, *Am. J. Pathol.,* 135, 27–32, 1989.
70. **Cooper, B. J., Valentine, B. A., Wilson, S., Patterson, D. F., Concannon, P. W.,** Canine muscular dystrophy: confirmation of X-linked inheritance, *J. Hered.,* 79, 405–408, 1988.
71. **Valentine, B. A., Cooper, B. J., de Lahunta, A., O'Quinn, R., and Blue, J.,** Canine X-linked muscular dystrophy: clinical studies, *J. Neurol. Sci.,* 88, 69–81, 1988.
72. **Cooper, B. J., Winand, N. J., Stedman, H., Valentine, B. A., Hoffman, E. P., Kunkel, L. M., Scott, M., Fishbeck, K. H., Kornegay, J. N., Avery, R. J., Williams, J. R., Schmickel, R. D., and Sylvester, J. E.,** The homologue of the Duchenne locus is defective in X-linked muscular dystrophy of dogs, *Nature,* 334, 154–156, 1988.

Chapter 17

NEUROLEPTIC MALIGNANT SYNDROME

Chester A. Pearlman

TABLE OF CONTENTS

I. Introduction 388

II. Relationship of NMS to Malignant Hyperthermia 389

III. Pathogenesis of NMS 389

IV. Effects of Neuroleptics on Membranes: Relation to NMS 390

References 391

I. INTRODUCTION

The neuroleptic malignant syndrome (NMS) is a rare, idiosyncratic reaction to neuroleptic drugs.[1-3] The term neuroleptic refers to centrally acting dopamine blocking agents which are widely used in the treatment of schizophrenic, manic (bipolar), and other psychotic disorders, in the behavioral management of delirium and dementia, and to prevent nausea and vomiting. NMS was first described in 1961 by Delay, who was also the first psychiatrist to use these drugs. He emphasized the paradoxical occurrence of fever in conjunction with a pale skin, in contrast to the expected vasodilation. To call attention to the potentially disastrous consequences of failure to stop administration of the drug, he coined the term *syndrome malin des neuroleptiques* because of its superficial resemblence to toxic shock, which is called *syndrome malin* in France.

Subsequent research has indicated that the most common features of NMS are severe (lead pipe) rigidity of limb and/or trunk muscles and fever. Accompanying signs are tachycardia, lability of blood pressure, copious sweating, and disturbances of consciousness. The most common laboratory finding is an elevated serum creatine kinase thought to be related to the muscular rigidity. These abnormalities can lead to serious complications, such as renal failure from myoglobinuria and dehydration, pneumonia and respiratory failure from compromised respiration, and pulmonary emboli so that the mortality in cases of NMS prior to 1980 was over 20%. In recent years, the prognosis has improved considerably due to wider recognition and treatment with dopamine agonists and muscle relaxants.[1,4]

While all clinically used neuroleptics have been implicated in NMS,[1,2] the number of cases involving drugs with high D2 dopamine-receptor blocking potency, such as haloperidol, is much greater than that with low potency agents, such as chlorpromazine, but the implications are unclear because of the much more widespread use of high potency agents. About 60% of cases are male with an average age of about 40, but cases have been reported from infants to age 78. There is no evidence of genetic susceptibility. Some cases have suggested predisposing effects of brain damage, influence of dose or change in dose, or other environmental factors, such as exhaustion or dehydration, but most cases did not involve these elements.[1-3] Estimations of population incidence have varied from 0.1 to 1.4%, but the populations with high incidence had many bipolar patients.[5,6] Patients with affective disorders seem to be more susceptible to NMS just as they are more likely to have other dopamine-related side effects from neuroleptic treatment, such as Parkinsonian motor signs and tardive dyskinesia.[1]

Some puzzling aspects of NMS are the marked variability in onset ranging from a single dose to several years of treatment, although a majority of cases occurred after 3 to 9 days, and the intermittent vulnerability. Many cases had previously received the same or a different neuroleptic without incident, and only about one third of cases subsequently treated with neuroleptics had a recurrence of NMS.[1,7,8]

II. RELATIONSHIP OF NMS TO MALIGNANT HYPERTHERMIA

With respect to the role of membrane pathology, Caroff first noted the similarities between NMS and malignant hyperthermia (MH), a condition involving abnormal muscular contractility which is further discussed in Part V. Severe rigidity and hyperthermia are found in both NMS and MH. Vulnerability to precipitating agents is intermittent and often stress related, and dantrolene, which is the principal treatment for the rigidity of MH,[9] has also been helpful in NMS.[1,2] While studies with neuromuscular blocking agents have demonstrated the CNS source of NMS rigidity in contrast to the muscular origin of MH, several patients with NMS have shown susceptibility to MH according to the muscle contracture test.[10,11] Although none developed actual MH during general anesthesia or with other common MH precipitants, such as succinylcholine, a muscle relaxant which is used during electroconvulsive therapy (ECT),[12] one patient had a reaction suggestive of MH with a curariform agent used for muscle paralysis during ECT.[13] This incident was somewhat puzzling because such drugs are regularly used in patients with MH susceptibility because of their supposed lack of direct muscle effects.[9] The significance of these findings is unclear, however, because no episode of MH has ever been reported during the millions of ECT involving use of succinylcholine despite the statistical likelihood that hundreds of patients with MH susceptibility were at risk.[10] One patient with evidence of MH susceptibility had a slight febrile response to ECT with dantrolene used for muscle relaxation which did not recur when the dose of dantrolene was increased.[14] It seems likely that pharmacokinetic aspects of the use of succinylcholine for ECT, or even some aspect of ECT itself, are responsible for the absence of episodes of MH.[15] Thus, clarification of the implications of MH susceptibility in NMS patients will require further study.

III. PATHOGENESIS OF NMS

Research on the pathogenesis of NMS has focused on dopaminergic mechanisms of temperature regulation in the hypothalamus and dopaminergic involvement in regulation of muscle tone by the basal ganglia.[1] For example, blockade of post-synaptic dopamine receptors in the anterior hypothalamus of the rat by local injection of a neuroleptic prevented the normal vasodilation of the tail in response to a heat load with resultant hyperthermia. This effect could account for the pale fever described by Delay as well as the copious sweating resulting from utilization of sweat glands for heat dissipation. This factor is also partially responsible for the susceptibility to heatstroke of patients treated with neuroleptics, but since NMS occurs at ordinary temperatures, an endogenous heat load is apparently involved. Obvious sources are the muscular rigidity and, in the absence of rigidity, the severe motor agitation of some psychotic, especially manic, patients. While the precise cause of neuroleptic-induced rigidity remains unknown, some suggestive mechanisms have been proposed involving dopaminergic regulation of a feedback motor-control loop from cortex to basal ganglia to thalamus and cortex.[16] Disturbance of dopaminergic function, as in Parkinson's disease or with neuroleptic treatment, results in simultaneous facilitation of agonist and antagonist muscles, which is manifested as rigidity. Intracellular recordings in animals have shown that treatment with neuroleptics for a few days actually produces a depolarization inactivation of the dopaminergic innervation of the basal ganglia,[17] so that compensatory mechanisms must protect most patients from severe rigidity and potential NMS. While several peptides, such as neurotensin, cholecystokinin, substance P, and endogenous opioids, have been shown to modulate various animal models of this mechanism of rigidity, none has helped explain the intermittent vulnerabiltiy.[1] One suggestive clinical report found a low serum iron in 19 of 20 patients with NMS which resolved spontaneously with improvement in the symptoms of NMS.[18] Similar iron deficiency (without hematologic signs) has been shown to impair dopaminergic function in some animal models[19] and is associated with akathisia, a motoric restlessness which is a side effect of neuroleptic treatment.[20] Another, more speculative, hypothesis about vulnerability of NMS is derived from the evidence[16,21] of differential localization and function of D1 and D2 dopamine receptors in the basal ganglia. Cellular developmental research has shown that the basal ganglia contain discrete areas, called striosomes, intermingled with a relatively undifferentiated matrix. D1 receptors appear to be primarily located in striosomes. Behavioral studies have suggested a synergistic role of D1 and D2 pharmacologic agonists in regulation of motor function. While the antipsychotic and most motor effects of neuroleptics appear to be mediated by D2 receptors, vulnerability to NMS might involve deficient compensatory response of the striosomal D1 function to D2 inactivation by neuroleptics.

A recent study of muscle pathology in a case of NMS[22] concluded that the evidence of uncoupled phosphorylation suggested a muscular origin of the hyperthermia. Work summarized by Britt,[23] however, has indicated that such excitation-contraction uncoupling can result from high fever as well as contribute to its continued presence. The beneficial effect of dantrolene in NMS appears to derive from its interruption of this vicious cycle by facilitation of reabsorption of calcium from the myoplasm into the sarcoplasmic reticulum with restoration of normal muscular relaxation.[1]

IV. EFFECTS OF NEUROLEPTICS ON MEMBRANES: RELATION TO NMS

Other research has provided more direct evidence of effects of neuroleptics on cell membranes. One study found evidence of enhanced lipid peroxidation in the cerebrospinal fluid of patients taking phenothiazines, including one case of NMS,[24] that suggested possible membrane damage. Some patients have had an elevation of serum creatine kinase associated with neuroleptic treatment without rigidity or other evidence of NMS.[25-27] This leakage of creatine kinase before the smaller myoglobin molecule suggested some factor besides increased membrane porosity. Another study found alterations in structural order (the reciprocal of fluidity) of brain membranes from rats receiving various neuroleptics.[28] Several *in vitro* studies have shown inhibition by neuroleptics of calmodulin function and other enzymatic processes.[29] The relevance of these findings to NMS, however, is unclear. The *in vitro* studies involved micromolar concentrations in contrast to the nanomolar concentrations observed clinically.[30] The lipid peroxidation and membrane order findings were more prominent with low potency phenothiazines and appeared to be dose related in contrast to the greater association of NMS with high potency agents without dose relationship. Nevertheless, the possibility that vulnerability to NMS is associated with some membrane effect of neuroleptics remains an intriguing hypothesis for future research. From this perspective, it is of interest that psychotic states are frequently associated with an elevated serum creatine kinase of muscular origin without neuroleptic exposure,[31] suggesting the possibility of a similar membrane defect in the central nervous system and the muscles of such patients.

Another potential line of research might focus on the dynamics of intracellular iron. Ordinarily, such iron is stored by binding to the protein ferritin, but it may be released under certain conditions and could contribute to the previously mentioned enhanced lipid peroxidation.[32] Moreover, studies using chelating agents have indicated that alteration of dopamine function by iron deficiency is not related to membrane order but may result from some involvement of iron in the structure of dopamine receptors.[33]

A final speculation comes from reports of reinduction of symptoms of NMS by lithium treatment.[34] While this might possibly have resulted from interference by lithium with striatal dopamine synthesis[35] or some other process related to induction of Parkinsonian rigidity,[36] the time course was more consistent with a second-messenger effect of lithium on dephosphorylation of a phosphate,[37] which might be related to one of the previously discussed membrane mechanisms.

REFERENCES

1. **Pearlman, C.,** Neuroleptic malignant syndrome: a review of the literature, *J. Clin. Psychopharmacol.,* 6, 257–273, 1986.
2. **Lazarus, A., Mann, S. C., and Caroff, S. N.,** *The Neuroleptic Malignant Syndrome and Related Conditions,* American Psychiatric Press, Washington, D.C., 1989.

3. **Caroff, S. N. and Mann, S. C.**, Neuroleptic malignant syndrome, *Psychopharmacol. Bull.*, 24, 25–29, 1988.
4. **Shalev, A., Hermesh, H., and Munitz. H.**, Mortality from neuroleptic malignant syndrome, *J. Clin. Psychiatry*, 50, 18–25, 1989.
5. **Keck, P. E., Jr., Pope, H. G., Jr., and McElroy, S. L.**, Frequency and presentation of neuroleptic malignant syndrome: a prospective study, *Am. J. Psychiatry*, 144, 1344–1346, 1987.
6. **Friedman, J. H., Davis, R., and Wagner, R. L.**, Neuroleptic malignant syndrome: the results of a 6-month prospective study of incidence in a state psychiatric hospital, *Clin. Neuropharmacol.*, 11, 373–377, 1988.
7. **Susman, V. L. and Addonizio, G.**, Recurrence of neuroleptic malignant syndrome, *J. Nerv. Ment. Dis.*, 176, 234–241, 1988.
8. **Wells, A. J., Sommi, R. W., and Crimson, W. L.**, Neuroleptic rechallenge after neuroleptic malignant syndrome: case report and literature review, *Drug Intell. Clin. Pharm.*, 22, 475–479, 1988.
9. **Gronert, G. A.**, Malignant hyperthermia, *Anesthesiology*, 52, 395–423, 1980.
10. **Caroff, S. N., Rosenberg, H., Fletcher, J. E., Heiman-Patterson, T. D., and Mann, S. C.**, Malignant hyperthermia susceptibility in neuroleptic malignant syndrome, *Anesthesiology*, 67, 20–25, 1987.
11. **Araki, M., Takagi, A., Higuchi, I., and Sugita, H.**, Neuroleptic malignant syndrome: caffeine contracture of single muscle fibers and muscle pathology, *Neurology*, 38, 297–301, 1988.
12. **Hermesh, H., Aizenberg, D., Lapidot, M., and Munitz, H.**, Risk of malignant hyperthermia among patients with neuroleptic malignant syndrome and their families, *Am. J. Psychiatry*, 145, 1431–1434, 1988.
13. **Grigg, J. R.**, Neuroleptic malignant syndrome and malignant hyperthermia, *Am. J. Psychiatry*, 145, 1175, 1988.
14. **Franks, R. D., Aoueille, B., III, and Mahowald, M. C.**, ECT use for a patient with malignant hyperthermia, *Am. J. Psychiatry*, 139, 1065–1066, 1982.
15. **Johnson, G. C. and Santos, A. B.**, More on ECT and malignant hyperthermia, *Am. J. Psychiatry*, 140, 266–267, 1983.
16. **Penney, J. B., Jr. and Young, A. B.**, Striatal inhomogeneities and basal ganglia function, *Movement Disorders*, 1, 3–15, 1986.
17. **Bunney, B. S.**, Antipsychotic drug effects on the electrical activity of dopaminergic neurons, *Trends Neurosci.*, 7, 212–215, 1984.
18. **Rosebush, P. I. and Mazurek, M. F.**, Serum abnormalities in neuroleptic malignant syndrome, *Am. Psychiatr. Assoc. Annu. Meet.*, New Research 77, 1988.
19. **Youdim, M. B. H.**, Brain iron metabolism: biochemical and behavioral aspects in relation to dopaminergic neurotransmission, in *Handbook of Neurochemistry*, Vol. 10, Lajtha, A., Ed., Plenum Press, New York, 1985, 731–755.
20. **Brown, K. W., Glen, S. E., and White. T.**, Low serum iron status and akathisia, *Lancet*, I, 1234–1236, 1987.
21. **Clark, D. and White, F. J.**, Review: D1 dopamine receptor — the search for a function: critical evaluation of the D1/D2 dopamine receptor classification and its functional implications, *Synapse*, 1, 347–388, 1987.
22. **Martin, D. T. and Swash, M.**, Muscle pathology in the neuroleptic malignant syndrome, *J. Neurol.*, 235, 120–121, 1987.
23. **Britt, B. A.**, Etiology and pathology of malignant hyperthermia, *Fed. Proc.*, 38, 44–48, 1979.
24. **Pall, H. S., Williams, A. C., Blake, D. R., and Lunec, J.**, Evidence of enhanced lipid peroxidation in the cerebrospinal fluid of patients taking phenothiazines, *Lancet*, II, 596–599, 1987.

25. **Pearlman, C., Wheadon, D., and Epstein, S.**, Creatine kinase elevation after neuroleptic treatment, *Am. J. Psychiatry,* 145, 1018–1019, 1988.
26. **Naganuma, H. and Fujii, I.**, The effect of intramuscular injection of neuroleptics on serum CPK, *Kyushu N-psych.,* 32, 46–48, 1986.
27. **Naganuma, H., Sasaki, I., ad Fujii, I.**, Oral administration of neuroleptics and CPK, *Kyushu N-psych.,* 34, 209–212, 1988.
28. **Cohen, B. M. and Zubenko, G. S.**, In vivo effects of psychotropic agents on the physical properties of cell membranes in the rat brain, *Psychopharmacology,* 86, 365–368, 1985.
29. **Weiss, B., Prozialeck, W. C., and Wallace, T. L.**, Interaction of drugs with calmodulin, *Biochem. Pharmacol.,* 31, 2217–2226, 1982.
30. **Seeman, P.**, Anti-schizophrenic drugs — membrane receptor sites of action, *Biochem. Pharmacol.,* 26, 1741–1748, 1977.
31. **Meltzer, H. Y.**, Neuromuscular dysfunction in schizophrenia, *Schizophr. Bull.,* 2, 106–135, 1976.
32. **Samokyszyn, V. M., Thomas, C. E., Reif, D. W., Saito, M., and Aust, S. D.**, Release of iron from ferritin and its role in oxygen radical toxicities, *Drug Metab. Rev.,* 19, 283–303, 1988.
33. **Ben-Shachar, D., Finberg, J. P. M., and Youdim, M. B. H.**, Effect of iron chelators on dopamine D_2 receptors, *J. Neurochem.,* 45, 999–1005, 1985.
34. **Susman, V. L. and Addonizio, G.**, Reinduction of neuroleptic malignant syndrome by lithium, *J. Clin. Psychopharmacol.,* 7, 339–341, 1987.
35. **Engel, J. and Berggren, U.**, Effects of lithium on behaviour and brain monoamines, *Acta Psychiatr. Scand.,* 61 (Suppl. 280), 133–142, 1980.
36. **Sansone, M. E. G. and Ziegler, D. K.**, Lithium toxicity: a review of neurologic complications, *Clin. Neuropharmacol.,* 8, 242–248, 1985.
37. **Menkes, H. A., Baraban, J. M., Freed, A. N., and Snyder, S. H.**, Lithium dampens neurotransmitter response in smooth muscle: relevance to action in affective illness, *Proc. Natl. Acad. Sci. U.S.A.,* 83, 5727–5730, 1986.

Index

Dependence phenomena, Ca^{2+} channel regulation in, 33
Depolarization, spreading depression-like, in neuronal damage, 192–194
Dermatan sulfate, HIV replication not inhibited by, 345
Desferoxamine
  in isolated reperfused heart tissue, 258, 260
  in renal ischemia, 293–296, 305
Dextrans, in renal ischemia, 305
Dextran sulfate, against AIDS, 344–346
Diabetes, see also Hyperglycemia
  antioxidants against, 16
  membrane ion channels and
    ATP-sensitive K^+ channel, 89–91
    Ca^{2+}-activated K^+ channel, 87–90
    characterization of diabetes, 81–82
    electrophysiology of pancreatic β-cell, 82–83
    insulin secretion, 80–81
    ion channels in pancreatic β-cell, 82–83
    pancreatic morphology, 80
    planar lipid bilayer, 86–87
    sulfonylureas, 83, 85
  reduced affinity of insulin receptor to insulin in, 10
Diacylglycerol
  in asbestos-associated lung cancer, 137
  in cerebral vasospasm, 229, 231, 234–237
  in gastric acid secretion, 114
  in mitogenic signal transduction, 315, 317
  phospholipase C in production of, 11–12
  in renal ischemia, 301–302
2,4-Diamino-6-piperidineylpyridine 3-oxide, see Minoxidil
Diazoxide
  ATP-sensitive, glibenclamide-sensitive K channel in smooth muscle affected by, 63, 65–66
  for hair growth, 66
  against hypertension, 46, 66
Dibucaine, in renal ischemia, 299
Dielectric strength, of cell membrane, 6
Diene conjugates, in renal ischemia, 293, 300
Diffuse pleural fibrosis, genesis of, 135
1,4-Dihydropyridine antagonists, in voltage-dependent Ca^{2+} channel blockage, 30–32, 34, 36–37
Dihydropyridine derivatives
  against hypertension, 44
  as K channel openers, 46
Diltiazem
  against hypertension, 44
  structural formula for, 28
  in voltage-dependent Ca^{2+} channel blockage, 31
5,5′-Dimethyl-1-pyrroline-N-oxide (DMPO), as spin trap molecule, 249–250, 254–256, 258–259, 262–263
Dimethylthiourea, asbestos toxicity prevented by, 131
Diphenhydramine, in cerebral vasospasm, 232
Diphenylhydantoin
  structural formula of, 33
  in T channel characterization, 29
Disease processes, biological membranes in
  Ca^{2+} channels and, 7–10
  causes for illness, 4
  diseases which do not appear to have membrane involvement, 16–18
  evolution of cell membrane structure, 4–6
  features of cell membranes, 6–7
  ion pumps and, 7, 9–10
  pathology of cell membranes and, 11–15
  pharmacologic approach to membrane-linked diseases, 15–16
  regulation through membrane receptors and, 10
DMPO, see 5,5′-Dimethyl-1-pyrroline-N-oxide
DNA damage, in asbestos-induced membrane damage, 132–133
Donnan equilibrium, osmotic effects of, 292
Dopamine
  in cerebral vasospasm, 213
  in fluorescence between smooth muscle cells, 102–103
  in gastric microcirculatory system, 107, 110
  in neuroleptic malignant syndrome, 388–391
Dopamine-β-hydroxylase-negative nerves, catecholaminergic, identification of, 103
Drug withdrawal, Ca^{2+} channel regulation in, 33
Duchenne muscular dystrophy
  clinical features of, 368–369
  gene in, 369–371
  hypercontracted fiber incidence in, 377
  immunohistochemical analysis of DMDP expression in, 372
  membrane hypothesis of, 376–378
Dysmenorrhea, Ca^{2+} channel antagonists against, 30
Dystrophin
  defect of in Duchenne/Becker muscular dystrophies
    clinical features of DMD/BMD muscular dystrophies, 368–369
    cloning of DMD/BMD gene, 369, 370
    localization of DMD/BMD gene to Xp21, 369
    mdx mouse, 379–380
    membrane hypothesis of DMD, 376–378
  discovery of, 370–371
  expression pattern of, 374
  family of, 379
  membrane localization of, 371–375
  molecular structure of, 371
  structure and subcellular organization of, 378
  tissue specific isoforms of, 378–379

E

Early genes, see Oncogenes
Eclampsia, Ca^{2+} channel antagonists against, 30
EDHF, see Endothelium-derived hyperpolarizing factor
EDRF, see Endothelium-derived relaxing factor
EDTA, in renal ischemia, 294
EEG, see Electroencephalogram
EGF, see Epidermal growth factor
EGTA, in renal ischemia, 302–303
Ehrlich's ascites tumor model, anticancer activity of prostaglandin derivatives in, 356–358
Eicosanoids
  in cerebral vasospasm, 213, 226

in renal ischemia, 303–304
Elastase, in tumor local invasion, 324
Electroencephalogram (EEG), in hypoxia and ischemia, 186
Electron paramagnetic resonance (EPR) spectroscopy
  absorption shapes in, 244
  applications of
    direct measurements on isolated reperfused hearts, 251–254
    in vivo methods, 263–266
    spin-trapping measurements on cells, 257–263
    spin-trapping measurements on isolated reperfused hearts, 254–258
  background information on, 242–246
  line shapes in, 245
  oxygen free radicals and, 247–250
  spectral shapes in, 246
  Zeeman splitting in, 243, 245
Electron transport, microsomal, schematic representation of, 329
Emboli, heterotypic, in tumor migration to secondary sites, 326
Endoplasmic reticulum, in cancer, 328–331
Endothelin
  in cerebral vasospasm, 231
  in gastric microcirculatory system, 107–108, 113
Endothelium-derived hyperpolarizing factor (EDHF), hyperpolarization induced by, 59
  in cerebral vasospasm, 228, 237
  in stomach, 107
Endothelium-derived relaxing factor (EDRF)
Energy depletion, in cell injury, 184
Energy metabolism, in living creatures, 4–5
Energy shortage, metabolic, pathways of cell destruction caused by, 185
Enteric nervous regulation, of gastrointestinal tract, 120
Envelope glycoproteins, HIV-1, inhibition of maturation of, 339–340
env gene, in AIDS, 338
Enzymes, see also specific enzymes
  in protection against oxygen toxicity, 6
  in protein phosphorylation, 10
EP-A-0321273, EP-A-0321274, EP-A-0326297, EP-A-0354553, as K channel opener, 46
Epidermal growth factor (EGF)
  in inositol lipid metabolism, 152–156
  in mitogenic signal transduction, 316, 317
  in tumor local invasion, 324
Epilepsy, Ca^{2+} channel antagonists against, 30
Epinephrine, in cerebral vasospasm, 213
EPR, see Electron paramagnetic resonance
*erb*B oncogene, in expression of growth factors and growth factor receptors, 318, 320–321
Esophageal spasm, Ca^{2+} channel antagonists against, 30
Estrogen, in voltage-dependent Ca^{2+} channel regulation, 35, 37
Evans Blue, against AIDS, 347
Evolution, of cell membrane structure, 4–6
Excitable cells
  ATP-dependent K channel in, 52–54

Ca^{2+} in, 24
signals produced by membranes of, 7
Excitotoxic hypothesis, of neuronal damage, 196–197
Extracellular matrix, in tumor cell spreading to secondary sites, 320, 322
Extravasation, in tumor migration to secondary sites, 328

F

Falck-Hillarp's method, in catecholaminergic nerve location, 99, 101–102, 118–119
Fascia dentata, in hypoxia, 193
Fenton reaction
  in electron paramagnetic resonance spectroscopy, 248
  in isolated reperfused heart tissue, 255, 257
  reactive oxygen metabolites and, 129
Feroxamine, in renal ischemia, 295
Ferrioxamine, in renal ischemia, 300
Ferritin, in renal ischemia, 295–296
FFAs, see Free fatty acids
Fibrin degradation products, in cerebral vasospasm, 213
Fibronectin receptor, in tumor local invasion, 325–326
Fibrosis
  in asbestos-induced diseases, 133–135
  in cerebral vasospasm, 211, 215
Flavin mononucleotide, in renal ischemia, 296
Flunarizine, against hypertension, 44
5-Fluorouracil, OC-5186 with, against cancer, 357–359
Formaldehyde-induced fluorescence method, in catecholaminergic nerve location, 99, 101–102, 118–119
FR-C8653, against hypertension, 44
Free fatty acids (FFAs), in renal ischemia, 299–300
Free radicals, see Oxygen free radicals
Fucoidan, as HIV reverse transcriptase inhibitor, 345

G

gag gene products, in HIV-1, inhibition of myristylation of, 340
Gallopamil, against hypertension, 44
Gangliosides, in shortening time spent in spreading depression-like polarization, 193
Gastric acid, mechanism of secretion of, 114
Gastric mucosal injury
  arachidonic acid cascade in, 279–281
  background information on, 272
  ischemia-reperfusion in, 275–278
    microcirculatory disturbances and autonomic nervous receptors in
      alteration of autonomic nervous activity and receptor distribution in gastric ulcer formation, 118–119
    autonomic nervous system distribution in stomach, 96–104
    enteric nervous regulation of gastrointestinal tract, 120

vascular effector effector receptor distribution in gastric microcirculatory system, 105–117
oxygen free radicals in, 272–273, 278–281
platelet-activating factor in, 274
stress in, 274–275
superoxide dismutase inhibitor in, 278
Gastrin-releasing peptide (GRP), in stomach, 103, 107, 111, 114
Gastrointestinal tract, enteric nervous regulation of, 120
Gelatinase, in tumor local invasion, 324
Gelsolin, in tumor migration to secondary sites, 327
G-factor, in electron paramagnetic resonance spectroscopy, 243
Glaucoma, Ca^{2+} channel antagonists against, 30
Glibenclamide
 ATP-sensitive K channel in smooth muscle and, 53–54, 130
 ion channels in diabetes and, 83
Glomerulus, diagrammatic representation of, 289
Glucose, as extracellular antioxidant, 130
Glutamate, in hypoxic spreading depression-like depolarization, 190, 197
Glutamic acid, excitatory amino acid receptors for, Ca^{2+} channels and, 32
Glutathione
 as cellular antioxidant, 6, 130
 in cerebral vasospasm, 217, 223
 in renal ischemia, 293
Glutathione peroxidase
 in cerebral vasospasm, 217, 222
 in renal ischemia, 291
Glycine, in renal ischemia, 295, 300
Glycolipid sulfonic acid, against AIDS, 347
Glycolysis
 anaerobic, in hypoxia, 182
 in ATP synthesis, 5
 in mitochondrial membranes, in cancer, 332
Glycyrrhizn, against AIDS, 346
Glycyrrhizn sulfate, against AIDS, 346
gp41 transmembrane glycoprotein, encoding of, 338
gp120 envelope glycoprotein
 encoding of, 338
 inhibition of maturation of, 339–340
G-proteins
 Ca^{2+} channels and, 29, 31, 38
 in mitogenic signal transduction, 316–317
 permeability of ion channels regulated by, 10–11
Granulation tissue, formation of, in asbestos-induced diseases, 133
Green algae, appearance of on Earth, 5
Growth factor receptors, in mitogenic signal transduction, 315–317
Growth factors, gene expression of, in cancer, 318–321
Growth stimuli, inositol lipid metabolism and, 149–157
GRP, see Gastrin-releasing peptide
GTP, see Guanosine triphosphate
Guanosine triphosphate (GTP)
 binding of to G-protein, 10–11
 in renal ischemia, 300

Guanylate cyclase, in renal ischemia, 303
GVIA, in N channel characterization, 29

H

H^+
 in mitogenic signal transduction, 317–318
 in renal ischemia, 291
H-7, in cerebral vasospasm, 232–233, 235
Haber-Weiss reaction
 in EPR spectroscopy, 248
 reactive oxygen metabolites and, 16, 129, 131
 in renal ischemia, 294
Hair growth
 diazoxide for, 66
 minoxidil for, 45, 66
Haloperidol, in neuroleptic malignant syndrome, 388
Heart
 isolated reperfused, free radical generation in, 251–257
 voltage-dependent Ca^{2+} channel regulation in, 35–37
Heart attack, see Myocardial infarction
Hematoma, in destruction of local brain parenchyma, 210
Hemoglobin, structure of, 5
Heparin
 against AIDS, 344–345
 in Ca^{2+} release, 48–49
Hepatoma, endoplasmic reticulum membranes in, 330
Hexokinase, in mitochondrial membranes, in cancer, 332–333
High-energy state, required by calcium pump mechanism, 9
Hippocampus, in hypoxia, 187, 189, 192–195, 198
Histamine
 in cerebral vasospasm, 213
 in gastric microcirculatory system, 108–109, 114–115
HIV-1, see Human immunodeficiency virus 1
Hormones, see also specific hormones
 arachodonic acid in synthesis of, 11
 in Ca^{2+} metabolism and mobilization, 24–26
 in cAMP regulation, 10
12-HPETE, in cerebral vasospasm, 213, 221, 226–228, 235
15-HPETE, in cerebral vasospasm, 213, 217–218, 220, 225–226, 231
Human immunodeficiency virus 1 (HIV-1)
 background information on, 17, 338–339
 inhibition of maturation of HIV-1 envelope glycoproteins, 339–340
 inhibition of myristylation of gag gene products, 340
 MR-356 against, 360
 OC-3186 against, 360
 OC-5186 against, 359–360
 polyanionic compounds active at cell surface against anionic compounds, 346–347
 dextran sulfate and related molecules, 344–346
 phosphorothioate oligodeoxynucleotides, 347–348

sulfonic acids, 347
sCD4 as therapy for, 341–344
Hydrogen peroxide
  in cerebral vasospasm, 213
  in EPR spectroscopy, 247
Hydroperoxide, in cerebral vasospasm, 213
6-Hydroxydopamine, in voltage-dependent Ca^{2+} channel regulation, 35
Hydroxyeicosatetraenoic acids, in cerebral vasospasm, 219–220, 224–228, 235, 237
Hydroxyl radical
  in electron paramagnetic resonance spectroscopy, 247–251
  free iron in production of, 16
  in renal ischemia, 293–294
  reoxygenation after deoxygenation in production of, 13
4-Hydroxynonenal, in renal ischemia, 303
Hydroxyquinone, against cerebral vasospasm, 217
5-Hydroxytryptamine, see Serotonin
Hyperfine coupling constant, in EPR spectroscopy, 245–246, 249
Hyperglycemia, in retardation of hypoxic spreading depression, 194, see also Diabetes
Hyperparathyroidism, in hypertension, 25
Hypertension
  Ca^{2+} channel antagonists against, 28–30, 44
  cromakalim against, 45, 65
  diazoxide against, 46, 66
  essential
    Ca^{2+} in, 25
    hyperparathyroidism in, 25
    parathyroid hormone in, 25
    vitamin D in, 25
  minoxidil against, 45
  nicorandil against, 44–45
  pinacidil against, 65–66
  pulmonary, Ca^{2+} channel antagonists against, 30
  voltage-dependent Ca^{2+} channel regulation in, 36–37
Hyperthyroidism, voltage-dependent Ca^{2+} channel regulation in, 35, 37
Hypertrophic cardiomyopathy
  Ca^{2+} channel antagonists against, 29–30
  voltage-dependent Ca^{2+} channel regulation in, 36
Hypoglycemia, in acceleration of hypoxic spreading depression, 194
Hypothyroidism, voltage-dependent Ca^{2+} channel regulation in, 35, 37
Hypoxanthine, in renal ischemia, 294
Hypoxia
  neuronal membranes affected by
    background information on, 182–183
    Ca^{2+} concentration in cell survival, 184–186
    Ca^{2+} elevation in neuronal damage, 195–196
    delayed neuronal damage compared to, 197
    general consequences of tissue hypoxia, 183–184
    NMDA receptor role in, 196–197
    prolonged spreading depression-like depolarization damages neurons, 192–194
    reversible arrest of function in hypoxic central nervous system, 186–188

spreading depression-like membrane response of hypoxic neurons, 188–194
pathological disorders in, 12
sickle hemoglobin and, 16

I

IGF, see Insulin-like growth factor
IGF-1, see Insulin-like growth factor 1
IL-1, see Interleukin 1
IL-6, see Interleukin 6
Illness, causes for, 4
Indomethacin, in renal ischemia, 295, 299, 305
Inflammatory response
  in asbestos-induced diseases, 133–135
  in cerebral vasospasm, 215, 226
  in myocardial ischemia, 16
Inorganic substances, in cerebral vasospasms, 213
Inositol lipids
  calpain and, in mitogenic signaling, 166–170
  metabolism of, 143–157
  in renal ischemia, 302
Inositol triphosphate (IP_3)
  in Ca^{2+} release, 48–49
  in cerebral vasospasm, 229
  in gastric acid secretion, 114
  in mitogenic signal transduction, 317
  phospholipase C in synthesis of, 11–12
  pinacidil-induced inhibition of synthesis of, 65
  in renal ischemia, 301–302
Insulin
  in diabetes, 10, 80–82, 85–86, 89
  growth factor stimulation of, 155–156
  in obesity, 10
  scheme for secretion of, 85
  in voltage-dependent Ca^{2+} channel regulation, 35, 37
Insulin-like growth factor (IGF), in mitogenic signal transduction, 316–317
Insulin-like growth factor 1 (IGF-1), inflammatory macrophage release of, in asbestos-induced diseases, 134
Integrins, in tumor local invasion, 325
Interleukin 1 (IL-1), inflammatory macrophage release of, in asbestos-induced diseases, 134
Interleukin 6 (IL-6), inflammatory macrophage release of, in asbestos-induced diseases, 134
Interstitial connective tissue, in tumor cell spreading to secondary sites, 322
Intestinal hypermotility, Ca^{2+} channel antagonists against, 30
Intracranial pressure, elevated, in local destruction of brain parenchyma, 210
Ion channels, diabetes and, 80–91
Ion fluxes, changes in, in smooth muscle, 54–58
Ion pumps
  in disease processes, 7, 9–10
  regulation of, 9–10
IP_3, see Inositol triphosphate
Iron
  in hydroxyl radical production, 16
  in isolated reperfused heart tissue, 256–257

in neuroleptic malignant syndrome, 390–391
in renal ischemia, 291, 294–296
Ischemia
  arrest latency in, 186
  Ca^{2+} concentration in, 13–14
  cerebral, Ca2+ channel antagonists against, 30
  hypoxia distinguished from, 182
  myocardial
    Ca^{2+} in, 27
    free radical generation and, 251–252, 254–258, 264, 266
    inflammatory response after, 16
  pathological disorders in, 12–14
  renal, 288–305
  subarachnoid hemorrhage in, 210
  transient, interstitial ion levels and extracellular potential of rat cerebral neocortex during, 191
  voltage-dependent Ca^{2+} channel regulation in, 36
Ischemia-reperfusion injury
  antioxidants against, 16
  Ca^{2+} channel blockers against, 15–16
  characterization of, 13
  of gastric mucosa, 275–278
  K efflux inhibitors against, 15–16
  oxygen free radicals in, 14
  phospholipase inhibitors against, 15–16
  proteinase inhibitors against, 15–16
  schematic representation of, 15

J

Juxtaglomerular apparatus, diagramatic representation of, 289

K

K^+
  in cerebral vasospasm, 213
  depolarization, in voltage-dependent Ca^{2+} channel regulation, 35
  distribution of across cell membrane, 7
  in hypoxic spreading depression-like depolarization, 190, 199
  in mitogenic signal transduction, 317
  in renal ischemia, 290, 292
Kainic acid, in voltage-dependent Ca^{2+} channel regulation, 35, 37
K^+ channels
  in diabetes, 87–91
  in disease processes, 8
  schematic representation of, 9
  in vascular smooth muscles
    background information on, 44–46
    in excitable cells, 52–54
    K channel opener effects on, 54–65
    macroscopic K current in smooth muscles, 47–50
    unitary K current in smooth muscles measured using patch-clamp procedures, 50–52
K^+ efflux inhibitors, against ischemia-reperfusion injury, 15–16

2-Ketoisocapoate, effects of on ATP-sensitive, glibenclamide-sensitive K channel in smooth muscle, 63
Kidneys
  anatomy and physiology of, 286–289
  oxidative stress in, 292–294
    calcium role in, 297–301
    iron role in, 294–296
  renal ischemia in, 289–290
    cold, 291–292
    warm, 290–291
  second messengers in, 301–303
$KMnO_4$ fixation, of cholinergic nerves, 99
Krebs-Henseleit solution, in cerebral vasospasm, 233, 235
KRN 1391, as K channel opener, 46
Kynurenine, in cerebral vasospasm, 213

L

Labor, premature, Ca^{2+} channel antagonists against, 30
Lactic acid, in hypoxia, 194
Lähmungszeit, in anoxia and ischemia, 186
Lambert-Eaton myasthenic syndrome, voltage-dependent Ca^{2+} channel regulation in, 36–37
Laminin receptor, in tumor local invasion, 325–326
Latch bridges, in cerebral vasospasm, 229
Late spasm, protein kinase C in, 233–237
L-band bridge, in electron paramagnetic resonance spectroscopy, 263
L channels, Ca^{2+}, 29–30, 32, 36–38, 81
Lead, in voltage-dependent Ca^{2+} channel regulation, 35
Lectins, in tumor migration to secondary sites, 326, 328
Lemakalim, see Cromakalim
Lentinan sulfate, against AIDS, 344
Leukocyte adhesion, in cerebral vasospasm, 210
Leukotrienes
  arachodonic acid in synthesis of, 11
  in cerebral vasospasm, 213, 220, 225, 228
  in renal ischemia, 299, 304
Lipid hydroperoxides, in cerebral vasospasm, 221, 223, 226, 228–229
Lipid peroxidation
  in asbestos-induced membrane damage, 132
  in cerebral vasospasm, 215–227
  phenothiazines and, 391
  in renal ischemia, 293–296, 300, 303
Lipoxins, arachadonic acid in synthesis of, 11
Lipoxygenase pathway
  in cerebral vasospasm, 220–221, 223, 225–226, 228–229, 237
  in hormone synthesis, 11
  in renal ischemia, 304
Lithium, in neuroleptic malignant syndrome, 391
Lung cancer, asbestos in, 137–138
Lungs, entry and retention of asbestos fibers in, 126–127
LY211808, LY222675, as K channel openers, 46
Lysophosphatide residues, in renal ischemia, 299
Lysosomal changes, in cell injury, 184

M

Macrophages, lung, biology and function of, 127–128
Magnesium deficiency, in cerebral vasospasm, 213
Malaria
　possible cell membrane importance in, 17
　prostaglandin derivatives against, 362–363
Malignant hyperthermia, neuroleptic malignant syndrome relationship to, 389
Malignant mesothelioma, asbestos in, 124–125, 136–138
Malonaldehyde, in cross-linking of proteins, 132
Manic syndrome, Ca^{2+} channel antagonists against, 30
Manidipine, against hypertension, 44
Mannitol, in renal ischemia, 291, 293, 305
Mannoheptulose, in inhibition of glucose metabolism, 84
mdx mouse, muscular dystrophy treatment and, 379–380
Membrane potential, changes in, in smooth muscle, 54–58
Membrane receptors, in tumor cells, 142–170
Messengers, cellular, see also specific messengers
　phospholipase C in production of, 11
　in protein phosphorylation by protein kinase, 10
Metastasis
　characterization of, 320
　local invasion and, 322–326
　migration to secondary sites and, 326–328
　schematic representation of, 321
　steps of, 322–323
Methemoglobin, oxyhemoglobin conversion to, 216
Methionine, oxidation of, 132
Methysergide, in cerebral vasospasm, 232
Microcirculatory disturbances, in acute gastric mucosal lesions, 105–117
Migraine, Ca^{2+} channel antagonists against, 30
Minoxidil
　ATP-sensitive, glibenclamide-sensitive K channel in smooth muscle and, 64
　for hair growth, 45, 66
　against hypertension, 45
Mitochondria
　Ca^{2+} uptake by, 184
　in cancer, 331–334
　dysfunction of, in cell injury, 184
　energy transduction in, 6
　membrane structure of, 6
　oxygen free radical production by, 13
Mitogenic pathways, in mammalian cells, schematic representation of, 316
Mitogenic signals
　in normal and transformed cells, 166–170
　transduction of, in cancer, 315–318
Monoclonal antibodies, to receptor proteins, 16
Morphine, in voltage-dependent Ca^{2+} channel regulation, 35, 37
Motion sickness, Ca^{2+} channel antagonists against, 30
Mouse leukemia virus, against HIV-1, 360–361
MR-256, synthesis of, 356
MR-356
　against HIV-1, 360
　synthesis of, 356–357
Multicellular organisms, appearance of on Earth, 5
Multidrug resistance, in cancer chemotherapy, 17, 359
Muscle cells
　Ca^{2+} in contraction of, 13–14
　protein kinase C-mediated contraction of, 228–231
　signals produced by membranes of, 7
　voltage-dependent Ca^{2+} channel regulation in, 3, 37
Muscular dysgenesis, voltage-dependent Ca^{2+} channel regulation in, 36
Muscular dystrophy, possible cell membrane importance in, 17
MVIIA, in N channel characterization, 29
Myasthenia gravis, Ca^{2+} channel regulation in, 33
Myocardial infarction
　antioxidants against, 16
　Ca^{2+} channel antagonists against, 30
　ischemia-reperfusion injury in, 13
Myonecrosis, in cerebral vasospasm, 211, 228
Myosin, phosphorylation of, and light chain kinase, in cerebral vasospasm, 214, 228–230, 232, 234, 237
Myristate, in inhibition of HIV-1 replication, 340
N-Myristoyl glycinal diethylacetal, in inhibition of myristoylation of p17, 340

N

Na⁻
　distribution of across cell membrane, 7
　in hypoxic spreading depression-like depolarization, 188, 197
　in mitogenic signal transduction, 317–318
　in renal ischemia, 290
Na-Ca exchanger
　in disease processes, 8
　schematic representation of, 9
Na⁺ channels
　in disease processes, 8
　schematic representation of, 9
Na⁺K⁺-ATPase, depression of membrane-bound, in cell injury, 184
N channels, Ca^{2+}, 29, 32–33
Necrosis, in cerebral vasospasm, 211, 228
nef gene, in AIDS, 338–339
Nephron, diagramatic representation of, 288
neu family oncogenes, in expression of growth factors and growth factor receptors, 320
Neuroleptic malignant syndrome
　background information on, 388–389
　effects of neuroleptics on membranes, 390–391
　malignant hyperthermia relationship to, 389
　pathogenesis of, 389–390
Neuronal membranes
　hypoxia effects on, 182–199
　signals produced by, 7

Neuropeptide Y (NPY), in stomach, 103–104, 107, 111
Neurotransmitter antagonists, in cerebral vasospasm, 233
Neurotransmitters
  Ca^{2+} in release of, 13–14
  in gastric microcirculatory system, 105–109
Neutrophil adhesion inhibitors, in ischemia-reperfusion injury, 15
Neutrophils, self-eating reaction of, 16
Nicaraven (AVS), against cerebral vasospasm, 217, 223
Nicardipine
  in cerebral vasospasm, 232–233
  against hypertension, 44
Nicorandil
  against angina, 65
  ATP-sensitive, glibenclamide-sensitive K channel in smooth muscle and, 55, 57, 60–63, 65
  against hypertension, 44, 45
2-Nicotineamidoethyl nitrate, see Nicorandil
Nifedipine
  against hypertension, 44
  structural formula for, 28
  in voltage-dependent Ca^{2+} channel regulation, 35, 37
Niguldipine, as K channel opener, 46
Nimodipine
  against hypertension, 44
  in voltage-dependent Ca^{2+} channel regulation, 35
Nisoldipine, against hypertension, 44
Nitrendipine
  against hypertension, 44
  in voltage-dependent Ca^{2+} channel regulation, 35–37
Nitroblue tetrazolium reduction, in superoxide assay, 249
Nitro-compounds, against hypertension, 44
Nitrogen, spectral shapes of, in electron paramagnetic resonance spectroscopy, 246
Nitroglycerin, against angina, 65
N-methyl-D-aspartate (NMDA)
  excitatory amino acid receptors for, 32
  glutamate receptors and, 29
  in hypoxic spreading depression-like depolarization, 190, 192, 194
  in neuronal damage, 196–199
NMDA, see N-methyl-D-aspartate
Norepinephrine
  in cerebral vasospasm, 213, 218
  dopaminergic receptor binding to, 110
  in gastric acid secretion, 114
No-reflow phenomenon, in decrease of radical generation, 252
Noxiustoxin, as K channel inhibitor, 66
NPY, see Neuropeptide Y

O

Obesity, reduced affinity of insulin receptor to insulin in, 10
Obstructive lung disorder, Ca^{2+} channel antagonists against, 30

OC-2186, synthesis of, 356
OC-3186
  against AZT toxicity, 361–362
  against HIV-1, 360
  synthesis of, 356–357
OC-5186
  against cancer, 357–359
  5-fluoracil with, 357–359
  against HIV-1, 359–360
  against malaria, 363
  synthesis of, 356–357
OdC28, against AIDS, 348
Octan-1-ol, structural formula of, 33
Oncogenes, see also specific genes
  in expression of growth factors and growth factor receptors, 318–321
  in mitogenic signal transduction, 317–318
  in plasma membrane interactions, 321
  in tumor cells, 157–165
Organelles, membrane structure of, 6, see also specific structures
Ornithine decarboxylase, in asbestos-associated lung cancer, 137
Oxidative phosphorylation, in ATP synthesis, 5
Oxidative stress, renal, 292–294
Oxygen
  in evolution of life on Earth, 5
  in pathology of cell membranes, 12–15
  spectral shapes of, in electron paramagnetic resonance spectroscopy, 246
  toxicity of, 6, 15
Oxygen free radicals
  in cell injury, 184
  in electron paramagnetic resonance spectroscopy, 242–266
  in gastric mucosal injury, 272–273, 278–281
  in ischemia-reperfusion injury, 14
  reoxygenation after deoxygenation in production of, 13
  scavengers of
    in cerebral vasospasm, 217, 221–223
    in ischemia-reperfusion injury, 1–16
    in renal ischemia, 304
Oxygen metabolites, reactive, in acute asbestos cytotoxicity
  biological damage caused by, 131–133
  biological sources of, 129–130
  chemistry of, 128–129
  intracellular defenses against, 130–131
Oxygen paradox, characterization of, 13
Oxyhemoglobin
  in cerebral vasospasm, 213
  in lipid peroxidation, 216
Oxypurinol, isolated reperfused heart tissue and, 259, 263
Ozone, formation of from oxygen, 5

P

p17 protein, in AIDS, inhibition of myristylation of, 340
P170 glycoprotein

in membrane exposed to anticancer agents, 17
in multi-drug resistance, 359
P1060, see Pinacidil
PAF, see Platelet-activating factor
Pancreas, β-cells in, in diabetes, 80–87, 90
Papaverine derivatives, against hypertension, 44
Paraquat, in induction of lipid peroxidation, 132
Parasympathetic nervous system (PNS), in innervation of gastric mucosa, 99
Parathyroid hormone, in Ca^{2+} metabolism and mobilization, 24–26
Parenchyma, brain, local destruction of, 210
Parkinson's disease, in voltage-dependent Ca^{2+} channel regulation, 36
Paroxysmal supraventricular tachyarrhythmias, Ca^{2+} channel antagonists against, 29
Patch-clamp procedures
  K channel opener effects on K channels in vascular smooth muscle measured via, 58–65
  unitary K current in smooth muscles measured via, 50–52
Pathology, of cell membranes, in disease processes, 11–15
PDA, see Phorbol-1,2-acetate
PDGF, see Platelet-derived growth factor
Pentosan, as HIV reverse transcriptase inhibitor, 345–346
Pentosan polysulfate, against AIDS, 344–345
Peptide-peptide histidine isoleucine, in peptidergic nerve endings, 103
Peptidergic nerves, in stomach, 103–104
Peripheral vascular diseases, Ca^{2+} channel antagonists against, 28, 30
Permeabilities, of ion channels, sequential changes in, 7
Peroxidases, in renal ischemia, 293
Peroxidization, of phospholipids, 11
Peroxylamine, nitrogen coupling and, 252–253
PGOs, see Prostaglandin derivatives
Phagocytosis, frustrated, 130–131
Phallodine, as K channel inhibitor, 66
Pharmacologic approach, to membrane-linked diseases, 15–16
Phenothiazines, in neuroleptic malignant syndrome, 391
Phentolamine, in cerebral vasospasm, 232
Phenylalanine, oxidation of, 132
Phenylalkylamine
  binding site for, at L class of Ca^{2+} channels, 34
  derivatives, against hypertension, 44
Phenylephrine, in voltage-dependent Ca^{2+} channel regulation, 35
Phorbol-1,2-acetate (PDA), in cerebral vasospasm, 231–234
Phorbol esters
  in contraction of canine basilar artery, 231–234
  in mitogenic signal transduction, 317
  in tumor local invasion, 324
Phorbol myristate acetate (PMA), as tumor promoter, 136
Phosphatases, in cerebral vasospasm, 229

Phosphate, in renal ischemia, 292, 305
Phosphatidylcholine, in tumor migration to secondary sites, 327
Phosphatidylinositol
  in cerebral vasospasm, 229
  in mitogenic signal transduction, 315–318
  in renal ischemia, 301–303
Phosphatidylinositol biphosphate, in gastric acid secretion, 114
Phospholipase A_2
  in arachidonic acid production, 11–12
  in phospholipid decomposition, 11
  in renal ischemia, 299, 301, 303
  in tumor migration to secondary sites, 326–327
Phospholipase C
  in arachidonic acid production, 11–12
  in cerebral vasospasm, 229
  in diacylglycerol production, 11–12
  in gastric acid secretion, 114
  in inositol triphosphate production, 11–12
  in mitogenic signal transduction, 316
  in phospholipid decomposition, 11
  in renal ischemia, 303
  in tumor migration to secondary sites, 326–327
Phospholipases
  Ca^{2+} in triggering of action of, 11
  inhibitors of, against ischemia-reperfusion injury, 155–16
  in renal ischemia, 299
Phospholipid bilayer
  in diabetes, 86–87, 90
  as insulating material for cell membrane, 6–8
Phospholipids
  degradation of, in cell injury, 184
  peroxidized, phospholipase A_2 in preferential decomposition of, 11
Phosphorothioate oligodeoxynucleotides, against AIDS, 347–348
Phosphorylation, in regulation of pumps and channels, 9–10, 12
Pinacidil
  against hypertension, 65–66
  ATP-sensitive, glibenclamide-sensitive K channel in smooth muscle and, 57, 61, 64–65
  side effects of, 65
Piperazine derivative, against hypertension, 44
Pirenzepine, in visualization of muscarinic acetylcholine receptors, 107–108
Planar lipid bilayers, in diabetes, 86–87, 90
Plasma membranes, in cancer, 314–328
Plasminogen activators, in tumor local invasion, 324–325
Platelet-activating factor (PAF), in gastric mucosal injury, 274
Platelet adhesion, in cerebral vasospasm, 210
Platelet-derived growth factor (PDGF)
  in inositol lipid metabolism, 149–152, 155–156
  inflammatory, 134
  in mitogenic signal transduction, 316–317
  in tumor local invasion, 324
Pleural plaques, characterization of, 124, 135
PMA, see Phorbol myristate acetate

PN 200-110
  against hypertension, 44
  in voltage-dependent Ca^{2+} channel regulation, 35–36
PNS, see Parasympathetic nervous system
Podosomes, in tumor migration to secondary sites, 326
pol gene, in AIDS, 338
Polymorphonucleocytes, in renal ischemia, 294, 304
Polypeptides, in cerebral vasospasm, 213
Polyunsaturated fatty acids (PUFAs)
  in endoplasmic reticulum membranes, in cancer, 330
  in renal ischemia, 293, 300
Porin, in mitochondrial membranes, in cancer, 332–333
Post-ischemic neuron damage, compared to hypoxic neuron damage, 197–198
Prostacyclins
  arachodonic acid in synthesis of, 11
  in renal ischemia, 299, 304–305
Prostaglandin derivatives (PGOs)
  background information on, 356
  in chemotherapy
    against cancer, 356–359
    against retroviral diseases, 359–362
    against malaria, 362–363
  mechanism of action of, 364
  synthesis of, 356
Prostaglandins
  arachodonic acid in synthesis of, 11
  in cerebral vasospasm, 213, 217–219, 228, 231, 234
  in mitogenic signal transduction, 317
  in renal ischemia, 299
Protease inhibitors
  in inhibition of neutrophil self-eating reaction, 16
  against ischemia-reperfusion injury, 1–16
Proteinases, matrix-degrading, in tumor local invasion, 324
Protein-decomposing enzymes, calcium-activated, Ca^{2+} in triggering of action of, 11
Protein kinase A, in gastric acid secretion, 114
Protein kinase C
  in cerebral vasospasm, 229–237
  in contraction of smooth muscles, 228–231
  in gastric acid secretion, 114
  in mitogenic signal transduction, 317–318
  in phosphorylation of channel proteins, 11–12
  in renal ischemia, 302
Protein kinases, in protein phosphorylation, 10
Protein oxidation, in asbestos-induced membrane damage, 132
Proteins
  Ca^{2+} in synthesis of, 13
  in cerebral vasospasms, 213
Protein-tyrosine kinase pathway, in mitogenic signal transduction, 315–316, 318
Proteoglycanases, in tumor local invasion, 324
Proteoglycans, cell surface, in tumor local invasion, 323
Protooncogenes, see Oncogenes
Protoporphyrin structures, of cytochromes, 5
Psychosomatic diseases, possible cell membrane importance in, 18
PUFAs, see Polyunsaturated fatty acids
Pulmonary hypertension, Ca^{2+} channel antagonists against, 30
Pumps, regulation of, 9–10
Pyrethroids, in T channel characterization, 29
Pyridine nucleotides, reduced, in renal ischemia, 290, 292

Q

Quinones, in renal ischemia, 296
^{3}H-Quinuclidinyl benzilate, in visualization of muscarinic acetylcholine receptors, 105–106
Quisqualate, excitatory amino acid receptors for, Ca^{2+} channels and, 32

R

R24571 calmodulin inhibitor, in cerebral vasospasm, 232
Raffinose, in renal ischemia, 305
ras family oncogenes, in expression of growth factors and growth factor receptors, 319–321
Raynaud's phenomenon, Ca^{2+} channel antagonists against, 29
Receptor-operated channels, Ca^{2+}, 28–29, 31
Receptors, membrane, in disease processes, regulation via, 10
Red blood cell ghosts, in cerebral vasospasm, 213, 216, 221
Reserpine, in voltage-dependent Ca^{2+} channel regulation, 35
Respiratory burst system
  in renal ischemia, 294
  tumor promoters as stimulants of, 136
rev gene, in AIDS, 339
RGD recognition receptors, in tumor local invasion, 325
Ribofuranan sulfate, against AIDS, 344
RLV mouse leukemia virus, against HIV-1, 360–361
Ro 31-6930, as K channel opener, 46
RP 52891, as K channel opener, 46
Ruthenium red, in renal ischemia, 298
Ryanodine, in Ca^{2+} release, 47–49

S

S0121, as K channel opener, 46
Sarcoplasmic reticulum membrane, K channel opener effects on, 57–58
sCD4, against AIDS, 341–344
^{3}H-SCH23390, dopaminergic receptor binding to, 110
Schiff bases, in renal ischemia, 293, 300
SdC5, SdC15, and SdC28, against AIDS, 348
Second messengers
  in neuroleptic malignant syndrome, 391
  in renal ischemia, 301–303
  in tumor cells, 157–165
Selenium salts, in renal ischemia, 291

Semiquinone radical, in isolated reperfused heart tissue, 252–253
Serotonin, in cerebral vasospasm, 213, 218, 231, 234
Serpentine asbestos, classification and chemical composition of, 125
SG-75, see Nicorandil
Sickle cell anemia, possible cell membrane importance in, 16–17
Signal transduction, in tumor cells, 142–170
Singlet oxygen, reoxygenation after deoxygenation in production of, 13
Skeletal muscle, voltage-dependent Ca^{2+} channel regulation in, 35, 37
SKF 11197, as K channel opener, 46
Slow channel blockers, against hypertension, 44
SOD, see Superoxide dismutase
Somatostatin, in gastric acid secretion, 114
Spasmogens, in cerebral vasospasm, 213, see also specific spasmogens
Spinal cord injury
 antioxidants against, 16
 Ca^{2+} channel antagonists against, 30
Spin-trapping measurements, in electron paramagnetic resonance spectroscopy, 254–263
^{3}H-Spiperone, dopaminergic receptor binding to, 110
Spontaneous transient outward currents, in vascular smooth muscle, 47–49
Spreading depression-like membrane response, of hypoxic neurons, 188–194
SR 46142A, as K channel opener, 46
Staurospolin, in cerebral vasospasm, 232
Stimulus-response coupling, Ca^{2+} in, 24
Stomach, autonomic nervous system in, distribution of, 96–104
Stress, in gastric mucosal injury, 274–275
Stress proteins, in repair of ischemia-reperfusion injury, 15
Striosomes, dopamine receptors in, in neuroleptic malignant syndrome, 390
Strokes
 antioxidants against, 16
 ischemia-reperfusion injury in, 13
Strychnine, as K channel inhibitor, 66
Subarachnoid hemorrhage
 antioxidants against, 16
 Ca^{2+} channel antagonists against, 30
 in cerebral vasospasm, 210–212, 215, 217, 219–225, 231, 234–235, 237
Substrate metabolism, altered, in cell injury, 184
Succinylcholine, in precipitation of malignant hyperthermia, 389
Sucrose, in renal ischemia, 305
Sulfate, in renal ischemia, 292, 305
Sulfonic acids, against AIDS, 347
Sulfonylureas, ion channels in diabetes and, 83, 85–86
Superoxide anions
 in electron paramagnetic resonance spectroscopy, 247
 in lipid peroxidation, 216
 in renal ischemia, 296
Superoxide dismutase (SOD)
 in cerebral vasospasm, 222
 as extracellular antioxidant, 130
 inhibitor, in gastric mucosal injury, 278
 in isolated reperfused heart tissue, 255, 257–258, 263
 in protection against oxygen toxicity, 6
 reactive oxygen metabolites scavenged by, 130
 in renal ischemia, 291–292, 304–305
 in suppression of tumor promoter-induced DNA synthesis, 136
Superoxide radicals, neutrophil production of, 16
Superoxides, reoxygenation after deoxygenation in production of, 13
Suramin, incomplete inhibition of HIV-1 binding by, 345
Swainsonine, in inhibition of maturation of HIV-1 envelope glycoproteins, 340
Swelling, cellular, in cell injury, 184
Synaptosomes, signals produced by membranes of, 7

T

TA3090, against hypertension, 44
tat gene, in AIDS, 339
T channels, Ca^{2+}, 29, 32–33, 37
TEA, see Tetraethylammonium
TEMPO nitroxide spin label, in electron paramagnetic resonance spectroscopy, 264–266
Tetraethylammonium (TEA)
 in hypoxic spreading depression-like depolarization, 190
 as K channel inhibitor, 48–49, 51, 53, 58, 60–61, 63, 66
Tetramethrin
 structural formula of, 33
 in T channel characterization, 29
TGF-α, see Transforming growth factor α
TGF-β, see Transforming growth factor β
Thiobarbituric acid-positive substances
 in cerebral vasospasm, 216–217, 220–221
 in renal ischemia, 293
Thioformamides, as K channel openers, 46
Thrombin
 in cerebral vasospasm, 213
 in mitogenic signal transduction, 316
Thromboxane A_2, as spasmogen, in cerebral vasospasm, 213
Thromboxanes
 arachodonic acid in synthesis of, 11
 in renal ischemia, 299
Thyroid cell membrane, abnormalities of receptors on, in Basedow's disease, 10
Thyroid gland, voltage-dependent Ca^{2+} channel regulation in, 35, 37
Tinnitus, Ca^{2+} channel antagonists against, 30
TNF-α, see Tumor necrosis factor α
α-Tocopherol, in renal ischemia, 291
Tolbutamide

ATP-sensitive, glibenclamide-sensitive K channel in smooth muscle and, 53, 63, 65
ion channels in diabetes and, 83, 86
Tourette's disorder, Ca^{2+} channel antagonists against, 30
Transferrin
as extracellular antioxidant, 130
in renal ischemia, 295
Transforming genes, see Oncogenes
Transforming growth factor α (TGF-α)
in asbestos-induced diseases, 134
in mitogenic signal transduction, 317
Transforming growth factor β (TGF-β), inflammatory macrophage release of, in asbestos-induced diseases, 134
Transmembrane protein, in tumor migration to secondary sites, 327
Trehalose, in renal ischemia, 305
Tremolite, classification and chemical composition of, 125
Triplet hyperfine splitting, in electron paramagnetic resonance spectroscopy, 246
Troponin, as Ca^{2+} receptor, 24
Tryptamine, in cerebral vasospasm, 213
Tryptophan, oxidation of, 132
Tumor cells, membrane receptors and signal transduction in
background information on, 142–143
inositol lipid metabolism and, 143–149
epidermal growth factor, 152–155
growth factors that stimulate alternative pathways and insulin, 155–157
platelet-derived growth factor, 149–152
mitogenic signaling in normal and transformed cells, 166–170
oncogenes and second messengers, 157–166
Tumor necrosis factor α (TNF-α), inflammatory macrophage release of, in asbestos-induced diseases, 134
Tumor promoters, in asbestos-induced diseases, 135–136
Two-hemorrhage model, of cerebral vasospasm, 223, 233–234, 236
Tyrosine, oxidation of, 132
Tyrosine hydroxylase-positive nerves, catecholaminergic, identification of, 103
Tyrosine kinase, in mitogenic signal transduction, 315–316, 318

U

U74006F, against cerebral vasospasm, 217
Ulcers, gastric, alteration of autonomic nervous activity and receptor distribution in formation of, 118–119
Uric acid
as extracellular antioxidant, 130
in renal ischemia, 293
Urinary incontinence, Ca^{2+} channel antagonists against, 30
Uterus, voltage-dependent Ca^{2+} channel regulation in, 35, 37

V

v-abl oncogene, in expression of growth factors and growth factor receptors, 319–321
Vascular effectors, distribution of receptors of, in gastric microcirculatory system, 105–117
Vascular smooth muscle
ATP-sensitive K channel in, 44–67
Ca^{2+} in, 25, 27
Vasoactive intestinal peptide (VIP), in stomach, 103–104, 107, 111, 114
Vasocontractile substances, see Spasmogens
Vasopressin
in cerebral vasospasm, 213
in mitogenic signal transduction, 316
Verapamil
drug resistance in cancer and malaria reversed by, 17
against hypertension, 44
in renal ischemia, 298, 305
structural formula for, 28
in voltage-dependent Ca^{2+} channel regulation, 31, 37
Vertigo, Ca^{2+} channel antagonists against, 28, 30
v-fps oncogene, in expression of growth factors and growth factor receptors, 320
vif gene, in AIDS, 338
Vinblastine, prostaglandin derivatives with, against cancer, 359
VIP, see Vasoactive intestinal peptide
Vitamin C
as cellular antioxidant, 6, 130
in renal ischemia, 291
Vitamin D, in Ca^{2+} metabolism and mobilization, 24–25
Vitamin E
as cellular antioxidant, 6, 130
in cerebral vasospasm, 217, 222
in renal ischemia, 291, 293, 300
Voltage-clamp procedures, K channel opener effects on K channels in vascular smooth muscle measured via, 58–65
Voltage-dependent channels, Ca^{2+}, 28–31, 33, 35–38, 81
Voltage-gated channels, in disease processes, 8
vpr gene, in AIDS, 338
vpu gene, in AIDS, 338
v-ros, v-sis, and v-src oncogene, in expression of growth factors and growth factor receptors, 318, 320–321

W

W-7 calmodulin inhibitor, in cerebral vasospasm, 232
Warm ischemia, renal, 290–291
WAY-120, and WAY-491, as K channel openers, 46
W-conotoxins
N channels characterized by, 29, 32–33
structural formulas of, 33

X

Xanthine dehydrogenase, in renal ischemia, 293, 298
Xanthine oxidase
  blocker, in isolated reperfused heart tissue, 259, 261–262
  in renal ischemia, 293–294, 298–299
Xylofuranan sulfate, against AIDS, 344–345

Z

Zeeman splitting, of paramagnetic system, 243, 245

HEALTH SCIENCES LIBRARY
LUTHERAN COLLEGE
3024 FAIRFIELD AVE.
FORT WAYNE, IN 46807-1697